Felipe Cucker · Michael Shub (Eds.)

Foundations of Computational Mathematics

Springer

Berlin
Heidelberg
New York
Barcelona
Budapest
Hong Kong
London
Milan
Paris
Santa Clara
Singapur
Tokyo

Felipe Cucker · Michael Shub (Eds.)

Foundations of Computational Mathematics

Selected Papers of a Conference
Held at Rio de Janeiro, January 1997

Springer

Felipe Cucker
Department of Mathematics
City University of Hong Kong
83 Tat Chee Avenue, Kowloon
Hong Kong
e-mail: macucker@sobolev.ciyu.edu.hk
and:
Departament d'Economia
Universitat Pompeu Fabra
Ramon Trias Fargas 25-27
Barcelona 08005
Spain
e-mail (Spain): cucker@upf.es

Michael Shub
I.B.M. T.J. Watson Center
10598 Yorktown Heights, NY
USA
e-mail: shub@watson.ibm.com

Cataloging-in-Publication Data applied for

Die Deutsche Bibliothek - CIP-Einheitsaufnahme

Foundations of computational mathematics : selected papers of a conference held at Rio de Janeiro,
January 1997 / Felipe Cucker ; Michael Shub (ed.). - Berlin ; Heidelberg ; New York ; Barcelona ;
Budapest ; Hong Kong ; London ; Milan ; Paris ; Santa Clara ; Singapur ; Tokyo ; Springer, 1997
    ISBN 3-540-61647-0
NE: Cucker, Felipe [Hrsg.]

Mathematics Subject Classification (1991): 11Yxx, 13Pxx, 14Qxx, 65xx, 68Qxx, 68Q40

ISBN 3-540-61647-0 Springer-Verlag Berlin Heidelberg New York

The use of registered names, trademarks, etc., in this publication does not imply, even in the absence of
a specific statement, that such names are exempt from the relevant protective laws and regulations and
therefore free for general use.

Typeset with TEX: Data conversion by Lewis & Leins GmbH, Berlin
Cover design: design & production GmbH, Heidelberg

SPIN: 10543830   41/3020 - 5 4 3 2 1 0 - Printed on acid-free paper

Preface

This book contains a collection of articles corresponding to some of the talks delivered at the *Foundations of Computational Mathematics* conference held at IMPA in Rio de Janeiro in January 1997. Some of the others are published in the December 1996 issue of the Journal of Complexity. Both of these publications were available and distributed at the meeting. Even in this aspect we hope to have achieved a synthesis of the mathematics and computer science cultures as well as of the disciplines.

The reaction to the Park City meeting on Mathematics of Numerical Analysis: Real Number Algorithms which was chaired by Steve Smale and had around 275 participants, was very enthusiastic. At the suggestion of Narendra Karmarmar a lunch time meeting of Felipe Cucker, Arieh Iserles, Narendra Karmarkar, Jim Renegar, Mike Shub and Steve Smale decided to try to hold a periodic meeting entitled "Foundations of Computational Mathematics" and to form an organization with the same name whose primary purpose will be to hold the meeting.

This is then the first edition of FoCM as such. It has been organized around a small collection of workshops, namely

- Systems of algebraic equations and computational algebraic geometry
- Homotopy methods and real machines
- Information-based complexity
- Numerical linear algebra
- Approximation and PDEs
- Optimization
- Differential equations and dynamical systems
- Relations to computer science
- Vision and related computational tools

There were also twelve plenary speakers.

The variety of topics is reflected in the pages that follow. They also reflect a variety of styles since the articles run from one-page abstracts to complete mathematical papers.

We are indebted to several people. First of all to the workshop organizers: Lenore Blum, Felipe Cucker, Wolfgang Dahmen, Jim Demmel, Ron DeVore,

Alan Edelman, Clovis Gonzaga, Arieh Iserles, Herb Keller, Thomas Lickteig, Jean-Michel Morel, David Mumford, Erich Novak, Marie-Françoise Roy, Imre Simon, Andrew Stuart, and Henryk Wozniakowski. They selected the papers which appear in this volume. Then, to the local organizing committe: Alfredo Iusem, Eric Goles, Teresa Krick, Gregorio Malajovich, Benar Fux Svatier, Martin Tygel, and Jorge Zubelli.

We are greatly indebted to IMPA for making their lovely facility available and for financial support. SIAM provided us with help as a cooperating society and the meeting was endorsed by the IMU. The CNPq, CAPES, FAPERJ and FINEP, all state agencies from Brasil also supported the organization of the conference. And the same holds true for IBM and the NSF. To all of these institutions we want to express here our gratitude.

We are also indebted with Springer Verlag whose efficiency allowed the publication of this book within a short period of time.

Felipe Cucker, *Publications*
Mike Shub, *Chair*
FoCM

List of Contributors

Saugata Basu
Courant Institute of Mathematical
Sciences, New York University,
New York, NY 10012, USA

Jonathan M. Borwein
Centre for Experimental and
Constructive Mathematics,
Simon Fraser University,
Burnaby, BC V5A 1S6, Canada

Vicent Caselles
Dept. of Mathematics and Informatics,
University of Illes Balears,
07071 Palma de Mallorca, Spain
dmivca0@ps.uib.es
dmitcv0@ps.uib.es

Bartomeu Coll
Dept. of Mathematics and
Informatics,
University of Illes Balears,
07071 Palma de Mallorca, Spain
dmivca0@ps.uib.es
dmitcv0@ps.uib.es

Peter Dayan
Department of Brain and Cognitive
Sciences, E25-210, MIT,
Cambridge, MA 02139, USA

Jean-Pierre Dedieu
Laboratoire Approximation et
Optimisation,
Université Paul Sabatier,
31062 Toulouse, France

Alicia Dickenstein
Departamento de Matemática,
F.C.E. y N.,
Universidad de Buenos Aires,
Ciudad Universitaria,
Pabellón I, 1428 Buenos Aires,
Argentina
alidick@dm.uba.ar

Luca Dieci
School of Mathematics,
Georgia Institute of Technology,
Atlanta, GA 30332, USA
dieci@math.gatech.edu

Véronique Froidure, Jean-Eric Pin
LITP, CNRS, Université Paris VII,
2 Place Jussieu, 75251 Paris Cedex 05,
France

Jean-Sylvestre Gakwaya
Université de Mons-Hainaut,
Institut de Mathématique et
d'Informatique,

Avenue Maistriau, 15,
7000 Mons, Belgique
gakwaya@sun1.umh.ac.be

Shuhong Gao
Department of Mathematical Sciences,
Clemson University,
Clemson, SC 29634-1907,
USA
sgao@math.clemson.edu

Peter J. Giblin
Deptartment of Mathematics,
University of Liverpool,
Liverpool L69 3BX, UK
pjgiblin@liv.ac.uk

O. Gonzalez, A.M. Stuart
Division of Applied Mechanics,
Department of Mechanical
Engineering, Stanford University,
Stanford, CA 94305-4040, USA

Arieh Iserles
Department of Applied Mathematics
and Theoretical Physics,
University of Cambridge, Silver Street,
Cambridge CB3 9EW,
England

**Kh.T. Kholmurodov, I.V. Puzynin,
Yu.S. Smirnov**
Laboratory of Computing Techniques
and Automation,
Joint Institute for Nuclear Research,
Dubna, Russia
smirnov@lcta6.jinr.dubna.su

Peter Kirrinnis
Universität Bonn,
Institut für Informatik II,
Römerstr. 164,
53117 Bonn, Germany

Y. Kohayakawa
Instituto de Matemática e Estatística,
Universidade de São Paulo,
Rua do Matão 1010,
05508–900 São Paulo, SP,
Brazil

Sóstenes Lins
Universidade Federal de Pernambuco,
Departamento de Matemática,
CEP. 50.740-540, Recife-PE, Brazil

José Mario Martínez
Departamento de Matemática
Aplicada, IMECC-UNICAMP,
CP 6065, 13081-970 Campinas SP,
Brazil
martinez@ime.unicamp.br

Martín Matamala
Departamento de Ingeniería
Matemática, Facultad de Ciencias
Físicas y Matemáticas,
Universidad de Chile,
Beaucheff 850,
Casilla 170-correo 3,
Santiago-Chile

Guillermo Matera
Depto. de Matemáticas, F.C.E. y N.,
Univ. de Buenos Aires, Argentina
and
Inst. de Ciencias, Univ. Nac. Gral.
Sarmiento, Argentina
gmatera@dm.uba.ar

Sanjoy K. Mitter
Department of Electrical Engineering
and Computer Science,
Massachusetts Institute of
Technology,
Cambridge, MA 02139, USA

Cristopher Moore
Santa Fe Institute,
1399 Hyde Park Road,
Santa Fe, NM 87501, USA
moore@santafe.edu

Warren B. Moors
Department of Mathematics,
The University of Auckland,
Private Bag 92019, Auckland,
New Zealand

Jean-Michel Morel
CEREMADE,
Universite Paris-Dauphine,
Place du Marechal Lattre de Tassigny
75775 Paris Cedex 16, France
morel@paris9.dauphine.fr

Bernard Mourrain
INRIA, SAFIR,
2004 route des Lucioles,
B.P. 93, 06902 Sophia Antipolis,
France
mourrain@sophia.inria.fr

Hans Munthe-Kaas
Institutt for Informatikk,
Universitetet i Bergen,
Høyteknologisenteret,
5020 Bergen, Norway
Hans.Munthe-Kaas@ii.uib.no

Victor Y. Pan
Mathematics and Computer Science
Department, Lehman College, City
University of New York,
Bronx, NY 10468, USA
VPAN@LCVAX.LEHMAN.CUNY.EDU

Suely Oliveira
Department of Computer Science,
Texas A&M University,
College Station, Texas 77843-3112,
USA

Wilson R. de Oliveira
Departamento de Informática
UFPE - CCEN,
Caixa Postal 7851 - CEP 50732-970
Recife - PE, Brazil
wrdo@di.ufpe.br
http://www.di.ufpe.br/~wrdo

Diego Pallara
Dipartimento di Matematica,
Università di Lecce,
C. P. 193, 73100, Lecce, Italy
pallara@le.infn.it

Daniel Panario
Department of Computer Science,
University of Toronto,
Toronto, Canada M5S-1A4
daniel@cs.toronto.edu

Robert J. Plemmons
Department of Computer Science,
Wake Forest University,
Winston-Salem, NC 27109, USA
plemmons@wfu.edu

Richard Pollack
Courant Institute of Mathematical
Sciences, New York University,
New York, NY 10012, USA

Sebastian Reich
Konrad-Zuse-Zentrum für
Informationstechnik (ZIB),
Takustr. 7,
14195 Berlin-Dahlem, Germany

J. Maurice Rojas
Massachusetts Institute of
Technology,
Mathematics Department,
77 Mass. Ave.,
Cambridge, MA 02139, USA

Ricardo Rosa
The Institute for Scientific
Computing & Applied Mathematics,
618 E. Third St.,
Indiana University,
Bloomington, IN 47405, USA
and
Departamento de Matemática
Aplicada, IM-UFRJ,
Caixa Postal 68530
Rio de Janeiro, RJ,
C.E.P. 21945-970, Brazil

Marie-Françoise Roy
IRMAR (URA CNRS 305),
Université de Rennes,
Campus de Beaulieu,
35042 Rennes Cedex, France

Guillermo Sapiro
Hewlett-Packard Labs,
1501 Page Mill Rd.,
Palo Alto, CA 94304, USA
guille@hpl.hp.com

Aron Simis
Instituto de Matemática,
Universidade Federal da Bahia,
40170–210 Salvador, Bahia, Brazil
aron@ufba.br

David E. Stewart
Mathematics Dept.,
Virginia Polytechnic Institute and
State University,
Blacksburg, VA 24061, USA

Roger Temam
The Institute for Scientific
Computing & Applied Mathematics,
618 E. Third St.,
Indiana University,
Bloomington, IN 47405, USA
and
Laboratoire d'Analyse Numérique,
Université de Paris-Sud,
Bâtiment 425, 91405 Orsay Cedex,
France

Jose Maria Turull Torres
Universidad Nacional de San Luis,
Argentina
and
Depto. de Sistemas, F.R.B.A.,
Univ. Tecnológica Nacional,
Argentina
turull@iamba.edu.ar

György Turán
Department of Mathematics,
Statistics and Computer Science,
University of Illinois at Chicago,
Chicago, IL 60607-7045, USA
and
Research Group on Artificial
Intelligence of the
Hungarian Academy of Sciences,
Szeged, Hungary

Erik S. Van Vleck
Department of Mathematical and
Computer Sciences,
Colorado School of Mines,
Golden, CO 80401, USA
erikvv@poincare.mines.edu

Farrokh Vatan
Electrical Engineering Department,
University of California, Los Angeles,
Box 951594,
Los Angeles, CA 90095, USA

Jorge R. Vera
Dept of Industrial Engineering,
University of Chile,
República 701, Santiago, Chile
jvera@dii.uchile.cl

Antonella Zanna
Department of Applied Mathematics
and Theoretical Physics,
University of Cambridge,
Silver Street,
Cambridge CB3 9EW, England

Contents

Computing Roadmaps of Semi-algebraic Sets on a Variety
(Extended Abstract)
Saugata Basu, Richard Pollack and Marie-Française Roy 1

Essentially Smooth Lipschitz Functions: Compositions and Chain Rules
Jonathan M. Borwein and Warren B. Moors 16

Junction Detection and Filtering
Vicent Caselles, Bartomeu Coll and Jean-Michel Morel 23

Recognition in Hierarchical Models
Peter Dayan 43

Continuity Σ−Algebras (Extended Abstract)
Wilson R. de Oliveira 63

Condition Number Analysis for Sparse Polynomial Systems
Jean-Pierre Dedieu 75

Residues in the Torus and Toric Varieties
Alicia Dickenstein 102

Piecewise Smooth Orthonormal Factors
for Fundamental Solution Matrices
Luca Dieci and Erik S. Van Vleck 104

Algorithms for computing finite semigroups
Véronique Froidure and Jean-Eric Pin 112

Extended Grzegorczyk Hierarchy in the BSS Model of Computability
Jean-Sylvestre Gakwaya 127

Affine-Invariant Symmetry Sets
Peter J. Giblin and Guillermo Sapiro 152

On the Qualitative Properties of Modified Equations
O. Gonzalez and A.M. Stuart 169

Numerical Methods on (and off) Manifolds
Arieh Iserles 180

On One Computational Scheme Solving the Nonstationary Schrödinger
Equation with Polynomial Nonlinearity
Kh.T. Kholmurodov, I.V. Puzynin, Yu.S. Smirnov 190

Newton Iteration Towards a Cluster of Polynomial Zeros
Peter Kirrinnis 193

Szemerédi's Regularity Lemma for Sparse Graphs
Y. Kohayakawa 216

Questions on Attractors of 3-Manifolds
Sóstenes Lins 231

A Trust-Region SLCP Model Algorithm for Nonlinear Programming
José Mario Martínez 246

On the height used by additives BSS machines
Martín Matamala 256

The Space Complexity of Elimination Theory: Upper Bounds
Guillermo Matera and Jose Maria Turull Torres 267

Global Stochastic Recursive Algorithms
Sanjoy K. Mitter 277

Dynamical Recognizers: Real-time Language Recognition by Analog
Computers (Extended Abstract)
Cristopher Moore 278

Solving special polynomial systems by using structured matrices and
algebraic residues
Bernard Mourrain and Victor Y. Pan 287

Numerical Integration of Differential Equations on
Homogeneous Manifolds
Hans Munthe-Kaas and Antonella Zanna 305

A Convergence proof of an Iterative Subspace Method
for Eigenvalues Problems
Suely Oliveira 316

Regularity of Minimizers of the Mumford-Shah Functional
Diego Pallara 326

Tests and Constructions of Irreducible Polynomials over Finite Fields
Shuhong Gao and Daniel Panario 346

Numerical Linear Algebra in Optical Imaging
Robert J. Plemmons 362

Explicit symplectic integration of rod dynamics
Sebastian Reich 368

Toric Laminations, Sparse Generalized Characteristic Polynomials,
and a Refinement of Hilbert's Tenth Problem
J. Maurice Rojas 369

Finite-Dimensional Feedback Control of a Scalar Reaction-Diffusion
Equation via Inertial Manifold Theory
Ricardo Rosa and Roger Temam 382

Computational aspects of jacobian matrices
Aron Simis 392

Rigid body dynamics and measure differential inclusions
David E. Stewart 405

Linear decision lists and partitioning algorithms for the construction
of neural networks
György Turán and Farrokh Vatan 414

Ill-Posedness and Finite Precision Arithmetic: A Complexity Analysis
for Interior Point Methods
Jorge R. Vera 424

Iterated Commutators, Lie's Reduction Method and Ordinary Differential
Equations on Matrix Lie Groups
Antonella Zanna and Hans Munthe-Kaas 434

Computing Roadmaps of Semi-algebraic Sets on a Variety (Extended Abstract)

Saugata Basu*, Richard Pollack** and Marie-Françoise Roy***

Abstract. We consider a semi-algebraic set S defined by s polynomials of degree at most d in k variables contained in an algebraic variety V of dimension k' defined as the zero set of a polynomial of degree d and two points defined by polynomials of degree t. We present an algorithm for computing a semi-algebraic path in S connecting two points if they happen to lie in the same connected component of S which works in time $(s^{k'+1} + k'st^{O(1)})d^{O(k^2)}$.

1 Introduction

1.1 The Problem

Let R be a real closed field and $V = Z(Q)$ be a real algebraic variety of real dimension k' defined as the zero set of a polynomial $Q \in R[X_1, \ldots, X_k]$ of degree d. Given a finite family of polynomials, $\mathcal{P} = \{P_1, \ldots, P_s\} \subset R[X_1, \ldots, X_k]$, and a semi-algebraic set $S \subset V$ defined by a Boolean formula with atoms of the form $P_i \sigma 0$, $\sigma \in \{<, >, =\}$, we construct $R(S) \subset S$, called a *roadmap* of S. The set $R(S)$ is a semi- algebraic set of dimension at most one, satisfying

1. for every semi-algebraically connected component C of S, $C \cap R(S)$ is non-empty and semi-algebraically connected.
2. for every $x \in R$, and for every semi-algebraically connected component C' of S_x, $C' \cap R(S) \neq \emptyset$. (Here, and everywhere else in this paper π is the projection on the first coordinate, S_X is $S \cap \pi^{-1}(X)$, where $X \subset R$ and we abbreviate $S_{\{v\}}$, $S_{(-\infty,v)}$ and $S_{(-\infty,v]}$ by S_v, $S_{<v}$ and $S_{\leq v}$ respectively.)

Denote by D the ring generated by the coefficients of Q and of $P_1, \ldots, P_s$. If Q and the polynomials in $\mathcal{P}$ have total degrees bounded by d, then we compute $R(S)$ using $s^{k'+1}d^{O(k^2)}$ arithmetic operations in D. Moreover, the degrees of the polynomials defining the roadmap are bounded by $d^{O(k^2)}$.

* Supported in part by NSF grants CCR-9402640 and CCR- 9424398
** Supported in part by NSF grants CCR-9402640 and CCR- 9424398
*** Supported in part by the project ESPRIT-LTR 21024 FRISCO and by European Community contract CHRX-CT94- 0506

We also describe a *connecting subroutine*, connecting a point $x \in S$ to the roadmap $R(S)$. This algorithm constructs a *connecting curve*: a connected semi-algebraic subset of S which has dimension at most one connecting x to $R(S)$. If x is described by univariate polynomials of degree at most t then the complexity of this algorithm is $k'st^{O(1)}d^{O(k^2)}$.

Thus the roadmap algorithm together with the connecting subroutine enables us to decide whether two points x and y of S described by univariate polynomials of degree at most t are in the same connected component of S in time $(s^{k'+1} + k'st^{O(1)})d^{O(k^2)}$. Furthermore, if x and y are in the same connected component, we obtain a semi-algebraic path joining the two points which consists of certain branches of the roadmap and of the connecting curves.

1.2   Prior Results

The problem of deciding connectivity properties of semi-algebraic sets has attracted much attention during the last ten years. The original motivation came from robot motion planning where the free space of a robot can be modeled as a semi-algebraic set. In this context it is important to know whether a robot can move from one configuration to another (see [16]). This is equivalent to deciding whether the two corresponding points in the configuration space are in the same connected component of the free space.

In robotics, the configuration space of a robot is often embedded as a lower dimensional algebraic variety in a higher dimensional real Euclidean space (see for example [12] page 65.) In such a situation it is of interest to design algorithms which take advantage of this fact and whose complexity depends on the dimension of this variety rather than that of the ambient space.

In 1987, John Canny [5] introduced the notion of a roadmap of a semi-algebraic set. Similar notions were also considered by Grigor'ev and Vorobjov [8], Heintz, Roy, and Solerno [11] and Gournay and Risler [9].

Canny [5] gave an algorithm that after subsequent modifications [6] constructed a roadmap for a semi-algebraic set defined by polynomials whose sign invariant sets give a Whitney regular stratification of R^k whose complexity is $s^k(\log s)d^{O(k^4)}$. For an arbitrary semi-algebraic set he perturbs the defining polynomials and is then able to decide if two points are in the same semi-algebraically connected component in the same time. This algorithm does not give a path joining the points. A Monte Carlo version of this algorithm has complexity $s^k(\log s)d^{O(k^2)}$.

Grigor'ev and Vorobjov [8] gave an algorithm with complexity $(sd)^{k^{O(1)}}$, for counting the number of connected components of a semi-algebraic set. Heintz, Roy, and Solerno [11] and Gournay and Risler [9] gave algorithms which compute a roadmap for any semi-algebraic set whose complexity was also $(sd)^{k^{O(1)}}$. Note that unlike the complexity of Canny's algorithm, the complexities of these algorithms are not separated into a combinatorial part (the part depending on s) and an algebraic part (the part depending on d). Since the given semi-algebraic set might have $(sd)^k$ different connected components the combinatorial complexity of Canny's algorithm is nearly optimal. On the other hand an algebraic complexity of $d^{O(k^2)}$ is unavoidable as long as the basic technique of recursing down the dimension is used as it is in all known algorithms for this problem. Canny's algorithm makes use of Thom's isotopy lemma for stratified sets, and as a result it requires the use of generic projections, as well as perturbations to put the input polynomials into general position in a very strong sense. In order to do this in a deterministic fashion, $O(s + k^2)$ different transcendentals are introduced, requiring the algebraic operations to be performed over an extended ring, which raises the algebraic complexity of the algorithm to $d^{O(k^4)}$.

In [2] a deterministic algorithm constructing a roadmap for any semi-algebraic set in time $s^{k+1}d^{O(k^2)}$ is given. This complexity is again nearly optimal in its combinatorial complexity and moreover it does not require the regular stratification condition, or generic projections, which helps to reduce the algebraic complexity to $d^{O(k^2)}$. The algorithm computes a semi-algebraic path between the input points if they happen to lie in the same connected component and hence this solves the extended version of the problem.

1.3 Our work

In this paper, we give an algorithm for computing a roadmap of a semi-algebraic set contained in an algebraic set of dimension k'. The exponent of s in the combinatorial complexity of this algorithm depends only on k' rather than on k, the dimension of the ambient space.. Our result relies heavily on the construction of approximating varieties [15] as well as on a new algorithm with complexity $s^{k'+1}d^{O(k)}$ for constructing points in every connected component of $S \subset V$ [3], which also uses [15]. It also uses several algorithms from [2].

The rest of the paper is organized as follows. In section 2 we state a topological lemma that is crucial for our algorithm. Next, we develop some algorithmic tools that allow us to compute curves in a given algebraic set parametrized by the first coordinate and to identify certain *distinguished slices* where it is necessary to make recursive calls in one lower dimension.

In section 3 we give an algorithm which constructs a roadmap of an algebraic set which is required to pass through a specified set of points. This will serve as a base case for our recursion (see section 4).

In section 4 we introduce the notion of *combinatorial level* which we use in our algorithm. We give an algorithm, that given a family of polynomials $\mathcal{P}$ constructs a *uniform* roadmap which has the property that its intersection with any set defined by a single weak sign condition on $\mathcal{P}$ is a roadmap for that set. We also describe a *connecting subroutine* to the uniform roadmap.

Lastly in section 5, we give a roadmap algorithm for an arbitrary semi- algebraic set S, defined by the family of polynomials $\mathcal{P}$. The idea is to construct uniform roadmaps for a perturbed family of polynomials on various approximating varieties close to V, for which the combinatorial level is k' and to take the limit of the curves so constructed when the parameters of perturbation tend to 0. In order to ensure that the curve so constructed is connected in every connected component, we need to add curves connecting well chosen points to the various roadmaps on approximating varieties.

2   Preliminaries

2.1   A topological lemma

Our argument depends on perturbing the polynomials by various *infinitesimals* and working over the field of Puiseux series in these infinitesimals. We write $R\langle\delta\rangle$ for the real closed field of Puiseux series in δ with coefficients in R (see [4]). The sign of an element in this field agrees with the sign of the coefficient of the lowest degree term in δ. This order makes δ positive and smaller than any positive element of R. The map eval_δ maps an element of $R\langle\delta\rangle$ which is bounded over R, i.e. one that has no negative powers of δ to its constant term.

Given a polynomial Q such that $Z(Q)$ is compact, let $\zeta \ll \frac{1}{\Omega}$ be infinitesimals, and let

$$Q_1 = (1 - \zeta)Q^2 + \zeta(X_1^{2d+2} + \cdots + X_k^{2d+2} - k\Omega^{2d+2}).$$

Then, $Z(Q_1)$ is a smooth hypersurface with a finite number of critical points with respect to π (for a proof, see [1]). Following an idea in [9], the *pseudo-critical values* of $Z(Q)$ are defined to be the first coordinates of the images of these critical points under the eval_ζ map.

Given a family of polynomials $\mathcal{P} = \{P_1, \ldots, P_S\}$ such that for every $P \in \mathcal{P}$, $Z(P)$ is bounded and S is a closed and bounded semi-algebraic set defined by $P_1 \geq 0, \ldots, P_s \geq 0$, we call $v \in R$ a pseudo-critical value for S if it is a pseudo-critical value for any of the non-empty algebraic sets defined by $P_{i_1}^2 + \cdots + P_{i_l}^2 = 0$, $1 \leq i_1 < i_2 \cdots < i_l \leq s$.

The following lemma, which generalizes a result in [9], allows us to characterize a certain finite number of values of the first coordinate so that it suffices to make recursive calls at the corresponding slices when constructing the roadmap for a basic semi-algebraic set.

Lemma 1 *Let the sets $Z(P_i), 1 \leq i \leq s$ be compact and C be a non-empty semi-algebraically connected component of $S_{[a,b]}$. If $v \in (a,b)$ is the only pseudo-critical value for S in $[a,b]$, or if $v \in [a,b]$ and there are no pseudo-critical values for S in $[a,b]$, then the set C_v is non-empty and semi-algebraically connected.*

The proof appears in the full version of the paper.

2.2   Curve Segments Subroutine

The input is a polynomial $Q \in D[Y, X_1, \ldots, X_k]$, whose degree is at most d and a set of m points, say $M \subset Z(Q)$, such that each point of M is described by univariate polynomials of degree at most t. This means precisely that the points of M are given by a pair (u, σ) where u is a univariate representation (i.e. a $k+2$-tuple of univariate polynomials $u = (f, g_0, \ldots, g_k),)$ and $\sigma \in \{0, 1, -1\}^{\deg f}$ for which there exists a root t of f with Thom encoding σ (i.e. $\mathrm{sign}(f^{(i)}(t)) = \sigma_i$) such that $x = \frac{g_i}{g_0}(t)$.

We assume that $Z(Q)$ is bounded. This is no loss of generality since when $Z(Q)$ is not bounded, we first introduce a new variable X_{k+1} and a new infinitely large variable Ω and replace Q by the polynomial $Q^2 + (X_1^2 + \cdots + X_{k+1}^2 - \Omega^2)^2$. At the end of the computation we will replace Ω by a large enough element $M \in R$ (see [1] for details) and extend the roadmap to infinity to obtain the roadmap over all of R^k.

The output contains a partition of the Y axis into points, called *distinguished values*, and open intervals. The distinguished values (and hence the intervals) are ordered. Among the distinguished values are the first coordinates of the points in M. Over each distinguished value the subroutine also outputs a finite set of points, called *distinguished points* (the points of M are all distinguished points), and over each interval a finite set of *curve segments*: i.e. continuous semi-algebraic curves parameterized by the interval along the Y axis. The distinguished points and the output output curve segments are contained in $Z(Q)$. The set of distinguished values and the sets of distinguished points and curve segments satisfy the following properties.

P_1: Let $[a, b]$ be an interval, and C be a non-empty semi-algebraically connected component of $Z(Q)_{[a,b]}$. If $v \in (a, b)$ is the only distinguished value in $[a, b]$, or if $v \in [a, b]$ and there are no distinguished values in the interval $[a, b]$, then the set C_v is non-empty and semi-algebraically connected.

P_2: For every $y \in R$ the set of distinguished points and curve segments output intersect every semi-algebraically connected component of $Z(Q)_y$.

P_3: For each curve segment output over an interval with endpoint a given distinguished value, there exists a distinguished point over this distinguished value which belongs to the closure of the curve segment.

The adjacency relations between curve segments and distinguished points are computed in the subroutine.

The subroutine makes use of the parametrized cell representatives subroutine and the univariate representation subroutine in [1]. The parametrized cell representatives subroutine allows one to construct curves in $Z(Q)$ satisfying P_2, and using the univariate representation subroutine we compute the pseudo-critical values of Q. This ensures property P_1 by lemma 1.

The cost of computing the curves and the pseudo-critical values is bounded by $d^{O(k)}$ (see [1] for details). However, there is the additional cost of sorting the distinguished values computed, comparing them to the first coordinates of points in M, and intersecting the curves constructed with the hyperplanes corresponding to the first coordinates of points in M, and this adds a factor of $m \log m t^{O(1)}$. The total cost is bounded by $(m \log m) t^{O(1)} d^{O(k)}$.

3   The roadmap algorithm for an algebraic set

The input is a polynomial $Q \in D[X_1, \ldots, X_k]$ whose total degree is at most d and a set of m points M such that each point in M is defined by polynomials of degrees not greater than t. The output is a roadmap $R(Z(Q), M)$ of $Z(Q)$ containing M. According to subsection 2.2 we can suppose without loss of generality that $Z(Q)$ is bounded.

Using the curve segments subroutine taking X_1 as parameter, we obtain a partition of the X_1 axis and sets of semi-algebraic continuous curves parametrized along the X_1 axis, with properties P_1, P_2 and P_3.

At every distinguished value v, make a recursive call to the algorithm in $R^{k-1} = \pi^{-1}(v)$, with the following input: the polynomial, $Q(v, X_2, \ldots, X_k)$, and the set of points which are the points of M and the distinguished points belonging to the fiber $\pi^{-1}(v)$. Note that the polynomial $Q(v, X_2, \ldots, X_k)$, has its coefficients in the ordered ring $D[v]$.

The correctness of the algorithm follows from the correctness of the curve segments subroutine. The cost of a call to the curve segments subroutine is $(m \log m) t^{O(1)} d^{O(k)}$ arithmetic operations in the ring D.

We make a recursive call at $m + d^{O(k)}$ distinguished values. Note that in a recursive call at depth ℓ the computations are performed over a ring $D[v_1, \ldots, v_\ell]$, and a single arithmetic operation in $D[v_1, \ldots, v_\ell]$ costs $t^{O(1)} d^{O(k\ell)}$ operations in D. Thus, the total number of arithmetic operations made in the ring D during the algorithm is bounded by $(m \log m) t^{O(1)} d^{O(k^2)}$.

4   An Algorithm for a Uniform Roadmap

4.1   Combinatorial Level

Given a polynomial Q and a set of polynomials $\mathcal{P} = \{P_1, \ldots, P_s\}$ we define the *combinatorial level* of the system $(Q, \mathcal{P})$ to be the minimum number ℓ (if it exists) satisfying the following:

1. no more than ℓ of the polynomials in $\mathcal{P}$ have a real zero in common with Q, and
2. any real zero common to ℓ polynomials of $\mathcal{P}$ and Q is isolated.

Note that if the combinatorial level of $(Q, \mathcal{P})$ is bounded by ℓ, then the combinatorial level of $(Q^2 + P_1^2, \{P_2, \ldots, P_s\})$ is bounded by $\ell - 1$.

We say that a family of polynomials $\mathcal{P} = \{P_1, \ldots, P_s\}$ is in *general position* over $Z(Q)$ of dimension k' if no $k' + 1$ polynomials of this family have a real zero in common with Q and if the only real zeroes common to k' polynomials of the family and Q are isolated. Thus, if the set of polynomials $P_1, \ldots, P_s$ is in general position over $Z(Q)$ of dimension k', then the combinatorial level of the system $(Q, \mathcal{P})$ is at most k'.

4.2   Uniform Roadmap

In subsection 4.5 we describe an algorithm that takes as input a polynomial Q for which $Z(Q)$ is compact and a set of polynomials $\mathcal{P} = \{P_1 \ldots, P_s\}$, where the combinatorial level of the system $(Q, \mathcal{P})$ is at most ℓ, and the degrees of the polynomials are bounded by d.

The algorithm outputs a *uniform roadmap* i.e. a one-dimensional semi-algebraic set, $R(Q, \mathcal{P})$, which is a union of open curve segments and points satisfying the following two properties:

(a) The signs of the polynomials, $Q, P_1, \ldots, P_s$ are constant on each curve segment,

(b) The intersection of this set with any basic closed semi- algebraic set S_σ, defined by a sign condition $\sigma = (0, \sigma_1, \ldots, \sigma_s)$, i.e. $Q = 0, P_1\sigma_1 0, \ldots, P_s\sigma_s 0$, where $\sigma_i \in \{\geq, \leq\}$, (we call such a σ a *weak* sign condition on $\mathcal{P}$) is a roadmap $R(S_\sigma)$.

The complexity of the algorithm is $s^{\ell+1} d^{O(k^2)}$.

4.3   Basic idea

Consider a sweep hyperplane orthogonal to the X_1 axis, defined by $X_1 = v$. Consider too, a weak sign condition, say σ, of the polynomials $\mathcal{P}$, and let S_σ be the set defined by σ. Consider the semi-algebraically connected components $C_1(v), \ldots, C_\ell(v)$ of $S_{\leq v}$. The key fact is that, as the hyperplane sweeps from left to right, if two of these connected components, $C_i(v), C_j(v)$, *join together* and become one connected component at some point, then there exists an algebraic set $Z(P_{i_1}, \ldots, P_{i_m})$, such that a semi-algebraically connected component of the intersection of this algebraic set with S belongs to this single connected component. Moreover, the X_1 coordinate of the sweep hyperplane at this point is a pseudo-critical value for $Z(P_{i_1}, \ldots, P_{i_j})$.

The algorithm constructs curves in the connected components, $C_i(v)$, such that to the left of the sweep hyperplane the curves satisfy the roadmap conditions for $C_i(v)$. When two connected components join together to form a single connected component, the partial roadmaps of these two components computed so far, get linked by means of a recursive call to the algorithm. The recursive call computes a roadmap in the slice, restricted to the algebraic set mentioned above.

4.4   Finger Subroutine

We first describe a subroutine which, given a polynomial Q, a set of polynomials $\mathcal{P} = \{P_1, \ldots, P_s\}$ and a point p in $Z(Q)$, with $\pi(p) = v$ constructs a finite number of continuous semi-algebraic curves starting at p so that every semi-algebraically connected component of every realizable sign condition of $\mathcal{P}$ in $Z(Q)$ sufficiently near and to the left of p, contains one of these curves without the point p.

The subroutine takes as input a polynomial Q and a family of polynomials $\mathcal{P} = \{P_1, \ldots, P_s\}$ with combinatorial level of the system $(Q, \mathcal{P})$ bounded by ℓ and a point $p \in Z(Q)$ which is defined by polynomials of degree at most $d^{O(k)}$.

It outputs a set of continuous curves starting at p with the property that every semi-algebraically connected component of non-empty sign conditions of $\mathcal{P}$ in $(Z(Q) \cap B_p(r))_{<v}$ for some sufficiently small r contains one of these curves without the endpoint p.

Let the set of polynomials (possibly empty) that are zero at p be $\mathcal{P}_p = \{P_j | j \in J\}$ and $B_p(\epsilon)$ be a ball of radius ϵ and center p, where ϵ is an infinitesimal. Using the sample points subroutine [1] with the polynomials defining $B_p(\epsilon)_{<v}$ along

with the polynomials $\mathcal{P}_p$ as input, we find a point $p_i(\epsilon)$ in every semi-algebraically connected component of every non-empty sign conditions of these polynomials in $B_p(\epsilon)_{<v}$.

Moreover, $\mathrm{eval}_\epsilon(p_i) = p$. For a small enough t_0, replacing ϵ by t where $0 \le t \le t_0$ gives for each $p_i(\epsilon)$, continuous curves $p_i(t)$, which join p to points in every semi- algebraically connected component of the non-empty sign conditions of $\mathcal{P}$ intersected with $(Z(Q) \cap B_p(\epsilon))_{<v}$.

The complexity of this subroutine is easily seen to be the cost of calling the sample points subroutine, with $O(\ell)$ polynomials, in k variables, $X_1, \ldots, X_k$.

Since at most ℓ polynomials can be zero at p and using the complexity bound of the sample points subroutine in [1], we conclude that the complexity of the finger subroutine is $\ell^{k+1} d^{O(k)}$.

4.5   The Uniform Roadmap Algorithm

Note that, if the combinatorial level of the system, $(Q, \mathcal{P})$ is 0 then we can use the algorithm for the algebraic case, since on every semi-algebraically connected component of $Z(Q)$ the signs of the polynomials in $\mathcal{P}$ are fixed and are not zero. Moreover, this algorithm has complexity $sd^{O(k^2)}$.

Also, if $k = 1$ then we can sort the roots of the univariate polynomials $Q, P_1, \ldots, P_s$ and output a description of the one dimensional set as the roadmap. The algorithm has complexity $(s \log s) d^{O(1)}$.

The remaining cases are described below.

We use the curve segment subroutine to compute a set of curves and distinguished values for the sets $Z(Q)$ and $Z(Q_J)$ for each $J \subset \{1, 2, \ldots, s\}$ of cardinality at most $\ell - 1$ where

$$Q_J = Q^2 + \sum_{j \in J} P_j^2.$$

Denote the set of curves constructed on $Z(Q_J)$ by $R(Q_J)$. We store the set of distinguished values and points computed by the subroutine and call the end points of these segments the set of *end points*.

Intersect each of these curve segments with the zero set of each polynomial in $\mathcal{P}$. Note that the intersection of a curve segment with the zero set of a polynomial is either the segment itself or a finite set of points (possibly empty). We check which case holds by substituting the univariate representation of the curve into each polynomial in $\mathcal{P}$ and checking whether or not the resulting univariate polynomial vanishes identically or not.

If the intersection is the curve segment itself, we ignore this intersection. Otherwise, the points of intersection yield a partition of the curve segment. We add these points of intersection to our list of end points.

The signs of the polynomials of $\mathcal{P}$ do not change on each segment of this partition. Moreover, we can maintain a data structure that stores the sign vector of the set of polynomials $\mathcal{P}$ on each curve segment and point computed above.

The X_1 coordinates of end points are called *distinguished values*. A *distinguished slice* is $\pi^{-1}(v)$ for a distinguished value v. Note that a distinguished value originates either from a call to the curve segments subroutine for a particular algebraic set $Z(Q_J)$, or as the X_1 coordinate of a point of intersection of a curve computed by the curve segment subroutine on a particular algebraic set $Z(Q_J)$. The algebraic set *associated* to v is $Z(Q_J)$ in the first two cases, and $Z(Q_\emptyset) = Z(Q)$ in the last case.

For every distinguished value v, compute the set M_J, of points of intersection of the curves constructed on $Z(Q_J)$ in the previous step and the slice $\pi^{-1}(v)$. Call the algorithm recursively to construct a uniform roadmap, $R(Q_{J_v}, \mathcal{P}_v, M_J)$, restricted to the slice $\pi^{-1}(v)$.

Note, that the combinatorial level of this system is $\ell - |J|$, and the number of variables is $k - 1$.

For each distinguished point p with $\pi(p) = v$ and the corresponding distinguished slice, we use the finger subroutine to construct curves joining p to points in every semi-algebraically connected component of every realizable sign condition of the family of polynomials $\mathcal{P}$ intersected with, $(Z(Q) \cap B_p(r_p)_{<v}$, for some small enough r_p. Let the other endpoints of these curves be $p_1, \ldots, p_j$.

For each $p_i, 1 \le i \le j$, in the previous step we compute a roadmap, $\Gamma(p_i)$, for the algebraic set $Z(Q)$ in the slice which passes through p_i as well as the set of points, $I(p_i) = R(Z(Q))_{\pi(p_i)}$.

We then compute the intersections of the points and curve segments defining $\Gamma(p_i)$ with $Z(P_i), 1 \le i \le s$, and retain the semi-algebraically connected component of $\Gamma(p_i)$ which contains p_i and over which the signs of the polynomials in $\mathcal{P}$ are constant. We call this set $\Gamma'(p_i)$.

For each p_i in the previous step, either $\Gamma'(p_i) \cap I(p_i) \ne \emptyset$ or for some $P \in \mathcal{P}$, $\Gamma'(p_i)$ intersects $Z(P)$, at a point q_i.

In the former case, continue with the next p_i in the list. In the latter case, repeat Step 4 with p_i replaced by q_i, and staying on the algebraic set $P = Q = 0$ instead of $Q = 0$

In other words, we compute a roadmap, $\Gamma(q_i)$, of $P = Q = 0$ passing through q_i and the set of points $I(q_i)$. We retain the semi-algebraically connected component, $\Gamma'(q_i)$ of $\Gamma(q_i)$ which contains q_i and over which the signs of the polynomials in $\mathcal{P}$ are constant.

Again, either $\Gamma'(q_i) \cap I(q_i)$ is non-empty, or $\Gamma'(q_i)$ intersects $Z(P_1)$ for some polynomial $P_1 \in \mathcal{P} \setminus P$, and so on. Since the combinatorial level of $(Q, \mathcal{P})$ is bounded by ℓ this procedure terminates, after at most ℓ iterations. When it terminates, we have connected the point p_i by a series of curves in the same weak sign condition as p_i and which lie in that slice, to some other point of our partially computed roadmap in this weak sign condition.

We output all the curves, and endpoints computed above with every curve labelled by the sign condition it satisfies.

4.6 The connecting subroutine

The connecting subroutine is the following: let $S \subset Z(Q)$ be a basic semi-algebraic set defined by weak sign conditions over $\mathcal{P}$ and an $x \in S$ described by polynomials of degree t. The connecting subroutine connects x to the intersection of S with the uniform roadmap of $R(Z(Q), \mathcal{P})$ inside S.

We compute the set M of points of intersection of $R(Z(Q))$ with $X_1 = x_1$ We start by constructing a road map $R(Z(Q)_{x_1}, M \cup \{x\})$. If x is connected to $R(Z(Q))$ inside S the connection is done.

Otherwise, let P_{i_1} be a polynomial in $\mathcal{P}$ such that there is a connected curve C contained in $R(Z(Q)_{x_1}, M \cup \{x\})$ going from x to y with $P_{i_1}(y) = 0$ and such that no P_{i_1} vanishes on C .

We start again the algorithm replacing Q by $Q_1 := Q^2 + (P_{i_1})^2$ and x by y.

Since the combinatorial level of $(Q, \mathcal{P})$ is bounded by ℓ, the algorithm terminates after at most ℓ iterations.

4.7   Proof of Correctness

We only give the idea behind the proof. The complete proof will appear in the full version of the paper. We prove that the set $R(Q, \mathcal{P})$ computed by the algorithm satisfies properties (a) and (b) of section 4.2.

Obviously property (a) of section 4.2 is satisfied by $R(Q, \mathcal{P})$. Let $v_1, \ldots, v_n$ be the set of distinguished values computed by the algorithm.

We fix a weak sign condition σ and prove that $R(Q, \mathcal{P}) \cap S_\sigma$ is a roadmap for S_σ. This fact follows by induction on i and the following two lemmas.

Lemma 2 *For* $1 \leq i \leq n$, *if* $R(Q, \mathcal{P})_{\leq v_i}$ *is a roadmap for* $S_{\leq v_i}$, *then* $R(Q, \mathcal{P})_{< v_{i+1}}$ *is a roadmap for* $S_{< v_{i+1}}$.

Lemma 3 *For* $1 \leq i \leq n$, *if* $R(Q, \mathcal{P})_{< v_i}$ *is a roadmap for* $S_{< v_i}$ *then* $R(Q, \mathcal{P}, M)_{\leq v_i}$ *is a roadmap for* $S_{\leq v_i}$.

4.8   Complexity

When $\ell = 0$, it follows from the analysis of the algebraic case, that the complexity is $s d^{O(k^2)}$. When $k = 1$, the complexity is $(s \log s) d^{O(1)}$.

The total number of calls to the curve segments subroutine is $\sum_{1 \leq j < \ell} \binom{s}{j}$, and each call costs $d^{O(k)}$. Thus, the total cost of the calls to the curve segments subroutine is bounded by $s^{\ell-1} d^{O(k)}$ arithmetic operations in D.

The cost of computing the intersection of the curves computed with the zero sets of the polynomials in $\mathcal{P}$ is bounded by by $s^\ell d^{O(k)}$.

The total cost of the calls to the finger subroutine and linking is bounded by $s^{\ell+1} d^{O(k)}$.

We next count the recursive calls. For each j, $0 \leq j < \ell$ we make $\binom{O(s)}{j} d^{O(k)}$ recursive calls, to the algorithm with the system having combinatorial level $\ell - j$ and geometric dimension $k - 1$.

Let $T(s, d, \ell, k)$ denote the number of arithmetic operations needed for the problem with these parameters. Since at any depth of the recursion the cost of a single arithmetic operations is bounded by $d^{O(k^2)}$ arithmetic operations in D, we ignore the fact that the ring changes as we go down in the recursion. Thus, we have the following recurrence,

$$T(s,d,\ell,k) \leq \sum_{0 \leq j < \ell} \binom{O(s)}{j} d^{O(k)} T(s,d,\ell-j,k-1) + O(s^{\ell+1}d^{O(k)}), \ell > 0, k > 1,$$

$$T(s,d,0,k) = sd^{O(k^2)}, k > 1,$$

$$T(s,d,\ell,1) = s \log sd^{O(1)}.$$

This recurrence solves to $T(s,d,\ell,k) = s^{\ell+1}d^{O(k^2)}$.

It follows immediately that the total cost is still bounded by $s^{\ell+1}d^{O(k^2)}$.

The complexity of the *connecting subroutine* is the following. If x is described by polynomials of degree bounded by t, the complexity of connecting x to $R(Z(Q),\mathcal{P})$ is $\ell st^{O(1)}d^{O(k^2)}$.

5   The Algorithm for the General Case

5.1   Basic idea

We give a roadmap algorithm for an arbitrary semi-algebraic set S, defined by the family of polynomials $\mathcal{P}$ which is contained in an algebraic variety V with given real dimension k'. Note that we do not compute the real dimension of V during the algorithm. According to subsection 2.2 we can suppose without loss of generality that $Z(Q)$ is bounded.

The idea is to construct uniform roadmaps for a perturbed family of polynomials which are in general position over approximating varieties which are close to V and of dimension k'. Thus the combinatorial level is k'. We then take the limit of the curves so constructed as the parameters of perturbation tend to 0. In order to ensure that the curve so constructed is connected in every connected component, we need to add curves connecting well chosen points to the various roadmaps on approximating varieties. These points are of two kinds:

1. *linking points* which ensure that curves coming from different approximating varieties whose images under the eval map intersect get connected and
2. *touching triples* ensuring that if the union of two connected components of two different sign conditions is connected, the union of the curves constructed in the two connected components is connected.

5.2   Linking points

We recall the construction of approximating varieties in [15] (see also [3]).

Let $V = Z(Q)$ be a real algebraic variety of dimension k'. Suppose that $V \subset B_0(r)$." We assume that Q has degree at most d and is non-negative non-negative everywhere on R^k. This assumption causes no loss of generality as we can replace Q with Q^2 at the cost of doubling the degree of Q.

For any index set $I = \{i_{k'+1}, \ldots, i_k\} \subset \{1, \ldots, k\}$ and an infinitesimal δ, let

$$a = (k - k')r^{-2(d+1)},$$

$$Q_I = (1 - \delta)Q - \delta(X_{i_{k'+1}}^{2(d+1)} + \cdots + X_{i_k}^{2(d+1)} - a),$$

$$Q_{I,T} = (1 - T)Q - T(X_{i_{k'+1}}^{2(d+1)} + \cdots + X_{i_k}^{2(d+1)} - a),$$

$$\mathcal{Q}_I = \{Q_I, \frac{\partial Q_I}{\partial X_{i_{k'+2}}}, \ldots \frac{\partial Q_I}{\partial X_{i_k}}\}$$

$$V_I = Z(\mathcal{Q}_I)$$

$$W = \cup_{I \subset [1,k], |I| = k - k'} V_I$$

With these notations, the following key results appear in [15].

Proposition 4 *[15] The real dimension of V_I is at most k'.*

Proposition 5 *[15] For every $x \in V$ there exists $y \in W$ such that $\mathrm{eval}_\delta(y) = x$.*

We now describe the construction of the linking points.
First we note that

$$\mathrm{eval}(V_I) = \mathrm{closure}(\{(x,t) \in R^{k+1} \mid Q_{I,t}(x) = 0\}) \cap \{T = 0\}$$

(this is a consequence of the proof that the image under eval of a bounded semi-algebraic set is semi-algebraic in [1]).
If S is a semi-algebraic set described by a quantifier-free formula, $\phi(X)$, then the closure of S is described by the following quantified formula:

$$\psi(X) = \forall(\epsilon)\exists(Y)(|X - Y|^2 < \epsilon^2) \wedge \phi(Y).$$

Note that, $\psi(X)$ is a first-order formula with two block of quantifiers, the first with one variable and the second with k variables. We denote by $\mathcal{R}_I$ the set of polynomials in $k + 1$ variables obtained after two steps of the block elimination subroutine in [1] applied to polynomials appearing in the first order formula describing closure($\{(x,t) \in R^{k+1} \mid Q_{I,t}(x) = 0\}$). These polynomials are such that closure($\{(x,t) \in R^{k+1} \mid Q_{I,t}(x) = 0\}$) is the union of semi-algebraically connected components of sets defined by sign conditions over $\mathcal{R}_I$ (note that we do not say that closure($\{(x,t) \in R^{k+1} \mid Q_{I,t}(x) = 0\}$) can be described by polynomials in $\mathcal{R}_I$). According to [1], the set $\mathcal{R}_I$ has $d^{O(k)}$ elements, and each of these elements has degree at most $d^{O(k)}$. We denote by $\mathcal{P}_I$ the set of polynomials in k variables obtained by substituting 0 for T in $\mathcal{R}_I$. Next we apply the sample point algorithm on a variety ([3]) with input V and $\mathcal{P} \cup_I \mathcal{P}_I$. The algorithm outputs a set L of $s^{k'} d^{O(k^2)}$ *linking points* defined by polynomials of degree at most $d^{O(k^2)}$. The cost is bounded by $s^{k'+1} d^{O(k^2)}$.

Note that each connected component, C, of every set defined by weak sign-conditions on the family $\mathcal{P}$ contained in V, is covered by the union of the approximating varieties V_I. Since, C is closed $\mathrm{eval}_\delta(V_I) \cap C$ is also closed. Moreover, $\cup_I \mathrm{eval}_\delta(V_I) \cap C = C$. Let $\mathcal{I}$ be the set of multi-indices I such that $\mathrm{eval}_\delta(V_I) \cap C$

is non-empty. Thus, if we connect, pairwise, the separate roadmaps arising from $V_I, V_J, I, J, \in \mathcal{I}$, for each pair I, J for which $\text{eval}_\delta(V_I) \cap \text{eval}_\delta(V_J) \cap C$ is non-empty, our roadmap will be connected inside C. The linking points allow us to do precisely this.

We use the ample point subroutine ([1]) to construct for every $p \in L$ and every I such that $p \in \text{eval}(V_I)$ a point p_I in V_I infinitesimally close to V and denote the points so obtained, L_I.

5.3 Touching triples

We compute a set N of *touching triples* $(p_1, p_2, \gamma_{1,2})$, where p_1, p_2 are points, and $\gamma_{1,2}$ is a semi-algebraic path joining, p_1 to p_2. The set N has the following property: For any two semi-algebraically connected components, C_1 and C_2, of sets S_{σ_1} and S_{σ_2}, where σ_1 and σ_2 are sign conditions on $\mathcal{P}$, such that $\bar{C}_1 \cap C_2 \neq \emptyset$, there exists $(p_1, p_2, \gamma_{1,2}) \in N$, such that $p_1 \in C_1$, $p_2 \in C_2$ and $\gamma_{1,2} \setminus \{p_2\} \in C_2$.

Thus, if C_1 and C_2 are two semi-algebraically connected components of two distinct sign conditions which are included in S and whose union is semi-algebraically connected then there exists $(p_1, p_2, \gamma_{1,2}) \in N$ such that $\gamma_{1,2}$ connects the point $p_1 \in C_1$ with the point $p_2 \in C_2$ by a semi-algebraic path lying in $C_1 \cup C_2$.

In order to compute the set N we introduce two infinitesimals $\eta' \ll \eta$, and consider the set $\mathcal{P}' = \cup_{P \in \mathcal{P}}\{P, P \pm \eta', P \pm \eta\}$.

We call the sample points subroutine on V [3] with the family $\mathcal{P}'$ as input, and construct a set of points in $(R\langle \eta, \eta' \rangle)^k$ which intersects every semi-algebraically connected component of the family $\mathcal{P}'$.

For each point p_1 constructed above, we construct the point $p_2 = \text{eval}_{\eta'}(p_1)$. The point p_1 is represented as a k-univariate representation (u, σ) with coefficients in $D[\eta, \eta']$. Replacing, η' in u by a new variable t, and letting t vary over the interval $[0, \eta']$ we get a semi-algebraic path $\gamma_{1,2}$ joining p_1 to p_2. We include the triple $(p_1, p_2, \gamma_{1,2})$ in the set N.

Next, we construct for every p such that there exists q and γ with $(p, q, \gamma) \in N$ and for every I such that $p \in \text{eval}(V_I)$ a point p_I in V_I infinitesimally close to V_I by the sample point subroutine ([1]) and denote by N_I the points of V_I so obtained.

We call M_I the union of N_I and L_I.

5.4 The Algorithm

We are finally in a position to describe the algorithm in the general case. We take a first infinitesimal δ. First we compute the set of linking points L and the corresponding L_I. Then, we compute the set of touching triples N and the corresponding set N_I and we output $M_I = L_I \cup N_I$.

We make a perturbation of the polynomials in $\mathcal{P}$ as follows.

We take two more infinitesimals ϵ and ϵ' with $\delta \ll \epsilon' \ll \epsilon$.

We replace the set $\mathcal{P}$ by the following set, $\mathcal{P}^*$, of $4s$ polynomials:

$$\mathcal{P}^* = \cup_{i=1,\ldots,s}\{(1 - \epsilon')P_i - \epsilon H_{4i-3}, (1 - \epsilon')P_i + \epsilon H_{4i-2},$$

$$(1 - \epsilon')P_i - \epsilon'\epsilon H_{4i-1}, (1 - \epsilon')P_i + \epsilon'\epsilon H_{4i}\},$$

where $H_i = (1 + \sum_{1 \le j \le k} i^j x_j^{d'})$ and d' is an even number greater than the degree of any P_i.

We have the following two lemmas which show that the family $\mathcal{P}^*$ is in general position over each V_I, and that there exists a correspondence between the cells of $\mathcal{P}$ and $\mathcal{P}^*$. Very similar lemmas occur in [3].

Lemma 6 *The combinatorial level of $(\mathcal{Q}_I, \mathcal{P}^*)$ is at most k'.*

Lemma 7 *Let σ be a realizable strict sign condition on the set of polynomials $\mathcal{P}$ on V given by $P_1 = \ldots = P_\ell = 0, P_{\ell+1} > 0, \ldots, P_s > 0$, and let C be a semi-algebraically connected component of the intersection of the realization $\mathcal{R}(\sigma)$ of σ with V. Let σ' be the weak sign condition on $\mathcal{P}^*$ given by,*

$$-\epsilon\epsilon' H_{4i} \le (1 - \epsilon')P_i \le \epsilon'\epsilon H_{4i-1}, 1 \le i \le \ell,$$

$$(1 - \epsilon')P_i \ge \epsilon H_{4i-3}$$

and $\mathcal{R}(\sigma')$ its realization. Then, for every I such that $\mathrm{eval}_{\epsilon'}(V_I) \cap \mathcal{R}(\sigma) \ne \emptyset$, there exists a unique semi-algebraically connected component C' of $\mathcal{R}(\sigma') \cap V_I$ such that $\mathrm{eval}_{\epsilon'}(C') \subset C_{R\langle\epsilon\rangle}$.
Moreover, if $x \in R^k$ is in C then $x \in C'$.

We now use the algorithm for constructing uniform roadmaps with input $(\mathcal{Q}_I, \mathcal{P}^*)$. We connect the points of M_I to the uniform roadmap of $(\mathcal{Q}_I, \mathcal{P}^*)$ using the connecting subroutine.

We then compute the image of the roadmap constructed above under the $\mathrm{eval}_{\epsilon'}$ map, and retain only those portions which are in the given set S.

In order to connect a point x to the roadmap, an I such that $x \in \mathrm{eval}_{\epsilon'}(V_I)$ is chosen and a point x_I infinitesimally close to x in V_I is constructed using the sample points subroutine ([1]). This point x_I is connected to the uniform roadmap $R(\mathcal{Q}_I, \mathcal{P}^*)$. Then we compute the image of the connecting curves under the $\mathrm{eval}_{\epsilon'}$ map.

5.5   Proof of Correctness and Complexity

The proof of correctness follows from the proof of correctness of the uniform algorithm and lemmas 6 and 7.

The combinatorial level of $(\mathcal{Q}_I, \mathcal{P}^*)$ is bounded by k'. Using the complexity bound of the uniform algorithm and noting that the cost of computing the various sets M_I is $s^{k'} d^{O(k^2)}$, we see that the complexity is bounded by $s^{k'+1} d^{O(k^2)}$.

If x is described by polynomials of degree at most t, the complexity of connecting x to the roadmap is $k' s t^{O(1)} d^{O(k^2)}$.

References

1. S. BASU, R. POLLACK, M.-F. ROY *On the combinatorial and algebraic complexity of Quantifier Elimination*, Proc. *35th IEEE Symp. on Foundations of Computer Science*, 632-642, (1994). (Journal version to appear in JACM.)
2. S. BASU, R. POLLACK, M.-F. ROY *Computing Roadmaps of Semi-algebraic Sets*, Proc. *28th Annual ACM Symposium on the Theory of Computing*, 168-173, 1996.
3. S. BASU, R. POLLACK, M.-F. ROY *On Computing a Set of Points meeting every Semi-algebraically Connected Component of a Family of Polynomials on a Variety* , submitted to Journal of Complexity.
4. J. BOCHNAK, M. COSTE, M.-F. ROY Géométrie algébrique réelle. Springer-Verlag (1987).
5. J. CANNY, The Complexity of Robot Motion Planning, MIT Press (1987).
6. J. CANNY, *Computing road maps in general semi-algebraic sets*, The Computer Journal, 36: 504–514, (1993).
7. G. E. COLLINS, *Quantifier elimination for real closed fields by cylindrical algebraic decomposition* Springer Lecture Notes in Computer Science 33, 515- 532.
8. D. GRIGOR'EV, N. VOROBJOV *Counting connected components of a semi-algebraic set in subexponential time.* Comput. Complexity, 2 :133-186 (1992).
9. L. GOURNAY, J. J. RISLER *Construction of roadmaps of semi-algebraic sets.* Technical Report, DMI, Ecole Normale Superieure, Paris, France (1992).
10. R. M. HARDT *Semi-algebraic Local Triviality in Semi-algebraic Mappings* American Journal of Math. 102, 1980, 291-302.
11. J. HEINTZ, M.-F. ROY, P. SOLERNÒ Single exponential path finding in semi-algebraic sets II : The general case. Algebraic geometry and its applications. C. L. Bajaj editor, Springer Verlag (1993) 467-481.
12. J.-C. LATOMBE *Robot Motion Planning*, Kluwer Academic Publishers.
13. J. MILNOR *Morse Theory*, Annals of Mathematical Studies, Princeton University Press.
14. J. RENEGAR *On the computational complexity and geometry of the first order theory of the reals*, J. of Symbolic Comput., (1992).
15. M.-F. ROY, N. VOROBJOV *Computing the Complexification of a Semi- algebraic Set*, ISSAC 96.
16. J. SCHWARTZ, M. SHARIR *On the 'piano movers' problem II. General techniques for computing topological properties of real algebraic manifolds.*, Advances in Applied Mathematics 4, 298-351.

Essentially Smooth Lipschitz Functions: Compositions and Chain Rules

Jonathan M. Borwein and Warren B. Moors

Abstract. We introduce a new class of real-valued locally Lipschitz functions, which we call *arc-wise essentially smooth*, and we show that if $g : R^n \to R$ is arc-wise essentially smooth on R^n and each function $f_j : R^m \to R$, $1 \leq j \leq n$ is strictly differentiable almost everywhere in R^m, then $g \circ f$ is strictly differentiable almost everywhere in R^m, where $f \equiv (f_1, f_2, ...f_n)$. We also show that all the semi-smooth and pseudo-regular functions are arc-wise essentially smooth. Thus, we provide a large and robust lattice algebra of Lipschitz functions whose generalized derivatives are extremely well-behaved.

Keywords: Lipschitz functions, Chain rule, Null sets, Differentiability, Essentially strictly differentiable

AMS (1991) subject classification: Primary: 49J520; 46N10 Secondary: 58C20.

1 Introduction and Preliminaries

In this extended abstract we indicate, in finite dimensions, that locally Lipschitz functions which are essentially smooth (strictly differentiable almost everywhere) possess extremely strong closure properties. The full, quite technical, details are given in [4]. Essentially strictly differentiable functions were studied in great detail in [3]. In particular, it was shown therein that all such functions possess very well-behaved Clarke generalized gradients. More specifically, each such function is:

- **generic**: generically Fréchet differentiable;
- **minimal**: the Clarke subgradient is reconstructable from *any* dense set of gradients;
- **integrable:** any function sharing the same Clarke subgradient must differ only by a constant (on any open connected set).

In consequence, this class provides an appropriate "pathology free" environment for nonsmooth analysis and optimization, both theoretical and algorithmic. For example, it is by now well appreciated that "semi-smoothness" plays an important role in the analysis of nonsmooth algorithms.

We recall some preliminary definitions regarding the Clarke subdifferential mapping, [6]. A real-valued function f defined on a non-empty open subset A of a Euclidean (finite dimensional) space X is *locally Lipschitz* on A, if for each $x_0 \in A$ there exists a $K > 0$ and a $\delta > 0$ such that

$$|f(x) - f(y)| \le K\|x - y\| \quad \text{for all} \quad x, y \in B(x_0, \delta)$$

($B(x_0, \delta)$ is the ball of radius δ around x_0.) For functions in this class, it is often instructive to consider the following three directional derivatives:

i. The *upper Dini derivative* at $x \in A$, in the direction y, is given by,

$$f^+(x; y) \equiv \limsup_{\lambda \to 0^+} \frac{f(x + \lambda y) - f(x)}{\lambda}.$$

ii. The *lower Dini derivative* at $x \in A$, in the direction y, is given by,

$$f^-(x; y) \equiv \liminf_{\lambda \to 0^+} \frac{f(x + \lambda y) - f(x)}{\lambda}.$$

iii. The *Clarke generalised directional derivative* at $x \in A$, in the direction y, is given by,

$$f^0(x; y) \equiv \limsup_{\substack{z \to x \\ \lambda \to 0^+}} \frac{f(z + \lambda y) - f(z)}{\lambda}.$$

It is immediate from these three definitions that for each $x \in A$ and each $y \in X$,

$$f^-(x; y) \le f^+(x; y) \le f^0(x; y).$$

We require several notions of differentiability associated with a locally Lipschitz function f:

iv. f is *differentiable at x, in the direction y*, if

$$f'(x; y) \equiv \lim_{\lambda \to 0} \frac{f(x + \lambda y) - f(x)}{\lambda} \quad \text{exists.}$$

v. f is *Gateaux differentiable at x*, if

$$\nabla f(x)(y) \equiv \lim_{\lambda \to 0} \frac{f(x + \lambda y) - f(x)}{\lambda} \quad \text{exists.}$$

for each $y \in X$ and $\nabla f(x)$ is a continuous linear functional on X.

We also require two slightly stronger notions of differentiability.

vi. f is *strictly differentiable at x, in the direction y*, if

$$\lim_{\substack{z \to x \\ \lambda \to 0^+}} \frac{f(z + \lambda y) - f(z)}{\lambda} \quad \text{exists.}$$

vii. f is *strictly differentiable* at x, if f is strictly differentiable at x, in every direction $y \in X$. We note that a function f is strictly differentiable at x, in the direction y if, and only if,

$$f^0(x; y) = f'(x; y) = -f^0(x; -y).$$

In [3] the present authors called a real-valued locally Lipschitz function f, defined on a non-empty open subset A, of a separable Banach space X, *essentially strictly differentiable* on A, if $\{x \in A : f$ is not strictly differentiable at $x\}$ is a null set, (Haar-null in the general setting, Lebesgue null in finite dimensions) and they denoted by $\mathcal{S}_e(A)$, the family of all real-valued essentially strictly differentiable locally Lipschitz functions defined on A. We shall also call such functions *essentially smooth* since strict differentiability is an appropriate localization of continuous differentiability.

We also have reason to consider vector-valued functions. Let A be a non-empty open subset of a Euclidean space X and let $f : A \to R^n$ be defined by

$$f(x) \equiv (f_1(x), f_2(x), ...f_n(x)) \quad \text{where } f_j : A \to R.$$

We say that the vector-valued function f is *essentially strictly differentiable* on A if $f_j \in \mathcal{S}_e(A)$ for each $1 \leq j \leq n$, and in this case we write: $f \in \mathcal{S}_e(A; R^n)$. In addition, we will say that f is *strictly differentiable* at $x \in A$, in the direction y if,

$$f_j^0(x; y) = -f_j^0(x; -y) \quad \text{for each } 1 \leq j \leq n.$$

In this case we write: $f^0(x; y) = -f^0(x; -y)$.

Let us now return to the purpose at hand: to show that the family of functions $\mathcal{S}_e(A)$, possesses very striking *closure properties* under composition. A first and optimistic guess might be, that if $f_1, f_2, \cdots, f_n \in \mathcal{S}_e(A)$ and $g \in \mathcal{S}_e(R^n)$, then $g \circ f \in \mathcal{S}_e(A)$, where $f \equiv (f_1, f_2, ...f_n)$. However, the following example shows that in full generality this is not true.

Example 1 Let A denote the open interval $(0, 1)$ in R and let C denote a Cantor subset of A with $\mu(C) > 0$. Define the real-valued functions f_1, f_2 and d_C on A by: $f_1 \equiv 0$, $f_2(t) \equiv t$ and $d_C(t) \equiv \text{dist}(t, C)$. Further, define $g : R^2 \to R$ by $g(x, y) \equiv \text{dist}((x, y), \{0\} \times C)$.

Clearly, f_1 and f_2 are strictly differentiable almost everywhere, in fact f_1 and f_2 are strictly differentiable everywhere. Moreover, by Theorem 8.5 in [3] we have that $g \in \mathcal{S}_e(R^2)$. We claim that $g \circ f \notin \mathcal{S}_e(A)$, where $f \equiv (f_1, f_2)$. To see this, observe that $g \circ f(x) = d_C(x)$ for each $x \in A$. Now, it is standard that d_C is not strictly differentiable at any point $x \in C$. Hence, it follows that $g \circ f$ is not strictly differentiable at any point of C, and so $g \circ f \notin \mathcal{S}_e(A)$.

We will see that imposing a tighter condition on g will repair this defect.

2   A chain rule for Lipschitz functions.

We now indicate that there is a large subclass of the essentially strictly differentiable functions, that is closed under composition. This class of functions is closely related to Valadier's *saine* functions, [9]. Let A be a non-empty open subset of R^n. We say that a real-valued locally Lipschitz function f defined on A is *arc-wise essentially smooth* on A if for each locally Lipschitz function $x \in \mathcal{S}_e((0,1); R^n)$,

$$\mu(\{t \in (0,1) : f^0(x(t); x^+(t)) \neq -f^0(x(t); -x^+(t))\}) = 0.$$

[Here, $x^+(t)$ denotes a vector of Dini derivatives. Valadier requires this to hold for *all* absolutely continuous arcs.] We shall denote by $\mathcal{A}_e(A)$, the family of all arc-wise essentially smooth functions on A.

Proposition 1 *For each non-empty open subset A of R^n, $\mathcal{A}_e(A) \subseteq \mathcal{S}_e(A)$.*

Proposition 2 *Let A be a non-empty open subset of R. Then $\mathcal{A}_e(A) = \mathcal{S}_e(A)$.*

Our next goal is to show that the class of functions $\mathcal{A}_e(A)$ is reasonably large. Let us begin with the obvious observation that $\mathcal{A}_e(A)$ contains all the C^1 functions defined on A. A useful adjunct is:

Lemma 3 *Let A be a non-empty open subset of R^n. Let f be a real-valued locally Lipschitz function defined on A. Then $f \in \mathcal{A}_e(A)$, if for each essentially strictly differentiable curve $x : (0,1) \to A$,*

$$\mu(\{t \in (0,1) : f^0(x(t); x'(t)) = f'(x(t); x'(t))\}) = 1.$$

The following lemma encapsulates the heart of our chain rule.

Lemma 4 *Let A be a non-empty open subset of a Euclidean space X. Let $f \equiv (f_1, f_2, ... f_n)$ be a locally Lipschitz function from A into R^n. Furthermore, let g be a real-valued locally Lipschitz function defined on a non-empty open subset B of R^n which contains $f(A)$. Suppose f is strictly differentiable at some point $x_0 \in A$, in the direction y and that either, (i) g is strictly differentiable at $f(x_0)$, in the direction $f'(x_0; y)$ or, (ii) $f'(x_0; y) = 0$. Then $g \circ f$ is strictly differentiable at x_0, in the direction y.*

We need a few more definitions before we can show that $\mathcal{A}_e(A)$ contains a significant class of functions, (above and beyond the C^1 functions). Let f be a real-valued locally Lipschitz function defined on a non-empty open subset A of a Euclidean space X. Then f is *upper semi-smooth (lower semi-smooth)* at a point $x \in A$, in the direction y if,

$$f^+(x; y) \geq \limsup_{\substack{t \to 0^+ \\ y' \to y}} f^+(x + ty'; y) \quad \left(f^-(x; y) \leq \liminf_{\substack{t \to 0^+ \\ y' \to y}} f^-(x + ty'; y) \right).$$

Moreover, we say that f is *semi-smooth* at a point $x \in A$, *in the direction* y if,

$$\limsup_{\substack{t \to 0^+ \\ y' \to y}} f^+(x + ty'; y) = f^+(x; y) = f^-(x; y) = \liminf_{\substack{t \to 0^+ \\ y' \to y}} f^-(x + ty'; y).$$

Additionally, we say that f is *arc-wise essentially upper semi-smooth* (*arc-wise essentially lower semi-smooth*) on A, if for each $x \in S_e((0, 1); A)$

$$\mu(\{t \in (0, 1) : f \text{ is upper (lower) semi-smooth at } x(t) \text{ in the direction } x'(t)\}) = 1.$$

The following technical result allows us to perform only "unilateral" analysis in our main theorem.

Theorem 5 *Let f be a real-valued locally Lipschitz function defined on a non-empty open subset A of a Euclidean space X. If f is arc-wise essentially upper semi-smooth (arc-wise essentially lower semi-smooth) on A, then $f \in \mathcal{A}_e(A)$.*

Now, we may show that the family of functions $\mathcal{A}_e(A)$ enjoys quite remarkable closure properties.

Theorem 6 *Let A be a non-empty open subset of R^m and suppose $f_1, f_2, ... f_n \in \mathcal{A}_e(A)$. Let g be a real-valued function locally Lipschitz function defined on a non-empty open subset U of R^n, which contains $f(A)$, where $f \equiv (f_1, f_2, \cdots, f_n)$. If $g \in \mathcal{A}_e(U)$, then $g \circ f \in \mathcal{A}_e(A)$.*

On relaxing the hypotheses on the $f_1, f_2, \cdots, f_n$ we may still recover a most satisfactory theorem:

Theorem 7 *Let A be a non-empty open subset of R^m, and suppose $f_1, f_2, ... f_n \in S_e(A)$. Suppose further, that U is a non-empty open subset of R^n, which contains $f(A)$, where $f \equiv (f_1, f_2, \cdots, f_n)$. Then for each $g \in \mathcal{A}_e(U)$, $g \circ f \in S_e(A)$.*

Corollary 8 *Let A be a non-empty open subset of R^m then $\mathcal{A}_e(A)$ and $S_e(A)$ are function algebras and vector lattices containing the C^1-functions and the convex functions.*

We should also observe that from the proof of the Theorem 7, it can be shown that for almost all x in A,

$$\nabla(g \circ f)(x) = \partial g(f(x)) \cdot \nabla f(x).$$

Here ∂g is the Clarke subgradient of g. Note, however, that g may fail to be strictly differentiable at $f(x)$, unless the range of $\nabla f(x)$ is all of R^n.

Finally, let us observe that the class of functions $\mathcal{A}_e(R^n)$ contains many distance functions. Indeed, one can show that if C is a regular subset of R^n, then the distance function generated by this set and any smooth norm on R^n, is in this class. To see this, we need to make the following observations:

(i) For any norm on R^n, the distance function d_C is regular at each point of C, see [1].

(ii) If the norm is smooth on R^n, then $-d_C$ is regular on $R^n \backslash C$, see [2].

We may now define an appropriate class of closed subsets of R^n as follows. A subset C of R^n is *arc-wise essentially smooth* if for each $x \in \mathcal{S}_e((0,1); R^n)$ the set

$$\{t \in (0,1) : x'(t) \in K_C(x(t)) \backslash T_C(x(t))\}.$$

has measure zero. Here $K_C(x)$ denotes the *contingent cone* of C at x, and $T_C(x)$ denotes the *Clarke tangent cone* of C at x. Recall that C is regular at x if $K_C(x) = T_C(x)$, [6]. Note that all sets that are regular except on a countable set are arc-wise essentially smooth. Thus, all closed convex sets and all smooth manifolds are arc-wise essentially smooth.

Proposition 9 *Let $\| \cdot \|$ be a smooth norm on R^n and let C be a non-empty closed subset of R^n. Then the distance function generated by the norm $\| \cdot \|$ and the set C is a member of $\mathcal{A}_e(R^n)$ if, and only if, C is arc-wise essentially smooth.*

It is immediate from Proposition 9 and Theorem 6 that, unlike the family of regular sets, the family of arc-wise essentially smooth sets is closed under finite unions. We should also observe that from Proposition 2 it follows that the boundary of every arc-wise essentially smooth set is a Lebesgue null set. We also note that the set of Example 1 is a null set but is not arc-wise essentially smooth.

Finally, observe that we may now provide general conditions to ensure that *exact penalty functions* of the form

$$f(x) + K d_C(x)$$

will be in $\mathcal{S}_e(A)$ or in $\mathcal{A}_e(A)$. Reasonable *marginal functions* also lie in $\mathcal{A}_e(R^n)$ or $\mathcal{S}_e(R^n)$ ([3]). Thus, these classes successfully exclude most of the "monsters" from nonsmooth optimization.

References

1. J.M. Borwein and M. Fabian, "A note on regularity of sets and of distance functions in Banach space," *J. Math. Anal. Appl*, **182** (1994), 566–560.
2. J. M. Borwein, S. P. Fitzpatrick and J. R. Giles, "The differentiability of real functions on normed linear spaces using generalized subgradients," *J. Math. Anal. and Appl.* **128** (1987), 512–534.
3. J. M. Borwein and W. B. Moors, "Essentially strictly differentiable Lipschitz functions," CECM Report 95-029.
4. J. M. Borwein and W. B. Moors, "A Chain Rule for Lipschitz Functions," CECM Report 96-057.
5. J. P. R. Christensen, "On sets of Haar measure zero in Abelian groups," *Israel J. Math.* **13** (1972), 255–260.
6. F. H. Clarke, Optimization and Nonsmooth Analysis, John Wiley, New York (1971).

7. W. B. Moors, "A characterisation of minimal subdifferential mappings of locally Lipschitz functions," *Set-Valued Analysis* **3** (1995), 129-141.
8. M. E. Munroe, Measure and Integration, Second edition, Addison-Wesley, Reading, 1971.
9. M. Valadier, "Entrainement unilateral, lignes de descente, fouctions Lipshitziennes non pathologiques," *C. R. Acad. Sci. Paris 308 Series I* (1989), 241-244.

Junction Detection and Filtering

Vicent Caselles, Bartomeu Coll and Jean-Michel Morel

Abstract. Models for image smoothing have not fully incorporated the restrictions imposed by the physics of image generation. In particular, any image smoothing process should respect the essential singularities of images (which we call junctions) and should be invariant with respect to contrast changes. We conclude that local image smoothing is possible provided singular points of the image have been previously detected and preserved. We define the associated degenerate partial differential equation and sketch the proof of its mathematical validity. Our analysis uses the phenomenological theory of image perception of Gaetano Kanizsa, whose mathematical translation yields an algorithm for the detection of the discontinuity points of digital images.

1 Introduction

Images, in a continuous model, are functions $u(x)$ defined on a domain Ω of the plane, generally a rectangle. Let us consider, without much loss of generality, grey level images, that is, images where $u(x)$ is a real number denoting the 'grey level' or brightness, tipically between 0 and 255. In practice, images are *digitized*, which means that the values of $u(x)$ are given only in a discrete grid of a rectangle. It is, however, very convenient to discuss the geometry of images in the continuous framework of functional analysis. This passage to a continuous model is valid provided it is shown that the algorithms defined in the continuous model can be made effective on their digitized trace.

If we accept as our primary goal to give a description of the objects present in an image we immediately face a difficulty, objects occur at different scales (and this leaving aside the question of what we mean by objects in real images). On the other hand, derivatives seem to be a nice tool to identify local variations of image intensity and we would like to compute them. The digitized image being discontinuous everywhere, this looks desperate unless some notion of scale is not introduced. The scale parameter, which we denote by t, measures the degree of smoothing we allow in an image. Scale $t = 0$ corresponds to the actual image, and we denote by $u(t, x)$ the image smoothed at scale t. As shown in [2], local smoothing processes are represented by solutions of parabolic partial differential equations whose initial datum $u(0, x) = u(x)$ is the image we want to analyze.

For a full discussion of it, we send the reader to [2]. Many smoothing equations have been proposed in the literature. For our purposes we recall three significant examples:

• The heat equation [21] :

$$\frac{\partial u}{\partial t} = \Delta u \,.$$

(1)

• The mean curvature motion (Osher-Sethian equation [18], see also [19])

$$\frac{\partial u}{\partial t} = |Du|curv(u) \,.$$

(2)

• The A.M.S.S. model ([2], [20])

$$\frac{\partial u}{\partial t} = |Du|(curv(u))^{\frac{1}{3}} \,.$$

(3)

where $\Delta u = \frac{\partial^2 u}{\partial x^2} + \frac{\partial^2 u}{\partial y^2}$ denotes the Laplacian of $u(t,x,y)$ with respect to the space variables (x,y), $Du = (\frac{\partial u}{\partial x}, \frac{\partial u}{\partial y})$ denotes the spatial gradient of u, $|Du|$ being the euclidean norm of Du and $curv(u) = div(\frac{Du}{|Du|})$, the curvature of the level lines of u, $div = \frac{\partial}{\partial x} + \frac{\partial}{\partial y}$ being the divergence operator. It is proven in [2] that the heat equation is the only linear isotropic image smoothing model and the A.M.S.S. model is the only contrast invariant and affine invariant image smoothing model (see also [20]). Contrast invariant means that the smoothing operator $T_t : u \rightarrow u(t) = T_t u$ commutes with any contrast change $u \rightarrow g(u)$, g being any nondecreasing real function. The Osher-Sethian equation is somehow an intermediate case, being both quasilinear and contrast invariant.

There is, however, a significant objection to the second and third models: in order to make sense mathematically, they assume the initial image to be continuous !. We shall see in the next section that this is not a realistic requirement for images. If we remove this requirement we can still draw the same conclusions as in [2], but only locally, at points where the image is continuous.

Our plan is as follows. In Sect. 2 we shall sketch the process of image formation and show how this process entails generation of singularities and leads to invariance requirements for the image operators. We shall be particularly interested in the singularities which are inherent to the image formation process. Then we shall discuss wether smoothing procedures which have been proposed in the past are compatible or not with the image formation process. In short: can a smoothing process be performed which does not destroy what it ought to detect, the image structure ?. We shall see that the answer is close to negative, and show that in fact *singularities in the image must be detected previously to any smoothing procedure*. Thus the rest of the paper is as follows. In Sect. 3, we deduce which are the largest invariant objects (or "atoms") in an image which can be taken as initial data for smoothing operations: we prove that they consist in pieces of level lines of the image joining singular points. Section 4 is devoted to the effective computation of "atoms" and singular points of the image and to first experiments. In Sect. 5 we propose a P.D.E. model with the right boundary conditions at image singular points. We display some experiments showing that the visual aspect of images is not altered by such structure preserving smoothing process. We display an experiment to show how, thanks to the smoothing thus defined, we can transform an image into a topographic map revealing the structure of the image and its essential singularities.

2 How Images Are Generated: Occlusion and Transparency as Basic Operations

We shall assume a simple model for image generation which is essentially geometric and only assumes that light propagates in straight lines, a physically approximately true law. According to this basic model, the *camera oscura* or *pinhole* model, every point of the surface of physical objects reflects solar or artificial light in all directions. Generation of images occurs when this light is focused by passing through a hole on the side of a black box and finally impressed on the face of the black box opposite to the hole. The black box can be a camera, in which case this face is a photosensitive paper or an array of electronic amplifiers. It also is the model of the eye, in which case the photosensitive surface is the retina. Calling x a generic point on the retina (or film) plane, we call $u(x)$ the result of this impression, which we shall throughout this paper assume to be a real value measuring intensity (or energy, or "grey level") of the light sent by x. We shall also assume that our images come from a natural, human or animal, environnment and therefore correspond to common retinal images. This may seem a very vague assertion but we shall give it a precise sense in the following.

2.1 Occlusion as Basic Operation on Images

Natural, human visual world is made of objects of many sizes and heights which are spread out on the ground and which tipically tend to occlude each other. As pointed out by the phenomenologist Kanizsa [15], we generally see parts of the objects in front of us because they occlude each other. Let us formalize this by assuming that objects have been added one by one to a scene. Given an object $\tilde{A}$ in front of the camera, we call A the region of the image onto which is projected by the camera. We call u_A the part of the image thus generated, which is defined in A. Assuming now that the object $\tilde{A}$ is added in a real scene $\tilde{R}$ of the world whose image was v, we observe a new image which depends upon which part of $\tilde{A}$ is in front of objects of $\tilde{R}$, and which part in back. Assuming that $\tilde{A}$ occludes objects of $\tilde{R}$ and is occluded by no object of $\tilde{R}$, we get a new image $u_{\tilde{R} \cup \tilde{A}}$ defined by

$$u_{\tilde{R} \cup \tilde{A}} = u_A \quad \text{in} \quad A$$
$$u_{\tilde{R} \cup \tilde{A}} = v \quad \text{in} \quad \mathbb{R}^2 \setminus A \ . \tag{4}$$

Of course, we do not take into account in this basic model the fact that objects in $\tilde{R}$ may intercept light falling on $\tilde{A}$, and conversely. In other terms, we have omitted the shadowing effects, which will now be considered.

2.2 Transparency (or Shadowing) as a Second Basic Operation on Images

Let us assume that the light source is a point in euclidean space, and that an object $\tilde{A}$ is interposed between a scene $\tilde{R}$ whose image is v and the light source. We call $\tilde{S}$ the shadow spot of $\tilde{A}$ and S the region it occupies in the image plane. The resulting image u is defined by

$$
\begin{aligned}
u_{\tilde{R},\tilde{S},g} &= v \quad \text{in} \quad \mathbb{R}^2 \setminus S \\
u_{\tilde{R},\tilde{S},g} &= g(v) \quad \text{in} \quad S .
\end{aligned}
\tag{5}
$$

Here, g denotes a contrast change function due to the shadowing, which is assumed to be uniform in $\tilde{S}$. Clearly, we must have $g(s) \leq s$, because the brightness decreases inside a shadow, but we do not know in general how g looks. The only assumption for introducing g is that points with equal grey level s before shadowing get a new, but the same, grey level $g(s)$ after shadowing. Of course, this model is not true on the boundary of the shadow, which can be blurry because of diffraction effects or because the light source is not really reducible to a point. Another problem is that g in fact depends upon the kind of physical surface which is shadowed so that it may well be different on each one of the shadowed objects. This is no real restriction, since this only means that the shadow spot S must be divided into as many regions as shadowed objects in the scene ; we only need to iterate the application of the preceding model accordingly. A variant of the shadowing effect which has been discussed in perception psychology by Gaetano Kanizsa is *transparency*. In the transparency phenomenon, a transparent homogeneous object $\tilde{S}$ (in glass for instance) is interposed between part of the scene and the observer. Since $\tilde{S}$ intercepts part of the light sent by the scene, we still get a relation like (2), so that transparency and shadowing simply are equivalent from the image processing viewpoint. If transparency (or shadowing) occurs uniformly on the whole scene, the relations (2) reduce to

$$
u_g = g(v) ,
\tag{6}
$$

which means that the grey-level scale of the image is altered by a nondecreasing contrast change function g.

2.3 Requirements for Image Analysis Operators

Of course, we do not know *a priori*, when we look at an image, what are the physical objects which have left a visual trace in it. We know, however, that the operations having led to the actual image are given by formulas (4, 5). Thus, any processing of the image should avoid to destroy the image structure resulting from (4, 5): this structure traduces the information about shadows, facets and occlusions which are clues to the physical organization of the scene in space. *The identity and shape of objects must be recovered from the image by means which should be stable with respect to those operations.* As a basic and important example, let us recall how the Mathematical Morphology school has insisted on the fact that image analysis operations must be invariant with respect to any contrast change g. In the following, we shall say that an operation T on an image u is *morphological* if

$$T(g(u)) = T(u) \qquad (7)$$

for any nondecreasing contrast change g.

3 Basic Objects of Image Analysis

We take into account the previous discussion to define the basic objects of image analysis. We call *basic objects* a class of mathematical objects, simpler to handle than the whole image, but into which any image can be decomposed and from which it can be reconstructed. Two classical examples of image decompositions are *additive decompositions into simple waves* (sine, cosine, wavelets, wavelet packets, etc) and *segmentations* (decomposition of the image into homogeneous regions separated by boundaries, or "edges"). The decomposition of an image into simple waves is an additive one and we have argued against it as not adapted to the structure of images, except for restoration processes. Indeed, operations leading to the construction of real world images are strongly nonlinear and the simplest of them, the constrast change, does not preserve additive decompositions. Concerning segmentations, the criterion for the creation of edges or boundaries is the strength of constrast on the edges, along with the homogeneity of regions. Both of these criteria are simply not invariant with respect to contrast changes. Indeed $\nabla g(u) = g'(u)\nabla u$ so that we can alter the value of the gradient by simply changing the constrast.

The Mathematical Morphology school offers a better alternative: to decompose an image into its binary shadows (or level sets), that is, we set $X_\lambda u(x) = white$ if $u(x) \geq \lambda$ and $X_\lambda u(x) = black$ otherwise. The white set is then called *level set* of u. It is easily seen that, under fairly general conditions, an image can be reconstructed from its level sets by the formula

$$u(x) = \sup\{\lambda, u(x) \geq \lambda\} = \sup\{\lambda, x \in X_\lambda u\} \ . \qquad (8)$$

The decomposition therefore is nonlinear, and reduces the image to a family of plane sets (X_λ). Notice that in the discrete framework, when the image is discretized in both space and grey level, the above formula is trivial to prove. In addition, if we assume that the level sets are closed Cacciopoli subsets of $\mathbb{R}^2$, that is, closed sets whose boundary has finite length, then a classical theorem asserts that its boundary is a union of closed Jordan curves (see [16], Chapter 6). In the discrete framework, all this is obvious and we can associate with each level set a finite set of Jordan curves which define its boundary. Conversely, the level set is uniquely defined from those Jordan curves. We shall call them *level curves of the image*. Are level sets and level curves the sought for basic objects of image processing? In some sense they are better than all above discussed "basic objects" because they are invariant under contrast changes. To be more precise, if we transform an image u into gou, where g is an increasing continuous function (understood as a contrast change), then it is easily seen that the set of level sets of gou is equal to the set of level sets of u. However, level sets are drastically altered by occlusion or shadowing. Let us discuss this point. One can see in Fig. 1 an elementary example of image generated by occlusion. A grey disk is partly occluded by a darker square (a). In (b) we display a perspective view of the image graph. In (c) and (d) we see two of the four distinct level sets of the image, the other ones being the empty set and the whole plane. It is easily seen that none of the level sets (c) and (d) corresponds to physical objects in the image. Indeed, one of them results from their union and the other one from their set difference. The same thing is true for the level lines (e) and (f) : they appear as the result of some "cut and paste" process applied to the level lines of the original objects. It is nontheless true, because of the invariance with respect to constrast changes, that the basic objects of image processing must be somehow parts of the level lines. This leads us to what will be our proposition.

• **Our proposition : Basic objects are all junctions of level lines, (particularly : T-junctions, X-junctions) and parts of level lines joining them.**

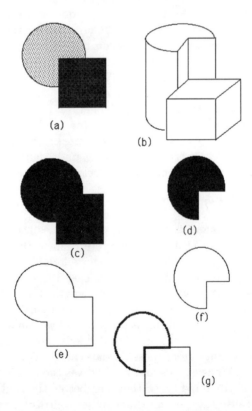

Fig. 1. Image, level sets, level lines and T-junctions

Let us explain the terms contained in this proposition starting with T-junctions. We refer to Fig. 1 for a first simple (but, in fact, general) example. The level lines (e) and (f) represent two level lines at two different levels and in (g) we have superposed them in order to put in evidence how they are organized with respect to each other. We have displayed one of them as a thin line, the other one as a bold line and the common part in grey. Thus, *we say that there is a T-junction at a point x if two level lines meet at point x*. The half level lines (or branches) starting from x are thus four in number; two coincide and two take opposite directions. (The branches which coincide belong to different level lines while the branches which take opposite directions may also belong to the same level line or not).

The next case of singularity is caused by the transparency phenomenon and we shall call it X-junctions. In the transparency phenomenon, we have assumed that inside a spot S, an original image v is changed into $u = g(v)$, where g is a contrast change. We keep $u = v$ outside the spot S. As a consequence, a level line with level λ outside S becomes a level line with level $g(\lambda)$ inside the spot, so that the level lines inside S are continuations of level lines outside S. Fuchs showed this results in the *continuation* perceptual phenomenon : we tend to see the level line of v crossing the boundary of S [11]. Of course, this illusion is based on the continuity of direction. In the same way, of course, the boundary of S is seen as crossing the level lines of v. In fact, level lines cannot cross and the apparent crossing, which we shall call X-junction, consists in the meeting at a point x of two level lines wich locally create four angular regions. The grey levels need not be constant inside each region, but we know that the ranges of the four angular region are ordered as a consequence of the fact that each pair of regions is separated by a level line. We refer to [7] for a detailed discussion of the transparency phenomenon.

4   Mathematical-Physical Discussion

Let us summarize our proposition for basic objects, or "atoms" of image processing. We argue about the invariance of the image analysis with respect to the accidents of the image generation. The reasoning has two stages.

Argument 1 (*Invariance Argument*)

• *Since the image formation may include an unknown and non recoverable contrast change : We reduce the image to its parts invariant with respect to contrast changes, that is, the level lines.*

• *Since, every time we observe two level lines (or more) joining at a point, this can be the result of an occlusion or of a shadowing, we must break the level lines at this point : indeed, the branches of level lines arriving at a junction are likely to be parts of different visual objects (real objects or shadows). As a consequence, every junction is a possible cue to occlusion or shadowing.*

A remarkable point about Argument 1 is that it needs absolutely no assumption about the physical objects, but only on the "final" part of image generation by contrast changes, occlusions and shadowing. The conclusion of Argument 1 coincides with what is stated by phenomenology [15]. Indeed, Gaetano Kanizsa proved the main structuring role of T and X-junctions in our interpretation of images. Of course, he does not give indications on how these junctions should be detected, but Argument 1 shows that there is a canonical way to do the detection.

5   Experimental Kanizsa Programme

5.1   Computation of Basic Visual Events

In this section, we discuss how the basic perceptual-physical events discussed above can be detected in digital images and we present experimental results. The above description of T-, X-junctions is based on the assumptions that

• Level sets and level lines can be computed.

• The meeting of two level lines with different levels can be detected.

In a digital image, the level sets are computed by simple thresholding. A level set $\{u(x) \geq \lambda\}$ can be immediately displayed in white on black background. In the actual technology, $\lambda = 0, 1, ..., 255$, so that we can associate with an image 255 level sets. The Jordan curves limiting the level sets are easily computed by a straightforward algorithm : they are chains of vertical and horizontal segments limiting the pixels. In the numerical experiments, these chains are represented as chains of pixels by simply inserting" boundary pixels" between the actual image pixels.

We define "junctions" in general as every point of the image plane where two level lines (with different levels) meet. As we shall see in experiments, all contrasted edges present in an image tend to generate 'junctions' if the image is slightly blurry : indeed, the smoothing of an edge generates a lot of parallel level lines which collapse at many points, thus generating tiny X-junctions. As a consequence, the meeting of two level lines can be considered as physically meaningful if and only if the level lines diverge significantly from each other. In order to distinguish between the true junctions and the ones due to quantization effects, we decide to take into account T-junctions if and only if *the area of the occulting object, the apparent area of the occulted object and the area of background are large enough.* This threshold on area must be of course as low as possible, because there is a risk to loose T-junctions between small objects. Small size significant objects appear very often in natural images : they may simply be objects seen at a large distance. In any case, the area threshold only depends on the quantization parameters of the digital image and tends ideally to zero when the accuracy of the image tends to infinite. The algorithm for the elimination of dubious junctions is as follows.

Junction Detection Algorithm

• Fix an area threshold n (in practice, $n = 40$ pixels seems sufficient to eliminate junctions due to quantization effects) and a grey level threshold b (in practice : $b = 2$ is sufficient to avoid grey level quantization effects).

• At every point x where two level lines meet : define $\lambda_0 < \mu_0$ the minimum and maximum value of u in the four neighboring pixels of x.

• We denote by L_λ the connected component of x in the set $\{y, u(y) \leq \lambda\}$ and by M_μ the connected component of x in the set $\{y, u(y) \geq \mu\}$. Find the smallest $\lambda \geq \lambda_0$ such that the area of L_λ is larger than n. Call this value λ_1. Find the largest $\mu, \lambda_1 \leq \mu \leq \mu_0$, such that the area of M_μ is larger than n. We call this value μ_1. If $\mu_1 - \lambda_1 \geq 2b$, and if the set $\{y, \mu_1 - b \geq u(y) \geq \lambda_1 + b\}$ has a connected component containing x with area larger than n, then retain x as a valid junction.

6   Image Filtering Preserving Kanizsa Singularities

6.1   Image and Level Curve Smoothing Models

In this section, we push a little further the exploration of invariant image or level lines filtering methods which have been recently proposed in [2] and [20]. According to the axiomatic presentation of image iterative smoothing (or "pyramidal filtering", or "scale space") proposed in [2], all iterative filtering methods, depending on a scale parameter t, can be modelized by a diffusive partial differential equation of the kind

$$\frac{\partial u}{\partial t}(t, x, y) = F(D^2 u(t, x, y), Du(t, x, y), x, y, t)$$
$$u(0, x, y) = u_0(x, y),$$

(9)

where $u_0(x, y)$ is the initial image, $u(t, x, y)$ is the image filtered at scale t and Du, D^2u denote the first and second derivatives with respect to (x, y).

The most invariant "scale space" proposed in [2], is the Affine Morphological Scale Space (AMSS), that is, an equation of the preceding kind which is invariant with respect to contrast changes of the image and affine transforms of the image plane,

$$\frac{\partial u}{\partial t} = |Du|(curv(u))^{\frac{1}{3}}$$
$$u(0) = u_0,$$

(10)

where

$$curv(u)(x, y) = \frac{u_{xx}(u_y)^2 - 2u_{xy}u_x u_y + u_{yy}(u_x)^2}{(u_x^2 + u_y^2)^{\frac{3}{2}}}(x, y)$$

(11)

is the curvature of the level line passing by (x, y). The interpretation of the AMSS model is given in [2] and corresponds to the following evolution equation for curves, proposed independently in [20] and which we call (ASS) (Affine Scale Space of curves)

$$\frac{\partial C}{\partial t}(t, s) = (curv(C(t, s)))^{\frac{1}{3}}\mathbf{n}(t, s)$$
$$C(0, s) = C_0(s),$$

(12)

where $C(0, s)$ is an initial curve embedded in the plane and parameterized by s, $\mathbf{n}(t, s)$ is the normal unit vector at $C(t, s)$ and $curv(C(t, s))$ the curvature of the curve at $C(t, s)$. The relation between (AMSS) and (ASS) is formally the following : (AMSS) moves all level lines of $u(t, x, y)$ as if each one were moving by (ASS). At the mathematical level, however, existence and uniqueness results are not at the same stage. (AMSS) has been proved to have a unique solution in the viscosity sense when the initial image is continuous. (ASS) is proved so far to have a unique smooth solution when the initial curve is convex [20]. *The main drawback of the classification proposed in [2] is the (necessary) assumption that the initial image $u_0(x)$ is continuous with respect to x*. Without this assumption, the (AMSS) model and the related morphological models loose mathematical consistency. This drawback can be avoided by using the Osher-Sethian [18] method : If we wish to apply (AMSS) to (e.g.) a binary and therefore discontinuous image $\mathbf{1}_X$, the characteristic function of X, we substitute to $\mathbf{1}_X$ the signed distance function u_0 to X, which is Lipschitz and has X as zero level set : $X = \{x, u_0(x) \leq 0\}$. We apply to u_0 the (AMSS) model and define the evolution of X as the zero level set of $u(t)$. Evans and Spruck [10] have shown that such an evolution does not depend upon the chosen distance function and coincides with (ASS) when the evolution of the boundary of X is well defined and smooth. In other terms, we can use (AMSS) either for moving continuous functions, or for moving a single set, or a single Jordan curve, by moving its distance function. In order to deal with general discontinuous functions, we deduce that the (AMSS) evolution will be well defined only if we can reduce it to the joint motion of single level curves.

6.2   A Kanizsa Model for Image Evolution

Here, we refer to our discussion on the basic objects of Image Analysis, in Sect. 3. According to this discussion, the decomposition of u_0 into its level lines followed by a further independent analysis of each is not correct : indeed, level lines interact at all image singularities (T-, X-junctions,...) and we have argued that atoms of the image must be the pieces of level lines joining junctions and not the level lines themselves. In other terms, a correct atomic decomposition of the image implies a previous *segmentation* of the level lines. After this segmentation only, the smoothing described by (AMSS) or (ASS) can be applied. This means that we are allowed to move the level lines (for smoothing, or multiscale analysis purposes) but with *internal* boundary conditions : the level lines being during their motion constrained to pass by their endpoints, the junctions. The image model which arises from the preceding discussion is the following.

Definition 1 (Kanizsa image model). We call image a function $u(x)$ whose level lines have all a finite length and meet only on a closed set without interior $\mathcal{Z}$, which we call the set of junctions of u.

When the junction set is finite (which is the case for digital images), we set $\mathcal{Z} = \{z_1, z_2, ..., z_n\}$. According to the preceding discussion, the correct adaptation of the (AMSS) model to images having junctions is

Algorithm 1 : Junctions-preserving, affine invariant and contrast invariant smoothing
• Compute the junctions $z_1, z_2, ..., z_n$ (by the Junction Detection algorithm, which is affine invariant).
• Move each piece of level curve C by the (ASS) model. If the curve has endpoints, they remain fixed and the (ASS) evolution is not allowed to let the curve cross other junctions. This yields $C(t)$.
• Reconstruct a smoothed image $u(t, x, y)$ which has the curves $C(t)$ as level lines and the z_i as junctions. This reconstruction is possible because the (ASS) model preserves the inclusion of curves (see [4])

Mathematical and Numerical Discussion of Algorithm 1. Algorithm 1 is well defined provided the (ASS) model is mathematically correctly posed ; this point is, however, still not completely solved. In addition, the independent processing of all pieces of level lines looks computationally heavy. So we prefer to consider a variant of (AMSS) implementing Algorithms 1 and 2. We consider the characteristic function $\mathbf{1}_Z(x, y) = 0$ if $(x, y) \in Z$ and $\mathbf{1}_Z(x, y) = 1$ if $(x, y) \notin Z$ of the junctions and apply to the original image the equation

$$\frac{\partial u}{\partial t} = \mathbf{1}_Z |Du| (curv(u))^\alpha$$
$$u(0) = u_0. \tag{13}$$

with $\alpha = 1/3$, which we call the $(AMSS1)$ model. We conjecture that such an equation makes sense, and has a unique viscosity solution (in the sense defined in [9]) preserving the singularities and moving each piece of level line by the (ASS) model. In addition, we conjecture that the pieces of level lines joining junctions keep the same endpoints by such a process. We shall discuss in the next section in which way well-posedness can be proved for (AMSS1).

Let us now consider a well-posed approximation of (AMSS1). We replace $\mathbf{1}_Z$ by a function $\mathbf{1}_{Z,\varepsilon}$ which is smooth (e.g. C^∞) and satisfies

$$\mathbf{1}_{Z,\varepsilon}(x, y) = 1 \tag{14}$$

if the distance of (x, y) to Z is larger than ε and

$$\mathbf{1}_{Z,\varepsilon}(x, y) = 0 \tag{15}$$

if the distance of (x, y) to Z is smaller than $\frac{\varepsilon}{2}$ and $0 \leq \mathbf{1}_{Z,\varepsilon} \leq 1$ everywhere. The considered model therefore is

$$(AMSS_\varepsilon) \quad \frac{\partial u}{\partial t} = \mathbf{1}_{Z,\varepsilon} |Du| (curv(u))^{\frac{1}{3}}. \tag{16}$$

and it is an equation of the kind (9) where F is continuous with respect to all variables, nondecreasing with respect to its first argument and satisfying the condition

$$\forall x \in \mathbb{R}^2, \quad F(0, 0, x) = 0, \tag{17}$$

and some additional continuity conditions for F which are easily checked in our case. Then, as it is proved in [12], $(AMSS_\varepsilon)$ has a unique solution in the viscosity sense. Therefore, it also satisfies a local comparison principle. Indeed, by Theorem 4.2 in [12], we have a comparison principle between sub- and supersolutions of $(AMSS_\varepsilon)$. Moreover, because of (17), constant functions are solutions of $(AMSS_\varepsilon)$, and, as in [8] Theorem 4.5, by using [8] Proposition 2.3 and the comparison principle mentioned before, one can construct by the Perron method a (unique) viscosity solution for $(AMSS_\varepsilon)$. In this argument, we have assumed that the image is defined in all of $\mathbb{R}^2$. This does not matter, since we can impose without loss of consistency in our model that all points on the boundary of our image are T-junctions. Then the image can be extended by 0 to the whole plane. Another point to make clear in the application of the above theorems is following : the existence and uniqueness results apply only if the initial datum u_0 is continuous, which is not our case. So we can argue in the following way. Let us call $\mathcal{Z}_\varepsilon$ the ε-neighborhood of $\mathcal{Z}$. We know that outside $\mathcal{Z}_{\frac{\varepsilon}{2}}$ the function u_0 is continuous. So we can, by Tietze Theorem, extend u_0 inside $\mathcal{Z}_{\frac{\varepsilon}{2}}$ as a continuous function, which we call $\tilde{u}_0$. Then the existence and uniqueness theorems mentioned above apply. It only remains to prove that the solution thus obtained does not depend upon the choice of the extension $\tilde{u}_0$. Indeed, it is easily seen that inside $\mathcal{Z}_{\frac{\varepsilon}{2}}$ solutions of $(AMSS_\varepsilon)$ with continuous initial data does not move. We deduce, as an easy consequence of the comparison principle, that all continuous extensions of u_0 inside $\mathcal{Z}_{\frac{\varepsilon}{2}}$ yield the same solution outside.

The interpretation of $(AMSS_\varepsilon)$ is essentially the same as for $(AMSS1)$: the level lines of u with endpoints on $\mathcal{Z}$ are constrained to keep the same endpoints and the only difference with respect to $AMSS1$ is an alteration of their speed in an ε-neighborhood of $\mathcal{Z}$. Of course, it is more than likely that the solution of $(AMSS_\varepsilon)$ converges to the (to be proved) solution of $(AMSS1)$, which is the right model. We conjecture that this convergence is true and we can give hints in the next subsection about existence for $(AMSS1)$ which also suggest in which sense this convergence can be true. Returning to $(AMSS_\varepsilon)$, we would like to make clear that it is in fact the right model for what we can numerically implement. Any numerical implementation of $(AMSS1)$ is done on a grid and derivatives of u at a pixel (x, y) computed with the neighbooring pixels. We refer to [3] where a very easy and most invariant such numerical scheme (due to Luis Alvarez and Frederic Guichard) is proposed for $(AMSS)$. The only alteration which we do for this scheme is following : If (x, y) is a vertex of the grid belonging to $\mathcal{Z}$, we simply fix in the scheme the four pixels surrounding (x, y). In other terms, we call u^n the successive discretized values of u and write the Alvarez-Guichard scheme for $(AMSS_\varepsilon)$ in the form

$$u^{n+1}(x, y) = u^n(x, y) + \Delta t F(D^2 u^n(x, y), D u^n(x, y), (x, y)) \qquad (18)$$

where of course (x, y) are discrete values and Δt a small enough time (or scale) increment. Then the new scheme, associated with $(AMSS_\varepsilon)$ coincides with (18) if (x, y) are the coordinates of a pixel not touching a $T - junction$, and

$$u^{n+1}(x, y) = u^n(x, y) \tag{19}$$

otherwise. This slight change in the scheme has dramatic consequences in the evolution of the image in scale space, as we shall see in the numerical experiments.

6.3 Existence and Uniqueness Proof for the Junction Preserving Smoothing

We give a sketch of the proof of existence and uniqueness of viscosity solutions for model (13) with $\alpha = 1$, the Osher-Sethian classical mean curvature adapted to keep the the junctions fixed,

$$\frac{\partial u}{\partial t} = \mathbf{1}_Z |Du|(curv(u)) \tag{20}$$

with initial condition $u_0(x, y)$. We adapt the notion of viscosity solutions to this problem. Let Z be the set of singularities described in previous sections. Given $p \in Z$, let us denote by $\tau_p(x) \in S^1$, $x \neq p$, the direction of x taking p as the origin of coordinates. Let X be the space of functions $f \in C(\mathbb{R}^2 \setminus Z)$ such that for any sequence $x_n \to p, p \in Z$ with $\tau_p(x_n) \to v \in S^1$, then the $\lim_n f(x_n)$ exists and let $\mathcal{R}$ be the space of functions of the form $\phi(\tau_p(x))$, $p \in Z$, $x \in \mathbb{R}^2 \setminus \{p\}$, ϕ being a smooth function on S^1, i.e., the set of smooth radial functions with respect to some point of Z. In the next definition $()^*, ()_*$ denote the upper and lower semicontinuous rearrangement of the function in the argument, respectively.

Definition 2 Let u be an upper (lower) semicontinuous function on $[0, T] \times \mathbb{R}^2 \setminus Z$. We say that u is a viscosity subsolution (supersolution) of if $\forall (\phi, \psi, x_0, t_0)$, $\phi \in C^2(\mathbb{R}^2)$, $\psi \in C^1([0, T])$, $x_0 \notin Z$, $t_0 \in]0, T]$ such that $u(t, x) - \phi(x) - \psi(t)$ has a local maximum (minimum) at (t_0, x_0), then

$$\frac{\partial \psi}{\partial t}(t_0) \leq |D\phi(x_0)| curv(\phi)(x_0) \quad (resp. \geq) \quad if \quad D\phi(x_0) \neq 0 \tag{21}$$

or

$$\frac{\partial \psi}{\partial t}(t_0) \leq 0 \quad (resp. \geq) \quad if \quad D\phi(x_0) = 0, D^2\phi(x_0) = 0 \tag{22}$$

and $\forall (\phi, \psi, x_0, t_0)$, $\phi \in \mathcal{R}$, $\psi \in C^1([0, T])$, $x_0 \in Z$, $t_0 \in]0, T]$ such that $(u(t, x) - \phi(x))^* - \psi(t)$ $((u(t, x) - \phi(x))_* - \psi(t))$ has a local maximum (minimum) at (t_0, x_0),

$$\frac{\partial \psi}{\partial t}(t_0) \leq 0 \quad (resp. \geq). \tag{23}$$

A function $u \in C([0, T], X)$ is a viscosity solution if it is a viscosity sub- and supersolution.

Then we prove

Theorem 3 Let $u_0 \in X$. Then there exists a unique viscosity solution $u \in C([0, T], X)$ of (20) with initial datum u_0.

The proof being too long to be included here we only give a sketch of it.

Step 1. Let $F = F_0 \cup F_1 \cup F_2$ where F_0 (resp. F_1, F_2) is the set of smooth Jordan curves $C : [0,1] \to \mathbb{R}^2$: with $C(0) = C(1)$ and no point in $\mathcal{Z}$ (resp. with $C(0) = C(1) \in \mathcal{Z}$, $C(0) \neq C(1)$ and both in $\mathcal{Z}$). In this step we study the motion of curves in F by the so-called intrinsic heat equation

$$(IHE1) \quad \frac{\partial C(t,s)}{\partial t} = curv(C)\mathbf{n}(t,s), \tag{24}$$

When $C(0,s) \in F_0$, the curve moves according to (24). In this case existence, uniqueness and regularity results were proved in [13]. When $C(0,s) \in F_1 \cup F_2$, we add to (24) the boundary conditions

$$C(t,0) = z_0, \quad C(t,1) = z_1 \tag{25}$$

where z_0, z_1 denote the fixed endpoints of the curve $C(t,s)$ which belong to $\mathcal{Z}$ ($z_0 = z_1$ if $C(0,s) \in F_1$). In order to generalize Grayson's result to the case where the tips of the curve are fixed, one may proceed as follows. Assuming that the initial curve $C(0,s)$ is parameterized by a parameter s ranging from 0 to 1, we embed C_0 into a periodic curve by setting

$$\tilde{C}(0,s) = 2C(0,0) - C(0,-s) \quad \text{if} \quad -1 \leq s \leq 0 \tag{26}$$

and then by embedding $\tilde{C}$ into a 2-periodic curve. If, by this process, $\tilde{C}$ has remained a simple curve (a curve without self-intersection), then we can apply the main theorem in [13] and assert that $\tilde{C}(t,s)$ is uniquely defined, independent of the initial parametrization, and analytic for every $t > 0$. In addition, because of the uniqueness of the evolution, $\tilde{C}(t,s)$ is invariant by the symmetry $x \to -x$ in $\mathbb{R}^2$, so that z_0 belongs to $\tilde{C}$ for every t. In the same way, z_1 belongs to $\tilde{C}$ and we define the solution of (IHE1) as the part of $\tilde{C}$ joining z_0 and z_1. In the case where $\tilde{C}_0$ does present self-intersections, the results in [5], [6], [1] can be applied to get local existence, uniqueness and smoothness of a solution in some $[0, T_0]$, $T_0 > 0$. Moreover this motion can be extended until the part of the evolving curve corresponding to the fundamental domain touches a point of $\mathcal{Z}$. This obliges to impose the following set of rules

Rule 1. When an evolving curve touches a point of $\mathcal{Z}$ then we subdivide the curve in pieces belonging again to F and let them evolve as such.

Rule 2. The evolving curve will never cross a point of $\mathcal{Z}$. Thus, either the evolving curve contains a new point of $\mathcal{Z}$ or it is homotope in $\mathbb{R}^2 \setminus \mathcal{Z}$ to the initial curve.

Rule 3. If the point $p \in \mathcal{Z}$, not in $C(0,s)$, is in $C(t,s)$, $t > 0$ and the directional tangents to $C(t,s)$ from both sides at p make an angle of π then the curve continues its evolution as if p were not on it, in as much as we respect Rule 2.

Using this set of rules and using the results of [5], [6], [13], [1] we may define globally in time a smooth evolution of curves in F and prove that no singularities appeared in the fundamental domain of the curve. Let $H(t)C$ denote this evolution for any curve $C \in F$. This evolution satisfies a comparison principle, if the region inside C_1 is contained in the region inside C_2, $C_1, C_2 \in F$, then the region inside $H(t)C_1$ is contained in the region inside $H(t)C_2$. Moreover this motion is stable with respect to perturbations of the initial conditions. If they do not encounter another point in $\mathcal{Z}$, curves in $F_0 \cup F_1$ collapse to a point in finite time while curves in F_2 tend to the straight line joining the two points in $\mathcal{Z}$ belonging to the curve as $t \to \infty$.

Step 2. Our purpose is to define the corresponging evolution operator on X. To simplify our notation, let D be the region determined by a finite set of curves in F, either inside or outside. Then we define $H(t)D$ as the region determined by the evolution of the corresponding set of curves. Given a function $\phi \in X$, smooth in $\mathbb{R}^2 \setminus \mathcal{Z}$, with, at most, a finite number of nondegenerate critical points (i.e., it is a Morse function) we define

$$H(t)\phi(x) = \sup\{\lambda : \lambda \quad \text{a regular value of} \quad \phi, \quad x \in H(t)[\phi \geq \lambda]\}. \qquad (27)$$

$H(t)$ acts as a order preserving semigroup of contractions which can be extended to all X. Then, following the strategy of [14] we prove that for any $\phi \in X$, $H(t)\phi$ is a viscosity solution of (20). Then, adapting the usual proof, we prove a comparison result between sub- and supersolutions of (20).

Theorem 4 *Let u be upper semicontinuous in $\mathbb{R}^2 \setminus \mathcal{Z}$, $v \in C([0, T], X)$. Suppose that u is a viscosity subsolution of (20) and v a viscosity supersolution. Then*

$$\sup_{x \in \mathbb{R}^2 \setminus \mathcal{Z}} (u(t, x) - v(t, x))^+ \leq \sup_{x \in \mathbb{R}^2 \setminus \mathcal{Z}} (u(0, x) - v(0, x))^+. \qquad (28)$$

The same estimate also holds if we assume $u \in C([0, T], X)$, v be lower semicontinuous in $\mathbb{R}^2 \setminus \mathcal{Z}$.

Then uniqueness follows. Another important consequence of the comparison result is the following

Theorem 5 *Let $u_0 \in X$. Then the solution $u_\varepsilon(t, x)$ of $(AMSS_\varepsilon)$ converges to the viscosity solution of (20).*

A full account of these results can be seen in [7].

Remark. We conjecture that the result proved in Step 1 can be also proved for the equation (ASS) plus the boundary conditions $C(t, 0) = z_0, C(t, 1) = z_1$, with the difference that the convergence to the straight line when the initial curve is in F_2 should take a finite time.

6.4   Experiments in Junction-Preserving Image Filtering

The experiments have been made on WorkStation SUN IPC, with an image processing environment called MEGAWAVE II, whose author is Jacques Froment.

- **Experiment 1 : detection of T-junctions as meeting points of level lines.** The original grey-level image *El caminante* is Fig. 2, call it u_1. In Fig. 3(a) we see *El caminante* without skirting-board : $u_1 > 60$ and Fig. 3(b) is *El caminante* with skirting-board. We display in Fig. 4(a) the level lines of Fig. 3(a) in white and the level lines of Fig. 3(b) in black with their common parts in dark grey. Candidates to T-junctions generated by the main occlusions can be seen. They are characterized as meeting points of a black, a white and a grey line. In Fig. 4(b), we see all T-junctions detected from those two level sets, after a filtering of spurious T-junctions has been made, with area threshold $n = 80$.

- **Experiment 2. Image filtering preserving Kanizsa singularities on a real image.** Figure 5 displays the result of the Junction Detection Algoritm applied to *El caminante* with area threshold $n = 40$ and grey level threshold $b = 2$ with (in white) small "T"s indicating locations of detected junctions. All physically plausible junctions seem detected. As expected, the boundary of the image generates an almost continuous line of junctions, as well as the main occlusive structures and shadows of the image. Figure 6(a) : (AMSS) analysis of Fig. 2 at scale $t = 12$. Figure 6(b) : Algorithm 1 applied to Fig. 2, at the same scale. The fixed junctions are those of Fig. 5.

- **Experiment 3. Topographic maps.** The application of Algorithm 1 simplifies the shape of level lines and permits to better visualize the whole topographic organization of level lines and junctions (that is, all of the proposed "atoms" for image analysis). Figure 7(a) : original level lines of *El caminante* displayed for all levels multiple of 5. Figure 7(b) shows the same level lines corresponding to Fig. 6(b).

- **Experiment 4. Sensibility to noise.** The following experiments do not pretend to exhaust the subject of sensibility to noise of the junctions. Figures 8(a) and 8(b) : two details of a synthetic image which for display reasons we zoomed by a factor 2. The detail images have 40% of the pixels destroyed and replaced by an uniformly distributed random value between 0 and 255. In white, one can see the obtained T-junctions and X-junctions with $n = 40$ and $b = 2$.

Acknowledgement. We gratefully acknowledge partial support by EC Project "Mathematical Methods for Image Processing", reference ERBCHRXCT930095, DGICYT project, reference PB94-1174 and CNES. Also, we thank Jean Petitot, Bernard Teissier, Marcella Mairota, Denis Pasquignon, Jacques Froment, Jean-Pierre d'Ales, Antonin Chambolle and Frédéric Guichard for valuable information and discussions.

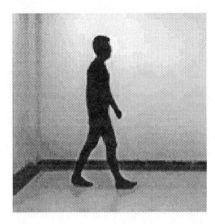

Fig. 2. Original image, u_1, El caminante

Fig. 3. Left: Figure 3(a), $[u_1 > 90]$. Right: Figure 3(b), $[u_1 > 140]$

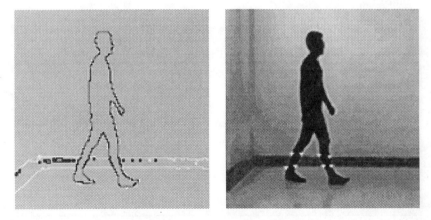

Fig. 4. Left: Figure 4(a), level lines of Fig.3(a) in white and level lines of Fig. 3(b) in black with their common parts in dark grey. Right: Figure 4(b), detected T-junctions with area threshold $n = 40$

Fig. 5. Dectected junctions corresponding to Fig. 2 with $n = 40$ and $b = 2$

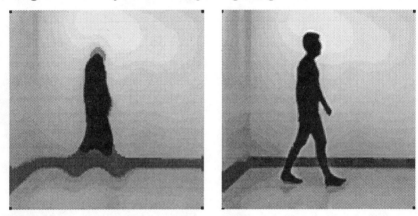

Fig. 6. Left: Figure 6(a), (AMSS) analysis of Fig. 2 at scale $t = 12$. Rigth: Figure 6(b), Algorithm 1 applied to Fig. 2 with $t = 12$ fixing the junctions of Fig. 5

Fig. 7. Left: Figure 7(a), all level lines of Fig. 5 which are multiple of 5. Right: Figure 7(b), same level lines corresponding to Fig. 6(b)

Fig. 8. Left: Figure 8(a). Right: Figure 8(b). Junctions detected in synthetic images with added noise

References

1. Altschuler, S.J., Grayson, M.: Shortening space curves and flow through singularities. J. Diff. Geometry **35** (1992) 283–298
2. Alvarez, L, Guichard, F., Lions, P.L., Morel, J. M.: Axioms and fundamental equations of image processing. Arch. Rational Mechanics and Anal. **16, IX** (1993), 200–257
3. Alvarez, L., Morel, J.M.: Formalization and Computational Aspects of Image Analysis. Acta Numerica 1994, Cambridge University Press.
4. Alvarez, L., Morales, F.: Affine Morphological Multiscale Analysis of Corners and Multiple Junctions. Ref 9402, August 1994, Dept. Informatica y Systemas, Universidad de Las Palmas de Gran Canaria. To appear in International Journal of Computer Vision.
5. Angenent, S.: Parabolic equations for curves on surfaces. Part I. Curves with p-integrable curvature. Ann. Math. **132** (1990) 451–483
6. Angenent, S.: Parabolic equations for curves on surfaces. Part II. Intersections, blow-up and generalized solutions. Ann. Math. **133** (1991) 171–215
7. Caselles, V., Coll, B., Morel, J.M.: A Kanizsa programme. Preprint CEREMADE, Univ. Paris-Dauphine, 1995. Submitted to Int. J. Comp. Vision.
8. Chen, Y.G., Giga Y., Goto, S.: Uniqueness and existence of viscosity solutions of generalized mean curvature flow equations. J. Diff. Geometry **33** (1991) 749–786.
9. Crandall, M.G., Ishii, H., Lions, P.L.: User's guide to viscosity solutions of second order partial differential equation. Bull. Amer. Math. Soc., **27** (1992) 1–67.
10. Evans, L.C., Spruck, J.: Motion of level sets by mean curvature. J. Differential Geom., **33** (1991) 635–681.
11. Fuchs, W.: Experimentelle Untersuchungen ueber das simultane Hintereinandersehen auf der selben Sehrichtung. Zeitschrift fuer Psychologie, **91** (1923) 154–253.
12. Giga, Y., Goto, S., Ishii, H., Sato, M.M.: Comparison principle and convexity preserving properties for singular degenerate parabolic equations on unbounded domains. Indiana Univ. Math. J., **40** (1991) 443–470.
13. Grayson, M.: The heat equation shrinks embedded plane curves to round points. J. Differential Geometry **26** (1987) 285–314.
14. A Generalization of the Bence, Merriman and Osher algorithm for motion by mean curvature. Preprint.
15. Kanizsa, G.: Vedere e pensare. Il Mulino, Bologna, 1991.
16. Morel, J.M., Solimini, S.: Variational methods in image processing. Birkhäuser, 1994.

17. Nitzberg, M., Mumford, D., Shiota, T.: Filtering, Segmentation and Depth. Lecture Notes in Computer Science, 662, Springer-Verlag 1993.
18. Osher, S., Sethian, J.: Fronts propagating with curvature dependent speed: algorithms based on the Hamilton-Jacobi formulation. J. Comp. Physics **79** (1988) 12–49.
19. Rudin, L.I.: Images, numerical analysis of singularities and shock filters. PhD. dissertation, Caltech, Pasadena, California n 5250:TR, 1987.
20. Sapiro, G., Tannenbaum, A.: On affine plane curve evolution. Journal of Functional Analysis, **119** (1994) 79–120.
21. Witkin, A.P.: Scale-space filtering. Proc. of IJCAI, Karlsruhe 1983, 1019–1021.

Recognition in Hierarchical Models

Peter Dayan

Abstract. Various proposals have recently been made which cast cortical processing in terms of hierarchical statistical generative models (Mumford, 1994; Kawato, 1993; Hinton & Zemel, 1994; Zemel, 1994; Hinton *et al*, 1995; Dayan *et al*, 1995; Olshausen & Field, 1996; Rao & Ballard, 1995). In the case of vision, these claim that top-down connections in the cortical hierarchy capture essential aspects of how the activities of neurons in primary sensory areas are generated by the contents of visually observed scenes. The counterpart to a generative model is its statistical inverse, called a *recognition* model (Hinton & Zemel, 1994). This takes low-level activities and produces probability distributions over the entities in the world that could have led to them, expressed as activities of neurons in higher visual areas that model the image generation process. Even if a generative model is computationally tractable, its associated recognition model may not be. In this paper, we study various different types of exact, sampling-based and approximate recognition models in the light of computational and cortical constraints.

I Introduction

There are two popular notions as to the major on-line (as opposed to learning) mode of cortical processing. One concentrates on *discrimination* or *classification* of input from the sensory epithelium. For instance, if images contain a single handwritten digit, then the task for cortex in recognising or interpreting an image is to produce a probability distribution reporting which digit might be present. Invariances of various sorts are key – successive layers are taken as ignoring ever more information present in the image but irrelevant to the digit class, such as the style of the digit (*eg* italic or roman), the thickness of the strokes, the position on the page, *etc.* Purely bottom-up processing in the cortical hierarchy is typically thought of as implementing classification, justified by results such as Perrett *et al's* (1982) on the speed of face recognition, whose calculations imply that there little, if any, time for lateral or top-down influences. There is also substantial statistical theory underlying discrimination. However, its cortical instantiation is somewhat problematical. First, it is not clear how the sort of supervised training that underlies most classification systems could be arranged. Second, it is not clear how relevant prior information (such as that the particular writer for an image tends to favour particular curly strokes) can be properly incorporated into the recognition process.

The contending notion suggests that cortex builds a *model* (usually a probability density model) of the input it receives (Grenander, 1976; Mumford, 1994). The model captures the statistical structure of the observed input, according probabilities to particular inputs commensurate with their frequency in the world. In ideal circumstances, the model will reflect accurately the actual process by which images are created – *eg* for the images of handwritten digits, the model will include explicit choices for the identity of the digit, the style, the thickness of the strokes, *etc*. In cortex, these choices should be instantiated in the activities of particular amongst groups or populations of neurons. Recent statistical models for cortex have taken note of its layered structure (Felleman & Van Essen, 1991), and suggested that top-down and/or lateral connections contain the generative model. The model represents a (possibly complicated) probabilistic *prior* over observable scenes.

Given such a model, the most general task in interpreting a particular image is to blend information from the senses with this prior information (consistent with Bayes theorem) to report a posterior distribution over the various generative choices – *ie* analysis by synthesis. Alternatively, given some loss function, single values might be produced that summarise the posterior probability distribution. If one of the generative choices is the identity of the handwritten digit, then interpreting an image entails reporting the distribution over the digits it might contain. This is a characteristic inverse problem (Marroquin, 1985) – regularisation theory is an alternative way of describing the same operations (Poggio & Torre, 1984) – and is also the conventional way that maximum likelihood models (strictly maximum *a posteriori* models) are used for classification or discrimination. The generative mode for cortex therefore requires the discrimination mode too. This paper studies the discriminative phase, called *recognition* (Hinton & Zemel, 1994), that emerges as the Bayesian inverse to top-down generation.

If top-down and/or lateral weights in cortex are involved in the generative model, it is natural to conclude that the bottom-up weights are concerned with recognition. Of course, the other weights could also be involved – there are many cases for which top-down influences over perception are strong, and others such as binocular rivalry (Leopold & Logothetis, 1996; Logothetis *et al*, 1996) for which there appears to be an on-going interaction between bottom-up and top-down processing (Dayan, 1996). This paper studies different ways that recognition, or approximations to recognition, can be implemented for various sorts of generative model.

In some cases the recognition phase is computationally straightforward. Two important examples are when it is linear, which is the case of factor analysis (FA) discussed in detail in the next section, and when it involves a one-of-n operation, as in a mixture model such as the popular mixture of Gaussians (Nowlan, 1991) and mixture of experts (Jacobs *et al*, 1991) architectures. Even if exact recognition is tractable, we will see that there are different ways of implementing it, mixing combinations of bottom-up, top-down and lateral processing.

For many other generative models, even ones that are simple to specify, recognition is computationally challenging. Two examples are the unsupervised version of the Boltzmann machine (BM; Hinton & Sejnowski, 1986) and causal belief networks (*eg* Pearl, 1988). For the BM, both the generative and the recognition distributions are computationally intractable to calculate. For directed belief networks, their structure makes it easy to calculate the prior probability over a set of generative choices. However, calculating the posterior probability distribution given an observation is again difficult.

If exact recognition is intractable, one has two options. Markov chain Monte-Carlo methods can be used to collect samples that (at least asymptotically) reflect the exact recognition inverse (Neal, 1992;1993).[1] In cases like the BM or belief nets, Gibbs sampling specifies a Markov chain whose stationary distribution is the true posterior. Typically, one needs to run the chain for a while until transients are sure to have decayed, and then take samples of the states of the chain as being samples from the true recognition distribution. The disadvantage of using stochastic simulation is the time it takes for transients to decay, and also the number of independent runs necessary if there are large energy barriers between states (or, equivalently, the high variance in the samples). Various methods for overcoming these problems have been suggested, in particular forms of annealing.

If Monte-Carlo sampling based on the true generative model is not to be used, then some form of approximation to the true recognition inverse is needed. Various different such schemes have been suggested, each with its own characteristics. The Helmholtz machine (HM; Hinton *et al*, 1995; Dayan *et al*, 1995) has a top-down belief-net generative model leading to a lowest layer, which represents the direct sensory report of images. The HM uses a bottom-up belief net to instantiate an approximation to the recognition inverse. In one version, the recognition distribution is described through samples that are generated stochastically (which is computationally easy) – and parameters of this bottom-up net are learnt during a training phase to make its samples appropriate. An alternative is to use mean-field methods (Saul *et al*, 1996; Jaakkola *et al*, 1996). Here, a parameterised form is chosen for the approximation to the whole inverse distribution for a particular image, where the particular parameterisation is chosen to make calculations easy. The parameters are then updated to minimise a Kullback-Leibler based measure of the difference between the approximate and the actual recognition distributions. A further alternative to the Helmholtz machine or mean field methods is to abandon the requirement of finding the true posterior distribution, and rather look for just its maximum.

This paper studies aspects of these different recognition choices. We are particularly interested in the relationship between the information contained in the bottom-up weights and that contained in the top-down or generative weights. Iterative recognition schemes that employ top-down weights turn out to require that the bottom-up weights are essentially the transpose of the generative weights (as is also true for principal components analysis). Recognition schemes that concentrate on feedforward processing require bottom-up weights that are not purely the transpose of the top-down weights. In general, there remains an

---

[1] The inverse is a whole distribution, and therefore can be specified by samples as in Monte-Carlo methods, or through a parameterisation, as in mean-field methods.

unresolved conflict between having fast and feedforward recognition, as inspired by Perrett *et al's* (1982) results, and having iterative recognition that blends bottom-up and top-down information in an appropriate way, including handling so-called explaining away effects in non-linear generative models in which either one generative cause for an input is active, or another one is active, but probably not both. The role of lateral processing is also unclear. There is neither the experimental evidence nor the computational compulsion to adopt one scheme in particular at present. Other issues are also important, particularly the way in which the efficacies of the connections are specified through experience, but they are not the current focus.

The next section studies in depth the factor analysis case of a two-layer and linear generative model with Gaussian noise, and comments on its multi-layer extension, as in Chou, Willsky & Benveniste (1994) and Chou, Willsky & Nikoukhah (1994); and section 3 looks briefly at causal belief networks with binary units, which raise such problems as explaining away.

II   Factor Analysis

A simple two-layer generative model is shown in figure 1 and is given by:

$$\mathbf{y} \sim \mathcal{N}[0, \Phi] \qquad \mathbf{x} \sim \mathcal{N}\left[\mathcal{G}^T \mathbf{y}, \Psi\right] \qquad \Psi = \text{diag}\left(\tau_1^2, \ldots, \tau_n^2\right) \qquad (1)$$

where $\mathcal{N}(\mathbf{a}, \Gamma)$ is a multivariate Gaussian distribution with mean $\mathbf{a}$ and covariance matrix Γ, $\mathbf{y} \in \Re^m$, $\mathbf{x} \in \Re^n$, and $\mathcal{G}$ is an $m \times n$ matrix of generative weights. Equations 1 imply that the components x_i of $\mathbf{x}$ are mutually independent, given the values of their belief net parents $\mathbf{y}$. Components y_i are called the *factors* underlying the observed examples $\mathbf{x}$.

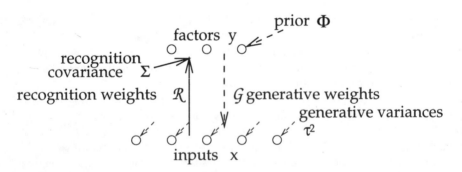

Fig. 1. Factor Analysis. The dotted lines on the right show the elements of the linear and Gaussian generative model in equations 1, involving factors $\mathbf{y}$ and observables $\mathbf{x}$. The solid lines on the left show the elements of the recognition model of equations 3 that is the inverse of the generative model.

This generative model is exactly that underlying the statistical technique of factor analysis (see Everitt 1984; Jolliffe, 1986 for introductions; Dempster, Laird

& Rubin, 1977, and Rubin & Thayer, 1982 for more proximal analysis). It was pointed out as being a linear version of the Helmholtz machine by Neal (personal communication; Neal & Dayan, 1996) and was used for to model images of digits by Hinton *et al* (1996). Given this generative model, and a particular example **x**, the role of recognition is to calculate the posterior distribution $\mathcal{P}[\mathbf{y}|\mathbf{x}]$ over the generators **y** given the observed image **x**, or perhaps instead to calculate some particular value $\mathbf{y}^*$ that summarises this posterior distribution.

The joint distribution over **x** and **y** is Gaussian

$$\mathcal{P}[\mathbf{x},\mathbf{y}] \propto \exp\{-\frac{1}{2}\left[\mathbf{y}^T\Phi^{-1}\mathbf{y} + \left(\mathbf{x} - \mathcal{G}^T\mathbf{y}\right)^T \Psi^{-1}\left(\mathbf{x} - \mathcal{G}^T\mathbf{y}\right)\right]\}, \qquad (2)$$

therefore the posterior distribution $\mathcal{P}[\mathbf{y}|\mathbf{x}]$ is also Gaussian $\mathcal{N}[\mathcal{R}^T\mathbf{x}, \Sigma]$, where

$$\mathcal{R} = \Psi^{-1}\mathcal{G}\left(\Phi^{-1} + \mathcal{G}\Psi^{-1}\mathcal{G}^T\right)^{-1} \qquad \Sigma^{-1} = \Phi^{-1} + \mathcal{G}\Psi^{-1}\mathcal{G}^T \qquad (3)$$

The maximum *a posteriori* value $\mathbf{y}^{\mathrm{MAP}}$ comes from minimising $-\log\mathcal{P}[\mathbf{x},\mathbf{y}]$:

$$\mathcal{E}[\mathbf{y}] = \mathbf{y}^T\Phi^{-1}\mathbf{y} + \left(\mathbf{x} - \mathcal{G}^T\mathbf{y}\right)^T \Psi^{-1}\left(\mathbf{x} - \mathcal{G}^T\mathbf{y}\right), \quad ie \quad \mathbf{y}^{\mathrm{MAP}} = \mathcal{R}^T\mathbf{x} \qquad (4)$$

as it must be, since the posterior distribution is Gaussian and therefore unimodal. The maximum likelihood value $\mathbf{y}^{\mathrm{ML}}$ comes from setting $\Phi^{-1} = 0$ in equation 4, which is dimensionally reasonable in the case in which there are fewer factors than input variables, *ie* $m < n$.

Top-down information for a particular case could be seen as changing the prior over **y** in equation 1 in two ways. If it changes the prior covariance matrix for the factors, then $\mathcal{R}$ must change too. If it just specifies a non-zero unconditional mean for the factors: $\mathbf{y} \sim \mathcal{N}[\bar{\mathbf{y}}, \Phi]$, then the new posterior is just

$$\mathbf{y}|\mathbf{x} \sim \mathcal{N}\left[\mathcal{R}^T\mathbf{x} + \left(\Phi^{-1} + \mathcal{G}\Psi^{-1}\mathcal{G}^T\right)^{-1}\Phi^{-1}\bar{\mathbf{y}}, \Sigma\right] \qquad (5)$$

We are now in a position to describe some of the various proposals for recognition. For none of them is it quite clear how the posterior covariance matrix Σ might be represented in cortex (see Neal & Dayan, 1996). In the case that Φ is rotationally invariant, the factors are infamously rotationally underspecified. This allows us the cortically convenient option of taking Σ as being just a diagonal matrix. An additional problem with this linear factor analysis model is that it is only well determined (even up to rotation) if there are sufficiently fewer factors than input dimensions. This is not true of cortex, and is not necessary for non-linear factor analysis models (*eg* Olshausen & Field, 1996) or models that include temporal effects (*eg* Rao & Ballard, 1995).

Bottom-Up Method

The factor analysis version of the Helmholtz machine (Neal & Dayan, 1996; Hinton *et al*, 1996) is similar to the standard version of the Helmholtz machine in that it devotes a set of parameters to a feedforward belief net structure that is intended to approximate the recognition inverse using only bottom-up processing.

This uses the feedforward weights $\mathcal{R}$ shown in figure 1, and an explicitly parameterised feedforward recognition covariance matrix Σ. This covariance matrix specifies the parameters of the noise corrupting the mean value $\mathcal{R}^T\mathbf{x}$, and so can be used to generate samples from the recognition model.

Note that recognition only requires a linear operation on the input pattern $\mathbf{x}$, but that the relationship between the top-down weights $\mathcal{G}$ and the bottom-up weights $\mathcal{R}$ is obscured by the priors Φ and Ψ. This is characteristic of methods that calculate the posterior distribution (or samples from it) based purely on bottom-up processing. The obscuring factor, $(\Phi^{-1} + \mathcal{G}\Psi^{-1}\mathcal{G}^T)^{-1}$ balances the prior variability of a factor (from Φ^{-1}) with the extent to which that factor might be responsible for inputs, modulated by the extent to which noise might be responsible instead (from $\mathcal{G}\Psi^{-1}\mathcal{G}^T$).

In principal components analysis (PCA), the value y_j would be the projection of the input $\mathbf{x}$ onto the jth orthonormal eigenvector of the covariance matrix of all the inputs. For PCA, the generative and recognition weight matrices are just transposes of each other ($\mathcal{R} = \mathcal{G}^T$), both containing the relevant eigenvectors. Most of the methods discussed below that employ top-down weights during processing also have $\mathcal{R} = \mathcal{G}^T$, but they do not contain the eigenvectors.

The advantage of the one-shot method in giving the mean of the posterior distribution in a single feedforward operation is offset by the disadvantage that it is not clear what to do if some particular input value x_k is not available on a particular case, for instance due to occlusion. $\mathcal{R}$ is tailored to the fact that all the inputs will be available. Also, $\mathcal{R}$ implicitly incorporates knowledge about the prior Φ over the factors, and so if top-down information can specify a different prior covariance matrix $\Phi' \neq \Phi$ on some particular occasion, then the bottom-up weights $\mathcal{R}$ will be incorrect. If top-down information only changes the unconditional mean, then the expression for the posterior mean in equation 5 shows that $\mathcal{R}$ is still appropriate.

Top-Down Method

More in the spirit of mean field methods, it is also possible to derive the posterior distribution of equations 3 in an iterative manner, using constitutively the top-down weights that define the generative model in the first place. A good way to understand this is through the same minimum description length (MDL; Rissanen, 1989) coding argument that motivates the Helmholtz machine. Consider using a distribution $\hat{\mathcal{N}} \equiv \mathcal{N}[\hat{\mathbf{y}}, \hat{\Sigma}]$ as a stochastic code for example $\mathbf{x}$. This means that a sample $\mathbf{y}^s$ is drawn from $\hat{\mathcal{N}}$, is coded itself using the prior $\mathcal{N}[\mathbf{0}, \Phi]$ over $\mathbf{y}$, and is used to provide a conditional prior $\mathcal{N}[\mathcal{G}^T\mathbf{y}^s, \Psi]$ over the actual image $\mathbf{x}$. The net mean description length for example $\mathbf{x}$ using this code has two additive components. The first is the cost of coding the sample $\mathbf{y}^s$ from $\hat{\mathcal{N}}$ (minus the bits back, Hinton & Zemel, 1994), and, on average, is:

$$\mathcal{F}_1 = KL\left\{\mathcal{N}\left[\hat{\mathbf{y}}, \hat{\Sigma}\right], \mathcal{N}\left[\mathbf{0}, \Phi\right]\right\} = \frac{1}{2}(\log\frac{|\Phi|}{|\hat{\Sigma}|} + \text{tr}\left(\hat{\Sigma}\Phi^{-1}\right) + \hat{\mathbf{y}}^T\Phi^{-1}\hat{\mathbf{y}}) \quad (6)$$

where $KL\{\mathcal{P}, \mathcal{Q}\}$ is the Kullback-Leibler divergence from $\mathcal{P}$ to $\mathcal{Q}$. The second component is the cost of coding the image $\mathbf{x}$ given $\mathbf{y}^s$, which, on average, is:

$$\mathcal{F}_2 = \frac{1}{2}((\mathbf{x} - \mathcal{G}^T\hat{\mathbf{y}})^T \Psi^{-1} (\mathbf{x} - \mathcal{G}^T\hat{\mathbf{y}}) + \text{tr}\left(\mathcal{G}\Psi^{-1}\mathcal{G}^T\hat{\Sigma}\right) + \sum_j \log \tau_j^2) + \mathcal{K} \quad (7)$$

where $\mathcal{K}$ is a constant that does not depend on $\hat{\mathbf{y}}$ or $\hat{\Sigma}$.

Shannon's theorem guarantees that the description length $\mathcal{F} \equiv \mathcal{F}_1 + \mathcal{F}_2$ for any choice of $\hat{\mathbf{y}}$ and $\hat{\Sigma}$ is greater than or equal to $-\log \mathcal{P}[\mathbf{x}]$ under the generative model, and equality holds when the coding distribution $\hat{\mathcal{N}}$ is the true recognition distribution (*ie* the true probabilistic inverse to the generative distribution). Consider, therefore, minimising $\mathcal{F}$ with respect to $\hat{\mathbf{y}}$ and $\hat{\Sigma}$. The linear and Gaussian nature of the generative model makes the two minimisations independent. As might be expected from equation 3, the optimisation of $\hat{\Sigma}$ is also independent of the input $\mathbf{x}$ and therefore can be done once and for all. Optimising $\hat{\mathbf{y}}$ is more interesting. The parts of $\mathcal{F}$ that depend on $\hat{\mathbf{y}}$ comprise a quadratic form, with $\nabla_{\hat{\mathbf{y}}}\mathcal{F} = \Phi^{-1}\hat{\mathbf{y}} - \mathcal{G}\Psi^{-1}(\mathbf{x} - \mathcal{G}^T\hat{\mathbf{y}})$. We can therefore either read off the optimal $\hat{\mathbf{y}}$ as implied by equation 3 (by solving for $\nabla_{\hat{\mathbf{y}}}\mathcal{F} = \mathbf{0}$), or implement gradient descent in $\mathcal{F}$ using the dynamical system (Olshausen & Field, 1996; Rao & Ballard, 1995):

$$\tau \frac{d\hat{\mathbf{y}}}{dt} = -\nabla_{\hat{\mathbf{y}}}\mathcal{F} = -\Phi^{-1}\hat{\mathbf{y}} + \mathcal{G}\Psi^{-1}(\mathbf{x} - \mathcal{G}^T\hat{\mathbf{y}}) \quad (8)$$

where τ is a time constant. The interesting aspect of this equation is its implications for the bottom-up weights and processing in the $\mathbf{x}$ layer. Equation 8 suggests calculating the prediction error $(\mathbf{x} - \mathcal{G}^T\hat{\mathbf{y}})$ in the $\mathbf{x}$ layer (this is the difference between the actual image $\mathbf{x}$ and the image that would be predicted from the mean top-down activities $\hat{\mathbf{y}}$), down-weighting this prediction error in the $\mathbf{x}$ layer by the noise magnitudes $1/\tau_j^2$ along each $\mathbf{x}$ dimension, and then propagating it through the transpose of the generative weights $\mathcal{G}$ to change $\hat{\mathbf{y}}$. Note the two major differences from the one-shot approach: a) the system is iterative, based on calculating the prediction errors; and b), as in PCA, the bottom-up weights are the transpose of the generative weights, rather than being dependent also on Ψ and Φ. The generative weights $\mathcal{G}$ that minimise $\mathcal{F}$ will nevertheless in general not be the same as the ones calculated by PCA, *ie* they will differ from the eigenvectors of the covariance matrix of the images.

If there is top-down information that changes the unconditional mean of $\mathbf{y}$, then this just adds an extra term $\Phi^{-1}\bar{\mathbf{y}}$ to the update equations. Changing the unconditional covariance matrix Φ requires just a change to the update within the $\mathbf{y}$ layer, and not a change to the bottom-up weights. Further, if the value of some input dimension is not specified on some particular occasion, then it should make no contribution in the term $\mathcal{G}\Psi^{-1}(\mathbf{x} - \mathcal{G}^T\hat{\mathbf{y}})$. This will be the case if the result of top-down influences *on the* $\mathbf{x}$ *layer* is to set the relevant component of $\mathbf{x}$ equal to its top-down mean, in the absence of any information from the scene. Top-down inference therefore avoids all the problems alluded to for bottom-up inference, at the expense of requiring iterations to satisfy $d\hat{\mathbf{y}}/dt = 0$.

Olshausen & Field (1996) developed a dynamical system like that of equation 8 from the starting point of minimising the cost:

$$\mathcal{C}(\hat{\mathbf{y}}, \mathcal{G}) = \sum_i f(\hat{y}_i) + \left(\mathbf{x} - \mathcal{G}^T \hat{\mathbf{y}}\right)^T \Psi^{-1} \left(\mathbf{x} - \mathcal{G}^T \hat{\mathbf{y}}\right) \tag{9}$$

which is closely related to $\mathcal{F}$.[2] In equation 9, $\hat{\mathbf{y}}$ is again the cortical representation of $\mathbf{x}$, and is also chosen to balance two costs. The first term is intended to encourage sparseness in the $\hat{\mathbf{y}}$, using a penalty term $f(y)$ such as $f(y) = \log(1 + y^2)$ which encourages $\mathbf{y}$ units to be silent. Just like component $\mathcal{F}_1$ in equation 6, this penalty term is essentially equivalent to that coming from a prior $\mathcal{P}[y] \propto e^{-f(y)}$ for the activities in the $\mathbf{y}$ layer, in which they are mutually independent. The term $-\Phi^{-1}\hat{\mathbf{y}}$ in equation 8 is replaced in $\nabla_{\hat{y}}\mathcal{C}(\hat{\mathbf{y}}, \mathcal{G})$ by a vector whose components are $-f'(\hat{y}_i)$. However, making $f(y)$ non-quadratic means that there is no longer a separation in the minimisations with respect to $\hat{\mathbf{y}}$ and $\hat{\Sigma}$ (just as certainty equivalence in control theory only holds in the linear case), and so minimising $\mathcal{C}$ with respect to $\hat{\mathbf{y}}$ becomes itself an approximation. One can write down the equivalent of $\mathcal{F}$, but such a simple dynamical system can only find (local) maximum *a posteriori* values and not the true Bayesian conditional mean.

Just like components $\mathcal{F}_2$ in equation 7, the second term in equation 9 encourages $\hat{\mathbf{y}}$ to provide a good model for the image $\mathbf{x}$, through the medium of the generative weights $\mathcal{G}$. Olshausen & Field (1996) showed that realistic generative receptive fields emerge for the $\mathbf{y}$ units when recognition is based on choosing $\hat{\mathbf{y}}^* = \text{argmin}_{\hat{y}}\mathcal{C}(\hat{\mathbf{y}}, \mathcal{G})$, and the generative weights are altered based on these values. Reasonable recognition receptive fields for the $\mathbf{y}$ units are also observed, but they depend on the images that are presented and require calculating $\hat{\mathbf{y}}^*$.

If $f(y)$ is quadratic, as in Rao & Ballard (1995), and effectively also in $\mathcal{F}_1$, then there is actually no incentive for $\hat{\mathbf{y}}^*$ to be sparse. For instance, if there were two units y_1 and y_2 with equal generative weights, then $\hat{y}_1^2 + \hat{y}_2^2 + (9 - [\hat{y}_1 + \hat{y}_2])^2$ is minimised with $\hat{y}_1^* = \hat{y}_2^* = 3$; whereas $\log(1 + \hat{y}_1^2) + \log(1 + \hat{y}_2^2) + (9 - [\hat{y}_1 + \hat{y}_2])^2$ is minimised at $(\hat{y}_1^*, \hat{y}_2^*) = (0.1, 8.8)$ or $(8.8, 0.1)$.

Lateral Method

Olshausen (personal communication) has pointed out an equivalent form of equation 8, for minimising $\mathcal{F}$ with respect to $\hat{\mathbf{y}}$:

$$\tau \frac{d\hat{\mathbf{y}}}{dt} = \mathcal{G}\Psi^{-1}\mathbf{x} - \left(\Phi^{-1} + \mathcal{G}\Psi^{-1}\mathcal{G}^T\right)\hat{\mathbf{y}}. \tag{10}$$

This suggests a slightly different calculation scheme from equation 8, in which the image $(\mathbf{x})$, rather than the prediction error for the image $\left(\mathbf{x} - \mathcal{G}^T\hat{\mathbf{y}}\right)$ is downweighted by Ψ^{-1} and propagated through a weight matrix $\mathcal{G}$ which again is just the transpose of the generative weights. Now, though, computation in the $\mathbf{y}$ layer is more complicated, including requiring connections that are sensitive to the noise magnitudes Ψ^{-1} in the $\mathbf{x}$ layer. This shares the disadvantage of the

---

[2] Olshausen & Field (1996) used $\Psi = \mathcal{I}$, based on the reasonable assumption that all errors in the $\mathbf{x}$ layer are equivalent.

bottom-up scheme in terms of requiring weights that include information about the priors Φ^{-1} and $\mathcal{G}\Psi^{-1}\mathcal{G}^T$, but does at least allow the non-specification of some particular input, provided, as also for equation 8 that the input is set at exactly its top-down predicted value at all times.

Combined Bottom-Up and Top-Down Method

The final method is based on multiplying the update in equation 10 by the positive definite matrix $\Sigma = (\Phi^{-1} + \mathcal{G}\Psi^{-1}\mathcal{G}^T)^{-1}$. In this case, one can also implement the dynamics:

$$\tau\frac{d\hat{\mathbf{y}}}{dt} = \left(\Phi^{-1} + \mathcal{G}\Psi^{-1}\mathcal{G}^T\right)^{-1}\mathcal{G}\Psi^{-1}\mathbf{x} - \hat{\mathbf{y}} = \mathcal{R}^T\mathbf{x} - \hat{\mathbf{y}} \tag{11}$$

This version has the attractive characteristic that if all the inputs are specified, then the feedforward information is instantly correct, and so iteration would in theory not be required (if the time constant $\tau = 1$). If some inputs are not specified, then, by the same reasoning as above, the system will still find the correct conditional mean for the factors.

In this simple linear and Gaussian case, there is therefore an update form that is based on: a) the one-shot method used in the conventional Helmholtz machine, and b) the iterative scheme employed in Rao & Ballard (1995) and also in mean field methods for the non-linear case (Jaakkola *et al*, 1996; Olshausen & Field, 1996).

Hierarchical Factor Analysis

Figure 2 shows a slightly more complicated, but still linear and Gaussian model which has three layers, two of which are spatially separated. This model, and yet more complicated versions of it, is due to Chou and Willsky and their colleagues (Chou, Willsky & Benveniste, 1994; Chou, Willsky & Nikoukhah, 1994; Luettgen & Willsky, 1995). These authors were interested in using Kalman filters in scale rather than in time to build tractable models of inputs that naturally live in multiple dimensions (such as images) rather than one dimension (such as auditory waveforms). The most natural generative model in two-dimensions is a Markov random field (MRF; Kinderman & Snell, 1980), but inference in MRFs is notoriously intractable. Chou *et al* (1994) develop a sophisticated theory of recognition in such structures, and we will mostly just cite the relevant results in our own terms, without writing out the more notionally gruesome ones.

One-Pass Method

The net prior distribution for $\mathbf{y}^a$ is Gaussian, with mean $\mathbf{0}$ and covariance matrix $\Xi^a + \mathcal{H}^{a^T}\Phi\mathcal{H}^a$. One can therefore use the results of the previous section to work out the conditional mean and variance of the Gaussian distributions $\mathbf{y}^a|\mathbf{x}^a$, and equivalently $\mathbf{y}^b|\mathbf{x}^b$. Only the means depend on the images $\mathbf{x}^a$ and $\mathbf{x}^b$. It turns out that one can also combine the means of these conditional distributions in a linear manner to work out the mean of the Gaussian distribution $\mathbf{z}|\mathbf{x}^a, \mathbf{x}^b$, including information from both halves of the input. These linear feedforward operations are closely related to those in equation 3, only with added complexity because of the separated inputs.

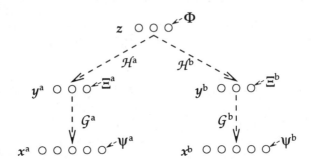

Fig. 2. Hierarchical factor analysis, after Chou *et al* (1994). The three-layer linear Gaussian generative model for inputs $\mathbf{x}^a$ and $\mathbf{x}^b$ via factors $\mathbf{z}$ and $\mathbf{y}^a$ and $\mathbf{y}^b$. Φ is the covariance of $\mathbf{z}$, Ξ^a is the covariance of the noise corrupting $\mathbf{y}^a$ from $\mathcal{H}^{a^T}\mathbf{z}$, and Ψ^a the noise corrupting $\mathbf{x}^a$ from $\mathcal{G}^{a^T}\mathbf{y}^a$. Only the generative weights are shown.

Although one can calculate $\mathbf{z}|\mathbf{x}^a, \mathbf{x}^b$ using purely bottom-up calculations, and $\mathbf{y}^a$ is conditionally independent of $\mathbf{y}^b$ and therefore $\mathbf{x}^b$ *given* $\mathbf{z}$,, the structure of the generative model makes it clear that $\mathbf{y}^a$ is conditionally *dependent* on $\mathbf{x}^b$ in the recognition circumstance in which we are given only $\mathbf{x}^a$ and $\mathbf{x}^b$ and not $\mathbf{z}$. In fact, $\mathbf{y}^a, \mathbf{y}^b$ and $\mathbf{z}$ are joinly Gaussian, with a hyper-elliptical covariance structure. The Kalman filter framework can be used for smoothing as well as filtering, in this case, feeding information back from $\mathbf{z}|\mathbf{x}^a, \mathbf{x}^b$ to update the distribution $\mathbf{y}^a|\mathbf{x}^a$ to $\mathbf{y}^a|\mathbf{x}^a, \mathbf{x}^b$. Chou *et al* (1994) show how to do this efficiently using a single top-down pass, in a generalization of the Rauch-Tung-Striebel smoothing algorithm to this tree-like case. Chou *et al* (1994) also point out a slightly different variant on this two pass algorithm in which the bottom-up phase calculates terms such as $\mathbf{y}^{a^{ML}}$ ignoring the prior, and then the top-down phase applies all the information about the priors.

Iterative Methods

Although, just as for the case of the two-layer model, there is a substantially efficient non-iterative algorithm, Rao & Ballard (1995) pointed out that one can calculate the means of the distributions $\mathbf{y}^a|\mathbf{x}^a, \mathbf{x}^b$, $\mathbf{y}^b|\mathbf{x}^a, \mathbf{x}^b$ and $\mathbf{z}|\mathbf{x}^a, \mathbf{x}^b$ using a dynamical system closely related to those in equations 8 and 10. The quadratic form comprising the terms in the description length that depend on the means (written as $\hat{\mathbf{y}}^a$, *etc*) is given by

$$2\mathcal{F}' = \hat{\mathbf{z}}^T \Phi^{-1} \hat{\mathbf{z}} +$$

$$\left(\hat{\mathbf{y}}^a - \mathcal{H}^{a^T}\hat{\mathbf{z}}\right)^T \Xi^{a^{-1}} \left(\hat{\mathbf{y}}^a - \mathcal{H}^{a^T}\hat{\mathbf{z}}\right) + \left(\hat{\mathbf{y}}^b - \mathcal{H}^{b^T}\hat{\mathbf{z}}\right)^T \Xi^{b^{-1}} \left(\hat{\mathbf{y}}^b - \mathcal{H}^{b^T}\hat{\mathbf{z}}\right) +$$

$$\left(\mathbf{x}^a - \mathcal{G}^{a^T}\hat{\mathbf{y}}^a\right)^T \Psi^{a^{-1}} \left(\mathbf{x}^a - \mathcal{G}^{a^T}\hat{\mathbf{y}}^a\right) + \left(\mathbf{x}^b - \mathcal{G}^{b^T}\hat{\mathbf{y}}^b\right)^T \Psi^{b^{-1}} \left(\mathbf{x}^b - \mathcal{G}^{b^T}\hat{\mathbf{y}}^b\right).$$

Using gradient descent to solve for $\nabla\mathcal{F}' = \mathbf{0}$ leads to an analogue of equation 8 (Rao & Ballard, 1995):

$$\tau \frac{d\hat{\mathbf{z}}}{dt} = -\Phi^{-1}\hat{\mathbf{z}} + \mathcal{H}^a \Xi^{a^{-1}} \left(\hat{\mathbf{y}}^a - \mathcal{H}^{a^T}\hat{\mathbf{z}}\right) + \mathcal{H}^b \Xi^{b^{-1}} \left(\hat{\mathbf{y}}^b - \mathcal{H}^{b^T}\hat{\mathbf{z}}\right)$$

$$\tau \frac{d\hat{\mathbf{y}}^a}{dt} = -\Xi^{a^{-1}} \left(\hat{\mathbf{y}}^a - \mathcal{H}^{a^T}\hat{\mathbf{z}}\right) + \mathcal{G}^a \Psi^{a^{-1}} \left(\mathbf{x}^a - \mathcal{G}^{a^T}\hat{\mathbf{y}}^a\right)$$

$$(12)$$

These share with equation 8 the characteristic of how top-down prediction errors at the various levels ($\hat{\mathbf{y}}^a - \mathcal{H}^{a^T}\hat{\mathbf{z}}$ and $\mathbf{x}^a - \mathcal{G}^{a^T}\hat{\mathbf{y}}^a$) are downweighted by the noise covariances and propagated bottom-up through the transpose of the generative weight matrices. Since the overall joint distribution $\mathbf{z}, \mathbf{y}^a, \mathbf{y}^b | \mathbf{x}^a, \mathbf{x}^b$ is elliptical, the use of these means is slightly tricky. Again, in a non-linear or non-Gaussian case, such as a multilayer analogue of Olshausen & Field's (1996) sparsity prior, the separation between means and covariances would disappear, and it would no longer be possible to use these equations to work out the true posterior means.

There is also an analogue of equation 10:

$$\tau \frac{d\hat{\mathbf{z}}}{dt} = \mathcal{H}^a \Xi^{a^{-1}} \hat{\mathbf{y}}^a + \mathcal{H}^b \Xi^{b^{-1}} \hat{\mathbf{y}}^b - \left(\Phi^{-1} + \mathcal{H}^a \Xi^{a^{-1}} \mathcal{H}^{a^T} + \mathcal{H}^b \Xi^{b^{-1}} \mathcal{H}^{b^T}\right) \hat{\mathbf{z}}$$

$$\tau \frac{d\hat{\mathbf{y}}^a}{dt} = \mathcal{G}^a \Psi^{a^{-1}} \mathbf{x}^a + \Xi^{a^{-1}} \mathcal{H}^{a^T} \hat{\mathbf{z}} - \left(\Xi^{a^{-1}} + \mathcal{G}^a \Psi^{a^{-1}} \mathcal{G}^{a^T}\right) \hat{\mathbf{y}}^a$$

$$(13)$$

which implies the use of lateral operations. The equivalent of equation 11 is more complicated, however, because the bottom up estimate of $\hat{\mathbf{y}}^a$ given just $\mathbf{x}^a$ is different from the final estimate of $\hat{\mathbf{y}}^a$ given both $\mathbf{x}^a$ and $\mathbf{x}^b$. Unlike equation 10, it is not clear how to specify a single set of bottom-up weights that can conveniently be used for both one-pass and iterative inference.

III   Non-linear Models

The previous section considered the case of linear and Gaussian generative models for which there are computationally tractable ways of calculating the exact recognition inverse distributions. Even though there may be purely bottom-up ways of doing this in regular cases, for which there is no occlusion and no top-down information relevant to inference for a particular image, iterative methods can also be used, and have certain demonstrable advantages. However, linear and Gaussian models are unlikely to suffice, even for the most primitive datasets. One class of non-linear models that has been studied in some depth is that of binary belief networks with sigmoidal activation functions (Neal, 1992; Hinton *et al*, 1995; Saul *et al*, 1996). These preserve the top-down structure of the generative model, but, for a single unit x_1, they have (*cf* equation 1):

$$x_1 \sim \mathcal{B}\left[\sigma\left(\left[\mathcal{G}^T \mathbf{y}\right]_1\right)\right] \tag{14}$$

where $\mathcal{B}[p]$ is the binomial distribution with mean p and σ is a sigmoid function whose output lies between 0 and 1. The activities in layer $\mathbf{y}$ are set similarly, except, for a two-layer network, that the only input to the binomial distribution is a bias term. An extra component of $\mathbf{y}$ is treated as a bias for determining $\mathbf{x}$. A three layer network is shown in figure 3a.

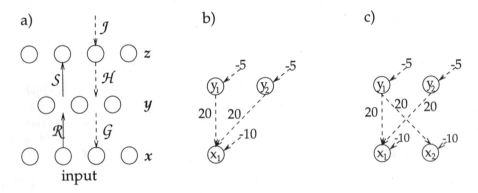

a)

b)

c)

input

Fig. 3. a) Three layer sigmoidal belief net. The layered structure shows the conditional independencies in the net $P[\mathbf{x}, \mathbf{y}, \mathbf{z}] = P[\mathbf{z}]P[\mathbf{y}|\mathbf{z}]P[\mathbf{x}|\mathbf{y}]$ where the units within a layer are mutually independent given the activities in the layer above and are set according to equation 14 using the generative weights $\mathcal{H}$, $\mathcal{G}$ (and the biases $\mathcal{J}$ for the $\mathbf{z}$ layer that receives no other input). $\mathcal{R}$ and $\mathcal{S}$ are bottom-up parameters for recognition. For the Helmholtz machine, recognition is based on a bottom-up belief network which specifies $P[\mathbf{y}, \mathbf{z}|\mathbf{x}] = P[\mathbf{y}|\mathbf{x}]P[\mathbf{z}|\mathbf{y}]$ (which is true) with an approximation of mutual independence within the $\mathbf{y}$ layer and within the $\mathbf{z}$ layer given the $\mathbf{y}$ activities (which is generally not true). b,c) Two non-linear generative models. Both figures give the generative connections allowing $\mathbf{y}$ units to describe activity patterns over $\mathbf{x}$ units. All units are stochastic binary units, with activation functions as given in the text. The weights with no originating nodes are biases. b) Explaining away. $x_1 = 1$ requires explaining by either $y_1 = 1$ or $y_2 = 1$ but not both (*ie* $P[\mathbf{y} = \{0,0\}|x = 1] = 0.0033; P[\mathbf{y} = \{1,1\}|x = 1] = 0.0033$ and $P[\mathbf{y} = \{0,1\}|x = 1] = P[\mathbf{y} = \{1,0\}|x = 1] = 0.4967$. c) if $y_1 = 1$ also generates $x_2 = 1$, then at question is whether or not there should be a direct negative influence from x_2 to y_2 in the recognition model.

Examples of layered belief networks with this structure are given by Hinton *et al* (1995), in particular the comparatively large networks (with $4-16-16-64$

units in three hidden layers and one input layer) that model 8×8 binary images of handwritten digits. A key aspect of these generative models is that the activities of units *within* a layer are mutually independent, given the activities of the units in the layer above. This makes specifying the generative probabilities very simple. In such cases, it is again necessary to calculate the recognition inverse to this generative model – now the recognition distribution assigns probabilities to the 2^n binary states of the n hidden units in the network, given the activities in the lowest input layer. For general networks, this distribution is not tractably computable, and therefore sampling or approximations are necessary. Even though in the generative model, activities of units within a layer are independent given the activities in the layer above, activities of units within a layer in the recognition model need not be mutually independent given the activities in the layer below.

The non-linear belief net model makes for much richer generative and recognition models than the linear Gaussian models of factor analysis. Two very simple but revealing generative models are shown in figure 3b;c. The left example shows a case of explaining away (Pearl, 1988). All the units are most likely to be off (0). However, the occurrence of x being on (1) requires explanation by either $y_1 = 1$ or $y_2 = 1$. The *a priori* unlikelihood that the **y** units are active makes it unlikely that $y_1 = y_2 = 1$; having $y_1 = 1$ *explains away* $x = 1$ and so obviates the need for $y_2 = 1$ as well. Figure 3c is a modification of this example in which y_1 tends to generate $x_1 = x_2 = 1$.

Bottom-Up Method

The Helmholtz machine (see figure 3a) suggests building a bottom-up recognition model that is also a sigmoidal belief net and (at least in the stochastic version) drawing samples from this net to approximate the full recognition distribution. Two simplifications are made: a) despite the caveat above, the activities of units within a layer are forced to be independent given the activities in the layer below, to avoid having to parameterise and manipulate the full conditional distribution within each layer,[3] and b) the connections of the recognition belief net are set using an incorrect training procedure that chooses them to minimise (locally) a *wrong* error measure that is nonetheless computationally convenient. Write the states of all the units other than the inputs (**x**) as α. Then, if the probability accorded to a particular α by the recognition model is $\mathcal{Q}_\alpha(\mathcal{R})$, where $\mathcal{R}$ are the weights of the non-linear recognition model, then $\mathcal{R}$ should correctly be chosen to minimise $KL\{\mathcal{Q}_\alpha(\mathcal{R}), \mathcal{P}[\alpha|\mathbf{x}]\}$ (Hinton & Zemel *et al*, 1994). Instead, in the sleep learning procedure (Hinton *et al*, 1995), they are chosen to minimise $KL\{\mathcal{P}[\alpha|\mathbf{x}], \mathcal{Q}_\alpha(\mathcal{R})\}$. Since the Kullback-Leibler divergence is not symmetric, these quantities are not the same. If the divergence cannot be forced to be 0 (as is likely given the approximations employed), then the main difference is that minimising the first requires that $\mathcal{P}[\alpha|\mathbf{x}]$ be small whenever $\mathcal{Q}_\alpha(\mathcal{R})$ is small, whereas minimising the second requires that $\mathcal{Q}_\alpha(\mathcal{R})$ be small whenever $\mathcal{P}[\alpha|\mathbf{x}]$

---

[3] Note, however, that this does not mean that *all* the units are mutually independent.

is small.

This bottom-up method for approximating the recognition model works quite well in practice (see Frey *et al*, 1996). However, in cases such as explaining away (including the simple example of figure 3b) it fails. It assigns independent bottom-up probabilities of 0.5 to $y_1 = 1$ and $y_2 = 1$, and so 50% of the time chooses settings for **y** that are incorrect.

The bottom-up recognition inverse to the generative model in figure 3c is correct. The point is that if $x_1 = x_2 = 1$ for a particular case, then it must be that $y_1 = 1$ rather than $y_2 = 1$. Even though the generative weight from y_2 to x_2 is 0, the recognition weight from x_2 to y_2 is negative. A more interesting case of this is seen in the weight patterns for the 8-bit shifter problem in Dayan *et al* (1995). Units in the first hidden layer generate the activity of single pixels within both eyes, shifted by one pixel left or right with respect to each other. The recognition weights for these units have positive values for the pixels whose activities are actually generated, but inhibitory side-lobes from the neighbouring pixels, to avoid spurious activation.

It has been suggested that lateral connections within a layer might be used as in a Boltzmann machine solely for the recognition model (Dayan & Hinton, 1996). The generative model would still involve only top-down influences as in figure 3a (to ensure that the the generative model is still tractable), but the recognition model would employ lateral links to circumvent the requirement that the units within a layer be independent. The disadvantage is that computationally complex Gibbs sampling within a layer has to be used to instantiate recognition. Also, in cases such as explaining away, there would have to be an explicit negative lateral connection between y_1 and y_2.

Top-Down Methods

Mean-field   The major alternatives to the bottom-up recognition model described above are mean field methods, pioneered for belief nets by Saul, Jaakkola & Jordan (1996) and Jaakkola, Saul & Jordan (1996). These effectively choose a parameterisation for an approximation to the recognition distribution for a particular case, and optimise the parameters to minimise the equivalent of the correct Kullback-Leibler divergence. Most approximations force *all* the units to be mutually independent (not just those units within a single layer, given the activities in the layer below). One can treat the linear and Gaussian case from the previous section exactly in mean-field terms, using a parameterisation with means for all the units and a particular covariance structure. Minimising the Kullback-Leibler divergence turns out to require satisfying a set of self-consistency equations at each unit, and there are algorithms that descend monotonically in the divergence whilst updating units asynchronously using only information local to a unit and its incoming and outgoing connections. In the linear Gaussian case, solving these self-consistency equations is exactly solving for the correct mean values.

The mean field theory of Jaakkola *et al* (1996) is a suitable non-linear counterpart. It uses as its sigmoid activation function the normal distribution function

$\sigma(a) = \frac{1}{\sqrt{2\pi}} \int_{-\infty}^{a} e^{-b^2/2} db$. This leads to a cost function $\mathcal{C}$ for minimisation for which the contribution from the terms in the $\mathbf{y}$ layer is $2\mathcal{C}_{\hat{\mathbf{y}}} =:$

$$\sum_{ij} \mathcal{G}_{ji}^2 \sigma(\hat{y}_j)\sigma(-\hat{y}_j) + \sum_j (\hat{y}_j - \sum_k \mathcal{H}_{kj}\sigma(\hat{z}_k))^2 + \sum_i (\hat{x}_i - \sum_j \mathcal{G}_{ji}\sigma(\hat{y}_j))^2 \quad (15)$$

summing for i over units in $\mathbf{x}$, for j over $\mathbf{y}$, for k over $\mathbf{z}$, where $\sigma(\hat{y}_j)$ is the mean activity of unit y_j, and $\hat{x}_i$ for input unit i is set so that $\sigma(\hat{x}_i)$ is very close to 0 or 1 as appropriate. Equivalent contributions to $\mathcal{C}$ come from the other layers in the network. This cost function is related to that for the linear and Gaussian case (equations 6 and 7) and for Olshausen & Field (1996) (equation 9). The contribution to $-\nabla_{\hat{y}_j}\mathcal{C}_{\hat{\mathbf{y}}}$ from the second and third (*ie* the prediction error) terms of equation 15 is

$$-(\hat{y}_j - \sum_k \mathcal{H}_{kj}\sigma(\hat{z}_k)) + \sum_i \mathcal{G}_{ji}\sigma'(\hat{y}_j)(\hat{x}_i - \sum_{j'} \mathcal{G}_{j'i}\sigma(\hat{y}_{j'}))$$

which has a very close parallel in the equivalent term for the linear and Gaussian hierarchical model as in $d\hat{y}^a/dt$ in equation 12. Minimising $\mathcal{C}$ can either be accomplished using gradient descent, as in equation 12, or by updating $\hat{y}_j$ such that $\nabla_{\hat{y}_j}\mathcal{C} = 0$. In both cases only local information is used at each unit, and the bottom-up connections from units in a layer to units in the layer above must be the transpose of the generative weights in the opposite direction.

Mean field approximations such as this can cope well with inverting the generative model of figure 3b when $x = 1$, since there are two stable minima, one with $y_1 = 1; y_2 = 0$ and the other with $y_1 = 0; y_2 = 1$. Moreover, unlike the lateral Boltzmann machine, they do not require the existence of an explicit connection between y_1 and y_2.

Stochastic simulation An alternative approach to calculating the recognition distribution is to use a Markov-chain Monte-Carlo method. Neal (1992) discusses stochastic simulation in sigmoidal belief nets in some detail, including the use of such samples for learning the generative weights. Here, we point out a simple approximation in the limit of small generative weights that shows the similarity with the other top-down uses of bottom-up connections.

In the case in figure 3a, with $\sigma(a) = 1/(1+e^{-a})$, we are interested in sampling from $\mathcal{P}[\mathbf{y}, \mathbf{z}|\mathbf{x}]$. In the simplest version, we visit each of the units in layers $\mathbf{y}$ and $\mathbf{z}$ in some random sequence, and choose a new state stochastically, taking into account influences from its parents (just the generative biases for the $\mathbf{z}$ units) and its children. Take unit y_1, writing $\bar{\mathbf{y}} = \{y_2, \ldots, y_n\}$. We should set its new state to 1 according to $\mathcal{P}[y_1 = 1|\bar{\mathbf{y}}, \mathbf{z}, \mathbf{x}]$. It turns out to be most convenient to calculate instead

$$\rho_1 = \log \frac{\mathcal{P}[y_1 = 1|\bar{\mathbf{y}}, \mathbf{z}, \mathbf{x}]}{\mathcal{P}[y_1 = 0|\bar{\mathbf{y}}, \mathbf{z}, \mathbf{x}]} = \log \frac{\mathcal{P}[\mathbf{z}]\mathcal{P}[y_1 = 1|\mathbf{z}]\mathcal{P}[\bar{\mathbf{y}}|\mathbf{z}]\mathcal{P}[\mathbf{x}|y_1 = 1, \bar{\mathbf{y}}]}{\mathcal{P}[\mathbf{z}]\mathcal{P}[y_1 = 0|\mathbf{z}]\mathcal{P}[\bar{\mathbf{y}}|\mathbf{z}]\mathcal{P}[\mathbf{x}|y_0 = 1, \bar{\mathbf{y}}]}$$

$$= \sum_k \mathcal{H}_{k1}z_k + \sum_i \log \frac{\mathcal{P}[x_i|y_1=1,\bar{\mathbf{y}}]}{\mathcal{P}[x_i|y_1=0,\bar{\mathbf{y}}]} \text{ as } \mathcal{P}[y_1 = 1|\mathbf{z}] = \sigma\left(\sum_k \mathcal{H}_{k1}z_k\right) \quad (16)$$

Since

$$\mathcal{P}[x_i|y_1 = 1, \bar{\mathbf{y}}] = \frac{x_i + (1 - x_i)e^{-\mathcal{G}_{1i} - \sum_{j\neq 1}\mathcal{G}_{ji}y_j}}{1 + e^{-\mathcal{G}_{1i} - \sum_{j\neq 1}\mathcal{G}_{ji}y_j}}$$

then, if $\mathcal{G}_{1i}$ is small compared with $\sum_{j\neq1}\mathcal{G}_{ji}y_j$, it turns out that

$$\log \frac{\mathcal{P}[x_i|y_1 = 1, \bar{\mathbf{y}}]}{\mathcal{P}[x_i|y_1 = 0, \bar{\mathbf{y}}]} = \mathcal{G}_{1i}\left(x_i - \mathcal{P}[x_i = 1|y_1 = 0, \bar{\mathbf{y}}]\right).$$

Since $\mathcal{G}_{1i}$ is small, to zeroth order

$$\mathcal{P}[x_i = 1|y_1 = 0, \bar{\mathbf{y}}] = \mathcal{P}[x_i = 1|y_1 = 1, \bar{\mathbf{y}}] = \mathcal{P}[x_i = 1|\mathbf{y}]$$

is the top down prediction that $x_i = 1$ whatever the state of y_1. Using this in equation 16, we have:

$$\rho_1 \approx \sum_k \mathcal{H}_{k1}z_k + \sum_i \mathcal{G}_{1i}\left(x_i - \mathcal{P}[x_i = 1|\mathbf{y}]\right) \tag{17}$$

which consists of the obvious top-down influence from $\mathbf{z}$ and a bottom-up influence that tends to reduce the prediction error $(x_i - \mathcal{P}[x_i = 1|\mathbf{y}])$ for the state of x_i. Just as in the other top-down cases (such as the dynamic system in equation 8 or the mean field theory of Jaakkola *et al*, 1996) the prediction error is propagated bottom-up through the transpose of the generative weights $\mathcal{G}$. Stochastic simulation then requires that y_1 be set to 1 with probability $\sigma(\rho_1)$. As for mean field methods, explaining away can be handled (although not quite as in equation 17, since the weights are not insubstantial), again without recourse to direct connections between y_1 and y_2.

For the generative model in figure 3c, note how making the recognition weight from x_2 to y_2 zero rather than negative makes it more complicated to work out that if $x_1 = x_2 = 1$, then $y_1 = 1$ rather than $y_2 = 1$.

IV   Discussion

In this paper, we have studied the issue of inverting various sorts of directed belief net generative models. Such generative models are attractive as ways of capturing the essence of the hierarchical structure of cortex, where the activities of neurons in successively higher cortical areas represent the generation of successively more abstract entities in scenes. Inverting the generative models, *ie* going from sensory input to the activities of the neurons that represent its likely generators, is essential to interpret scenes, and also (though this has not been stressed here) to learn appropriate generative models. The inverse operation, called recognition, is akin to discrimination or classification, and, in many cases, doing it exactly is computationally intractable. We assume that the bottom-up weights in the cortex (from V1 to V2, *etc*) are important for recognition, but that they may operate in conjunction with the top-down and lateral weights. We have also seen some of the relationships amongst recent suggestions as to how cortex might implement a generative model.

Various schemes for performing recognition have been suggested. One class uses only bottom-up connections, either for exact recognition, as for factor analysis, or for approximate recognition, as for the Helmholtz machine. In the latter case, it is tractable to draw samples stochastically from the bottom-up model. Although purely bottom-up models are fast, a strong requirement suggested by evidence on the speed of processing images of objects, they suffer from a number of disadvantages in terms of the difficulty of incorporating top-down information, coping with occlusion, and integrating information from disparate parts of a scene further than the credible spread of feedforward connections. Further, in important cases such as explaining away, bottom-up models that make reasonable approximations, such as that the activities of units in a layer are mutually independent given the activities in the layer below, are incompetent.

An alternative scheme is suggested by mean field methods (Saul *et al*, 1996). Here a parameterised form is chosen for the recognition distribution for a particular case, and the parameters are set by an optimisation process. For suitable parameterisations, optimisation is achieved by a set of local operations. Although there are no bottom-up weights as such for mean field methods, we saw various cases in which the influence of the units in one layer on those in the layer above is calculated by passing some form of prediction error (*ie* the difference between their actual activation and the activation predicted on the basis of the states of the units in the layer above) through the transpose of the generative weights. Optimisation is iterative – this solves the problems mentioned for the purely bottom-up method, but raises questions as to the time required. We saw simple cases in which there is a difference between the bottom-up weights implied by purely bottom-up recognition and the bottom-up weights implied by this top-down scheme.

A further alternative was to use lateral weights within each layer. This was either to eliminate the requirement for iteration between layers (as in equation 8), or to fix problems with approximations made for bottom-up inference, as in explaining away.

Different generative models impose different requirements on their recognition inverses. Linear models with Gaussian noise are particularly simple – even in the case of Chou *et al* (1994) with multiple layers and only partial connectivity, there are algorithms for working out the true recognition inverses which require nothing more than one bottom-up and one top-down pass. Non-linear models, such as the layered belief nets of figure 3a do not possess such tractable inverses, and so approximations are necessary.

This analysis is unsatisfyingly incomplete. Foremost, the interaction between bottom-up, top-down and lateral connections is still open. Given the structural differences between lateral and bottom-up weights, one might expect to find a computational difference too. Top-down influences must clearly be felt during recognition; however it is not apparent how to have bottom-up weights that implement one-shot recognition in ideal circumstances, but can be used for interacting bottom-up and top-down processing in less ideal cases. The role of the lateral weights is also mysterious. One suggestion is that they help repair problems with too restrictive approximations in the bottom-up recognition model, but this use imposes strong requirements on the presence of dense connections (so that all cases of explaining away in the generative model can be properly handled) and/or on the time available for Gibbs sampling. If the lateral weights are also involved in specifying the generative model, then the whole system becomes a form of Boltzmann machine.

An issue raised by the Kalman filter models of section 2 is the various covariance matrices for the activities at different layers. In the linear cases with Gaussian noise that we have considered, the covariance matrices are fixed once the parameters of the generative model are fixed, and they do not depend on the input for a particular scene. In non-linear cases, and in cases which include temporal effects (Rao & Ballard, 1995), this is not true. Retaining just the diagonal terms of the covariance matrix of the activities is simple (Sutton, 1992). Retaining the off-diagonal terms is more complicated because of their numerosity and the complexity of the calculations that lead to them.

Apart from capturing its general hierarchical characteristics, none of the models in this paper is very faithful to the real details of cortical processing. Apart from the many complexities of the structure of lower and higher visual processing areas, important issues are that the anatomical spread of top-down connections is *broader* than that of the bottom-up connections, and that there appears to be a difference over developmental time in their specification. For instance, the top-down connections from V2 to V1 in humans wait in the lowest cortical layers in V1 for a few months, and only migrate to what becomes their main targets in layer 2 at the same time that the intracortical lateral connections within layer 2 are also maturing (Burkhalter, 1993). This might favour a different class of models (Luttrell, 1995) from the ones discussed here in which the bottom-up recognition weights are actually primary, heeding some other developmental call, and the top-down and lateral weights merely build the generative model that is most consistent with whatever activities the recognition model ultimately specifies.

Acknowledgements

I am most grateful to Brendan Frey, Geoff Hinton, Tommi Jaakkola, Mike Jordan, Zhaoping Li, Radford Neal, Bruno Olshausen and Lawrence Saul for helpful discussions. This work was funded by NIMH grant R29-MH55541-01. All opinions expressed are those of the author.

References

1. Burkhalter, A (1993). Development of forward and feedback connections between areas V1 and V2 of human visual cortex. *Cerebral Cortex, 3*, 476-87.
2. Chou, KC, Willsky, AS & Benveniste, A (1994). Multiscale recursive estimation, data fusion, and regularization. *IEEE Transactions on Automatic Control, 39*, 464-478.
3. Chou, KC, Willsky, AS & Nikoukhah, R (1994). Multiscale systems, Kalman filters, and Riccati equations. *IEEE Transactions on Automatic Control, 39*, 479-492.
4. Dayan, P (1996). A Hierarchical model of visual rivalry. Submitted to *Neural Information Processing Systems, 9.*

5. Dayan, P & Hinton, GE (1996). Varieties of Helmholtz Machine. *Neural Networks,* in press.

6. Dayan, P, Hinton, GE, Neal, RM & Zemel, RS (1995). The Helmholtz machine. *Neural Computation,* **7**, 889-904.

7. Dempster, AP, Laird, NM & Rubin, DB (1977). Maximum likelihood from incomplete data via the EM algorithm. *Proceedings of the Royal Statistical Society,* **B 39**, 1-38.

8. Everitt, BS (1984). *An Introduction to Latent Variable Models.* London: Chapman and Hall.

9. Felleman DJ & Van Essen DC (1991) Distributed hierarchical processing in the primate cerebral cortex. *Cerebral Cortex,* **1**, 1-47.

10. Frey, BJ, Hinton, GE & Dayan, P (1995). Does the wake-sleep algorithm produce good density estimators? *Advances in Neural Information Processing Systems, 8,* forthcoming.

11. Grenander, U (1976-1981). *Lectures in Pattern Theory I, II and III: Pattern Analysis, Pattern Synthesis and Regular Structures.* Berlin: Springer-Verlag., Berlin, 1976-1981).

12. Hinton, GE, Dayan, P, Frey, BJ & Neal, RM (1995). The wake-sleep algorithm for unsupervised neural networks. *Science,* **268**, 1158-1160.

13. Hinton, GE, Dayan, P & Revow, M (1996). Modeling the manifolds of images of handwritten digits. *IEEE Transactions on Neural Networks,* forthcoming.

14. Hinton, GE & Sejnowski, TJ (1986). Learning and relearning in Boltzmann machines, In DE Rumelhart, JL McClelland and the PDP research group, editors, *Parallel Distributed Processing: Explorations in the Microstructure of Cognition. Volume 1: Foundations,* Cambridge, Massachusetts: MIT Press, 282-317.

15. Hinton, GE & Zemel, RS (1994). Autoencoders, minimum description length and Helmholtz free energy. In JD Cowan, G Tesauro and J Alspector, editors, *Advances in Neural Information Processing Systems 6.* San Mateo, CA: Morgan Kaufmann, 3-10.

16. Jaakkola, T, Saul, LK & Jordan, MI (1996). Fast learning by bounding likelihoods in sigmoid type belief networks. *Advances in Neural Information Processing Systems, 8,* forthcoming.

17. Jacobs, RA, Jordan, MI, Nowlan, SJ & Hinton, GE (1991). Adaptive mixtures of local experts, *Neural Computation,* **3**, 79-87.

18. Jolliffe, IT (1986) *Principal Component Analysis,* New York: Springer-Verlag.

19. Kawato, M, Hayakama, H & Inui, T (1993). A forward-inverse optics model of reciprocal connections between visual cortical areas. *Network,* **4**, 415-422.

20. Kinderman, R & Snell, JL (1980). *Markov Random Fields and their Applications.* American Mathematical Society.

21. Leopold, DA & Logothetis, NK (1996). Activity changes in early visual cortex reflect monkeys' percepts during binocular rivalry. *Nature,* **379**, 549-554.

22. Logothetis, NK, Leopold, DA & Sheinberg, DL (1996). What is rivalling during binocular rivalry. *Nature,* **380**, 621-624.

23. Luettgen, MR & Willsky, AS (1995). Likelihood calculation for a class of multiscale stochastic models, with application to texture discrimination. *IEEE Transactions on Image Processing,* **4**, 194-207.

24. Luttrell, SP (1995). A componential self-organizing neural network. Submitted to *Advances in Neural Information Processing Systems, 8.*

25. Marroquin, JL (1985). *Probabilistic Solution of Inverse Problems.* PhD Thesis, AI Lab, MIT, Cambridge, MA.

26. Mumford, D (1994). Neuronal architectures for pattern-theoretic problems. In C Koch and J Davis, editors, *Large-Scale Theories of the Cortex*. Cambridge, MA: MIT Press, 125-152.

27. Neal, RM (1992). Connectionist learning of belief networks. *Artificial Intelligence*, **56**, 71-113.

28. Neal, RM (1993). *Probabilistic Inference using Markov Chain Monte Carlo Methods*. Technical Report CRG-TR-93-1. Department of Computer Science, University of Toronto.

29. Neal, RM & Dayan, P (1996). *Factor Analysis using Delta-Rule Wake-Sleep Learning*. TR-9607, Department of Statistics, University of Toronto.

30. Nowlan, SJ (1991). *Soft Competitive Adaptation: Neural Network Learning Algorithms based on Fitting Statistical Mixtures*. CMU Technical Report CMU-CS-91-126, Carnegie-Mellon University, Pittsburgh PA.

31. Olshausen, BA & Field, DJ (1996). Emergence of simple-cell receptive field properties by learning a sparse code for natural images. *Nature*, **381**, 607-609.

32. Pearl, J (1988). *Probabilistic Reasoning in Intelligent Systems: Networks of Plausible Inference*. San Mateo, CA: Morgan Kaufmann.

33. Perrett, DI, Rolls, ET & Caan W (1982). Visual neurones responsive to faces in the monkey temporal cortex. *Experimental Brain Research*, **47**, 329-342.

34. Poggio, T & Torre, V (1984). Ill-posed problems and regularization analysis in early vision. *Proceedings of ARPA Image Understanding Workshop*, 257-263.

35. Rao, PNR & Ballard, DH (1995). *Dynamic Model of Visual Memory predicts Neural Response Properties in the Visual Cortex*. Technical report 95.4, Department of Computer Science, Rochester, NY.

36. Rissanen, J (1989). *Stochastic Complexity in Statistical Inquiry*. Singapore: World Scientific.

37. Rubin, DB & Thayer, DT (1982). EM algorithms for maximum likelihood factor analysis. *Psychometrika*, **47**, 69-76.

38. Saul, LK, Jaakkola, T & Jordan, MI (1996). Mean field theory for sigmoid belief networks. *Journal of Artificial Intelligence Research*, **4**, 61-76.

39. Sutton RS (1992). Gain adaptation beats least squares? In *Proceedings of the 7th Yale Workshop on Adaptive and Learning Systems*.

40. Ullman, S (1994). Sequence seeking and counterstreams: A model for bidirectional information flow in the cortex. In C Koch and J Davis, editors, *Large-Scale Theories of the Cortex*. Cambridge, MA: MIT Press, 257-270.

41. Zemel, RS (1994). *A Minimum Description Length Framework for Unsupervised Learning*. PhD Dissertation, Computer Science, University of Toronto, Canada.

Continuity $\Sigma-$Algebras (Extended Abstract)

Wilson R. de Oliveira*

Abstract. We generalize the notion of ordered and metric $\Sigma-$algebras to algebras where the carriers are continuity spaces [9]. The main motivation for this generalization comes from the subject of Real Number Computation. We aim at providing the scientific computing programmer with tools from the formal specification theory in order to achieve a rigorous development and analysis of numerical programs.

Classification of topics covered: abstract data types, program specification, exact real arithmetic, lazy functional languages, continuity spaces.

1 Introduction

In contrast to seminumerical computation, numerical computation has been relatively neglected by the formal methods researchers despite the fact that numerical programs are being used in vital activities on our day-life. We propose here to use one of the most widely used and successful approach to the specification of data types: the algebraic specification method.

The main reason for this neglect certainly is the very nature of the data type in question: the reals. A real number is an infinity entity and their collection is uncountable. Besides there are a plethora of topological notions such as continuity, convergence, limits, *etc.* which should be taken in account.

Algebraic specification methods provide techniques for *data abstraction* and the *structured specification, validation* and analysis of data structures. An excellent explanation of the ideas involved in the field is given by Martin Wirsing in [22]:

> The basic idea of the algebraic approach consists in describing data structures by just giving the names of the different sets of data, the names of the basic functions and their characteristics properties. The

---

*Supported by the Brazilian Research Council CNPq-RHAE, grant no. 610293192.3.

properties are described by formulas (mostly equations) which are invariant under isomorphism. An *abstract data type* is defined to be an isomorphism class of data structures, *e.g.* an isomorphism class of many-sorted algebras. An *algebraic specification* is a description of one or more such abstract data types. The methodological use of abstract data types in programming is grouping together in one specification all the basic functions that manipulate particular sorts of data, and then "hiding" the representation, in the sense that the *defined data* can only be manipulated by calling the functions provided by the specification.

The data type being specified is then the term algebra quotient out by the equations describing its properties. The initial algebra approach to specification of data types has proposed solutions to deal with data of infinity elements ([12,20]) *via* completion of the finite ones but their solutions does not cover the **lazy reals**. The mainstream approaches come then in two flavors: ordered and metric. The ordered approach uses continuos cpos [12] and ideal completion; and the metric approach uses ultrametric spaces [20] and Cauchy sequence completion.

One drawback of the ordered approach to continuous data types is that domains (Scott domains, algebraic cpos, continuous cpos) are not generally closed under quotients. In its turn the metric approach is not able to deal with partial information. Besides, to date, the metric approach is restricted to ultrametric spaces.

The author has suggested a common generalization of the above two approaches by using quasimetric spaces [6], spaces on which a non symmetric real distance function is defined. Most of the drawbacks of the classical approaches (see below) to continuous data types disappear but the category of continuity spaces has a richer structure and a wider collection of morphisms. The (natural) category for quasimetric spaces asks as morphism the non-increasing maps while for continuity spaces the maps are just continuous. Thus, as a category, the continuity spaces are indeed a generalization of the cpo and metric approaches. The exploration of the richer type structure of the continuity spaces is left for another occasion [4].

The troubles with the classical approaches (ordered and metric) come to the fore when addressing the problem of specifying the reals as an abstract data type. As we shall see in Section 2 the space obtained is not a domain (algebraic nor continuous cpos) of any kind and can not be seen as an ultrametric space. The reals as data type is an important issue in the analysis and semantics of numeric programs [3].

In its full generality Continuity Spaces are quantales valued distances. A quantale is sort of continuos lattice on which a notion of "addition" is defined so that it makes sense, say, to propose that $d(x,y) \leq d(x,z) + d(z,y)$.

The continuity spaces are sufficiently general do deal with most of the topological structures met in mathematical and computational practice [9].

After introducing the notion of a continuity Σ-algebra and obtaining initiality results a specification of the reals as an abstract data type is given. In the full version of this extended abstract detailed investigation on computability, efficiency, representations of continuous functions, *etc.* are envisaged. Preliminary results on this direction is presented in [5] using program schemes. It has to be pointed out, however, that a notion of non-deterministic data type must be employed. It is known ([3]) that, in order to represent every continuous and computable real function, we must have either a representation-dependent basic function or ,a non-deterministic basic operations. Thus the notion of hyperspaces in Section 3 can not be avoided.

In what follows a variant of the redundant balanced radix notation [14] is employed. In this representation we allow digits to be negative as well as positive. For a more complete study of this representation in contrast to others see [3]. What it is important for the work here is its simplicity. None of the results depend on the chosen representation. The redundant balanced radix notation is also known as signed digits representation. In its simplest form it is also known as *tritstreams* representation or modified binary representation, since the only digits allowed are in the set $T = \{-1, 0, 1\}$. We shall, w.l.o.g, limit ourselves to the discussion of reals in $[-1, 1]$.

2   Motivations

One of the most widely used representation of the reals in both practical and theoretical computer science is the signed digit representation (*q.v.* [3,11,15,21,13]). The reason comes from its simplicity, computability of the arithmetic operations, and easy implementation in lazy functional languages. Here a real number is represented by an infinite list or string. The reals represented in this way are usually called Lazy Reals since they can be implemented in a lazy functional language as a lazy list.

Numbers in the interval $[-1, 1]$ can be represented as signed binary digits *i.e.* string composed of digits in $T = \{-1, 0, 1\}$ or $T = \{-, 0, +\}$.

Let T^ω be the set of *streams* (infinite strings) on $T = \{-, 0, +\}$. The real number represented by the stream $\alpha = a_1 a_2 \dots$ is defined as $[\![\alpha]\!]_R = \sum_{i>0} a_i \times 2^{-i}$.

A string $\alpha = a_1 \dots a_n \in V^*$ represents the set of possible continuations of α; hence, an interval. This defines a map $[\![\ ]\!]_I$ from T^* to the closed intervals

$$[\![a_1 \dots a_n]\!]_I = [(\sum_{i=1}^{n} a_i \times 2^{-i}) - 2^{-n}, (\sum_{i=1}^{n} a_i \times 2^{-i}) + 2^{-n}]$$
$$= [[\![a_1 \dots a_n -^\omega]\!]_R, [\![a_1 \dots a_n +^\omega]\!]_R]$$

Notation 1   *We use* $[\![\ ]\!]$ *to denote the function which behaves as* $[\![\ ]\!]_I$ *on the strings and as* $[\![\ ]\!]_R$ *on streams.*

We have now at our disposal two ways of looking at the *real domain*. As the set of finite and infinite strings on T; and as the set of closed *dyadic* intervals[1] plus the real line. Actually, the set T^∞ $(= T^* \cup T^\omega)$ needs to be quotiented out by the equivalence relation $\alpha \equiv \beta$ iff $[\![\alpha]\!] = [\![\beta]\!]$. An information order $\sqsubseteq$ can then be defined on the equivalence classes. Indicating by $[\alpha]$ the equivalence class containing α, $[\alpha] \sqsubseteq [\beta]$ iff $[\![\beta]\!] \subseteq [\![\alpha]\!]$.

The equivalence on strings can be described more operationally as:

$$0 + x = + - x \tag{0}$$

$$0 - x = - + x \tag{1}$$

where $x \in T^\infty$, which generalizes the obvious arithmetic fact that $0 + \frac{1}{4} = \frac{1}{2} - \frac{1}{4}$ and that $0 - \frac{1}{4} = -\frac{1}{2} + \frac{1}{4}$.

The reader familiar with data type specification would suggest to approach the specification of the "real domain" by starting with finite lists of $\{-, 0, +\}$, subject them to the equations (0) and (1).Quotienting out the (prefix) order on the **finite** lists by these equations, in a standard way, gives us in effect the appropriate closed intervals with the intended order.

The approach mostly used in the literature coming from the continuous data type theory ([12]), would take the ideal completion of the poset of closed dyadic intervals. This yields a Scott domain [19] but gives us three versions of each dyadic rational.

For example, for 0 we have 0^ω, $-+^\omega$ and $+-^\omega$ which respectively correspond to the ideals generated by $\{[-\frac{1}{2^k}, \frac{1}{2^k}] | k \geq 0\}$, $\{[-\frac{1}{2^k}, 0] | k \geq 0\}$ and $\{[0, \frac{1}{2^k}] | k \geq 0\}$. In the completion, the ideal generated by the first set approximates the last two and the three can not be identified.

Another approach ([16]) is to take a continuous quotient and use a notion of normalizer. But, for a start our equivalence relation is not continuous. In the completed space we have *e.g.* $-+^n \equiv 0^n-$ and $-+^\omega = \bigsqcup_n -+^n \not\equiv \bigsqcup_n 0^n- = 0^\omega$ (recall that $\bigsqcup_n -+^n$ is the least upper bound of the sequence $(-+^n)_{n \geq 0}$ w.r.t. $\sqsubseteq$ order). There obviously exists (*q.v.* [12]) a least continuous extension of our equivalence which would identify the three copies of each dyadic rational. The trouble is that the resulting space is a partial order which is far from being a domain of any kind. It is not even a continuous cpo. For example, we have that, for any $k \geq 0$, $-+^k \sqsubseteq \bigsqcup_n +-^n = 0^\omega$ and $-+^k \not\sqsubseteq +-^n$ for any n. Which implies that $-+^k, k \geq 0$, is not compact. Similar argument shows that no finite string is compact. Thus the resulting space can not be obtained as a completion of any poset of compact elements and, by definition, is not an algebraic nor a continuous cpo.

When dealing with infinite objects one usually works on the finite portions of them and get the results on the infinite *via* some sort of completion. That is, in a nutshell, the methodology of Computability Theory, Domain Theory, Denotational Semantics, Constructive Mathematics and some approaches to continuous data types. We have seen above that we

---

[1]Of the form $[\frac{(m-1)}{2^n}, \frac{(m+1)}{2^n}]$ where n is the length of the string, m is the integer $\sum_{i=1}^{n} a_i \times 2^{n-i}$. Rationals of the form $m/2^n$, $m, n \in \mathbb{Z}$ are called *dyadic*.

can not follow this methodology when dealing with the real domain in the cpo approach. It is worth mentioning that the use of metric space is also not adequate, firstly by the fact that we can not represent partial or finite information (metric spaces are necessarily Hausdorff) and secondly that the approach to metric data types in the literature (*q.v.* [20]) uses *ultrametrics* (those space are necessarily zero dimensional). Our proposal is then to use *continuity space*.

3   Continuity Spaces

We quickly survey the main results and definitions of continuity spaces needed in Section 4. The reader interested in more details may consult [9,8,10]

The bottom element of a complete lattice V is denoted by 0, the top element by ∞ and for A a subset of V, the least upper bound of A is denoted by $\sup A$ and the greatest lower bound of A by $\inf A$.

A *value distributive lattice* is a completely distributive lattice V satisfying the following two conditions: (1) $\infty \succ 0$ and (2) if $p \succ 0$ and $q \succ 0$, then $p \wedge q \succ 0$. A *quantale* $\mathcal{V} = \langle V, \leq, + \rangle$ consists of a complete lattice $\langle V, \leq \rangle$ and an associative and commutative binary operation $+$ on V satisfying: (1) for all $p \in V$, $p + 0 = p$ and (2) for all $p \in V$ and all families $\{q_i\}_{i \in I}$ of elements of V, $p + \inf_{i \in I} q_i = \inf_{i \in I}(p + q_i)$. A *value quantale* is a quantale $\mathcal{V} = \langle V, \leq, + \rangle$ such that $\langle V, \leq \rangle$ is a value distributive lattice.

Definition 2 *A $\mathcal{V}$-continuity space is a pair $\mathcal{X} = (X, d)$ consisting of a set X and a function $d : X \times X \to V$ satisfying the following conditions: (1) for all $x \in X$, $d(x, x) = 0$ and (2) for all x, y, $z \in X$, $d(x, z) + d(z, y) \geq d(x, y)$.*

If $\mathcal{X} = (X, d)$ is a $\mathcal{V}$-continuity space, then the *dual* of $\mathcal{X}$ is the pair $\mathcal{X}^* = (X, d^*)$, where for all x, $y \in X$, $d^*(x, y) = d(y, x)$ and the *symmetrization* of $\mathcal{X}$ is the pair $\mathcal{X}^s = (X, d^s)$, where for all x, $y \in X$, $d^s(x, y) = d(x, y) \vee d(y, x)$.

There is a natural topology on a $\mathcal{V}$-continuity space $\mathcal{X} = (X, d)$, which is defined in a way completely analogous to the definition of the metric topology on a metric space, and is called the *induced topology* on $\mathcal{X}$. The induced topology on $\mathcal{X}^*$ is called the *dual topology* on $\mathcal{X}$ and induced topology on $\mathcal{X}^s$ the *symmetric topology* on $\mathcal{X}$. For A a subset of X, $cl(A)$ is the closure of A in the induced topology, $* - cl(A)$ is the closure of A in the dual topology, and $s - cl(A)$ is the closure of A in the symmetric topology. Similar notations will be used for other familiar topological notions. In particular, a function is said to be symmetrically continuos if it is continuous with respect to the symmetric topology.

Recall for use below that the *specialization order* on a topological space (X, τ), denoted by $\leq_\tau$, is defined by: $x \leq_\tau y$ iff $x \in cl\{y\}$. Then $\leq_\tau$ is always reflexive and transitive, and it is a partial order on X iff X is T_0.

A map $f : X \to Y$ between $\mathcal{V}$-continuity spaces X and Y is *nonexpansive* if for all x_1, $x_2 \in X$, $d_X(x_1, x_2) \geq d_Y(f(x_1), f(x_2))$. A $\mathcal{V}$-continuity space

$\mathcal{X} = (X,d)$ is said to be T_0 if $d(x,y) = 0$ and $d(y,x) = 0$ imply $x = y$, for all x, $y \in X$. Assume (X,d) is a $\mathcal{V}$-continuity space. Then X is called a $\mathcal{V}$-*domain* if it is T_0 and compact in its symmetric topology. Let $\mathcal{V}$-**Dom** denote the category with objects the $\mathcal{V}$-domains and morphisms the continuous functions. The proof that $\mathcal{V}$-**Dom** is closed under elementary type forming operations is found in [9].

A net $(x_\lambda)_{\lambda \in \Lambda}$ in a $\mathcal{V}$-continuity space X is *Cauchy* if for every $\varepsilon \succ 0$ there is a λ_0 such that for all μ, $\nu \geq \lambda_0$, $\varepsilon \geq d(x_\mu, x_\nu)$. X is complete if every Cauchy net in X has a limit in the symmetric topology on X. Cauchy completeness in continuity spaces and its relationship to natural order theoretic notions of completeness are studied in [8]. X is *totally bounded* if for all $\varepsilon \succ 0$ there is a finite $F \subseteq X$ such that $X = \cup_{y \in F} N_\varepsilon^s(y)$.

The proof of the next result is just a translation of the usual argument for the corresponding result on uniform spaces.

Theorem 3 *Assume $\mathcal{X} = (X,d)$ is a T_0 $\mathcal{V}$-continuity space. Then X is a $\mathcal{V}$-domain iff X is complete and totally bounded.*

Powerdomains provide domain-theoretic analogs of the power set. Their consideration is motivated by the need to model nondeterministic constructs. In this paper they are especially important, since they nondeterministic operations must be used.

The standard results of the theory of the *upper* powerdomain can be adapted to the setting of $\mathcal{V}$-domains. The lower and convex powerdomains can also be adapted to this setting. Assume X is a $\mathcal{V}$-continuity space. Define the *upper Hausdorff distance*, $d_{\mathcal{U}}$, on the power set of X as

$$d_{\mathcal{U}}(A,B) = \sup_{b \in B} \inf_{a \in A} d(a,b), A, B \subseteq X.$$

The *upper powerdomain* of X, $\mathcal{U}(\mathcal{X})$ is defined to be the collection of nonempty $*-$closed subsets of X with the upper Hausdorff distance.

Notation 4 *Denote by* **2** *the value quantale* $\{0,1\}$ *with* $0 \leq 1$ *and* $1 + 1 = 0 + 0 = 0$ *and* $1 + 0 = 0 + 1 = 1$; *by* **2**$^\bullet$ *the value quantale* $\{2^{-n} \mid n \in \mathbb{N}\} \cup \{0\}$ *with the usual order and* max; *by* $\mathbb{I}$ *the value quantale of real numbers in* $[0,1]$ *with the usal order and the truncated addition* $x + y = \min\{x+y, 1\}$.

Example 5 *On the algebraic cpo Σ^∞, define a* **2**$^\bullet$*-valued distance by:*

$$d(x,y) = 2^{-\max\{n \mid x[n] \leq y[n]\}}$$

where $x[n]$ denotes the n-truncation of x, that is, the result of deleting all terms of x after the first n. Notice that $d(x,y) = 0$ iff $x \leq y$. We see that d induces the Scott topology on Σ^∞, and is a **2**$^\bullet$*-domain iff Σ is finite.*

Example 6 *Let D be any Scott domain, and $r : B_D \to \mathbb{N}$ a map (a 'rank function') such that $r^{-1}(n)$ is a finite set for each $n \in \mathbb{N}$. Define a* **2**$^\bullet$*-valued distance by:*

$$d(x,y) = 2^{-\max\{n \mid e \sqsubseteq x \Rightarrow e \sqsubseteq y \text{ for every } e \text{ of rank } \leq n\}}$$

Then d is a $\mathbf{2^\bullet}$-domain and induces the Scott topology of D.

The two examples above could as well be defined as a $\mathbb{I}$-valued distance given rise to different kinds of spaces, as we shall see in a moment.

As we have hinted at above, an important part in the specification of an abstract data type is to obtain a quotient of a space. As usual the quotient space X/R of a space X by a given equivalence relation R is defined in such a way that the structure on X/R is *final* with respect to the canonical map $x \mapsto [x]$ of X onto X/R (in a suitable category). For (quasi)metric spaces the suitable category for constructions has as morphisms the non-expansive maps (*q.v.* [1]). The situation is surprisingly similar here.

Definition 7 *Given a continuity space (X, d) and an equivalence relation R we define the quotient continuity space $X/R = (Y, d_Y)$ as the greatest distance on Y for which $[.]$ is non-increasing, i.e. $d_Y([x], [y]) \le d(x, y)$.*

Concretely we have the following. Let $\pi = \langle x_0, x_1, \cdots, x_{2n+1} \rangle$ be a *path*, i.e. $x_{2i-1} R x_{2i}$ $(i = 1, \cdots, n)$, whose *length* is $l(\pi) = \sup_{i=0}^{n} d(x_{2i}, x_{2i+1})$. The triangle inequality requires that $d_Y([x_0], [x_{2n+1}]) \le l(\pi)$. Thus we define d_Y by: $d_Y(u, v) = \inf\{l(\pi) | \pi \text{ is a path from } x \text{ to } y \text{ where } [x] = u, [y] = v\}$.

Observation 8 *It can easily be shown that the definition of quotient space above is coherent with the categorical notion of quotient structure of [1] which, by applying their general definitions to our case, says that X/R is the quotient continuity space of (X, d) if the canonical map $\eta_R : (X, d) \to X/R$ has a co-optimal lift d'. A function $f : (X, d) \to Y$ has co-optimal lift if there exists a distance d' such that $f : (X, d) \to (Y, d')$ is non-expansive and whenever (Z, d'') and $g : Y \to Z$ are such that $g.f : (X, d) \to (Z, d'')$ is non-expansive,*

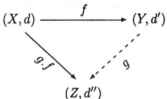

then $g : (Y, d') \to (Z, d'')$ is non-expansive.

4   Continuity Σ—algebras

The trouble with the discussion in Section 2 is that we have forgotten the *quantitative* structure. A continuity can be defined on T^∞ as in the Example 5:

$$d(x, y) = 2^{-\max\{n | x[n] \le y[n]\}} .$$

0^ω, $+-^\omega$ are to be identified because their *distance* is (to be made) 0:

$$d(0^\omega, +-^\omega) = d(\lim_{n \to \infty} 0^n, \lim_{n \to \infty} +-^n) \equiv 0$$

We assume familiarity with the basics notions of [12] and in particular with [17], where the notion of *non-deterministic* data type is studied. For the sake of simplicity, let us restrict ourselves w.l.o.g. to the one sorted case. So, in the terminology of [12], $\langle \Sigma_i \rangle_{i \in \mathbb{N}}$ is a one sorted operator domain and $\Sigma = \bigcup_{i \in \mathbb{N}} \Sigma_i$. The elements of Σ_i are the *operators symbols* of *arity i*. A nondeterministic operation $\sigma : A \to B$ is equally interpreted as a multivalued function $f : A \to 2^B$ or a function $f : A \times B \to \mathbb{B}$ where $\mathbb{B}$ is the flat domain of truth values with distance $d(x,y) = 0 \iff x = \bot$.

Definition 9 *Let U be a set, whose elements are called* variables *or* indeterminates, *disjoint from Σ. The set $FT(\Sigma, U)$ of all Σ-terms in the variables of U is defined inductively as:*

1. $U \cup \Sigma_0 \subset FT(\Sigma, U)$; *and*
2. *if $\sigma \in \Sigma_k$ and $t_i \in FT(\Sigma, U)$, then $\sigma(t_1, \ldots, t_k) \in FT(\Sigma, U)$.*

We define the term algebra $FT(\Sigma, U)$ *of all Σ-terms in the variables of U and the set of Σ-equations as usual [12].*

For the purpose of defining a distance function a Σ-term is better seen as a (finite) Σ-*tree*.

Definition 10 *A Σ-tree is a partial function $t : \mathbb{N}^* \longrightarrow \Sigma$ such that, for all $u \in \mathbb{N}^*$ and $i \in \mathbb{N}$*

1. $ui \in \mathsf{dom}(t) \Rightarrow u \in \mathsf{dom}(t);$
2. $ui \in \mathsf{dom}(t) \Rightarrow \exists n > 0 . t(u) \in \Sigma_n$ *and* $i < n$

t is a *labeling* function and an integer i in ui is meant to be the root of the ith subtree after arriving at the node $t(u)$.

Notation 11 $- CT_\Sigma$ denotes the set of all Σ-trees.
- FT_Σ the finite Σ-trees (i.e. all t such that $\mathsf{dom}(t)$ is finite).
- $\mathbb{N}^{(n)} = \{u \in \mathbb{N}^* | \mathsf{length}(u) \le n\} = \{u[k] \mid u \in N^*, k \le n\}$
- $t^{(n)} = t \mid \mathbb{N}^{(n)}$ i.e. $t^{(n)} : \mathbb{N}^{(n)} \to \Sigma$ and $t^{(n)}(u) = t(u)$, for all $u \in \mathbb{N}^{(n)}$

In what follows, unless otherwise stated, $\mathcal{D}$ will always denote one of the $2^\bullet$ or $\mathbb{I}$ value quantale.

Definition 12 $d : CT_\Sigma \times CT_\Sigma \to \mathcal{D}$ is defined as

$$d(t_1, t_2) = 2^{\max\{n | t_1^{(n)} \subseteq t_2^{(n)}\}} .$$

Observations 13   1. $\langle CT_\Sigma, d \rangle$ is a complete continuity space which is $\mathcal{D}$-domain if $\bigcup_{i \in \mathbb{N}} \Sigma_i$ is finite (as a set). From now on we assume $\bigcup_{i \in \mathbb{N}} \Sigma_i$ is finite.
2. $\langle CT_\Sigma, d \rangle$ is the completion of FT_Σ (with the $\mathcal{D}$-valued distance induced from that of CT_Σ).

We make CT_Σ into a Σ-algebra in the following way:

1. For $\sigma \in \Sigma_0$, let $\sigma_{CT} = \{\langle \lambda, \sigma \rangle\};$
2. For $\sigma \in \Sigma_n$, $n > 0$, and $t_1, \ldots, t_n \in CT_\Sigma$ let $\sigma_{CT} = \{\langle \lambda, \sigma \rangle\} \cup \bigcup_{i < n} \{\langle iu, \sigma' \rangle | \langle u, \sigma' \rangle \in t_{i+1}\}$

Observations 14   – if $t_i \in CT_\Sigma$, $1 \le i \le n$, and $\sigma \in \Sigma_n$, then $\sigma_{CT}(t_1, \ldots, t_n) \in CT_\Sigma$
– FT_Σ can be made into a Σ-algebra similarly.

Before arriving at what would be a reasonable definition of a continuity Σ–algebra we observe the following simple result:

Lemma 15 *The operations of CT_Σ are contractive.*

Proof (sketch). Just observe in the simple case σ is unary that $d(\sigma t, \sigma t') = \frac{1}{2}d(t, t')$ $\square$

We conclude then that the operations must be at least contractive functions.

We want to define a class of complete continuity Σ–algebras $CQAlg_\Sigma$ for which CT_Σ is initial. We then start with a class of continuity Σ–algebras $QAlg_\Sigma$ for which FT_Σ is initial and obtain the initiality result for $CQAlg_\Sigma$ from that of FT_Σ from the extension of the unique $h : FT_\Sigma \to \mathcal{A}$ to $\hat{h} : CT_\Sigma \to \mathcal{A}$. This being guaranteed by the fact that CT_Σ is the completion of FT_Σ. Thus $\mathcal{D}$-**Dom** is the natural candidate.

Definition 16 *A function $h : \mathcal{A} \to \mathcal{B}$ between Σ–algebras is a homomorphism if*

$$h(\bot_\mathcal{A}) = \bot_\mathcal{B}$$
$$h(\sigma_\mathcal{A}) = \sigma_\mathcal{B}$$
$$h(\sigma_\mathcal{A}(t_1, \ldots, t_n)) = \sigma_\mathcal{B}(h(t_1), \ldots, h(t_n))$$

Theorem 17 *FT_Σ is initial in the category of continuity Σ–algebras with nonexpansive operations and continuous homomorphisms.*

Proof We just need to prove that the uniquely defined homomorphism is continuous. Let $\varepsilon > 0$ be given. By induction hypothesis there is δ such that $d_\Sigma(t, t') \le \delta \Rightarrow d_\mathcal{A}(h(t), h(t')) \le \varepsilon$. As $d_\Sigma(\sigma(t), \sigma(t')) \le \frac{1}{2}d_\Sigma(t, t')$, we have that $d_\Sigma(\sigma(t), \sigma(t')) \le \frac{1}{2}\delta \Rightarrow d_\Sigma(t, t') \le \delta$. Thus, making $\delta' = \frac{1}{2}\delta$ we have:

$$
\begin{aligned}
d_\Sigma(\sigma(t), \sigma(t')) \le \delta' &\Rightarrow d_\Sigma(t, t') \le \delta \\
&\Rightarrow d_\mathcal{A}(h(\sigma(t)), h(\sigma(t'))) \le \varepsilon && \text{since} \\
&\le d_\mathcal{A}(h(t), h(t')) && \sigma_\mathcal{A} \text{ is nonexpansive} \\
&\le \varepsilon
\end{aligned}
$$

$\square$

Corollary 18 *CT_Σ is initial in the category of complete continuity Σ–algebras with nonexpansive operations and continuous homomorphisms.*

Theorem 19 *Let E be a set of equations and $CT_{\Sigma,E}$ be the quotient of CT_Σ by the equivalence relation induced by E. $CT_{\Sigma,E}$ is initial in the class of (Σ, E)-algebras by considering both categories of algebras above.*

Proof This automatically comes from the initiality of CT_Σ and the fact that the canonical morphism

$$\eta_E : (CT_\Sigma, d) \to CT_{\Sigma,E}$$

has a co-optimal lift d_E in the category of continuity spaces and nonexpansive morphisms (*q.v.* Observation 8). From that we conclude that any g making the diagram below commute is unique:

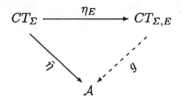

$\square$

We conclude this section with a very naïve specification of the abstract data type **Lazy Reals**. Due to space limitations we postpone the full specification to the full version.

- $\Sigma_0 = \{\lambda\}, \Sigma_1 = \{+, 0, -\}, \Sigma_3 = \{\text{cond}\}, \Sigma_i = \emptyset, i > 2$.
- $T_\Sigma \approx \{-, 0, +\}^*$, with the distance as in Example 5.
- $E = \{(0 + x, + - x), (0 - x, - + x)\}$ over $T_\Sigma(\{x\})$.
- $T_{\Sigma/E}$ is isomorphic to the continuity space T^* of Section 2.
- $\text{cond}(x, y, z) = \begin{cases} y & [\![x]\!] \leq 0 \\ z & 0 \leq [\![x]\!] \\ w & w \sqsubseteq y \& w \sqsubseteq z, \text{ otherwise} \end{cases}$
- The completion of the totally bounded space $T_{\Sigma/E}$ has as maximal elements a set which is homeomorphic to real line when $\mathcal{D} = \mathbb{I}$ and the real line with the discrete topology when $\mathcal{D} = \mathbf{2}^\bullet$.
- The function cond is a parallel test function and cannot be implemented sequentially. In [3] it is proved that no set of sequential primitives are sufficient to define the reals as an abstract data type.

5   Conclusions

A very simple space can be defined from which the reals can be obtained as the maximal elements. This space, named the **Lazy Reals**, is a cpo but is not a domain of any kind. Thus we can not use the available theory of continuous data types and the methodology of obtaining the "infinite" points as the completion of the "finite" ones. Our solution was then to use continuity spaces and we defined the notion of continuity Σ-algebras.

The problem of representing all continuous functions on the reals is being tackled in the full version of this extended abstract, where a notion of non-deterministic quasimetric Σ-algebras is fully developed. The need for non-determinism comes from the fact that one can not represent all continuous functions deterministically and keeps the abstractness of the representation by hiding it from the users of the data type [3].

Our approach the specification of reals as an algebraic data type is new. Despite the use of similar real number representation found in the literature one should stress the differences. In [3] is shown that no set of sequential primitives are sufficient to define the reals as an abstract data type but is left open the use of non-sequential primitives. [11] restricts to algebraic cpo case (and as a result the reals are obtained indirectly *via* retracts). [18] uses quasiuniformities and do obtain the real line as the subspace of total elements but no notion of continuous algebras is introduced. The same can be said to [7] which uses continuous cpos and uses the ĉpos of intervals; it is similar in spirit to [11] (in the sense that extends a λ-calculus based language with a type of real numbers).

While all the approaches cited above, including mine, obtain the reals as completions those of [2,21,13] start with the reals already given in way or another. [2] assumes the reals are available for manipulations and defines computations and its complexity over the ordered field of real numbers by Random Access Machine model; this is a departure from the standard effectivity theory of computations in the sense that assumes non (Turing) computable primitives such as the equality relation over the reals and does no take as computable Turing computable functions such as the exponential function. [21] by using representations applies his well developed Type 2 Theory of Effectivity to the real numbers represented as infinity strings (exactly the representation used in this work) where computations are performed by oracle Turing machines. [13] proposes a modified real RAM model which is polynomially equivalent to the Turing model of [21].

References

1. M. A. Arbib and E. G. Manes. *Arrows, structures and Functors: The categorical Imperative.* Academic Press, Inc., 1975.
2. Leonore Blum, Mike Shub, and Steve Smale. On a theory of computation and computational complexity over the real numbers: Np-completeness, recursive functions and universal machines. *Bulletin of the Americam Mathmatical Society*, 21:1–46, 1989.
3. Hans Boehm and Robert Cartwright. Exact real arithmetic: Formulating reals numbers as functions. Technical Report Rice COMP TR88-66, Department of Computer Science, Rice University, April 1988.
4. W. R. de Oliveira. Inverse limits continuous algebras. manuscript, Departamento de Informática, UFPE, 1996.
5. W. R. de Oliveira. Progam schemes over continuous algebras (extended abstract). accepted for publication in the *2nd. Workshop on Computability and Complexity in Analysis*, August 22-23, 1996, Universitat Trier, Germany.
6. W. R. de Oliveira and M. B. Smyth. Quasimetric Σ-algebras. In *SEMISH'94, XXI Seminario Integrado de Software e Hardware*, pages 547–561, Caxambu, Minas Gerais, Brazil, 31/07 to 05/08 1994. Sociedade Brasileira de Computação.
7. Martín H. Esacardó. PCF extended with real numbers. To appear in Theor. Computer. Science, 1996 (available electronically at http://theory.doc.ic.ac.uk:paper/Escardo/papers), 1996.

8. Robert C. Flagg. Completeness in continuity spaces. In R. A. G. Seely, editor, *Category Theory 1991*, pages 183–200. AMS, Providence, 1992.
9. Robert C. Flagg and Ralph Kopperman. Continuity spaces: Reconciling domains and metric spaces – Part I. Submitted to Theoretical Computer Science.
10. Robert C. Flagg and Ralph Kopperman. Fixed points and reflexive domain equations in categories of continuity spaces. In *Proc. of MFPS XI, Electronic Notes in Theoretical Computer Science 1*, pages 1–17, New Orleans, LA, March 29 to April 1 1995. Tulane University, Elsevier Science B.V.
11. P. Di Gianantonio. Real numbers computability and domain theory. Submitted to Information and Computation. Available electronically at http://www.dimi.uniud/~pietro, 1996.
12. J. Goguen, J.W. Thatcher, E.G. Wagner, and J.B. Wright. An initial algebra semantics and continuous algebras. *Journal of the ACM*, 24(1):68–95, January 1977.
13. Peter Hertling and Vasco Brattka. Feasible real randon access machines. In Ker-I Ko and Klaus Weihrauch, editors, *Proceedings of the Workshop on Computability and Complexity in Analysis*, pages 71–82, Hagen, Germany, August 19-20 1995. FernUniversität, Informatik Berichte.
14. Donald Knuth. *The Art of Computer Programming - Seminumerical Algorithms*, volume I. Addison-Wesley, 1969. section 4.5.
15. K. I. Ko and H. Friedman. Computational complexity of real functions. *Theoretical Computer Science*, 20, 1982.
16. M. R. Levy and T. S. E. Maibaum. Continuous data types. *SIAM Journal of. Computing*, 11(2):201, May 1982.
17. Tobias Nipkow. Non-deterministics data types: Models and implementations. *Acta Informatica*, 22:629–661, 1986.
18. Philipp Sünderhauf. A faithful computational model of the real numbers. *Theoretical Computer Science*, 151:277–294, 1995.
19. D. S. Scott. Domains for denotational semantics. In M. Nielson and E. M. Schmidt, editors, *Automata, Languages and Programming: Proceedings 1982*. Springer-Verlag, Berlin, 1982. Lecture Notes in Computer Science **140**.
20. V. Stoltenberg-Hansen and J. V. Tucker. Algebraic and fixed point equations over inverse limits of algebras. *Theoretical Computer Science*, 87:1–24, 1991. Fundamental Study.
21. Klaus Weihrauch. A foundation of computable analysis. In Ker-I Ko and Klaus Weihrauch, editors, *Proceedings of the Workshop on Computability and Complexity in Analysis*, pages 25–40, Hagen, Germany, August 19-20 1995. FernUniversität, Informatik Berichte.
22. Martin Wirsing. Algebraic specification. In J. van Leeuwen, editor, *Handbook of Theoretical Computer Science*, volume B, chapter 13, pages 675–788. Elsevier Science Publishers B. V., 1990.

Condition Number Analysis
for Sparse Polynomial Systems

Jean-Pierre Dedieu

Abstract. Is a sparse polynomial system more easy to solve than a non sparse one? In this paper we introduce a new invariant, the sparse condition number, and we study sparse polynomial system analysis in terms of this invariant.

1   Introduction

The condition number associated with a numerical problem measures the sensitivity of the considered problem to variations of the input. We study here polynomial systems $f : \mathbb{C}^n \to \mathbb{C}^n$. In this case, if $x \in \mathbb{C}^n$ is a simple zero of f and if $\dot{f}$ is a first order variation of the system, the corresponding first order variation $\dot{x}$ of x is equal to $\dot{x} = K(f,x)\dot{f} = -Df(x)^{-1}\dot{f}(x)$. The map $K(f,x)$ is linear and the condition number $C(f,x)$ of this problem is the operator norm of this map:

$$C(f,x) = \max_{\dot{f} \neq 0} \frac{\|K(f,x)\dot{f}\|}{\|\dot{f}\|}.$$

The condition number for polynomial systems has been introduced by Wilkinson [20], Wozniakowski [21] and, in the form given here, by Shub and Smale [14], [16]. These authors, in their famous serie "Complexity of Bézout Theorem", show that the complexity of some continuation algorithm for solving the polynomial system $f(x) = 0$ depends mainly on the condition number $C(f,x)$. Thus $C(f,x)$ appears as a crucial continuous invariant, related to f and x, which measures the complexity of solving polynomial systems in a model of computation over the complex numbers. This invariant has also a geometric interpretation: it is equal to the inverse of the distance to the critical set at x

$$\Sigma_x = \{g : \mathbb{C}^n \to \mathbb{C}^n \; : \; g(x) = 0 \; and \; rankDg(x) < n\}$$

see Shub-Smale [14], Dedieu [5].

If the condition operator $K(f,x)$ depends only on f and x it is important to remark that the condition number also depends on the class of perturbations we consider. Suppose that f belongs to some particular class of polynomial systems, say $\mathcal{K}$. Such a system is called here a *sparse system* since it is a priori defined by less parameters than a general (or dense) system. Let us list four important examples of sparse polynomial systems.

First Example: Fewnomials. Each equation in the system has a prescribed number of monomials:

$$f_i(z) = \sum_{\alpha \in A_i} a_\alpha z^\alpha$$

where A_i is a given subset in $\mathbb{N}^n$.

Second Example: SLP. Equations given by their computation scheme or by a straight line program. For example

$$(\dots((z + a_1)^{n_1} + a_2)^{n_2} \dots)^{n_{k-1}} + a_k.$$

Third Example: Eigenvalue Problem. These problems may be considered as a particular class of degree 2 equations, the parameters are here the entries of the considered matrix. It also may be considered as a fewnomial system.

Fouth Example: Polynomial Optimization Problems. To a polynomial program

$$\inf_{f_i(x)=0, 1 \le i \le p} f_0(x)$$

is classically associated a Langrangian system:

$$f_i(x) = 0, \ 1 \le i \le p,$$
$$Df_0(x) + \sum_{i=1}^p \lambda_i Df_i(x) = 0$$

where the unknowns are x and λ, the Lagrange multipliers.

We may define the *sparse condition number* by

$$C_\mathcal{K}(f,x) = \max_{\substack{\dot{f} \in \mathcal{K} \\ \dot{f} \ne 0}} \frac{\|K(f,x)\dot{f}\|}{\|\dot{f}\|}$$

so that, as an immediate property we have

$$C_\mathcal{K}(f,x) \le C(f,x).$$

If the complexity of some algorithm depends on the sparse condition number instead of the non sparse one, we obtain certainly a better bound in considering the sparse case. This is the program we try to develop in this paper. In Sections 2, 3 and 4 we develop the general machinery and we obtain an explicit description of the sparse condition number: Theorem 4.1. Here "f sparse" means f belongs to some Riemannian submanifold of polynomial systems. The second example shows clearly that the vector structure is too restrictive. In Section 5 we establish quantitative estimations for the variations of the zeros of a sparse polynomial system in terms of the variations of the system and its sparse condition number: Theorem 5.2, with an application to the eigenvalue problem: Theorem 5.6. In Sections 6 and 7 we give lower bounds for the separation number of f at x

that is the minimum distance from x to other zeros: Theorem 6.1 with also an application to the eigenvalue problem: Theorem 6.2. Section 8 is devoted to a Sparse Condition Number Theorem (Theorem 8.1) which relates the sparse condition number to the distance at the sparse discriminant. In Section 9 we study the complexity of a classical continuation algorithm based on Newton's iteration. When the homotopy path consists only in sparse polynomial systems the complexity can be bounded essentially by the square of the sparse condition number of the homotopy: Theorem 9.4.

The author started to study sparse condition numbers two years ago with Anne Bellido from Limoges University in France. We obtained the description of the sparse condition number. The other ideas contained here have mainly been developped when I was a visiting professor in City University of Hong Kong in May 1996. A lot of interesting discussions with Steve Smale stimulated this job.

2   Numerical Problems and Condition Number Theory

In this section we introduce some basic definitions: numerical problems, well-posed problems, ill-posed problems, condition operators, condition numbers. This material is taken from the paper Dedieu [5] and is included here in order to be self-contained.

We consider two (real or complex) Riemannian manifolds X with $dim\ X = n$ and Y with $dim\ Y = m$; X is the space of inputs and Y the space of outputs of the problem. We also consider a submanifold $V \subset X \times Y$ with $dim\ V = dim\ X = n$; V consists in the couples $(x, y) \in X \times Y$ where y is an output corresponding to the input x.

For example: $X = \mathbb{C}_d[x]$: complex polynomials with degree $\leq d$, $Y = \mathbb{C}$ and $V = \{(f, x) \in X \times Y\ :\ f(x) = 0\}$ represent the problem of finding the roots of a polynomial.

Let us denote by $T_x X$ the tangent space at x to X, by Π_1 and Π_2 the projections from V into X and Y respectively. We say that a problem $(x, y) \in V$ is *well-posed* when x is a regular value of the first projection that is when

$$D\Pi_1(x, y) : T_{(x,y)}V \to T_x X$$

is an isomorphism. Remenber that $dim\ V = dim\ X$. Otherwise we say that (x, y) is *ill-posed*. When $(x, y) \in V$ is well-posed the projection Π_1 is invertible in a neighborhood of (x, y) (inverse function theorem) and when Π_1^{-1} is composed with the second projection $\Pi_2 : V \to Y$ we obtain a C^1 operator $S : U_x \subset X \to U_y \subset Y$ satisfying

$$S(x) = y \quad and \quad graph\ S = V \cap (U_x \times U_y).$$

This operator associates to any input x' close to x the output $y' = S(x')$ close to $y = S(x)$.

The first order variations of the output y in terms of the first order variations of the input x are given by the derivative of S at x

$$K(x,y) = DS(x) : T_x X \to T_y Y.$$

This derivative is called the *condition operator* at (x, y) and can be computed via the equality

$$graph\ K(x,y) = T_{(x,y)} V.$$

The norm of this linear operator

$$C(x,y) = \max_{\substack{\dot{x} \in T_x X \\ \dot{x} \neq 0}} \frac{\|K(x,y)\dot{x}\|_{T_y Y}}{\|\dot{x}\|_{T_x X}}$$

is called the *condition number* at (x, y). It measures the sensitivity to input variations of the considered numerical problem.

Let E and F be Hermitian vector spaces and $L : E \to F$ a linear operator *onto*. The Moore-Penrose inverse of L is

$$L^+ = (L|_{(KerL)^\perp})^{-1}.$$

This operator satisfies $LL^+ = id_Y$ and $L^+L(x) = x$ for each $x \in (KerL)^\perp$.

Let $(x, y) \in V$ be a well-posed problem and $y' \in Y$ close to y. Let us assume that the second projection $\Pi_2 : V \to Y$ is a submersion at (x, y) or equivalently that $D\Pi_2(x, y) : T_{(x,y)} V \to T_y Y$ is onto. Notice that this condition requires $m \leq n$. Under this hypothesis we have the following

Theorem 2.1 *When $(x, y) \in V$ is a well-posed problem, there is a C^1 operator T defined on a neighborhood V_y of $y \in Y$ with values in a neighborhood of x in X and satisfying: $T(y) = x$, $ST(y') = y'$ for each $y' \in V_y$ with derivative $DT(y) = K(x,y)^+$.*

This derivative $DT(y) = K(x,y)^+$ is called the *inverse condition operator*. Its norm

$$C(x,y)^+ = \max_{\substack{\dot{y} \in T_y Y \\ \dot{y} \neq 0}} \frac{\|K(x,y)^+ \dot{y}\|_{T_x X}}{\|\dot{y}\|_{T_y Y}}$$

is called the *inverse condition number* at (x, y). The inverse condition operator can also be described by

$$K(x,y)^+ = D\Pi_1(x,y) D\Pi_2(x,y)^+$$

since $K(x, y) = D\Pi_2(x, y)D\Pi_1(x, y)^{-1}$. The graph of $K(x, y)^+$ is the orthogonal complement of $Ker D\Pi_2(x, y)$ in $T_{(x,y)} V$.

3   Representation Formulas for a Sparse Polynomial System

We use $\mathcal{H}_{d_i}$ to denote the set of homogeneous polynomial $f_i : \mathbb{C}^{n+1} \to \mathbb{C}$ of degree exactly d_i. We suppose 0 is such a polynomial so that $\mathcal{H}_{d_i}$ is a linear space. $\mathcal{H}_d$ is the set of homogeneous polynomial systems $f : \mathbb{C}^{n+1} \to \mathbb{C}^n$, $f =$

$(f_1, \ldots, f_n)$ with $f_i \in \mathcal{H}_{d_i}$ and $d = (d_1, \ldots, d_n)$. For homogeneous polynomials $f_i, g_i : \mathbb{C}^{n+1} \to \mathbb{C}$ of degree d_i let define

$$(f_i, g_i)_{d_i} = \sum_{|\alpha|=d_i} a_\alpha \bar{b}_\alpha \binom{d_i}{\alpha}^{-1}$$

where $f_i(z) = \sum_{|\alpha|=d_i} a_\alpha z^\alpha$, $g_i(z) = \sum_{|\alpha|=d_i} b_\alpha z^\alpha$, $\binom{d_i}{\alpha} = d_i!/\alpha_0! \ldots \alpha_n!$ and $z^\alpha = z_0^{\alpha_0} \ldots z_n^{\alpha_n}$. This scalar product has the following remarquable property: Kostlan [10],

Theorem 3.1 *This scalar product over $\mathcal{H}_{d_i}$ is unitarily invariant . In other words*

$$(f \circ u, g \circ u)_{d_i} = (f, g)_{d_i} \text{ for all } f, g \in \mathcal{H}_{d_i} \text{ and } u \in U(\mathbb{C}^{n+1}).$$

We now define an Hermitian scalar product over $\mathcal{H}_d$: for $f, g \in \mathcal{H}_d$

$$(f, g) = \sum_{i=1}^n (f_i, g_i)_{d_i}.$$

We also define the set of *sparse homogeneous polynomial systems* as a Riemannian subspace $\mathcal{K}$ of $\mathcal{H}_d$. More precisely, let $\mathcal{K}_i$ be a Riemannian subspace of $\mathcal{H}_{d_i}$ and $\mathcal{K}$ equal to the product of the spaces $\mathcal{K}_i$. When a system f belongs to $\mathcal{K}$ we say that f is sparse since it is a priori defined by less conditions than a general system in $\mathcal{H}_d$.

For all $f \in \mathcal{K}_i$ let us introduce the orthogonal projection

$$\Pi_{i,f} : \mathcal{H}_i \to T_f \mathcal{K}_i$$

onto the tangent space at f to $\mathcal{K}_i$. We give in the following some useful representation formulas and upper bounds for a polynomial and its derivatives: see Beauzamy-Dégot [2] and Shub-Smale [14].

Proposition 3.2 *For all $f \in \mathcal{H}_{d_i}$ and $x \in \mathbb{C}^{n+1}$ we have*

$$f(x) = (f(z), < z, x >^{d_i})_{d_i}$$

where $< z, x >$ denotes the usual scalar product in $\mathbb{C}^{n+1}$. Moreover

$$|f(x)| \leq \|f\|_{d_i} \|x\|^{d_i}.$$

Proof. The first formula is nothing else than the Riez Representation Formula for $f \to f(x)$. Write $f(z) = \sum_{|\alpha|=d_i} a_\alpha z^\alpha$. Since

$$< z, x >^{d_i} = \sum \binom{d_i}{\alpha} z^\alpha \bar{x}^\alpha$$

we have

$$(f(z), < z, x >^{d_i})_{d_i} = \sum \binom{d_i}{\alpha}^{-1} a_\alpha \overline{\binom{d_i}{\alpha}} \bar{x}^\alpha = \sum a_\alpha x^\alpha = f(x).$$

The second assertion is given by Cauchy-Schwarz Inequality:

$$|(f(z), < z, x >^{d_i})_{d_i}| \le \|f\|_{d_i} \| < z, x >^{d_i} \|_{d_i}$$

and the equality (easy to prove)

$$\| < z, x >^{d_i} \|_{d_i} = \|x\|^{d_i}.$$

In the sparse case we get

Corollary 3.3 *For all* $f \in \mathcal{K}_i$, $\dot{f} \in T_f \mathcal{K}_i$ *and* $x \in \mathbb{C}^{n+1}$ *we have*

$$\dot{f}(x) = (\dot{f}(z), \Pi_{i,f} < z, x >^{d_i})_{d_i}$$

so that

$$|\dot{f}(x)| \le \|\dot{f}\|_{d_i} N_{i,f}(x)$$

with $N_{i,f}(x) = \|\Pi_{i,f} < z, x >^{d_i} \|_{d_i}$.

We now study the case of derivatives.

Proposition 3.4 *For all* $f \in \mathcal{H}_{d_i-1}$ *and* $g \in \mathcal{H}_{d_i}$ *one has*

$$\left(\frac{\partial g}{\partial z_k}, f \right)_{d_i-1} = d_i \, (g, z_k f)_{d_i} \, .$$

Proof. By a direct compution this equality is easily checked for monomials.

Proposition 3.5 *For all* $f \in \mathcal{H}_{d_i}$ *and* $x, u \in \mathbb{C}^{n+1}$ *we have*

$$Df(x)(u) = (f(z), d < z, u >< z, x >^{d_i-1})_{d_i}.$$

Moreover

$$|Df(x)(u)| \le d\|f\|_{d_i} \|x\|^{d_i-1} \|u\|.$$

Proof. The first equality is an easy consequence of Propositions 3.2 and 3.4. The second assertion is given by Cauchy-Schwarz Inequality:

$$|Df(x)(u)| \le d\|f\|_{d_i} \| < z, u >< z, x >^{d_i-1} \|_{d_i}.$$

It remains to prove that

$$\| < z, u >< z, x >^{d_i-1} \|_{d_i} \le \|x\|^{d_i-1} \|u\|.$$

Let us denote $h(z) = < z, u >< z, x >^{d_i-1}$. Let U be a unitary transformation in $\mathbb{C}^{n+1}$ such that $U(x) = \|x\| e_0$. One has

$$h(U^{-1}z) = < U^{-1}z, u >< U^{-1}z, x >^{d_i-1} =$$

$$< z, Uu >< z, Ux >^{d_i-1} = < z, v > z_0^{d_i-1} \|x\|^{d_i-1}$$

with $v = Uu$. By Theorem 3.1 we have

$$\|h(z)\|_{d_i}^2 = \|h(U^{-1}z)\|_{d_i}^2 = \|x\|^{2(d_i-1)}\left(|v_0|^2 + \frac{1}{d_i}(|v_1|^2 + \ldots + |v_n|^2)\right) \leq$$

$$\|x\|^{2(d_i-1)}\|v\|^2 = \|x\|^{2(d_i-1)}\|u\|^2$$

and we are done.

In the sparse case we get:

Corollary 3.6 *For all integer $f \in \mathcal{K}_i$, $\dot{f} \in T_f\mathcal{K}_i$ and $x, u \in \mathbb{C}^{n+1}$ we have*

$$D\dot{f}(x)(u) = (\dot{f}(z), d\Pi_{i,f} < z, u >< z, x >^{d_i-1})_{d_i}.$$

Moreover

$$|D\dot{f}(x)(u)| \leq d\|\dot{f}\|_{d_i} N_{i,f}^{(1)}(x)\|u\|$$

with

$$N_{i,f}^{(1)}(x) = \max \|\Pi_{i,f} < z, u >< z, x >^{d_i-1} \|_{d_i}$$

where the max is taken for $\|u\| = 1$.

For higher derivatives we have the following

Proposition 3.7 *For all integer $k \geq 1$, $f \in \mathcal{H}_{d_i}$ and $x, u_1, \ldots, u_k \in \mathbb{C}^{n+1}$ we have $D^k f(x)(u_1, \ldots, u_k) =$*

$$(f(z), d(d-1)\ldots(d-k+1) < z, u_1 > \ldots < z, u_k >< z, x >^{d_i-k})_{d_i}.$$

Moreover

$$|D^k f(x)(u_1, \ldots, u_k)| \leq d(d-1)\ldots(d-k+1)\|f\|_{d_i}\|x\|^{d_i-k}\|u_1\| \ldots \|u_k\|.$$

We do not give here the proof of this result (induction over k). In the sparse case we get:

Corollary 3.8 *For all integer $k \geq 1$, $f \in \mathcal{K}_i$, $\dot{f} \in T_f\mathcal{K}_i$ and $x, u_1, \ldots, u_k \in \mathbb{C}^{n+1}$ we have $D^k \dot{f}(x)(u_1, \ldots, u_k) =$*

$$(\dot{f}(z), d(d-1)\ldots(d-k+1)\Pi_{i,f} < z, u_1 > \ldots < z, u_k >< z, x >^{d_i-k})_{d_i}.$$

Moreover

$$|D^k \dot{f}(x)(u_1, \ldots, u_k)| \leq d(d-1)\ldots(d-k+1)\|\dot{f}\|_{d_i} N_{i,f}^{(k)}(x)\|u_1\| \ldots \|u_k\|$$

with

$$N_{i,f}^{(k)}(x) = \max \|\Pi_{i,f} < z, u_1 > \ldots < z, u_k >< z, x >^{d_i-k} \|_{d_i}$$

where the max is taken for $\|u_1\| = \ldots \|u_k\| = 1$.

We use $\mathcal{P}_{d_i}$ to denote the set of polynomials $f_i : \mathbb{C}^n \to \mathbb{C}$ of degree less than or equal to d_i. There is a canonical isomorphism between $\mathcal{P}_{d_i}$ and $\mathcal{H}_{d_i}$ given by $f \in \mathcal{P}_{d_i} \to Hf \in \mathcal{H}_{d_i}$ with $Hf(x_0, x_1, \ldots, x_n) = x_0^{d_i} f(x_1/x_0, \ldots, x_n/x_0)$.

The inverse isomorphism is given by $H^{-1}g(x_0, x_1, \ldots, x_n) = g(1, x_1, \ldots, x_n)$. In order to avoid the introduction of new notations we often write f instead of Hf. We denote by $\mathcal{P}_d$ the set of polynomial systems $f : \mathbb{C}^n \to \mathbb{C}^n$, $f = (f_1, \ldots, f_n)$ with $f_i \in \mathcal{P}_{d_i}$ and $d = (d_1, \ldots, d_n)$. By the canonical isomorphism we obtain an induced Hermitian structure over $\mathcal{P}_{d_i}$ and $\mathcal{P}_d$. We will denote the corresponding norm by $\|f\|$ for each $\mathcal{P}_d$ and $\mathcal{H}_d$.

We define the set of *sparse polynomial systems* as a Riemannian submanifold $\mathcal{K}$ of $\mathcal{P}_d$, $\mathcal{K} = \prod_{i=1}^n \mathcal{K}_i$ with $\mathcal{K}_i$ a Riemannian submanifold of $\mathcal{P}_{d_i}$.

From the propositions 3.2 to 3.8 we deduce the followings:

Proposition 3.9 *For all $f \in \mathcal{P}_{d_i}$ and $x \in \mathbb{C}^n$ we have*

$$f(x) = (f(z), (1+ <z, x>)^{d_i})_{d_i}$$

where $<z, x>$ denotes the usual scalar product in $\mathbb{C}^n$. Moreover

$$|f(x)| \leq \|f\|_{d_i} \|x\|_1^{d_i}$$

with $\|x\|_1 = (1 + \|x\|^2)^{1/2}$.

This is an easy consequence of Proposition 3.2. In the sparse case we get:

Corollary 3.10 *For all $f \in \mathcal{K}_i$, $\dot{f} \in T_f\mathcal{K}_i$ and $x \in \mathbb{C}^n$ we have*

$$\dot{f}(x) = (\dot{f}(z), \Pi_{i,f}(1+ <z, x>)^{d_i})_{d_i}$$

so that

$$|\dot{f}(x)| \leq \|\dot{f}\|_{d_i} N_{i,f}(x)$$

with $N_{i,f}(x) = \|\Pi_{i,f}(1+ <z, x>)^{d_i}\|_{d_i}$ and $\Pi_{i,f}$ the orthogonal projection from $\mathcal{P}_{d_i}$ onto $T_f\mathcal{K}_i$.

Remark 3.11 The notation $N_{i,f}(x)$ is not ambiguous since the homogeneous and non-homogeneous versions of this number are equal. More precisely, for all $f \in \mathcal{K}_i$ we have $N_{i,f}(x) = N_{i,Hf}(1,x)$.

Proposition 3.12 *For all integer $k \geq 1$, $f \in \mathcal{P}_{d_i}$ and $x, u_1, \ldots, u_k \in \mathbb{C}^n$ we have $D^k f(x)(u_1, \ldots, u_k) =$*

$$(f(z), d(d-1)\ldots(d-k+1) <z, u_1> \ldots <z, u_k> (1+ <z,x>)^{d_i-k})_{d_i}.$$

Moreover

$$|D^k f(x)(u_1, \ldots, u_k)| \leq d(d-1)\ldots(d-k+1)\|f\|_{d_i}\|x\|_1^{d_i-k}\|u_1\| \ldots \|u_k\|.$$

In the sparse case we get:

Corollary 3.13 *For all integer $k \geq 1$, $f \in \mathcal{K}_i$, $\dot{f} \in T_f\mathcal{K}_i$ and $x, u_1, \ldots, u_k \in \mathbb{C}^n$ we have $D^k \dot{f}(x)(u_1, \ldots, u_k) =$*

$$(\dot{f}(z), d(d-1)\ldots(d-k+1)\Pi_{i,f} <z, u_1> \ldots <z, u_k> (1+ <z,x>)^{d_i-k})_{d_i}.$$

Moreover

$$|D^k \dot{f}(x)(u_1, \ldots, u_k)| \le d(d-1) \ldots (d-k+1) \|\dot{f}\|_{d_i} N_{i,f}^{(k)}(x) \|u_1\| \ldots \|u_k\|$$

with

$$N_{i,f}^{(k)}(x) = \max \|\Pi_{i,f} < z, u_1 > \ldots < z, u_k > (1+ < z, x >)^{d_i-k}\|_{d_i}$$

where the max is taken for $\|u_1\| = \ldots = \|u_k\| = 1$.

4 Condition Operator and Condition Number for a Sparse Polynomial System

Our aim, in this section, is to compute the condition number associated with a sparse polynomial system. We follow here the general scheme introduced in Section 2. We take $X = \mathcal{K}$, $Y = \mathbb{C}^n$ and

$$V = \{ (f,x) \in \mathcal{K} \times \mathbb{C}^n : f(x) = 0 \}.$$

Here $\mathcal{K}$ is a Riemannian submanifold of $\mathcal{P}_d$, $\mathcal{K} = \prod_{i=1}^n \mathcal{K}_i$ with $\mathcal{K}_i$ a Riemannian submanifold of $\mathcal{P}_{d_i}$.

If we differentiate the equation $f(x) = 0$ we obtain the following description of the tangent space at (f, x) to V

$$T_{f,x}V = \{(\dot{f}, \dot{x}) \in T_f\mathcal{K} \times \mathbb{C}^n : \dot{f}(x) + Df(x)\dot{x} = 0\}.$$

A problem (f, x) is *well-posed* if and only if $Df(x)$ is invertible. In such a case the condition operator

$$K_{\mathcal{K}}(f, x) : T_f\mathcal{K} \to \mathbb{C}^n$$

is given by

$$K_{\mathcal{K}}(f, x)\dot{f} = \dot{x} = -Df(x)^{-1}\dot{f}(x).$$

The condition number is the operator norm of $K_{\mathcal{K}}(f, x)$ that is

$$C_{\mathcal{K}}(f, x) = \|K_{\mathcal{K}}(f, x)\| = \max_{\substack{\dot{f} \in T_f\mathcal{K} \\ \dot{f} \ne 0}} \frac{\|K_{\mathcal{K}}(f, x)\dot{f}\|}{\|\dot{f}\|}.$$

Theorem 4.1 *Let* $(f, x) \in V$ *be a well-posed problem. Then*

$$C_{\mathcal{K}}(f, x) = \|Df(x)^{-1}Diag(N_{i,f}(x))\|$$

operator norm, with $Diag(N_{i,f}(x))$ *the* $n \times n$ *diagonal matrix with i-th entry equal to* $N_{i,f}(x) = N_{i,f_i}(x)$.

Proof. In virtue of the previous definitions we have

$$C_{\mathcal{K}}(f,x) = \max_{\substack{\dot{f} \in T_f \mathcal{K} \\ \dot{f} \neq 0}} \frac{\|Df(x)^{-1}\dot{f}(x)\|}{\|\dot{f}\|} = \max_{\|\dot{f}\| \leq 1} \|Df(x)^{-1}\dot{f}(x)\| =$$

$$\max_{\|\dot{f}\| \leq 1} \|Df(x)^{-1}Diag(N_{i,f}(x))Diag(N_{i,f}(x))^{+}\dot{f}(x)\|$$

where for any matrix $A \in \mathcal{M}_n(\mathbb{C})$ we denote by A^{+} the Moore-Penrose inverse of A. In our case, the i-th diagonal entry of $Diag(N_{i,f}(x))^{+}$ is equal to $N_{i,f}(x)^{-1}$ when $N_{i,f}(x) \neq 0$ and 0 when $N_{i,f}(x) = 0$. Remark that

$$Diag(N_i(x))Diag(N_{i,f}(x))^{+}\dot{f}(x) = \dot{f}(x)$$

since, by Corollary 3.10, $\dot{f}_i(x) = 0$ when $N_{i,f}(x) = 0$. By this corollary and the definition of $\|\dot{f}\|$ we remark that

$$\|Diag(N_{i,f}(x))^{+}\dot{f}(x)\| \leq 1$$

thus

$$\max_{\|\dot{f}\| \leq 1} \|Df(x)^{-1}\dot{f}(x)\| \leq \max_{\substack{u \in \mathbb{C}^n \\ \|u\| \leq 1}} \|Df(x)^{-1}Diag(N_{i,f}(x))u\|.$$

Let us prove that both max are equal. For any vector $u \in \mathbb{C}^n$ with $\|u\| \leq 1$ we take $\dot{f}_i(z) = \lambda_i \Pi_{i,f}(1+ < z,x >)^{d_i}$, with $\lambda_i = N_{i,f}(x)^{-1}u_i$ when $N_{i,f}(x) \neq 0$ and $\lambda_i = 0$ when $N_{i,f}(x) = 0$. We have $\dot{f}_i(x) = \lambda_i \Pi_i(1+ < z,x >)^{d_i}|_{z=x} = \lambda_i N_{i,f}(x)^2$ always by Corollary 3.10 so that $\dot{f}(x) = Diag(N_{i,f}(x))u$. Since $\|\dot{f}_i\| = |u_i|$ when $N_{i,f}(x) \neq 0$ we also have $\|\dot{f}\| \leq 1$. This yealds

$$\max_{\|\dot{f}\| \leq 1} \|Df(x)^{-1}\dot{f}(x)\| \geq \max_{\substack{u \in \mathbb{C}^n \\ \|u\| \leq 1}} \|Df(x)^{-1}Diag(N_{i,f}(x))u\|$$

and achieves the proof.

Remark 4.2 1. In the non-sparse case, that is when $\mathcal{K} = \mathcal{P}_d$, one has $N_{i,f}(x) = \|x\|_1^{d_i}$ by Proposition 3.1 and we obtain

$$C_{\mathcal{P}}(f,x) = \|Df(x)^{-1}Diag(\|x\|_1^{d_i})\|.$$

2. In Shub-Smale [14], [16] the authors study the condition number for homogeneous systems. They consider

$$V = \{ (f,x) \in \mathbb{P}(\mathcal{H}_d) \times \mathbb{P}(\mathbb{C}^{n+1}) : f(x) = 0 \}$$

and obtain

$$C_h(f,x) = \|f\|\|(Df(x)|_{x^\perp})^{-1}Diag(\|x\|^{d_i-1})\|.$$

This condition number is invariant under scaling for both f and x.

3. Another consequence of its definition is the inequality

$$C_{\mathcal{K}}(f,x) \leq C_{\mathcal{P}}(f,x).$$

The sparse condition number is smaller than the non-sparse one. Example 4.6 below shows that this difference can be large.

Example 4.3 Single Variable Polynomial Equations: General Case.
Let

$$f(z) = \sum_{i=0}^{d} a_i z^i, \ z \in \mathbb{C},$$

and $r \in \mathbb{C}$ be a simple root: $f(r) = 0$ and $f'(r) \neq 0$. Suppose now that this polynomial belongs to some subspace $\mathcal{K}$ of $\mathcal{P}_d$. In this case, we obtain

$$C_\mathcal{K}(f,r) = N(r)|f'(r)|^{-1}.$$

When $\mathcal{K} = \mathcal{P}_d$ we have $N(r) = (1+|r|^2)^{d/2}$ so that

$$C_\mathcal{H}(f,r) = (1+|r|^2)^{d/2}|f'(r)|^{-1}.$$

Example 4.4 Single Variable Polynomial Equations: Fewnomials.
We now consider the case of fewnomials, that is

$$f(z) = \sum_{i \in I} a_i z^i, \ z \in \mathbb{C},$$

where I is some subset of $\{0,1,\ldots,d\}$. The representation formula given in Corollary 3.10 is now the following

$$f(x) = (f(z), \sum_{i \in I} \binom{d}{i} z^i \bar{x}^i)$$

so that

$$N(x) = \left(\sum_{i \in I} \binom{d}{i} |x|^{2i} \right)^{1/2}.$$

This gives, when r is a simple root of f,

$$C_\mathcal{K}(f,r) = \left(\sum_{i \in I} \binom{d}{i} |r|^{2i} \right)^{1/2} |f'(r)|^{-1}.$$

The ratio of the sparse and non-sparse conditions numbers is given by

$$\left(\frac{C_\mathcal{K}(f,r)}{C_\mathcal{P}(f,r)} \right)^2 = \frac{\sum_{i \in I} \binom{d}{i} |r|^{2i}}{(1+|r|^2)^d}.$$

This number depends only on the modulus of the root r and on the set of monomials I.

In the case of quadrinomials:

$$f(z) = a_0 + a_1 z + a_{d-1} z^{d-1} + a_d z^d$$

this ratio is equal to

$$\left(\frac{C_{\mathcal{K}}(f,r)}{C_{\mathcal{P}}(f,r)}\right)^2 = \frac{1 + d|r|^2 + d|r|^{2d-2} + |r|^{2d}}{(1+|r|^2)^d}.$$

This last quantity is always less than or equal to 1, its minimum is reached for $|r| = 1$ and is equal to $(1+d)/2^{d-1}$.

About the location of the roots of a fewnomial we have the following:

Proposition 4.5 *Let* $f(z) = a_0 + a_1 z^{n_1} + \ldots + a_k z^{n_k}$ *with* $a_i \neq 0$. *Define* $F(t) = |a_k| - \sum_{i=0}^{k-1} |a_i| t^{n_k - n_i}$. *Let us denote by* r^{-1} *the unique positive root of* $F(t)$. *Then, for any root* z *of* f *we have* $|z| \leq r$.

Proof. Look at $F(t)$ for $t \geq 0$. At $t = 0$, $F(t)$ has a positive value: $|a_k|$. Then $F(t)$ decreases from this value until $-\infty$. So it possesses a unique positive root r^{-1}, $F(t)$ is positive before this root and negative after. If $f(z) = 0$ then, we can see easily that $F(|z|^{-1}) \leq 0$ so that $|z| \leq r$.

Corollary 4.6 (Kojima [9]) *All the zeros of the polynomial 4.5 lie on the disk* $|z| \leq s$ *where*

$$s = \max\{|a_0/a_1|^{1/n_1}, |2a_j/a_{j+1}|^{1/(n_{j+1}-nj)}, \ j = 1 \ldots k-1\}.$$

Proof. We have $F(1/s) \geq 0$ so that $s \leq r$. See also Marden [11], Sect. 30, Exerc. 11].

Example 4.7: The Eigenvalue Problem Let $A \in \mathcal{M}_n(\mathbb{C})$ be a $n \times n$ complex matrix and $b \in \mathbb{C}^n$ a given vector. We consider here the following version of the eigenvalue problem:

$$\lambda x = Ax \ and \ b_1 x_1 + \ldots b_n x_n = 1.$$

The normalization equation $b_1 x_1 + \ldots b_n x_n - 1 = 0$ is recommended by T.Y. Li. Let us look at this problem as a polynomial system. The inputs are the matrix A and the vector b, the outputs are x and λ. Thus $\mathcal{K}$ is the set of systems

$$f_i(z, \mu) = \mu z_i - (a_{i,1} z_1 + \ldots a_{i,n} z_n), \ 1 \leq i \leq n,$$

$$f_{n+1}(z, \mu) = b_1 z_1 + \ldots b_n z_n - 1.$$

The tangent space at (A, b) to $\mathcal{K}_i$ is

$$T_{f_i}\mathcal{K}_i = \{\dot{f}_i \ : \ \dot{f}_i(z) = \dot{a}_{i,1} z_1 + \ldots \dot{a}_{i,n} z_n, \ \dot{a}_i \in \mathbb{C}^n\} \subset \mathcal{P}_2,$$

when $1 \leq i \leq n$ and

$$T_{f_{n+1}}\mathcal{K}_{n+1} = \{\dot{f}_{n+1} \ : \ \dot{f}_{n+1}(z) = \dot{b}_1 z_1 + \ldots \dot{b}_n z_n, \ \dot{b} \in \mathbb{C}^n\} \subset \mathcal{P}_1.$$

Remark that these tangent spaces do not depend on (A, b). The representation formula for $\dot{f}_i(x, \lambda)$ is the following

$$\dot{f}_i(x, \lambda) = \dot{a}_{i,1} x_1 + \ldots \dot{a}_{i,n} x_n = (\dot{f}_i(z, \mu), 2(z_1 \bar{x}_1 + \ldots + z_n \bar{x}_n))_2$$

when $1 \leq i \leq n$ and

$$\dot{f}_{n+1}(x, \lambda) = \dot{b}_1 x_1 + \ldots \dot{b}_n x_n = (\dot{f}_{n+1}(z, \mu), z_1 \bar{x}_1 + \ldots + z_n \bar{x}_n))_1.$$

The numbers $N_{i,f}(x, \lambda)$ are equal to

$$N_{i,f}(x, \lambda) = \sqrt{2}\|x\| \text{ when } 1 \leq i \leq n \text{ and } N_{n+1,f}(x, \lambda) = \|x\|.$$

The derivative of f at (x, λ) is given by

$$Df(x, \lambda) = \begin{pmatrix} \lambda I_n - A & x \\ b^T & 0 \end{pmatrix}.$$

This matrix is invertible if and only if λ is a simple eigenvalue. The condition number of the eigenvalue problem is consequently equal to

$$C(A, b, x, \lambda) = \|x\| \left\| \begin{pmatrix} \lambda I_n - A & x \\ b^T & 0 \end{pmatrix}^{-1} \begin{pmatrix} \sqrt{2}I_n & 0 \\ 0 & 1 \end{pmatrix} \right\|$$

so that

$$C(A, b, x, \lambda) \leq \sqrt{2}\|x\| \left\| \begin{pmatrix} \lambda I_n - A & x \\ b^T & 0 \end{pmatrix}^{-1} \right\|.$$

5   Estimations for the Variations of the Zeros of a Sparse Polynomial System

Consider a sparse polynomial system $f \in \mathcal{K}$. Let g be another system, $g = f + \dot{f}$ with $\dot{f} \in T_f \mathcal{K}$. Let $x \in \mathbb{C}^n$ be a simple zero of f. If g is enough close to f there exists a root y of g close to x. This is a consequence of the implicit function theorem. Our objective, in this section, is to give a quantitative version of this fact: to estimate the distance from x to y in terms of the distance from f to g and the condition number $C_\mathcal{K}(f, x)$. There are a lot of such estimations in the single variable case: see for example Ostrowski [12], Bathia-Elsner-Krause [3], Beauzamy [1]. All these papers are based on the factorization of polynomials. Thus, they cannot be extended to the case of systems by similar arguments. We develop here an idea due to Steve Smale and based on α-theory. We first recall what is α-theory.

Let $F : \mathbb{C}^n \to \mathbb{C}^n$ be an analytic function. For any $z \in \mathbb{C}^n$ define the following numbers:

$$\beta(F, z) = \|DF(z)^{-1}F(z)\|$$

and

$$\gamma(F, z) = \sup_{k \geq 2} \left\| DF(z)^{-1} \frac{D^k F(z)}{k!} \right\|^{\frac{1}{k-1}}.$$

We give to these numbers the value ∞ when $DF(z)$ is singular. We also define

$$\alpha(F, z) = \beta(F, z)\gamma(F, z).$$

The following is originally due to Smale [18]. We give here a version by Wang Xinghua and Han Danfu with better constants Wang-Han [19].

Theorem 5.1 *If* $\alpha(F, z) \leq \dfrac{13 - 3\sqrt{17}}{4}$ *then there exists a zero* ζ *of* F *such that*

$$\|\zeta - z\| \leq \frac{7 + \sqrt{17}}{8}\beta(F, z).$$

Moreover, the sequence (z_k) *of Newton's iterates starting at* $z_0 = z$ *satisfies*

$$\|\zeta - z_k\| \leq \frac{7 + \sqrt{17}}{8}\left(\frac{1}{2}\right)^{2^k - 1}\|z_1 - z_0\|.$$

Our perturbation result is the following:

Theorem 5.2 *If*

$$\alpha(g, x) \leq \frac{13 - 3\sqrt{17}}{4} \quad \text{and} \quad \|Df(x)^{-1}(Df(x) - Dg(x))\| \leq \frac{9 - \sqrt{17}}{16}$$

then, there exists a zero y *of* g *such that*

$$\|x - y\| \leq 2C_\mathcal{K}(f, x)\|f - g\|.$$

Proof. By Theorem 5.1 we know that

$$\|y - x\| \leq \frac{7 + \sqrt{17}}{8}\beta(g, x)$$

for a certain zero y of g. We have $\beta(g, x) =$

$$\|Dg(x)^{-1}Df(x)Df(x)^{-1}Diag(N_{i,f}(x))Diag(N_{i,f}(x))^+(g(x) - f(x))\|$$

since $f(x) = 0$ and $g_i(x) - f_i(x) = \dot{f}_i(x) = 0$ when $N_{i,f}(x) = 0$. This gives $\beta(g, x) \leq$

$$\|Dg(x)^{-1}Df(x)\|\|Df(x)^{-1}Diag(N_{i,f}(x))\|\|Diag(N_{i,f}(x))^+(g(x) - f(x))\|$$

$$\leq \|Dg(x)^{-1}Df(x)\|C_\mathcal{K}(f, x)\|f - g\|$$

by the definition of the condition number and Corollary 3.7. We now use a classical lemma in linear algebra: let $A, B \in \mathcal{M}_n(\mathbb{C})$ with B invertible. If $\|B^{-1}(B - A)\| \leq \lambda < 1$ then A is invertible and $\|A^{-1}B\| \leq 1/(1 - \lambda)$. We take here $A = Dg(x)$, $B = Df(x)$ and $\lambda = (9 - \sqrt{17})/16$. From this lemma and the hypothesis of Theorem 5.2 we get $\|Dg(x)^{-1}Df(x)\| \leq 16/(7 + \sqrt{17})$ so that

$$\beta(g, x) \leq \frac{16}{7 + \sqrt{17}}C_\mathcal{K}(f, x)\|f - g\|$$

and we are done.

Remark 5.3 We have, in fact, proved the following more general estimation: if

$$\alpha(g,x) \le \frac{13 - 3\sqrt{17}}{4} \quad \text{and} \quad \|Df(x)^{-1}(Df(x) - Dg(x))\| \le \lambda < 1$$

then, there exists a zero y of g such that

$$\|x - y\| \le \frac{7 + \sqrt{17}}{8(1 - \lambda)} C_\mathcal{K}(f,x)\|f - g\|.$$

In Theorem 5.2 we consider the case of zeros of two polynomial systems f and g, $g = f + \dot{f}$ with $\dot{f} \in T_f\mathcal{K}$. We now give another estimation for two systems f and $g \in \mathcal{K}$. We consider here a path $F : t \in [0,1] \to (f_t, x_t) \in \mathcal{K} \times \mathbb{C}^n$ such that $f_t(x_t) = 0$, $rank\ Df_t(x_t) = n$, $f_0 = f$ and $f_1 = g$. We also denote by x and y the corresponding zeros of f and g, we introduce $L(f,g)$ the lenght of the path f_t in $\mathcal{K}$ and the condition number of the path

$$C_\mathcal{K}(F) = \max_{0 \le t \le 1} C_\mathcal{K}(f_t, x_t).$$

We will prove the following

Theorem 5.4

$$\|x - y\| \le C_\mathcal{K}(F)\ L(f,g).$$

Proof. Let us denote by $L(x,y)$ the lenght in $\mathbb{C}^n$ of the path from x to y. We have

$$\|x - y\| \le L(x,y) = \int_0^1 \|\dot{x}_t\|dt$$

with $\dot{x}_t$ the derivative with respect to t. Since $\dot{x}_t = K_\mathcal{K}(f_t, x_t)\dot{f}_t$ we get $\|\dot{x}_t\| \le C_\mathcal{K}(f_t, x_t)\|\dot{f}_t\|$ so that

$$\int_0^1 \|\dot{x}_t\|dt \le \int_0^1 C_\mathcal{K}(f_t, x_t)\|\dot{f}_t\|dt \le C_\mathcal{K}(F)L(f,g).$$

Example 5.5: The Eigenvalue Problem We study here the variation of the couple (eigenvector, eigenvalue) in term of the variation of the matrix. We use the same notations as in Example 4.7. We start with

$$f(x,\lambda) = \begin{pmatrix} (\lambda I - A)x \\ b^T x - 1 \end{pmatrix} = 0 \quad \text{and} \quad g(x,\lambda) = \begin{pmatrix} (\lambda I - B)x \\ c^T x - 1 \end{pmatrix}$$

so that $g \in f + T_f\mathcal{K}$. According to Theorem 5.2 we have to give estimates for $\alpha(g,x,\lambda)$ and $\|Df(x,\lambda)^{-1}(Df(x,\lambda) - Dg(x,\lambda))\|$.

Estimation for $\beta(g,x,\lambda)$.
We have

$$\beta(g,x,\lambda) = \|Dg(x,\lambda)^{-1}g(x,\lambda)\| =$$

$$\left\| \begin{pmatrix} \lambda I - B & x \\ c^T & 0 \end{pmatrix}^{-1} \begin{pmatrix} (\lambda I - B)x \\ c^T x - 1 \end{pmatrix} \right\| = \left\| \begin{pmatrix} \lambda I - B & x \\ b^H & 0 \end{pmatrix}^{-1} \begin{pmatrix} (A - B)x \\ (c - b)^T x \end{pmatrix} \right\| \le$$

$$\left\| \begin{pmatrix} \lambda I - B & x \\ c^T & 0 \end{pmatrix}^{-1} \right\| (\|A - B\|^2 + \|b - c\|^2)^{1/2} \|x\|.$$

Estimation for $\gamma(g, x, \lambda)$.
According to the definition of this number we have to compute the second derivative of $g(x, \lambda)$. It is given by

$$D^2 g(x, \lambda)(u, v) = v_{n+1} \begin{pmatrix} u_1 \\ \vdots \\ u_n \\ 0 \end{pmatrix} + u_{n+1} \begin{pmatrix} v_1 \\ \vdots \\ v_n \\ 0 \end{pmatrix}$$

and by an easy computation

$$\max_{\|u\| = \|v\| = 1} \|D^2 g(x, \lambda)(u, v)\| = 1.$$

This gives

$$\gamma(g, x, \lambda) \le \frac{1}{2} \left\| \begin{pmatrix} \lambda I - B & x \\ c^T & 0 \end{pmatrix}^{-1} \right\|.$$

Estimation for $\|Df^{-1}(Df - Dg)\|$.
We obtain easily

$$\|Df(x, \lambda)^{-1}(Df(x, \lambda) - Dg(x, \lambda))\| = \left\| \begin{pmatrix} \lambda I - A & x \\ b^T & 0 \end{pmatrix}^{-1} \begin{pmatrix} B - A & 0 \\ (b - c)^T & 0 \end{pmatrix} \right\|$$

$$\le \left\| \begin{pmatrix} \lambda I - A & x \\ b^T & 0 \end{pmatrix}^{-1} \right\| (\|B - A\|^2 + \|c - b\|^2)^{1/2}.$$

The conclusion of this study is resumed in the following

Theorem 5.6 *With the notations introduced previously if*

$$\left\| \begin{pmatrix} \lambda I - B & x \\ c^T & 0 \end{pmatrix}^{-1} \right\|^2 (\|A - B\|^2 + \|b - c\|^2)^{1/2} \|x\| \le \frac{13 - 3\sqrt{17}}{2}$$

and if

$$\left\| \begin{pmatrix} \lambda I - A & x \\ b^T & 0 \end{pmatrix}^{-1} \right\| (\|B - A\|^2 + \|c - b\|^2)^{1/2} \le \frac{9 - \sqrt{17}}{16}$$

then, there exist $y \in \mathbb{C}^n$ and $\mu \in \mathbb{C}$ such that

$$By = \mu y \quad and \quad c^T y = 1$$

with

$$\left(\|x-y\|^2 + |\lambda - \mu|^2\right)^{1/2} \leq$$

$$2\|x\| \left\|\begin{pmatrix} \lambda I - A & x \\ b^T & 0 \end{pmatrix}^{-1}\right\| \left(\|B-A\|_F^2 + 2\|b-c\|^2\right)^{1/2}.$$

Proof. The conclusion comes from Theorem 5.2 and from the different estimations given previously. We also use the upper bound for the condition number given in Example 4.7 and the following estimation

$$\|f-g\| = \left(\frac{1}{2}\|B-A\|_F^2 + \|b-c\|^2\right)^{1/2}.$$

6 Estimations for the Separation Number of a Sparse Polynomial System

Let $f \in \mathcal{P}_d$ be a polynomial system and $x \in \mathbb{C}^n$ any zero of f. The separation number of f at x is defined by

$$Sep(f,x) = \min_{\substack{f(y)=0 \\ y \neq x}} \|x-y\|$$

or 0 when x is not a simple zero. In this section we give a lower bound for $Sep(f,x)$ in terms of $\gamma(f,x)$. Lower bounds for $Sep(f,x)$ in terms of the condition number appear in Dedieu [6] and Blum-Cucker-Shub-Smale [4] Chap. 14, Theorem 3.

Theorem 6.1

$$\frac{1}{2\gamma(f,x)} \leq Sep(f,x).$$

Proof. We first suppose that x is a simple zero since, for a multiple zero, $Sep(f,x) = 0$ and $\gamma(f,x) = \infty$. Let y be another zero, distinct from x. By Taylor's Formula we have

$$Df(x)(y-x) = -\sum_{k=2}^{D} \frac{1}{k!} D^k f(x)(y-x)\dots(y-x)$$

with $D = \max d_i$, so that

$$\|y-x\| \leq \sum_{k \geq 2} \frac{1}{k!}\|Df(x)^{-1}D^k f(x)(y-x)\dots(y-x)\| \leq$$

$$\|y-x\| \sum_{k \geq 2} \frac{1}{k!}\|Df(x)^{-1}D^k f(x)\|\|y-x\|^{k-1}.$$

Since

$$\gamma(f,x) = \sup_{k \geq 2} \|Df(x)^{-1} \frac{D^k f(x)}{k!}\|^{\frac{1}{k-1}}$$

we obtain

$$1 \leq \sum_{k \geq 2} (\gamma(f,x)\|y - x\|)^{k-1}.$$

Let us now consider $P(t) = 1 - t - \ldots - t^{D-1}$. This polynomial decreases from 1 at $t = 0$ to $-\infty$ at $t = \infty$. Consequently it possesses a unique positive zero r and $P(t) > 0$ (resp. $P(t) < 0$) if and only if $t < r$ (resp. $t > r$). Since $P(1/2) > 0$ and $P(\gamma(f,x)\|y - x\|) \leq 0$ we obtain $1/2 < \gamma(f,x)\|y - x\|$ and we are done.

Example: The Eigenvalue Problem

Theorem 6.2 *With the same notations as in Example 5.4 suppose that*

$$Ax = \lambda x \quad and \quad b^T x = 1$$

for any other eigenvalue μ of A with corresponding eigenvector y, $b^T y = 1$, we have

$$\left\| \begin{pmatrix} \lambda I - A & x \\ b^H & 0 \end{pmatrix}^{-1} \right\|^{-1} \leq \left(\|x - y\|^2 + |\lambda - \mu|^2 \right)^{1/2}.$$

Proof. This a consequence of Theorem 6.1. We also use the upper bound for $\gamma(f,x,\lambda)$ given in Example 5.4.

7   Estimations for $\beta(\mathbf{g},\mathbf{x})$ and $\gamma(\mathbf{g},\mathbf{x})$

Let $f \in \mathcal{K} \subset \mathcal{P}_d$ as before and g be another system, $g = f + \dot{f}$ with $\dot{f} \in T_f\mathcal{K}$. Let $x \in \mathbb{C}^n$ be a simple zero of f. Our aim is here to give estimations for $\beta(g,x)$ and $\gamma(g,x)$ in terms of the condition number. We will use the numbers $N_{i,f}(x)$ and $N_{i,f}^{(k)}(x)$ defined in Corollaries 3.10 and 3.13. We have to make an hypothesis on the spaces $\mathcal{K}_i$: *either $\mathcal{K}_i$ is reduced to a single equation or $N_{i,f}(x) \neq 0$.* Notice that in the first case the tangent space $T_{f_i}\mathcal{K}_i$ is reduced to 0, the numbers $N_{i,f}(x)$ and $N_{i,f}^{(k)}(x)$ are also equal to 0. The set of indices i with $N_{i,f}(x) \neq 0$ is denoted by I and $D = \max d_i$. We also define

$$C_\mathcal{K}(g,x) = \|Dg(x)^{-1} Diag(N_{i,f}(x))\|$$

or ∞ when $rank Dg(x) < n$. The main results are the followings:

Theorem 7.1

$$\beta(g,x) \leq C_\mathcal{K}(g,x) \, \|f - g\|.$$

Theorem 7.2

$$\gamma(g,x) \le \frac{D(D-1)}{2} \max\{\, 1, \|g\| C_{\mathcal{K}}(g,x)\,\} \max_{\substack{k\ge 2 \\ i\in I}} \left(\frac{N_{i,f}^{(k)}(x)}{N_{i,f}(x)}\right)^{\frac{1}{k-1}}.$$

Theorem 7.3 *If*

$$D\, C_{\mathcal{K}}(f,x)\, \|f-g\| \max_{i\in I} \frac{N_{i,f}^{(1)}(x)}{N_{i,f}(x)} \le \lambda < 1$$

then

$$C_{\mathcal{K}}(g,x) \le \frac{1}{1-\lambda}\, C_{\mathcal{K}}(f,x).$$

Combining these theorems permits to give estimations for $\alpha(g,x)$ and $Sep\,(f,x)$ in terms of the condition number $C_{\mathcal{K}}(f,x)$ and the distance $\|f-g\|$. Such estimations may be useful in Theorems 5.2 and 6.1. We do not give here these estimations.

Proof of Theorem 7.1.

$$\beta(g,x) = \|Dg(x)^{-1}Diag(N_{i,f}(x))Diag(N_{i,f}(x))^{+}(g(x)-f(x))\|$$

since $f(x) = 0$ and $g_i(x) - f_i(x) = 0$ when $N_{i,f}(x) = 0$. This gives, by Corollary 3.10 and the definition of $C_{\mathcal{K}}(g,x)$

$$\beta(g,x) \le \|Dg(x)^{-1}Diag(N_{i,f}(x))\|\|Diag(N_{i,f}(x))^{+}(g(x)-f(x))\| =$$

$$C_{\mathcal{K}}(g,x)\|f-g\|.$$

Proof of Theorem 7.2. We have $\|Dg(x)^{-1}\dfrac{D^{(k)}g(x)}{k!}\| \le$

$$\|Dg(x)^{-1}Diag(N_{i,f}(x))Diag(N_{i,f}(x))^{+}\frac{D^{(k)}g(x)}{k!}\|$$

since $D^{(k)}g_i(x) = 0$ when $N_{i,f}(x) = 0$ by the hypothesis made in the beginning of this section. By Corollary 3.13 and the definition of $C_{\mathcal{K}}(g,x)$ we get

$$\le C_{\mathcal{K}}(g,x) \left(\sum_{i\in I}\left(N_{i,f}(x)^{-1}\frac{d_i(d_i-1)\dots(d_i-k+1)}{k!}\|g_i\|N_{i,f}^{(k)}(x)\right)^2\right)^{1/2}.$$

Since the quantity $(d_i(d_i-1)\dots(d_i-k+1)/k!)^{1/(k-1)}$, $k \ge 2$, is decreasing we get

$$\frac{d_i(d_i-1)\dots(d_i-k+1)}{k!} \le \left(\frac{d_i(d_i-1)}{2}\right)^{k-1} \le \left(\frac{D(D-1)}{2}\right)^{k-1}$$

so that

$$\|Dg(x)^{-1}\frac{D^{(k)}g(x)}{k!}\| \leq C_{\mathcal{K}}(g,x)\,\|g\| \left(\frac{D(D-1)}{2}\right)^{k-1} \max_{\substack{k\geq 2 \\ i\in I}} \frac{N_{i,f}^{(k)}(x)}{N_{i,f}(x)}.$$

Since

$$\gamma(g,x) = \sup_{k\geq 2}\left\|Dg(x)^{-1}\frac{D^k g(x)}{k!}\right\|^{\frac{1}{k-1}}$$

we obtain the following estimation

$$\gamma(g,x) \leq \frac{D(D-1)}{2}\,(C_{\mathcal{K}}(g,x)\,\|g\|)^{\frac{1}{k-1}} \max_{\substack{k\geq 2 \\ i\in I}} \left(\frac{N_{i,f}^{(k)}(x)}{N_{i,f}(x)}\right)^{\frac{1}{k-1}}$$

and we are done.

Proof of Theorem 7.3. We have $C_{\mathcal{K}}(g,x) =$

$$\|Dg(x)^{-1}Diag(N_i(x))\| = \|Dg(x)^{-1}Df(x)Df(x)^{-1}Diag(N_i(x))\| \leq$$

$$\|Dg(x)^{-1}Df(x)\|\|Df(x)^{-1}Diag(N_i(x))\| = \|Dg(x)^{-1}Df(x)\|C_{\mathcal{K}}(f,x)$$

We now use an argument similar to that given in the proof of Theorem 5.2: let $A, B \in \mathcal{M}_n(\mathbb{C})$ with B invertible. If $\|B^{-1}(B-A)\| \leq \lambda < 1$ then A is invertible and $\|A^{-1}B\| \leq 1/(1-\lambda)$. We take here $A = Dg(x)$, $B = Df(x)$. We get

$$C_{\mathcal{K}}(g,x) \leq \frac{1}{1-\lambda}C_{\mathcal{K}}(f,x)$$

if $\|Dg(x)^{-1}Df(x)\| \leq \lambda < 1$. The conclusion comes from the following estimation: $\|Df(x)^{-1}(Df(x) - Dg(x))\| =$

$$\|Df(x)^{-1}Diag(N_i(x))Diag(N_i(x))^{+}(Df(x) - Dg(x))\| \leq$$

$$\|Df(x)^{-1}Diag(N_i(x))\|\|Diag(N_i(x))^{+}(Df(x) - Dg(x))\| \leq$$

$$D\,C_{\mathcal{K}}(f,x)\,\|f-g\| \max_{i\in I}\frac{N_{i,f}^{(1)}(x)}{N_{i,f}(x)}$$

by Corollary 3.13 and since, by the hypothesis, $Df_i(x) - Dg_i(x) = 0$ when $i \notin I$.

8   The Condition Number Theorem

In this section we prove a Condition Number Theorem for sparse polynomial systems: the sparse condition number $C_{\mathcal{K}}(f,x)$ measures the distance of f from the set Σ_x of ill-posed problems at x. Such theorems have been proved by Eckart-Young [8] in the context of linear equations, Shub-Smale [14] for polynomial systems, Shub-Smale [16] for the eigenvalue problem, Dedieu [7] for the generalized eigenvalue problem. Dedieu [5] has put the Condition Number Theorem into a general setting.

Let us be more precise: $\mathcal{K}_i$ is a Riemannian submanifold of $\mathcal{P}_i$, $\mathcal{K} = \prod \mathcal{K}_i$ and

$$V = \{(f,x) \in \mathcal{K} \times \mathbb{C}^n \ : \ f(x) = 0\}.$$

Let $(f,x) \in V$ be a well-posed problem that is $rank\ Df(x) = n$. We define the fiber over x in V by

$$V_x = \{g \in \mathcal{K} \ : \ g(x) = 0\}$$

and the set of ill-posed problems at x by

$$\Sigma_x = \{g \in V_x \ : \ rank\ Dg(x) < n\}.$$

We have now to place conditions on $\mathcal{K}$:

$(H1)$ $\qquad\qquad\qquad N_{i,f}(x) > 0 \ for\ all\ i = 1 \ldots n,$

$(H2)$
$\quad$ *for each linear operator* $L : \mathbb{C}^n \to \mathbb{C}^n$ *there exits* $g \in V_x$ *with* $Dg(x) = L,$

$(H3)$ $\qquad\qquad\qquad V_x - f$ *is a vector subspace of* $\mathcal{P}_d.$

We will prove the following

Theorem 8.1

$$C_\mathcal{K}(f,x)^{-1} = \min_{g \in \Sigma_x} \|Diag(N_{i,f}(x))^{-1} D(f-g)(x)\|_F$$

The proof of this theorem is quite long and is shared in the following lemmas.

Lemma 8.2 *Define* $X_x = Ker\ K(f,x)^\perp$. *Then*

$$X_x = \{\dot{f} \in T_f \mathcal{K} \ : \ \dot{f}_i = \lambda_i \Pi_{i,f}(1 + < z, x >)^{d_i}, \ \lambda_i \in \mathbb{C}\}.$$

Proof. $K(f,x)\dot{f} = 0$ if and only if $\dot{f}(x) = 0$. By Corollary 3.10 this last equation may be read as orthogonality relations between $\dot{f}_i$ and the polynomial $\Pi_{i,f}(1 + < z, x >)^{d_i}$ for $i = 1 \ldots n$.

Lemma 8.3 $K(f,x) : X_x \to \mathbb{C}^n$ *is an isomorphism and its inverse is given by*

$$K(f,x)^+ \dot{x} = \left(-Df_i(x)\dot{x}\ \frac{\Pi_{i,f}(1 + < z, x >)^{d_i}}{N_{i,f}(x)^2} \right)_{1 \le i \le n}.$$

Proof. Let $\dot{x} \in \mathbb{C}^n$ be given. We want to write $K(f,x)\dot{f} = \dot{x}$ for some $\dot{f} \in X_x$. According to Lemma 8.2 we take $\dot{f}_i = \lambda_i \Pi_{i,f}(1 + < z, x >)^{d_i}$. The relation $K(f,x)\dot{f} = \dot{x}$ becomes $-Df(x)^{-1}(\lambda_i \Pi_{i,f}(1 + < z, x >)^{d_i})|_{z=x} = \dot{x}$. By Corollary 3.10 we get $(\lambda_i N_{i,f}(x)^2)_i = -Df(x)\dot{x}$ and we are done.

Remark 8.4 $K(f,x)^+$ is the Moore-Penrose inverse of $K(f,x)$. Its definition may be extended to any $g \in V_x$ by a similar formula:

$$K(g,x)^+\dot{x} = \left(-Dg_i(x)\dot{x}\ \frac{\Pi_{i,f}(1+ <z,x>)^{d_i}}{N_{i,f}(x)^2}\right)_{1\leq i\leq n}.$$

Lemma 8.5 *Each linear operator $L \in \mathcal{L}(\mathbb{C}^n, X_x)$ is equal to $K(g,x)^+$ for some $g \in V_x$.*

Proof. This lemma is a consequence of the definition of $K(g,x)^+$ and the hypothesis H2.

Lemma 8.6 *For all $g \in V_x$ we have $g \in \Sigma_x$ if and only if rank $K(g,x)^+ < n$.*

Proof. The proof is easy and left to the reader.

Remark 8.7 Since by Hypothesis H3 the set $V_x - f$ is a vector subspace of $\mathcal{P}_d$ we may define an orthogonal decomposition $V_x - f = E_x \oplus F_x$ with $g \in E_x$ if and only if $Dg(x) = 0$. For any $g \in V_x$ define $g_E - f_E$ the orthogonal projection of $g - f$ over E_x. We also define for $g, h \in V_x$

$$d_*(g,h) = \left(\|g_E - h_E\|_*^2 + \|K(g,x)^+ - K(h,x)^+\|_F^2\right)^{1/2}$$

where for a linear operator $L \in \mathcal{L}(\mathbb{C}^n, X_x)$ we denote by $\|L\|_F$ the Frobenius norm of L equal to $Trace(L^H L)^{1/2}$ and where $\|\ldots\|_*$ is any norm over E_x.

Lemma 8.8 *d_* is a distance over V_x.*

Proof. The only non trivial point to check is the implication $d_*(g,h) = 0$ implies $g = f$. If $d_*(g,h) = 0$ we have $Dg(x) = Dh(x)$ so that $g - h \in E_x$. Since we also have $g_E - h_E = 0$ we obtain $g = h$.

Lemma 8.9 *Let us denote by $\mathcal{S}(\mathbb{C}^n, X_x)$ the set of singular linear operators defined over $\mathbb{C}^n$ with values in X_x. We have*

$$C_K(f,x)^{-1} = d_F(K(f,x)^+, \mathcal{S}(\mathbb{C}^n, X_x)).$$

Proof. Let us write $K(f,x)|_{X_x} = VDU^H$ with U and V unitary over X_x and $\mathbb{C}^n$ and $D = (d_{i,j})$, $1 \leq i,j \leq n$, $d_{i,j} = 0$ if $i \neq j$ and $d_{1,1} \geq \ldots \geq d_{n,n} > 0$ the singular value decomposition of $K(f,x)|_{X_x}$. Then

$$C_K(f,x) = \|K(f,x)\| = \|K(f,x)|_{X_x}\| = d_{1,1}.$$

Now $K(f,x)^+ = UD^{-1}V^H$ and since the Frobenius norm is unitarily invariant, $d_F(K(f,x)^+, \mathcal{S}(\mathbb{C}^n, X_x))$ is equal to the F-distance of D to the set of $n \times n$ singular matrices. This distance is reached for $S = (s_{i,j})$, $s_{i,j} = 0$ if $i \neq j$, $s_{1,1} = 0$ and $s_{i,i} = d_{i,i}^{-1}$, $2 \leq i \leq n$ and is equal to $d_{1,1}^{-1}$: this achieves the proof.

Lemma 8.10 $C_K(f,x)^{-1} = \min_{g \in \Sigma_x} d_*(f,g) = d_*(f, \Sigma_x)$.

Proof. Let $g \in \Sigma_x$ be such that $d_*(f,g) = d_*(f, \Sigma_x)$, let us write $g - f = g_E + g_F$. Since $g - g_E = f + g_F \in f + V_x - f$ we have $g - g_E \in V_x$. Moreover, since $g_E \in E_x$ i.e $Dg_E(x) = 0$ one has $K(g - g_E, x)^+ = K(g,x)^+$ and $g - g_E \in \Sigma_x$. This gives

$$d_*(f, \Sigma_x)^2 = d_*(f, g)^2 = \|g_E\|_*^2 + \|K(f, x)^+ - K(g, x)^+\|_F^2 \leq$$

$$d_*(f, g - g_E)^2 = \|K(f, x)^+ - K(g - g_E, x)^+\|_F^2 = \|K(f, x)^+ - K(g, x)^+\|_F^2$$

so that $g_E = 0$ and $g - f \in F_x$. This yealds

$$d_*(f, \Sigma_x) = d_*(f, g) = \|K(f, x)^+ - K(g, x)^+\|_F$$

and since $K(g, x)^+$ is a singular operator

$$\geq d_F(K(f, x)^+, \mathcal{S}(\mathbb{C}^n, X_x)) = C_K(f, x)^{-1}$$

by Lemma 8.9. We now prove the inverse inequality. We have

$$C_K(f, x)^{-1} = d_F(K(f, x)^+, \mathcal{S}(\mathbb{C}^n, X_x)) = d_F(K(f, x)^+, L)$$

for some $L \in \mathcal{S}(\mathbb{C}^n, X_x)$. By Lemma 8.5 such an L can be written $L = K(g, x)^+$ with $g \in V_x$ and since L is singular, by Lemma 8.6, we have $g \in \Sigma_x$. We now write $g - f = g_E + g_F$ so that $g - g_E \in V_x$ and $K(g, x)^+ = K(g - g_E, x)^+$ as before. Since $g - g_E - f \in F_x$ we have

$$C_K(f, x)^{-1} = d_F(K(f, x)^+, L) = d_F(K(f, x)^+, K(g, x)^+) =$$

$$d_F(K(f, x)^+, K(g - g_E, x)^+) = d_*(f, g - g_E) \geq d_*(f, \Sigma_x)$$

and our Lemma is proved.

Lemma 8.11 $d_*(f, \Sigma_x) = \min_{g \in \Sigma_x} d_F(K(f, x)^+, K(g, x)^+)$.

Proof. The inequality $\geq$ is clear. Let us prove the other one. Let $\epsilon > 0$ and $g \in \Sigma_x$ be such that

$$\|K(f, x)^+ - K(g, x)^+\|_F - \epsilon \leq \min_{g \in \Sigma_x} \|K(f, x)^+ - K(g, x)^+\|_F.$$

We now write $g - f = g_E + g_F$ so that $g - g_E \in V_x$ and $K(g, x)^+ = K(g - g_E, x)^+$ as before. Since $g - g_E - f \in F_x$ we have $d_*(f, \Sigma_x) \leq d_*(f, g - g_E) = d_F(K(f, x)^+, K(g - g_E, x)^+) = d_F(K(f, x)^+, K(g, x)^+) \leq \epsilon + \min_{g \in \Sigma_x} \|K(f, x)^+ - K(g, x)^+\|_F$ and our lemma holds.

Lemma 8.12. *For any $g \in V_x$ we have*

$$\|K(g, x)^+\|_F = \|Diag(N_{i,f}(x)^{-1})Dg(x)\|_F.$$

Proof. For any $\dot{x}, \dot{y} \in \mathbb{C}^n$ we have $(K(g, x)^+\dot{x}, K(g, x)^+\dot{y}) =$

$$\sum_{i=1}^{n} Dg_i(x)\dot{x}\overline{Dg_i(x)\dot{y}} \left(\frac{\Pi_{i,f}(1 + <z, x>)^{d_i}}{N_{i,f}(x)^2}, \frac{\Pi_{i,f}(1 + <z, x>)^{d_i}}{N_{i,f}(x)^2} \right) =$$

$$\sum_{i=1}^{n} N_{i,f}(x)^{-2} Dg_i(x)\dot{x}\overline{Dg_i(x)\dot{y}} =$$

$$< Diag(N_{i,f}(x))^{-1} Dg(x)\dot{x}, Diag(N_{i,f}(x))^{-1} Dg(x)\dot{y} >_{\mathbb{C}^n} .$$

Consequently

$$K(g,x)^{+H}K(g,x) = \left(Diag(N_{i,f}(x))^{-1}Dg(x)\right)^{H} Diag(N_{i,f}(x))^{-1}Dg(x)$$

and

$$\|K(g,x)^{+}\|_F = \|Diag(N_{i,f}(x))^{-1}Dg(x)\|_F.$$

Remark 8.13 The proof of Theorem 8.1 is a consequence of Lemmas 8.10, 8.11 and 8.12.

9   Complexity of Sparse Continuation Methods

In this section we study the complexity of a classical continuation algorithm based on Newton's iteration. When the homotopy path consists only in sparse polynomial systems the complexity can be bounded essentially by the square of the sparse condition number of the homotopy. In [13] Renegar also relates the complexity of a continuation algorithm with the sparse structure: the number of nonzero coefficients occuring within the system.

Consider any curve $F : I \to V \setminus \Sigma$, $I = [0,1]$,

$$V = \{(f,x) \in \mathcal{K} \times \mathbb{C}^n : f(x) = 0\}$$

and Σ the critical set of V : $(f,x) \in \Sigma$ when $f(x) = 0$ and $rank\ Df(x) < n$. Write $F(t) = (f_t, x_t)$ so that $f_t(x_t) = 0$. The length of the curve $t \to f_t$ (resp. x_t) is denoted by L_f (resp. L_x). A partition of I is an increasing sequence t_i with $t_0 = 0$ and $t_k = 1$. We write f_i and x_i instead of f_{t_i} and x_{t_i}. Define a sequence $(z_i)_{0 \le i \le k}$ in $\mathbb{C}^n$ by

$$z_0 = x_0 \ and \ z_i = N_{f_i}(z_{i-1}), \ 1 \le i \le k,$$

with $N_{f_i}(z_{i-1}) = z_{i-1} - Df_i(z_{i-1})^{-1}f_i(z_{i-1})$ the Newton iterate.

Definition 9.1 (Smale [17]) *Let $f : \mathbb{C}^n \to \mathbb{C}^n$ be an analytic system. We say that $z \in \mathbb{C}^n$ is an approximate zero of f if the sequence $z_0 = z$, $z_{i+1} = N_f(z_i)$ is defined for all natural numbers i and there is an $x \in \mathbb{C}^n$ with $f(x) = 0$ and*

$$\|z_i - x\| \le \left(\frac{1}{2}\right)^{2^i - 1} \|z - x\|.$$

We say that x is the associated zero.

A criterion for z to be an approximate zero is given by

Theorem 9.2 (BCSS [4], Chap. 8, Theorem 1.] *Suppose that $f(x) = 0$ and that rank $Df(x) = n$. If*

$$\|z - x\|\gamma(f,x) \le \frac{3 - \sqrt{7}}{2}$$

then z is an approximate zero of f with associated zero x.

We return to our continuation method. When $\max_i |t_{i+1} - t_i|$ is small enough and hence k large, z_i is an approximate zero of f_i with associated zero x_i. Thus k is a good measure for the complexity of this continuation method.

Definition 9.3 *The sparse condition number of the homotopy F is*

$$C_\mathcal{K}(F) = \max_{0 \le t \le 1} C_\mathcal{K}(f_t, x_t).$$

The invariant γ for the homotopy F is (see section 7)

$$\gamma(F) = \max_{0 \le t \le 1} \gamma(f_t, x_t).$$

Theorem 9.4 *There is a partition $t_0 = 0 \ldots t_k = 1$ of $I = [0, 1]$ such that, for each i, z_i is an approximate zero of f_i relative to x_i with*

$$k = \lceil C_\mathcal{K}(F)\gamma(F)L_f \frac{4}{3 - \sqrt{7}} \rceil.$$

Remark 9.5. 1. A bound for $\gamma(F)$ in term of the condition number $C_\mathcal{K}(F)$ can be deduced from Theorem 7.2. It shows that k is bounded by the square of the sparse condition number of the homotopy.

2. When F is a linear homotopy: $f_t = (1 - t)f_0 + tf_1$, the length of the curve f_t is $L_f = \|f_0 - f_1\|$.

3. The non sparse case is studied in BCSS [4] Chap. 14, Theorem 4, for homogeneous polynomial systems and projective Newton's Method. Our proof reproduces a similar argument in an affine context. In the non sparse case more accurate bounds appear in Shub-Smale [14], [15] and [16].

Proof of Theorem 9.4. Since L_x is the length of the curve x_t one has

$$\sum_{i=0}^{k-1} \|x_i - x_{i+1}\| \le L_x \le C_\mathcal{K}(F)L_f$$

by Theorem 5.4. There exists a partition $t_0 = 0 \ldots t_1 = 1$ such that, for each i,

$$\|x_i - x_{i+1}\| \le \frac{C_\mathcal{K}(F)L_f}{k}$$

thus

$$\|x_i - x_{i+1}\|\gamma(F) \le \frac{3 - \sqrt{7}}{4}.$$

According to Theorem 9.2, Theorem 9.4 follows from the following assertion

$$\|z_i - x_i\|\gamma(f_i, x_i) \le \frac{3 - \sqrt{7}}{2}$$

or yet

$$\|z_i - x_i\|\gamma(F) \le \frac{3 - \sqrt{7}}{4}.$$

We will prove it by induction. It is true for $i = 0$ since $z_0 = x_0$. Now we have

$$\|z_i - x_{i+1}\|\gamma(F) \leq \|z_i - x_i\|\gamma(F) + \|x_i - x_{i+1}\|\gamma(F) \leq \frac{3 - \sqrt{7}}{2}.$$

By Theorem 5.2 z_i is an approximate zero of f_{i+1} with associated zero x_{i+1}. Thus

$$\|z_{i+1} - x_{i+1}\| \leq \frac{1}{2}\|z_i - x_{i+1}\|$$

and we are done.

10   Bibliography

1. B. Beauzamy, How the roots of a polynomial vary with its coefficients: a local quantitative result. Preprint, B. Beauzamy, Institut de Calcul Mathématique, 37 rue Tournefort, 75005 Paris, France.
2. B. Beauzamy, J. Dégot: Differential Identities, Transactions of the American Mathematical Society, Volume 347, Number 7, July 1995, pp 2607-2619.
3. R. Bhatia, L. Elsner, G. Krause, Bounds for the variation of the roots of a polynomial and the eigenvalues of a matrix. Lin. Alg. and its Apppl. 142(1990)195-209.
4. L. Blum, F. Cucker, M. Shub, S. Smale, Complexity and Real Computation. To appear.
5. J.P. Dedieu, Approximate Solutions of Numerical Problems, Condition Number Analysis and Condition Number Theorems. To appear in: Proceedings of the Summer Seminar on "Mathematics of Numerical Analysis: Real Number Algorithms", Park City, UT, July 17 - August 11, 1995, Lectures in Applied Mathematics.
6. J.P. Dedieu. Estimations for the separation number of a polynomial system. To appear in: Journal of Symbolic Computation.
7. J.P. Dedieu, Condition Operators, Condition Numbers and Condition Number Theorem for the Generalized Eigenvalue Problem. To appear in: Linear Alg. and its Appl.
8. C. Eckart, G. Young, A principal axis transformation for non-Hermitian matrices. Bull. Am. Math. Soc. 45(1939)118-121.
9. T. Kojima, On the limits of the roots of an algebraic equation. Tohoku Math. J. 11(1917)119-127.
10. E. Kostlan, On the distribution of the roots of random polynomials. In: M. Hirsch, J. Marsden, M. Shub Eds., From Topology to Computation: Proceedings of the Smalefeast, (1993)419-431, Springer-Verlag.
11. M. Marden, Geometry of Polynomials, AMS, 1966.
12. A. Ostrowski, Solutions of Equations in Euclidean and Banach Spaces. Academic Press, New-York, 1976.
13. J. Renegar, On the efficiency of Newton's Method in Approximating all Zeros of a System of Complex Polynomials. Math. Op. Research, 12(1987)121-148.

14. M. Shub, S. Smale, Complexity of Bézout's Theorem 1: Geometric Aspects. Journal of the Amer. Math. Soc. 6(1993)459-501.

15. M. Shub, S. Smale, Complexity of Bézout's Theorem 5: polynomial time. Theoretical Computer Science, 133(1994)141-164.

16. M. Shub, S. Smale, Complexity of Bézout's Theorem 4 : Probability of success ; Extensions. 1996. To appear in SIAM Journal of Num. Analysis.

17. S. Smale, The fondamental Theorem of Algebra and Complexity Theory. Bulletin of the Amer. Math. Soc., 4(1981)1-36.

18. S. Smale, Newton's method estimates with data at one point. In: The merging of disciplines: New directions in pure and applied mathematics. R. Ewing, K. Cross, C. Martin Eds. Springer Verlag, 1986.

19. Wang Xinghua, Han Danfu, On dominating sequence method in the point estimate and Smale theorem. Sci. China (Ser. A) 33(1990)135-144.

20. J. Wilkinson, Rounding Errors in Algebraic Processes. Her Majesty's Stationery Office, London, 1963.

21. H. Wozniakowski, Numerical Stability for Solving Nonlinear Equations, Num. Math., 27(1977)373-390.

Residues in the Torus and Toric Varieties

Alicia Dickenstein[**]

Let $f_1, \ldots, f_n$ be Laurent polynomials in n variables with a finite set V of common zeroes in the torus $T = (\mathbb{C}^*)^n$. The total residue of the differential form

$$\phi_q = \frac{q}{f_1 \cdots f_n} \frac{dt_1}{t_1} \wedge \cdots \wedge \frac{dt_n}{t_n},$$

i.e. the sum of the local Grothendieck residues of a Laurent polynomial q, at each of the points in V, with respect to $f_1, \ldots, f_n$, is a rational function of the coefficients of the f's which has interesting applications in a number of different contexts such as: integral representations formulae in complex analysis; explicit division and interpolation formulae and effective bounds for problems in commutative algebra (see [3] which contains an extensive bibliography); duality methods in the study of algebraic questions such as the ideal membership problem, computation of traces, elimination, complexity bounds, etc. ([2, 6, 8, 9, 10, 15]).

Given generic Laurent polynomials $f_1, \ldots, f_n$ with fixed Newton polytopes $\Delta_1, \ldots, \Delta_n$, the closures of the zeroes of the f_j's in the toroidal compactification X of T associated to the Minkowski sum $\Delta = \Delta_1 + \cdots + \Delta_n$, have no common intersection outside of T. This toric variety X may be made to play the same role that standard projective n-space plays for generic dense polynomials of fixed degrees. In a joint paper with E. Cattani and D. Cox ([4]), we showed that the total residue may be computed by a global object on X, namely the *toric residue* previously introduced by Cox [7]. This point of view is elaborated in [5] where we deduce symbolic algorithms for the computation of global residues as well as important properties of the total residue such as its polynomial dependence on coefficients corresponding to points in the interior of the Newton polytopes (with explicit bounds).

In current joint work with E. Cattani and B. Sturmfels, we study the behaviour of total residues with respect to boundary coefficients. Not surprisingly, the denominator of the global residue of ϕ_q, for a monomial $q = t^a$, is a product of powers of sparse resultants ([11, 16]) associated to all those facets of Δ relative to which a lies in the opposite side of Δ. The corresponding powers are the lattice distance from a to Δ suitably shifted.

In case $\Delta_1, \ldots, \Delta_n$ are "in general position", these facets resultants reduce to monomials, and we recover the corresponding results in [12] and [17].

[**] Supported by UBACYT and CONICET, Argentina

We deduce this behaviour of the denominators of global residues from the study of the denominators of the associated toric residues in X for multihomogeneous polynomials. In particular, we obtain new proofs of the known results in the case of homogeneous polynomials in projective space ([1, 13, 14]).

The same considerations that lead to the proof of the above results on residues and resultants, reveal the connection between resultants and toric jacobians and provide new determinantal algorithms for the computation of sparse resultants, which generalize the classical formula of Sylvester (ca. 1852) for the resultant of three ternary quadrics ([11],3.4.D).

References

1. Angeniol, Résidus et Effectivité. Unpublished manuscript, 1983.
2. E. Becker, J.-P. Cardinal, M.-F. Roy, and Z. Szafraniec, Multivariate Bezoutians, Kronecker Symbol and Eisenbud–Levine formula, in: *MEGA 94: Effective Methods in Algebraic Geometry* Progress in Mathematics 143 (Birkhäuser, 1996).
3. C. Berenstein, R. Gay, A. Vidras, and A. Yger, *Residue Currents and Bezout Identities*, Progress in Mathematics 114 (Birkhäuser, Basel Boston Berlin, 1993).
4. E. Cattani, D. Cox, and A. Dickenstein, Residues in Toric Varieties, *Compositio Mathematica*, to appear; alg-geom 9506024.
5. E. Cattani and A. Dickenstein, A Global View of Residues in the Torus, in: *MEGA 96: Effective Methods in Algebraic Geometry. Journal of Pure and Applied Algebra*, to appear.
6. E. Cattani, A. Dickenstein and B. Sturmfels, in: *MEGA 94: Effective Methods in Algebraic Geometry* Progress in Mathematics 143 (Birkhäuser, 1996) 135–164.
7. D. Cox, Toric residues, *Ark. Mat* **34** (1996) 73–96.
8. A. Dickenstein and C. Sessa, Duality methods for the membership problem, in: *MEGA 94: Effective Methods in Algebraic Geometry* , Progress in Mathematics 94, (Birkhäuser, 1991) 89–103.
9. M. Elkadi, Résidu de Grothendieck et Problèmes d' Effectivité , *Thèse*, Université de Bordeaux I, 1993.
10. N. Fitchas, M. Giusti, and F. Smietanski, Sur la complexité du théorème des zéros, in: *Proceedings 2nd Int. Conf. on Approximation and Optimization* (La Habana, Cuba, 1993).
11. I. Gel'fand, M. Kapranov, and A. Zelevinsky, *Discriminants, Resultants and Multidimensional Determinants* (Birkhäuser, Basel Boston Berlin, 1994).
12. O. A. Gelfond and A. G. Khovanskii, Newton polytopes and Grothendieck resiues, *Sov. Math. Dokl.*, to appear.
13. J.-P. Jouanolou, Singularités rationnelles du résultant, in: Springer LNM 732 (1979) 183–214.
14. J.-P. Jouanolou, Résultant et résidu de Grothendieck, Preprint IRMA, Strasbourg, 1992.
15. J. Sabia and P. Solerno, Bounds for Traces in Complete Intersections and Degrees in the Nullstellensatz, *AAECC Journal* **6** (1995) 353–376.6.
16. B. Sturmfels, Sparse Elimination Theory, in: D. Eisenbud, L. Robbiano, eds., *Computational Algebraic Geometry and Commutative Algebra*, Proceedings, Cortona, June 1991, (Cambridge University Press, Cambridge, 1993).
17. H. Zhang, Calculs de Résidus Toriques. Preprint, 1995.

Piecewise Smooth Orthonormal Factors for Fundamental Solution Matrices

Luca Dieci and Erik S. Van Vleck

Abstract. The purpose of this note is to report on some recent results concerning QR factorizations of smooth full rank matrices. The focus of this note is on the computation of the Q factor. In particular, unlike previous works on the subject, we report on two techniques based on reflectors and rotators which achieve piecewise smooth factorizations (see [7]). The benefits of the present techniques is that they can be implemented rather inexpensively, and that maintaining orthonormality for the Q factor is easy to guarantee even for rectangular matrices.

1 Introduction

There are many known instances in which one needs to compute a QR factorization of a smooth time varying (or, more generally, parameter dependent) full rank matrix $Y(t) \in \mathbb{R}^{n \times p}$, $n \geq p$. Since $\dot{Y} = [\dot{Y}(Y^T Y)^{-1} Y^T] Y$, we can always think of $Y(t)$ as being given by columns of a fundamental solution matrix for a linear system, a case to which we will restrict from now on:

$$\dot{Y} = A(t) Y, \quad A(t) \in \mathbb{R}^{n \times n}, \quad t \in [0, t_f]. \tag{1}$$

In particular, finding a time varying QR factorization of a fundamental solution matrix $Y(t)$ (or parts of it) is required in the so-called continuous orthonormalization technique for solving BVPs of ODEs (e.g., see [1]), or in the computation of Lyapunov exponents (e.g., see [2] and [5]). These applications provided our motivation, but there are many more situations in which one could conceivably need to obtain such factors, for example for updating techniques.

In what follows, we will seek the factorization $Y(t) = Q(t) R(t)$, where $Q(t) \in \mathbb{R}^{n \times p}$ and orthogonal, and $R(t) \in \mathbb{R}^{p \times p}$ upper triangular. In fact, we will only discuss ways to find Q. In principle, the problem is easily dealt with by: (i) Find $Y(t)$, and then (ii) perform its QR factorization whenever required. Smoothness of the factors can be easily guaranteed by monitoring the diagonal of R. But, aside from its lack of elegance, the trouble with this simple approach is the need for finding $Y(t)$. When $Y(t)$ is a solution of (1), it is not advisable to compute $Y(t)$, since (1) might have increasing/decreasing solution components, making the computation of Y possibly unstable, and quite likely hard. Rather, if we could compute the Q factor, and then change variables: $R := Q^T Y$, then we would be left with a conceptually and practically easier problem only involving quadratures to find R. Practically speaking, then, the emphasis has shifted on proper ways to find Q, in particular on how to ensure that the computed Q stays orthogonal. It should be realized that the Q equation will have to satisfy a skew-type differential equation (simply because $Q^T Q = I$), and hence one can anticipate some neutral stability characteristic for this equation.

A number of approaches to find numerically the Q factor have been proposed. One of the first we know of is in [8], and it is based on the differential equations satisfied by the continuous Gram-Schmidt procedure. Unfortunately, as was anticipated from the linear algebra context (see [9]), maintaining orthogonality numerically is difficult using the Gram-Schmidt technique, and one should follow an integration step by re-orthonormalization, much like what one does for the projected integrators (see below). An alternative to the viewpoint of the Gram-Schmidt procedure is obtained by thinking of Q as the orthogonal matrix performing the sought change of variables. Thus, by differentiating $Y = QR$ and using (1), one eventually obtains

$$\dot{R} = (Q^T A Q - S)R =: \tilde{A}(t)R, \qquad S := Q^T \dot{Q} \in \mathbb{R}^{p \times p} \qquad (2)$$

where the matrix $\tilde{A}$ is upper triangular, and the matrix S is skew-symmetric with

$$S_{ij} = \begin{cases} (Q^T A Q)_{ij} \,, i > j \\ \qquad\quad 0 \,, i = j \\ -(Q^T A Q)_{ji} \,, i < j \end{cases} \, , \quad i, j = 1, \ldots, p. \qquad (3)$$

The following equation for Q is also easily obtained

$$\dot{Q} = AQ - QQ^T AQ + QS$$
$$\text{or} \qquad\qquad\qquad\qquad\qquad (4)$$
$$\dot{Q} = (A - QQ^T AQQ^T + QSQ^T)Q =: H(Q,t)Q \,.$$

(In case $p = n$, then we simply have $\dot{Q} = QS$ of course.) Numerical techniques for (4) have been considered in [4], [6], [3], [10]. For the full case, $p = n$, Gauss RK (GRK) methods are known to be the only linear *automatic unitary integrators*, i.e. methods for which orthogonality of Q is automatically maintained at grid points. The main (and possibly only) drawback with this approach is the cost, since one needs to solve a nonlinear matrix equation at each step. In case $p < n$, no standard linear scheme can be expected to maintain orthogonality at gridpoints, when used to integrate (4); see [6]. However, in all cases, a simple family of schemes is provided by the so-called *projected unitary integrators*. These arise when we approximate the solution of (4) at mesh points t_k, call Z_k such an approximation, by an arbitrary integration scheme, and then factor $Z_k = Q_k R_k$, retaining the Q_k factor as an approximation of Q at t_k. In [10], the author uses the polar decomposition of Z_k to extract Q_k.

All above approaches are based on integration of the differential equation satisfied by Q, and attempt to recover orthogonality either by a direct numerical discretization, or by a simple post-processing operation. The alternative we consider next, instead, seeks the matrix Q as product of elementary orthogonal matrices, for us reflectors and rotators. We derive differential equations for these elementary factors, integrate these, and then recover the whole matrix Q. Preserving orthogonality becomes more or less trivial. However, the price we pay is that we cannot guarantee to obtain a globally smooth factorization, and only a piecewise smooth factorization is obtained. For all problems we know of, this is sufficient.

2   Piecewise Smooth Factorizations: Reflectors and Rotators

Here we report on two techniques recently analyzed in [7]. The basic idea is to seek Q as product of elementary orthogonal matrices, integrate for these elementary factors, and finally recover Q. The benefit of this approach is that integration of the elementary factors can be done in many ways, and to eventually preserve orthonormality is easy, if the process is well defined. In general, it is not, since there cannot be any guarantee that the sought smooth $Q(t)$ be the product of reflectors/rotators. However, for given t, any orthogonal matrix $Q(t)$ can always be written as product of reflectors or rotators, modulo a post-multiplication by a diagonal matrix of ± 1. These considerations lead to a procedure delivering a piecewise smooth Q.

Reflectors (Householder matrices).
   We want to find $Q^T(t) = P_p(t) \cdots P_1(t)$ such that $Q^T(t)Y(t) = R(t)$, where

$$P_i(t) = \begin{pmatrix} I_{i-1} & 0 \\ 0 & Q_i(t) \end{pmatrix}, Q_i(t) = I - \gamma_i(t)\mathbf{u}_i(t)\mathbf{u}_i(t)^T. \tag{5}$$

The idea is to derive differential equations for $\mathbf{u}_i$ and γ_i. First, consider the case of one vector: $\dot{\mathbf{y}} = A(t)\mathbf{y}$, $\mathbf{y} \in \mathbb{R}^n$, $t \in [0, t_f]$, $\mathbf{y}(0) = \mathbf{y}_0 \neq \mathbf{0}$. Want to find

$$Q(t) = I - \gamma(t)\mathbf{u}(t)\mathbf{u}^T(t), \quad \gamma(t) = \frac{2}{\mathbf{u}^T(t)\mathbf{u}(t)},$$

such that $Q(t)\mathbf{y}(t) = \mathbf{r}(t) = \begin{pmatrix} r_1(t) \\ 0 \\ \vdots \\ 0 \end{pmatrix}$, where of course $r_1 = \sigma\|\mathbf{y}\|$, $\sigma = \pm 1$. We

may take

$$\mathbf{u}(t) = \mathbf{y}(t) - \sigma\|\mathbf{y}\|, \tag{6}$$

by choosing

$$\sigma := \begin{cases} -1, & \text{if } y_1(0) \geq 0, \\ 1, & \text{if } y_1(0) < 0, \end{cases} \tag{7}$$

so that the above construction is well defined in the interval $[t_0, t_1)$, with $t_1 > 0$. With $A_s(t) = 1/2(A(t)+A^T(t))$, the following differential equations are obtained

(ICs may be obtained from (6))

$$\frac{d}{dt}\begin{pmatrix} \mathbf{u} \\ \sigma\|\mathbf{y}\| \end{pmatrix} = -\frac{\mathbf{u}^T A_s \mathbf{u}}{\sigma\|\mathbf{y}\|}\begin{pmatrix} \mathbf{e}_1 \\ -1 \end{pmatrix}$$

$$+\begin{pmatrix} A - 2\mathbf{e}_1\mathbf{e}_1^T A_s & (A - \mathbf{e}_1\mathbf{e}_1^T A_s)\mathbf{e}_1 \\ 2\mathbf{e}_1^T A_s & \mathbf{e}_1^T A_s \mathbf{e}_1 \end{pmatrix}\begin{pmatrix} \mathbf{u} \\ \sigma\|\mathbf{y}\| \end{pmatrix}. \tag{8}$$

For numerical purposes, it is better to work with the variables

$$\mathbf{v} := (\frac{\gamma}{2})^{1/2}\mathbf{u}, \qquad P = I - 2\mathbf{v}\mathbf{v}^T, \tag{9}$$

since now $\mathbf{v}^T\mathbf{v} = 1$ for all t for which $\mathbf{u}(t) \neq \mathbf{0}$. In this case, we get

$$\dot{\mathbf{v}} = \tfrac{\gamma}{2}(\dot{\mathbf{u}}\mathbf{v}^T - \mathbf{v}\dot{\mathbf{u}}^T)\mathbf{u} = (\gamma/2)^{1/2}(\dot{\mathbf{u}}\mathbf{v}^T - \mathbf{v}\dot{\mathbf{u}}^T)\mathbf{v}$$

$$\text{or} \quad \dot{\mathbf{v}} = (B - B^T)\mathbf{v}, \; B := (\gamma/2)^{1/2}\dot{\mathbf{u}}\mathbf{v}^T, \tag{10}$$

where (with $v_1 = \mathbf{e}_1^T\mathbf{v}$)

$$B = [\frac{1}{2v_1}(A - \mathbf{e}_1\mathbf{e}_1^T A_s)(2v_1\mathbf{v} - \mathbf{e}_1) + ((2v_1\mathbf{v}^T - \mathbf{e}_1^T)A_s\mathbf{v})\mathbf{e}_1]\mathbf{v}^T.$$

Remark. Clearly, (10) is a skew symmetric type equation, and it should thus be integrated with Gauss RK schemes so that the computed solution is of norm 1 at grid points. Integration of (8), instead, can be done with explicit schemes.

The situation for the general matrix case $Y(t) \in \mathbb{R}^{n \times p}$ goes similarly. We obtain

Algorithm $[0, h]$. Reflectors.

INPUT: $t = 0$, initial conditions $Y = Y(0) \in \mathbb{R}^{n \times p}$, and $h > 0$.
   - For $j = 1, \ldots, p$
      - Let $A = A(j : n, j : n)$, and $\mathbf{y}(0) = Y(j : n, j)$
      - Choose σ according to (7)
      - Integrate (8)-(10) on $[0, h]$
      - $A(t) = P_j A P_j - P_j \dot{P}_j$, $Y(0) = P_j(0)Y(0)$
   - Endfor j
OUTPUT: Have $Q^T(h) = P_p(h)\cdots P_1(h)$, and $\tilde{A}(h) = Q^T(h)A(h)Q(h) - Q^T(h)\dot{Q}(h)$, which is in $A(h)$, is triangular in its leading (n, p) submatrix. Likewise, $Q^T(h)Y(h)$ is triangular.

Notice that there is no guarantee that the underlying transformation $Q(t)$ is well defined for all $t \in [0, t_f]$. In particular, we might very well have a singular vector field when v_1 goes through 0. On the other hand, since $\mathbf{y}(t)$ cannot identically vanish, at any point $t \in [0, t_f]$ there is a sign choice like in (7) giving a Householder matrix for which the above representation is valid on some open interval about t, thus $[0, t_f]$ can be covered by a finite subset of these open intervals. In other words, a change of Householder frame (a *reimbedding*) must be contemplated. The resulting approach, leading to a piecewise smooth factorization of $Y(t)$, has been analyzed and implemented in [7], to which we refer for details. We simply mention that explicit construction of Y is not needed in order to decide when to change Householder frame and how to restart the computation.

Continuous Rotators.

Here we seek a representation for the matrix Q as product of elementary plane rotators. We first outline the basic process with no concern about its feasibility. Consider the elementary plane rotators for $i = 1, \ldots n - 1$, $j = i + 1, \ldots, n$,

$$
Q_{ij}(t) = \begin{pmatrix}
1 & \cdots & 0 & \cdots & 0 & \cdots & 0 \\
\vdots & \ddots & \vdots & & \vdots & & \vdots \\
0 & \cdots & c_{ij} & \cdots & -s_{ij} & \cdots & 0 \\
\vdots & & \vdots & \ddots & \vdots & & \vdots \\
0 & \cdots & s_{ij} & \cdots & c_{ij} & \cdots & 0 \\
\vdots & & \vdots & & \vdots & \ddots & \vdots \\
0 & \cdots & 0 & \cdots & 0 & \cdots & 1
\end{pmatrix}, \tag{11}
$$

where the rank-two block $\begin{pmatrix} c_{ij} & -s_{ij} \\ s_{ij} & c_{ij} \end{pmatrix}$ is in the i-th and j-th row/column. We have used the notation $c_{ij} = \cos(\theta_{ij}(t))$ and $s_{ij} = \sin(\theta_{ij}(t))$ for some function $\theta_{ij}(t)$, which is unknown. For the matrix $Y \in \mathbb{R}^{n \times p}$ we would want $Q = Q_{12} \cdots Q_{1n} \cdots Q_{pn}$ such that $Q^T Y(t)$ is upper triangular. First, consider the case of a single vector, that is we seek $Q_{12}, \ldots, Q_{1n}$ such that

$$
Q_{1n}^T(t) \cdots Q_{13}^T(t) Q_{12}^T(t) \mathbf{y}(t) = \sigma \|\mathbf{y}\| \mathbf{e}_1, \tag{12}
$$

and simply use the notation c_j, s_j for c_{1j}, s_{1j}. Differentiating (12), we obtain

$$
\begin{aligned}
\tfrac{d}{dt} \|\mathbf{y}\| \mathbf{e}_1 = \sigma \|\mathbf{y}\| [Q_{1n}^T \cdots Q_{12}^T A Q_{12} \cdots Q_{1n} \\
+ \dot{Q}_{1n}^T Q_{1n} + Q_{1n}^T \dot{Q}_{1,n-1}^T Q_{1,n-1} Q_{1n} + \cdots \\
+ Q_{1n}^T \cdots Q_{13}^T \dot{Q}_{12}^T Q_{12} Q_{13} \cdots Q_{1n}] \mathbf{e}_1.
\end{aligned} \tag{13}
$$

For $j = 2, \ldots, n$, let

$$
\alpha_j(t) = \mathbf{e}_j^T [Q_{1n}^T \cdots Q_{12}^T A Q_{12} \cdots Q_{1n}] \mathbf{e}_1, \tag{14}
$$

then obtain

$$
\begin{pmatrix}
c_3 \cdots c_n \dot{\theta}_2 \\
c_4 \cdots c_n \dot{\theta}_3 \\
\vdots \\
c_n \dot{\theta}_{n-1} \\
1 \dot{\theta}_n
\end{pmatrix} = \begin{pmatrix}
\alpha_2 \\
\alpha_3 \\
\vdots \\
\alpha_{n-1} \\
\alpha_n
\end{pmatrix}. \tag{15}
$$

Alternatively, we can formulate differential equations for the c_j, s_j factors directly, $j = 2, \ldots, n$. We obtain the system

$$
\left(\begin{array}{c} c_3 \cdots c_n \begin{pmatrix} \dot{c}_2 \\ \dot{s}_2 \end{pmatrix} \\ \vdots \\ \begin{pmatrix} \dot{c}_n \\ \dot{s}_n \end{pmatrix} \end{array} \right) = \left(\begin{array}{c} \begin{pmatrix} 0 & -\alpha_2 \\ \alpha_2 & 0 \end{pmatrix} \begin{pmatrix} c_2 \\ s_2 \end{pmatrix} \\ \vdots \\ \begin{pmatrix} 0 & -\alpha_n \\ \alpha_n & 0 \end{pmatrix} \begin{pmatrix} c_n \\ s_n \end{pmatrix} \end{array} \right). \tag{16}
$$

Remark. Integration of (16) should be done with Gauss Runge-Kutta schemes so that the computed factors will be orthogonal at gridpoints, that is $c_j^2 + s_j^2 = 1$ there. Integration of (15), instead, can be done with explicit schemes.

For the general case of a matrix $Y \in \mathbb{R}^{n \times p}$, if feasible, we can repeat the above process on the second column for the reduced problem, and so forth, until a full triangularization is achieved. However, there is no guarantee that systems such as (15)-(16) are solvable. The problem is that we have fixed an ordering in which we apply the elementary rotators. On the other hand, we can guarantee that the c_j do not vanish, $j = 2, \ldots, n$, on an interval $[0, t_1)$ by suitably choosing the ordering in which we apply them. For example, let $\pi_1 = [1, k, 2, \ldots, k-1, k+1, \ldots, n]$ be the index array specifying the order in which we apply the rotators. That is, let $Q_1 = Q_{1,\pi_1(2)} \cdots Q_{1,\pi_1(n)}$, and get ICs for the elementary rotators $Q_{1,\pi_1(2)}, \ldots, Q_{1,\pi_1(n)}$ by requiring that the rotated vectors $Q_{1,\pi_1(2)}^T(0)\mathbf{y}(0)$ up to $Q_1^T(0)\mathbf{y}(0)$ all have positive first entry. By re-defining α_j in (14) as $\alpha_{\pi_1(j)}$, $j = 2, \ldots, n$, and with similar indices replacements in (15)-(16), we can perform the integration of (15)-(16) up to some value $h > 0$. Repeating this reasoning for the remaining columns of Y gives the following algorithm.

Algorithm $[0, h]$. Rotators.

INPUT: $t = 0$, initial conditions $Y = Y(0) \in \mathbb{R}^{n \times p}$, and $h > 0$.
- For $j = 1, \ldots, p$
  - Let $A = A(j : n, j : n)$, $\mathbf{y}(0) = Y(j : n, j)$
  - Let π_j be the index array as defined above
  - Let $Q_j = Q_{j,\pi_j(j+1)} \cdots Q_{j,\pi_j(n)}$
  - Find ICs for $Q_j(0)$ as explained above
  - Find $Q_j(t)$ by integrating (15)-(16) on $[0, h]$
  - $A(t) = Q_j^T A Q_j + \dot{Q}_j^T Q_j$, $Y(0) = Q_j^T(0)Y(0)$
- Endfor j

OUTPUT: Have $Q(h) = Q_1(h) \cdots Q_p(h)$ such that $Q^T(h)Y(h)$ is triangular, and $\tilde{A}(h) = Q^T(h)A(h)Q(h) + \dot{Q}^T(h)Q(h)$, which is in $A(h)$, is triangular in its leading (n, p) submatrix.

As it was the case for reflectors, even now there is no guarantee that a chosen initial ordering in which we apply the rotators will suffice to cover the whole interval $[0, t_f]$. But, even in this case, a suitable reimbedding strategy, leading to a piecewise smooth factorization of Y, has been studied and implemented in [7], to which we refer for details.

3   An Example

Here we present a simple example to show the relative performance of the new approaches based on reflectors/rotators. More examples, discussion of reimbedding, and a more thorough analysis of the relative performances of all methods is in [7].

Example 3.1 This is Example 6.1 of [5]. We have $A(t) = Q(t)\tilde{A}(t)Q^T(t) + \dot{Q}(t)Q^T(t)$, where $\tilde{A}(t) = diag(\lambda_1, \cos(t), -\frac{1}{2\sqrt{t+1}}, \lambda_4)$. Here, $\lambda_1 = 1$, $\lambda_4 = -10$ and

$$Q(t) = diag(1, Q_\beta(t), 1) \cdot diag(Q_\alpha(t), Q_\alpha(t)),$$

where

$$Q_\gamma(t) = \begin{pmatrix} \cos(\gamma t) & \sin(\gamma t) \\ -\sin(\gamma t) & \cos(\gamma t) \end{pmatrix},$$

and $\alpha = 1, \beta = \sqrt{2}$. By construction, the exact matrix $Q(t)$ such that $Q(0) = I$, triangularizing $A(t)$ is precisely the above $Q(t)$. We have considered the following six methods: (a) projected and unitary integrators based on direct integration of (4) (in case $p = n$), henceforth labeled PRK and GRK; (b) Reflectors approaches based on (8) or (10), henceforth labeled U and V methods; and (c) Rotators approaches based on (15) and (16), henceforth labeled ANGLE and C/S methods. In Table 1, we report on the global error at $t_f = 1$ for all of these methods, and different stepsizes. Integration was done with the explicit midpoint rule (a Runge-Kutta method of order 2) for PRK, U, and ANGLE methods, and with the implicit midpoint rule (the Gauss RK rule of order 2) for GRK, V, and C/S methods.

Table 1. Example 3.1: Global Errors			
Method	$h = 0.1$	$h = 0.01$	$h = 0.001$
PRK	2.14E-2	1.83E-4	1.82E-6
GRK	1.62E-2	1.49E-4	1.49E-6
U	7.45E-3	6.20E-5	6.22E-7
V	3.08E-3	2.81E-5	2.81E-7
ANGLE	1.03E-2	6.63E-5	6.56E-7
C/S	7.25E-3	5.10E-5	5.10E-7

Although our implementations are presently far from being refined or efficient, we would like to mention that the ANGLE method was observed to be by far the least expensive of all those considered, while the GRK method was considerably more expensive than all others. We suspect that this will be a general trend.

Acknowledgements

Work supported in part under NSF Grant DMS-9306412 and NSF Grant DMS-9505049.

References

1. Ascher, U., Mattheij, R. M. and Russell, R. D.: *Numerical Solution of Boundary Value Problems for ODEs*, Prentice-Hall, Englewood Cliffs, N.J. (1988).
2. Benettin, G., Galgani, L., Giorgilli, A. and Strelcyn, J. M.: "Lyapunov Exponents for Smooth Dynamical Systems and for Hamiltonian Systems; A Method for Computing All of Them. Part 1: Theory ", and " ...Part 2: Numerical Applications ", *Meccanica* **15** (1980), pp. 9-20, 21-30.
3. Calvo, M.P., Iserles, A. and Zanna, A.: "Numerical Solution of Isospectral Flows", Technical Report DAMTP 1995 NA03, Cambridge University (1995).
4. Dieci, L., Russell, R. D. and Van Vleck, E. S.: "Unitary Integrators and Applications to Continuous Orthonormalization Techniques," *SIAM J. Numer. Anal.* **31** pp. 261-281.
5. Dieci, L., Russell, R. D., Van Vleck, E. S.: "On the Computation of Lyapunov Exponents for Continuous Dynamical Systems," to appear in *SIAM J. Numer. Anal.*.
6. Dieci, L. and Van Vleck, E.S.: "Computation of a Few Lyapunov Exponents for Continuous and Discrete Dynamical Systems," *Applied Numerical Math.* **17** (1995), pp. 275-291.
7. Dieci, L. and Van Vleck, E.S.: "Computation of Orthonormal Factors for Fundamental Solution Matrices," preprint (1996).
8. Goldhirsch, I., Sulem, P.L. and Orszag, S.A.: "Stability and Lyapunov Stability of Dynamical Systems: a Differential Approach and a Numerical Method," *Physica D* **27** (1987), pp. 311-337.
9. Golub, G.H. and Van Loan, C.F.: *Matrix Computations*, 2nd edition, The Johns Hopkins University Press (1989).
10. Higham, D.: "Time-stepping and preserving orthonormality", Technical Report NA 161, University of Dundee (1995).

Algorithms for computing finite semigroups

Véronique Froidure and Jean-Eric Pin

Abstract. The aim of this paper is to present algorithms to compute finite semigroups. The semigroup is given by a set of generators taken in a larger semigroup, called the "universe". This universe can be for instance the semigroup of all functions, partial functions, or relations on the set $\{1, \ldots, n\}$, or the semigroup of $n \times n$ matrices with entries in a given finite semiring.

The algorithm produces simultaneously a presentation of the semigroup by generators and relations, a confluent rewriting system for this presentation and the Cayley graph of the semigroup. The elements of the semigroups are identified with the reduced words of the rewriting system.

We also give some efficient algorithms to compute the Green relations, the local subsemigroups and the syntactic quasi-order of subsets of the semigroup.

1 Introduction

There are a number of complete and efficient packages for symbolic computation on groups, such as CAYLEY or GAP. Comparatively, the existing packages for semigroups are much less complete. Computers were used for finding the number of distinct semigroups of small order [10] or to solve specific questions on semigroups [16], but the main efforts were devoted to packages dealing with transformation semigroups. Actually, the primary goal of these packages is to manipulate finite state automata and rational expressions, and thus semigroups were uniquely considered as transition semigroups of finite automata. The first such package [5], written in APL, relied on algorithms developed by Perrot [18,14]. Given a finite deterministic automaton, it produced the list of elements of its transition semigroup and the structure of the regular $\mathcal{D}$-classes, including the Schützenberger group and the sandwich matrix. A much more efficient version, AUTOMATE, was written in C by Champarnaud and Hansel [3]. This interactive package comprised extended functions to manipulate rational languages, but the semigroup part did not include the computation of the Schützenberger groups. Another package, AMORE, was developed in Germany under the direction of W. Thomas (in particular by A. Potthoff) and can be obtained by anonymous ftp at ftp.informatik.uni-kiel.de:pub/kiel/amore/amore.ps.gz. It is comparable to AUTOMATE, since it is written in C and is also primarily designed to manipulate finite automata. However, it includes the computation of all $\mathcal{D}$-classes (regular or not, but without the Schützenberger groups). One can also test whether the transition semigroup is aperiodic, locally trivial, etc. A much less powerful, but convenient package was implemented by Sutner [27] using *Mathematica*. Other algorithms to compute Green's relations in finite transformation semigroups were also proposed in [15].

From the semigroup point of view, the main drawback of these packages lies in their original conception. Semigroups are always considered as transformation semigroups, and the algorithms used in these packages heavily rely on this feature. Although similar algorithms were designed for semigroups of boolean matrices by Konieczny [12,13], no general purpose algorithm was proposed so far in the literature. However, even in theoretical computer science, semigroups do not always occur as transformation semigroups. For instance, semigroups of boolean matrices [19,6,7] and more generally semigroups of matrices over commutative semirings [25,26] occur quite frequently, and therefore there is a strong need for a semigroup package similar to the existing ones on group theory. As a first step towards this goal, we present in this paper a general purpose algorithm to compute finite semigroups. Only a part of this algorithm has been implemented so far, but the first results are quite promising.

2 Definitions

2.1 Semigroups

A *semigroup* is a set equipped with an internal associative operation which is usually written in a multiplicative form. A *monoid* is a semigroup with an identity element (usually denoted by 1). If S is a semigroup, S^1 denotes the monoid equal to S if S has an identity element and to $S \cup \{1\}$ otherwise. In the latter case, the multiplication on S is extended by setting $s1 = 1s = s$ for every $s \in S^1$. The *dual* of a semigroup S is the semigroup $\tilde{S}$ defined on the set S by the multiplication $x \cdot y = yx$. We refer the interested reader to [9,14,20,8,1] for more details on semigroup theory.

Example 1. Let T_n be the monoid of all functions from $\{1, \ldots, n\}$ into itself under the multiplication defined by $uv = v \circ u$. This monoid is called the monoid of all transformations on $\{1, \ldots, n\}$. A *transformation semigroup* is simply a subsemigroup of some T_n. It is a well-known fact that every finite semigroup is isomorphic to a transformation semigroup. This is the semigroup counterpart of the group theoretic result that every finite group is isomorphic to a permutation group.

Example 2. A *semiring* is a set K equipped with two operations, called respectively addition and multiplication, denoted $(s,t) \to s+t$ and $(s,t) \to st$, and an element, denoted 0, such that:

1. $(K, +, 0)$ is a commutative monoid,
2. K is a semigroup for the multiplication,

3. for all $s, t_1, t_2 \in K$, $s(t_1 + t_2) = st_1 + st_2$ and $(t_1 + t_2)s = t_1 s + t_2 s$,
4. for all $s \in K$, $0s = s0 = 0$.

Thus the only difference with a ring is that inverses with respect to addition may not exist. Given a semiring K, the set $K^{n \times n}$ of $n \times n$ matrices over K is naturally equipped with a structure of semiring. In particular, $K^{n \times n}$ is a monoid under multiplication defined by

$$(rs)_{i,j} = \sum_{1 \leq k \leq n} r_{i,k} s_{k,j}.$$

Besides the finite rings, like $\mathbb{Z}/n\mathbb{Z}$, several other finite semirings are commonly used in the literature. We first mention the *boolean semiring* $\mathbb{B} = \{0, 1\}$, defined by the operations $0 + 0 = 0$, $0 + 1 = 1 + 0 = 1 + 1 = 1$ and $1 \cdot 1 = 1$, $0 \cdot 0 = 0 \cdot 1 = 1 \cdot 0 = 0$. Let us also mention the semiring $\mathbb{Z}_n = \{0, 1, \ldots, n\}$, with the operations $\oplus$ and $\otimes$ defined by $s \oplus t = \min\{s + t, n\}$ and $s \otimes t = \min\{st, n\}$. Other examples include the *tropical semiring* $(\mathbb{N} \cup \{\infty\}, \min, +)$ [25].

A relation $\mathcal{R}$ on a semigroup S is *stable on the right* (resp. *left*) if, for every $x, y, z \in S$, $x \mathcal{R} y$ implies $xz \mathcal{R} yz$ (resp. $zx \mathcal{R} zy$). A relation is *stable* if it is stable on the right and on the left. A *congruence* is a stable equivalence relation. Thus, an equivalence relation $\sim$ on S is a congruence if and only if, for every $s, t \in S$ and $x, y \in S^1$, $s \sim t$ implies $xsy \sim xty$. If $\sim$ is a congruence on S, then there is a well-defined multiplication on the quotient set $S/\sim$, given by

$$[s][t] = [st]$$

where $[s]$ denotes the $\sim$-class of $s \in S$.

Given two semigroups S and T, a *semigroup morphism* $\varphi : S \to T$ is a map from S into T such that for all $x, y \in S$, $\varphi(xy) = \varphi(x)\varphi(y)$. *Monoid morphisms* are defined analogously, but of course, the condition $\varphi(1) = 1$ is also required.

A semigroup (resp. monoid) S is a *quotient* of a semigroup (resp. monoid) T if there exists a surjective morphism from T onto S. In particular, if $\sim$ is a congruence on a semigroup S, then $S/\sim$ is a quotient of S and the map $\pi : S \to S/\sim$ defined by $\pi(s) = [s]$ is a surjective morphism, called the *quotient morphism* associated with $\sim$.

Let S be a semigroup. A *subsemigroup* of S is a subset T of S such that $t, t' \in T$ implies $tt' \in T$. Subsemigroups are closed under intersection. In particular, given a subset A of S, the smallest subsemigroup of S containing A is called the subsemigroup of S *generated* by A. The *Cayley graph* of S (relative to A) is the graph $\Gamma(S, A)$ having S^1 as set of vertices, and for each vertex s and each generator a, an edge labeled by a from s to sa.

An element e of a semigroup S is *idempotent* if $e^2 = e$. If s is an element of a finite semigroup, the subsemigroup generated by s contains a unique idempotent and a unique maximal subgroup, whose identity is the unique idempotent. Thus s has a unique idempotent power, denoted s^ω.

A *zero* is an element 0 such that, for every $s \in S$, $s0 = 0s = 0$. It is a routine exercise to see that there is at most one zero in a semigroup.

2.2   Free semigroups

An *alphabet* is a finite set whose elements are *letters*. A *word* (over the alphabet A) is a finite sequence $u = (a_1, a_2, \ldots, a_n)$ of letters of A. The integer n is the *length* of the word and is denoted $|u|$. In practice, the notation $(a_1, a_2, \ldots, a_n)$ is shortened to $a_1 a_2 \cdots a_n$. The empty word, which is the unique word of length 0, is denoted by 1. The (concatenation) *product* of two words $u = a_1 a_2 \cdots a_p$ and $v = b_1 b_2 \cdots b_q$ is the word $uv = a_1 a_2 \cdots a_p b_1 b_2 \cdots b_q$. The product is an associative operation on words. The set of all words on the alphabet A is denoted by A^*. Equipped with the product of words, it is a monoid, with the empty word as an identity. It is in fact the free monoid on the set A. This means that A^* satisfies the following universal property: if $\varphi : A \to M$ is a map from A into a monoid M, there exists a unique monoid morphism from A^* into M that extends φ. This morphism, also denoted φ, is simply defined by $\varphi(a_1 \cdots a_n) = \varphi(a_1) \cdots \varphi(a_n)$.

Let Σ be a subset of $A^* \times A^*$, and let $\sim_\Sigma$ be the least congruence $\sim$ on A^* such that $u \sim v$ for every pair $(u, v) \in \Sigma$. The quotient monoid $A^*/\sim_\Sigma$ is called the *monoid presented* by (A, Σ). A pair $(u, v) \in \Sigma$ is often denoted $u = v$. For instance, the monoid presented by $(\{a, b\}, ab = ba)$ is isomorphic to $\mathbb{N}^2$, the free commutative monoid on two generators. Presentations of semigroups are defined in the same way. Given a presentation Σ, the *word problem* is to know whether two given words are equivalent modulo $\sim_\Sigma$.

From the algorithmic point of view, presentations are in general intractable. See [11] for a survey. For instance, it is an undecidable problem to know whether a finitely presented semigroup is finite or not. There also exist finite presented semigroups with an undecidable word problem. To avoid these difficulties, we will follow a different approach, that goes back to Sakarovitch [23,24,17]. Some definitions are in order to describe it in a precise way.

Let A be a totally ordered alphabet. The *lexicographic order* is the total order used in a dictionary. Formally, it is the order $\leq_{lex}$ on A^* defined by $u \leq_{lex} v$ if and only if u is a prefix of v or $u = pau'$ and $v = pbv'$ for some $p \in A^*$, $a, b \in A$ with $a < b$. In the *military order*, words are ordered by length and words of equal length are ordered according to the lexicographic order. Formally, it is the order $\leq$ on A^* defined by $u \leq v$ if and only if $|u| < |v|$ or $|u| = |v|$ and $u \leq_{lex} v$.

For instance, if $A = \{a, b\}$ with $a < b$, then $ababb \leq_{lex} abba$ but $abba < ababb$. The next proposition summarizes elementary properties of the order $\leq$. The proof is straightforward and omitted.

Proposition 1 *Let $u, v \in A^*$ and let $a, b \in A$.*

1. *If $u < v$, then $au < av$ and $ua < va$.*
2. *If $ua \leq vb$, then $u \leq v$.*

An important consequence of Proposition 1 is that $\leq$ is a *stable order* on A^*: if $u \leq v$, then $xuy \leq xvy$ for all $x, y \in A^*$.
A *reduction* is a mapping $\rho : A^* \to A^*$ satisfying the following conditions:

1. $\rho \circ \rho = \rho$,

2. For all $u \in A^*$ and $a \in A$, $\rho(ua) = \rho(\rho(u)a)$ and $\rho(au) = \rho(a\rho(u))$.

Condition (2) can be extended as follows.

Lemma 2 *Let ρ be a reduction on A^*. Then for all $u, v \in A^*$, $\rho(uv) = \rho(\rho(u)v) = \rho(u\rho(v))$.*

Proof. We prove the equality $\rho(uv) = \rho(\rho(u)v)$ by induction on $|v|$. If $v = 1$, the result follows from condition (1). If the result holds for v, then for every letter $a \in A$, the following equalities hold by (2)

$$\rho(uva) = \rho((uv)a) = \rho(\rho(uv)a) = \rho(\rho(\rho(u)v)a) = \rho(\rho(u)va).$$

Similarly the equality $\rho(uv) = \rho(u\rho(v))$ is proved by induction on $|u|$. □

The set $R = \rho(A^*)$ is called the set of *reduced words* for ρ. The next proposition shows how reductions can be used to define monoids.

Proposition 3 *Let ρ be a reduction on A^* and let R be the set of its reduced words. Then R, equipped with the multiplication defined by $u \cdot v = \rho(uv)$, is a monoid.*

Proof. If $u, v, w \in A^*$,

$$(u \cdot v) \cdot w = \rho(\rho(uv)w) = \rho(uvw) = \rho(u\rho(vw)) = u \cdot (v \cdot w)$$

and thus the multiplication is associative. We claim that $\rho(1)$ is the identity of the multiplication. If $r \in R$, then $r = \rho(u)$ for some $u \in A^*$. Therefore $r \cdot \rho(1) = \rho(\rho(u)\rho(1)) = \rho(\rho(u)1) = \rho(\rho(u)) = \rho(u) = r$ and similarly, $\rho(1) \cdot r = r$. □

Conversely, given a monoid M generated by a set A, there is a natural morphism $\pi : A^* \to M$ defined by $\pi(a) = a$. Define a mapping $\rho : A^* \to A^*$ by setting, for each $u \in A^*$,

$$\rho(u) = \min\{v \in A^* \mid \pi(v) = \pi(u)\}$$

where the minimum is taken with respect to the military order. Then ρ is a reduction, called the *military reduction*, and the elements of M can be identified with the reduced words for ρ. This reduction also gives a presentation for M.

Theorem 4 *The monoid M is presented on A by the set of relations $\{(u = \rho(u)) \mid u \in MA \setminus M\}$.*

Proof. Let $\Sigma = \{(u = \rho(u)) \mid u \in MA \setminus M\}$. First, since $\pi(u) = \pi(\rho(u))$ by definition, M satisfies all relations of Σ and thus $u \sim_\Sigma v$ implies $\pi(u) = \pi(v)$. We claim that $u \sim_\Sigma \rho(u)$ for every word $u \in A^*$. Since $1 \in M$, every word $u \in A^*$ admits a unique factorization of the form $u = p(u)s(u)$ where $p(u)$ is the maximal prefix of u belonging to M. We prove the claim by induction on the length n of $s(u)$. If $n = 0$, then $p(u) = u$, $u = \rho(u)$ and the claim is trivial. Assume the claim holds for n and let u be a word such that $s(u) = a_1 \cdots a_{n+1}$. Then $p(u)a_1 \cdots a_n \sim_\Sigma \rho(p(u)a_1 \cdots a_n)$ by induction, and thus $u \sim_\Sigma \rho(p(u)a_1 \cdots a_n)a_{n+1}$. Now, by definition of Σ, $\rho(p(u)a_1 \cdots a_n)a_{n+1} \sim_\Sigma \rho(\rho(p(u)a_1 \cdots a_n)a_{n+1}) = \rho(u)$. Therefore, $u \sim_\Sigma \rho(u)$, proving the claim. Now $\pi(u) = \pi(v)$ implies $\rho(u) = \rho(v)$ and thus $u \sim_\Sigma v$. Thus Σ is a presentation of M. $\square$

3   The main algorithm

A semigroup S will be given as a subsemigroup of a given semigroup U, called the *universe*, generated by a subset A. This universe can be for instance the semigroup T_n or the semigroup of n by n matrices over a given semiring. We require the following information on the universe:

- the type of the elements (arrays, matrices over a semiring, etc.),
- an algorithm to compute the product of two elements of U,
- an algorithm to test equality of two elements of U,
- the set of generators A.

Given a subset A of a universe U, our main algorithm computes the submonoid of U generated by A. It is a little simpler to deal with monoids, so this point of view will be adopted in this presentation, but it is fairly easy to modify our algorithm to obtain the semigroup generated by A. The result of our computation can be formalized as follows:

Input: A universe U, a subset A of U and a total order on A.

Output: The military reduction $\rho : A^* \to M$ defining the submonoid M of U generated by A, the list of elements of M (sorted in military order) and the Cayley graphs $\Gamma(M, A)$ and $\Gamma(\tilde{M}, A)$.

3.1   A simplified version

We first present a simplified version of our algorithm, which just produces the sorted list of elements of M and the rewriting system. As was explained above, the elements of M are identified with reduced words of A^*. To each element $u \in M$ is associated its value $\nu(u)$ in the universe U.

The following pseudocode is our basic algorithm. The set of generators is given as a totally ordered set $A = \{a_1 < \ldots < a_k\}$. The first element of the list of elements is the empty word 1. The successor of the word u in the list is denoted $Next(u)$. The variable $Last$ denotes the last element of the current list.

Let $u := 1$ and $Last := 1$.
while true
for $i := 1$ to k,
compute $\nu(ua_i)$;
if $\nu(ua_i)$ is new
$Next(Last) := ua_i$;
$Last := ua_i$;
else if $\nu(ua_i) = \nu(u')$ for some $u' < ua_i$, produce the rule $ua_i \to u'$;
if u has a successor, $u := Next(u)$
else break;

The algorithm works as follows. For each element u of the list being completed and for each generator $a \in A$, the value ua is computed. If this value is the value of some element u' already in the list, a rule $ua \to u'$ is produced. Otherwise, a new element ua is created. The main properties of our algorithm are given in the following proposition.

Proposition 5 *The list of elements of M produced by the algorithm is sorted for the military order and the rules are all of the form $u \to v$ with $v < u$.*

Proof. A state of the program is determined by the values of the triple $(u, Last, i)$ at the beginning of the **for** loop. In a given state $(u, Last, i)$, the output is a list of the form $(1, \ldots, u, \ldots, Last)$. We claim that the interval $(u, \ldots, Last)$ is sorted for the military order and that $Last < ua_i$. This property is trivially satisfied at the initial state $(1, 1, 1)$. Passing from state $(u, Last, i)$ to state $(u, Last, i+1)$ (resp. $(u, ua_i, i+1)$) lets the property invariant, since $ua_i < ua_{i+1}$. Passing from state $(u, Last, k)$ to state $(Next(u), Last, 1)$ (resp. $(Next(u), ua_k, 1)$) also lets the property invariant. Indeed the interval $(Next(u), \ldots, Last)$ is a subinterval of $(u, \ldots, Last)$ and $Last < ua_k$. Thus $(u, \ldots, Last)$ (resp. $(u, \ldots, Last, ua_k)$) is sorted. Furthermore, $u < Next(u) \le Last < ua_k$ by assumption. Therefore, either $|Next(u)| = |Last| = |u| + 1$ and then $Last < ua_k < Next(u)a_1$, or $|Next(u)| = |u|$ and $Last < ua_k < Next(u)a_1$, since $u < Next(u)$. This proves the claim and shows that the output is sorted. The second part of the proposition is clear. $\square$

We illustrate our algorithm on an example.

Example 3. Let $U = T_6$ and let $A = \{a, b\}$ be the set of generators given in the following table

	1	2	3	4	5	6
a	2	2	4	4	5	6
b	5	3	4	4	6	6

We first calculate the value of the empty word 1 and of the words $1a = a$ and $1b = b$. Next the value of aa is equal to the value of a. This produces the rule $aa \to a$. The value of ab is new and thus ab is a new element. Similarly ba is a new element, but $bb \to ba$. Thus ab and ba are the only elements of length

2. Next, we calculate in this order the values of aba, abb, baa and baa. The first value is new, but the other three are not and produce the following rules: $abb \rightarrow aba$, $baa \rightarrow ba$, $bab \rightarrow ba$. Therefore, aba is the unique element of length 3 created at this step. It remains to calculate the values of $abaa$ and $abab$, which give the rules $abaa \rightarrow aba$ and $abab \rightarrow aba$. Finally, the elements of the monoid are represented in the following table

	1	2	3	4	5	6
1	2	3	4	5	6	7
a	2	2	4	4	5	6
b	5	3	4	4	6	6
ab	3	3	4	4	6	6
ba	6	4	4	4	6	6
aba	4	4	4	4	6	6

and the rewriting rules are

$$aa \rightarrow a \qquad bb \rightarrow ba \qquad abb \rightarrow aba \qquad baa \rightarrow ba$$
$$bab \rightarrow ba \qquad abaa \rightarrow aba \qquad abab \rightarrow aba$$

3.2   The extended version

Some computations are redundant in this algorithm. For instance, in the previous example, the computation of abb could have been avoided, since the rule $bb \rightarrow b$ infers $abb \rightarrow ab$. This example is generic. Let u be an element of M, and let $u = as$, where a is the first letter of u. If, for some generator $b \in A$, the word sb is not reduced, the word ub will not be reduced. Furthermore, if $s \rightarrow t$, then $u = as \rightarrow at$.

Example 4. Applying this improvement on the previous example would reduce the set of rules to the following set

$$aa \rightarrow a \qquad bb \rightarrow ba \qquad bab \rightarrow ba \,.$$

Thus $M = (\{a, b\} \mid aa = a, bb = ba, bab = ba)$.

However, there is a price to pay : we need to compute not only the elements of the form ua (for $u \in M$ and $a \in A$), but also those of the form au. doing so, we compute the Cayley graphs $\Gamma(M, A)$ and $\Gamma(\tilde{M}, A)$. Although the computation of the latter graph seems to be unnecessary, this information is actually needed to compute the Green relations, the local subsemigroups and the syntactic quasi orders (see the next sections).

A description of the data structure used to represent elements of M is in order. For a non-empty word u, denote by $f(u)$ (resp. $\ell(u)$) its first (resp. last) letters and by $p(u)$ (resp. $s(u)$) its prefix (resp. suffix) of length $|u| - 1$. Each element u of M (recall that u is a reduced word of A^*) is represented by a pointer on the following data:

1. The value $\nu(u)$ (an element of U).
2. The letters $f(u)$ and $\ell(u)$.
3. The addresses of $p(u)$ and $s(u)$.
4. The address Next of the successor of u, the minimal word of the set

$$\{v \in A^* \mid v \text{ is reduced and } u < v\}$$

5. For each generator $a \in A$, the address of $\rho(ua)$ and a flag to indicate whether ua is reduced.
6. For each generator $a \in A$, the address of $\rho(au)$.
7. a flag to indicate whether all the fields have been filled.

Note that the word u itself is not stored in this data structure, but can be easily retrieved, since we know its first letter and the address of its prefix $p(u)$. The $|M|$ addresses are stored in a sufficiently large table T, of size at least $\frac{10}{9}|M|s$, where s is the size of an address. Fast access is ensured through open addressing using a standard double hashing technique [4, pp. 235–237]. Two hash functions $h, h' : U \to \mathbb{N}$ are given. Let $v \in U$ be an element to be searched in the table T. The slots $T[h(v)]$, $T[h(v) + h'(v)]$, $T[h + 2h'(v)]$, $T[h(v) + 3h'(v)]$, ... are probed in this order and their values are compared to v. The search terminates successfully if the value v is found and terminates unsuccessfully if an empty slot is found. In the latter case, v is stored in the empty slot.

Note that this data structure also gives a representation of the Cayley graphs $\Gamma(S, A)$ and $\Gamma(\tilde{S}, A)$. Indeed, for each element $u \in M$ and each generator $a \in A$, we have stored the addresses of $\rho(ua)$ and $\rho(au)$.

The number of relations, the number of elements and the maximal length of the reduced words are stored in global variables. Initially, all these variables are set to 0. We now give the details of the algorithm.

Initialization. The data corresponding to the empty word 1 are filled. The value $\nu(1)$ is the identity of U. The fields corresponding to $f(u)$, $\ell(u)$, $p(u)$ and $s(u)$ are irrelevant. The successor of 1 is the letter a_1. For each generator a_i, the value $\nu(a_i)$ is computed. If $\nu(a_i) = \nu(a_j)$ for some $j < i$, the generator a_i is eliminated, and the rule $a_i \to a_j$ is created. Similarly, if $\nu(a_i) = 1$, the identity of U, the generator a_i is eliminated, and the rule $a_i \to 1$ is created. Otherwise, the value $\nu(a_i)$ is stored, its address is given to the field $\rho(a_i)$ of 1 and the flag is set to "reduced". The fields $f(a_i)$, $\ell(a_i)$, $p(a_i)$ and $s(a_i)$ are also filled. The variables giving the number of relations and the number of elements are updated.

Without lost of generality, we may suppose that no generator has been eliminated during this initialization. Thus the list of elements is $(1, a_1, \ldots, a_k)$. The pseudocode of the main loop is listed below.

```
let u := a_1 and Last := a_k;
while true
let s := s(u) and p := p(u);
for i := 1 to k,
(Computation of ua_i)
```

if sa_i is reduced,

compute $\nu(ua_i)$ and search this value in the table;

if $\nu(ua_i) = \nu(u')$ for some $u' < ua_i$, produce the rule $ua_i \to u'$;

else

$T[Last].Next := ua_i$;

$f(ua_i) := f(u)$; $\ell(ua_i) := a_i$; $p(ua_i) := u$; $s(ua_i) := s(u)a_i$;

$Last := ua_i$;

else   (sa_i is not reduced)

let $t := \rho(sa_i)$;   (note that $t < sa_i$)

if $t = 1$, **then** $\rho(ua_i) := f(u)$

else if $p(t) = s$,   (necessarily, $\ell(t) < a_i$ in this case)

then let $\rho(ua_i) := \rho(u\ell(t))$; create the rule $ua_i \to \rho(u\ell(t))$.

else   (the last case, $p(t) < s$)

$\rho(ua_i) = \rho\Big(\rho\big(f(u)p(t)\big)\ell(t)\Big)$;

(*Computation of* $a_i u$)

if the data relative to $\rho(a_i p)$ is available,

$\rho(a_i u) := \rho(\rho(a_i p)\ell(u))$;

else compute $\nu(a_i u)$ and search this value in the table;

if u has a successor, $u := Next(u)$

else break;

Some comments on this algorithm are in order. One could prove, in the spirit of the proof of Proposition 5, that at state $(u, Last, i)$, the data relative to all elements of M occurring before u has been filled. We use this information to compute $\rho(ua_i)$ and $\rho(a_i u)$.

To compute $\rho(ua_i)$, we first check whether $s(u)a_i$ is reduced. Since $s(u) < u$ in the military order, the data corresponding to $s(u)$ has been filled and thus, $t = \rho(s(u)a_i)$ is known. If $s(u)a_i$ is reduced, then the value of ua_i is computed. If this value is the value of some element u' already in the list, a rule $ua_i \to u'$ is produced. Otherwise, a new element ua_i is created. If, on the other hand, $s(u)a_i$ is not reduced, then $t < s(u)a_i$. If $t = 1$, then $\rho(ua_i) = \rho(f(u)s(u)a_i) = \rho(f(u)\rho(s(u)a_i)) = \rho(f(u))$. If $t \neq 1$, Proposition 1 shows that $p(t) \leq s(u)$. If $p(t) = s(u)$, then necessarily, $\ell(t) < a_i$. Therefore, $\rho(uf(t))$ has been already computed and $\rho(ua_i) = \rho(f(u)s(u)a_i) = \rho(f(u)\rho(s(u)a_i)) = \rho(f(u)t) = \rho(f(u)p(t)\ell(t)) = \rho(f(u)s(u)\ell(t)) = \rho(f(u)f(t))$. Finally, if $p(t) < s(u)$, then $\rho(ua_i) = \rho(f(u)p(t)\ell(t))$. Now, since $p(t) < s(u)$, $f(u)p(t) < f(u)s(u) = u$ by Proposition 1 and thus the data relative to $\rho(f(u)p(t))$ has been computed. In particular, $\rho(\rho(f(u)p(t))\ell(t)) = \rho(ua_i)$ is already known.

The computation of $\rho(a_i u)$ is easier. Observe that $\rho(a_i u) = \rho(a_i p(u)\ell(u))$. However, the data relative to $\rho(a_i p(u))$ may or may not have been computed at this stage. If this data is available, $\rho(a_i u)$ is already known. If it is not, then then the value of $\rho(a_i u)$ is computed. If this value happens to be the value of a new element, the value $\nu(a_i u)$ is the only field of the data associated with $\rho(a_i u)$ which is filled. The other fields will be filled later in the algorithm. For instance, in the example below, the list of elements is $\{a, b, aa, ab, ba, bb, aab, aba, abb\}$ when the computation of $\rho(b \cdot ab)$ takes place. A new value is found, the matrix $\left(\begin{smallmatrix} 3 & 3 \\ 3 & 3 \end{smallmatrix}\right)$, but the reduced word attached to this value will be unknown until the computation of $\rho(ba \cdot b)$.

Example 5. Let U the semigroup of 2×2 matrices with entries in $\mathbb{Z}_3$ and let $A = \{a, b\}$, where

$$a = \begin{pmatrix} 1 & 0 \\ 2 & 1 \end{pmatrix} \qquad b = \begin{pmatrix} 1 & 1 \\ 0 & 2 \end{pmatrix}.$$

The semigroup S generated by A contains 11 elements.

$$a = \begin{pmatrix} 1 & 0 \\ 2 & 1 \end{pmatrix} \quad b = \begin{pmatrix} 1 & 1 \\ 0 & 2 \end{pmatrix} \quad aa = \begin{pmatrix} 1 & 0 \\ 3 & 1 \end{pmatrix} \quad ab = \begin{pmatrix} 1 & 1 \\ 2 & 3 \end{pmatrix}$$
$$ba = \begin{pmatrix} 3 & 1 \\ 3 & 2 \end{pmatrix} \quad bb = \begin{pmatrix} 1 & 3 \\ 0 & 3 \end{pmatrix} \quad aab = \begin{pmatrix} 1 & 1 \\ 3 & 3 \end{pmatrix} \quad aba = \begin{pmatrix} 3 & 1 \\ 3 & 3 \end{pmatrix}$$
$$abb = \begin{pmatrix} 1 & 3 \\ 2 & 3 \end{pmatrix} \quad bab = \begin{pmatrix} 3 & 3 \\ 3 & 3 \end{pmatrix} \quad aabb = \begin{pmatrix} 1 & 3 \\ 3 & 3 \end{pmatrix}.$$

Rewriting rules

$$\begin{array}{lll} aaa \to aa & baa \to ba & bba \to bab \\ bbb \to bb & aaba \to aba & abab \to bab. \end{array}$$

Note that bab is a zero of S (it is easy to modify our algorithm to search for a zero). Thus $S = (\{a, b\} \mid aaa = aa, baa = ba, bba = 0, bbb = bb, aaba = aba, abab = 0)$.

Compared to the first version, the advantage of this algorithm is to avoid a number of computations inside the semigroup U.

4   Green relations

Green's relations on a semigroup S are defined as follows [14,20]. If s and t are elements of S, we set

$$\begin{array}{ll} s \,\mathcal{L}\, t & \text{if there exist } x, y \in S^1 \text{ such that } s = xt \text{ and } t = ys, \\ s \,\mathcal{R}\, t & \text{if there exist } x, y \in S^1 \text{ such that } s = tx \text{ and } t = sy, \\ s \,\mathcal{J}\, t & \text{if there exist } x, y, u, v \in S^1 \text{ such that } s = xty \text{ and } t = usv. \\ s \,\mathcal{H}\, t & \text{if } s \,\mathcal{R}\, t \text{ and } s \,\mathcal{L}\, t. \end{array}$$

For finite semigroups, there is a convenient representation of the corresponding equivalence classes. The elements of a given $\mathcal{R}$-class (resp. $\mathcal{L}$-class) are represented in a row (resp. column). The intersection of an $\mathcal{R}$-class and an $\mathcal{L}$-class is an $\mathcal{H}$-class. Each $\mathcal{J}$-class is a union of $\mathcal{R}$-classes (and also of $\mathcal{L}$-classes). It is not obvious to see that this representation is consistent: it relies in particular on the fact that, in finite semigroups, the relations $\mathcal{R}$ and $\mathcal{L}$ commute. The presence of an idempotent in an $\mathcal{H}$-class is indicated by a star. One can show that each $\mathcal{H}$-class containing an idempotent e is a subsemigroup of S, which is in fact a group with identity e. Furthermore, all $\mathcal{R}$-classes (resp. $\mathcal{L}$-classes) of a given $\mathcal{J}$-class have the same number of elements.

${}^{*}a_1, a_2$	${}^{*}a_3, a_4$	a_5, a_6
b_1, b_2	${}^{*}b_3, b_4$	${}^{*}b_5, b_6$

A $\mathcal{J}$-class.

In this figure, each row is an $\mathcal{R}$-class and each column is an $\mathcal{L}$-class. There are 6 $\mathcal{H}$-classes and 4 idempotents. Each idempotent is the identity of a group of order 2.

A $\mathcal{J}$-class containing an idempotent is called *regular*. One can show that in a regular $\mathcal{J}$-class, every $\mathcal{R}$-class and every $\mathcal{L}$-class contains an idempotent.

The computation of the $\mathcal{R}$-classes is fairly easy. It follows from the observation that the $\mathcal{R}$-classes of a semigroup S generated by A are the strongly connected components of the Cayley graph $\Gamma(S, A)$. The $\mathcal{L}$-classes can be computed in a similar way from the graph $\Gamma(\tilde{S}, A)$ corresponding to the left action of A on S. This fact is actually used in AMORE to compute the non regular $\mathcal{R}$-classes. Since our algorithm computes the graphs $\Gamma(S, A)$ and $\Gamma(\tilde{S}, A)$, Tarjan's algorithm [28] can now be used to compute its strongly connected components.

5   Local subsemigroups

If e is an idempotent of a finite semigroup S, the set

$$eSe = \{ese \mid s \in s\}$$

is a subsemigroup of S, called the *local subsemigroup* associated with e. This semigroup is in fact a monoid, since e is an identity in eSe. Local semigroups play an important role in the classification of rational languages [21]. Their computation is based on the following elementary lemma:

Lemma 6 *For every idempotent e, $eSe = eS \cap Se$.*

Proof. The inclusion $eSe \subset eS \cap Se$ is clear. For the opposite inclusion, let $s \in eS \cap Se$. Then $s = er = te$ for some $r, t \in S$. Therefore $ese = e(er)e = ere = tee = te = s$ and thus $s \in eSe$. $\square$

This gives a simple way to compute the local semigroup associated with e. Indeed, eS (resp. Se) is simply the set of vertices reachable from e in the graph $\Gamma(S, A)$ (resp. $\Gamma(\tilde{S}, A)$). Again this computation can be achieved by standard algorithms [4].

6   Syntactic quasi orders

Let P be a subset of a monoid M. The *syntactic quasi-order of* P is the quasi-order $\leq_P$ on M defined by $u \leq_P v$ if and only if, for every $x, y \in M$,

$$xvy \in P \Longrightarrow xuy \in P$$

The associated congruence $\sim_P$, defined by $u \sim_P v$ if and only if $u \leq_P v$ and $v \leq_P u$, is called the *syntactic congruence* of P. The quotient semigroup $S(P) = S/\sim_P$ is called the *syntactic semigroup* of P. See [22] for more details.

The computation of $\leq_P$ can be achieved as follows. Consider the graph G whose vertices are the pairs $(u, v) \in M \times M$ and the edges are of the form $(ua, va) \to (u, v)$ or $(au, av) \to (u, v)$, for some $a \in A$. This graph has $|M|^2$ edges and at most $2|A||M|^2$ vertices. Observe that for every $u, v \in M$, $u \not\leq_P v$ if and only if there exist $x, y \in M$ such that $xuy \notin P$ and $xvy \in P$. In other words, $u \not\leq_P v$ if and only if the vertex (u, v) is reachable in G from a vertex in $\bar{P} \times P$ (where $\bar{P}$ denotes the complement of P). Therefore, the computation of $\leq_P$ can be reduced to standard graph algorithms as follows:

1. First compute the graph G. This is easy from the knowledge of $\Gamma(S, A)$ and $\Gamma(\breve{S}, A)$.
2. Label each vertex (u, v) as follows:

$$\begin{cases} (0, 1) & \text{if } u \notin P \text{ and } v \in P \\ (1, 0) & \text{if } u \in P \text{ and } v \notin P \\ (1, 1) & \text{otherwise} \end{cases}$$

3. Do a depth first search in G (starting from each vertex labeled by $(0, 1)$) and set to 0 the first component of the label of all visited vertices.
4. Do a depth first search in G (starting from each vertex labeled by $(0, 0)$ or $(1, 0)$) and set to 0 the second component of the label of all visited vertices.
5. The label of each vertex now encodes the syntactic quasi-order of P in the following way:

$$\begin{cases} (1, 1) & \text{if } u \sim_P v \\ (1, 0) & \text{if } u \leq_P v \\ (0, 1) & \text{if } v \leq_P u \\ (0, 0) & \text{if } u \text{ and } v \text{ are incomparable} \end{cases} .$$

7   Conclusion

We have given several algorithms to compute finite semigroups. Contrary to most of the algorithms used in existing packages, our algorithms do not assume that semigroups are given as transformation semigroups. It would be interesting to analyze in a precise way the complexity of the different algorithms used in the existing packages.

References

1. J. Almeida, *Finite semigroups and universal algebra*, Series in Algebra Vol 3, Word Scientific, Singapore, (1994).
2. J. J. Cannon, Computing the ideal structure of finite semigroups, *Numer. Math.* **18**, (1971), 254–266.
3. J.M. Champarnaud and G. Hansel, AUTOMATE, a computing package for automata and finite semigroups, *J. Symbolic Computation* **12**, (1991), 197–220.
4. T. H. Cormen, C. E. Leiserson and R. L. Rivest, *Introduction to Algorithms*, MIT Press and McGraw-Hill, (1990).
5. G. Cousineau, J. F. Perrot and J. M. Rifflet, APL programs for direct computation of a finite semigroup, *APL Congres 73, Amsterdam, North Holand Publishing Co.*, (1973), 67–74.
6. V. Froidure, *Rangs des relations binaires et Semigroupes de relations non ambigus*, Thèse, Univ. Paris 6, France, (1995).
7. V. Froidure, Ranks of binary relations, *Semigroup Forum*, to appear.
8. P.M. Higgins, *Techniques of Semigroup Theory*, Oxford Univ. Press, (1992).
9. J.M. Howie, *An Introduction to Semigroup Theory*, Academic Press, London, (1976).
10. H. Jürgensen, Computers in semigroups, *Semigroup Forum*, **15**, (1977), 1–20.
11. O.G. Kharlampovich and M.V. Sapir, Algorithmic problems on varieties, *International Journal of Algebra and Computation*,
12. J. Konieczny, *Semigroups of Binary Relations*, Ph.D. Thesis, State Univ. of Pennsylvania (1992).
13. J. Konieczny, Green's equivalences in finite semigroups of binary relations, *Semigroup Forum* **48**, (1994), 235–252.
14. G. Lallement, *Semigroups and Combinatorial Applications*, John Wiley & Sons, New York, (1979).
15. G. Lallement and R. McFadden, On the determination of Green's relations in finite transformation semigroups, *J. Symbolic Comput.* **10**, (1990), 481–498.
16. E. Lusk and R. McFadden, Using Automated Reasoning Tools; A study of the Semigroup F_2B_2, *Semigroup Forum* **36**, (1987), 75–87.
17. M. Pelletier and J. Sakarovitch, Easy multiplications II. Extensions of Rational Semigroups, *Information and Computation/* **88**, (1990), 18–59.
18. J. F. Perrot, *Contribution à l'étude des monoïdes syntactiques et de certains groupes associés aux automates finis*, Thèse de doctorat, Univ. de Paris, France, (1972).
19. R. J. Plemmons and M. T. West, On the semigroup of binary relations, *Pacific Jour. of Math.* **35**, (1970), 743–753.
20. J.-E. Pin, *Variétés de langages formels*, Masson, Paris, (1984). English translation: *Varieties of formal languages*, Plenum, New-York, (1986).
21. J.-E. Pin, Finite semigroups and recognizable languages : an introduction, in NATO Advanced Study Institute *Semigroups, Formal Languages and Groups*, J. Fountain (ed.), Kluwer academic publishers, (1995), 1–32.
22. J.-E. Pin, A variety theorem without complementation, *Izvestiya VUZ Matematika* **39**, (1995), 80–90. English version, *Russian Mathem. (Iz. VUZ)* **39**, (1995), 74–83.
23. J. Sakarovitch, Description des monoïdes de type fini, *Elektronische Informationsverarbeitung und Kybernetik EIK* **17** (1981), 417–434.
24. J. Sakarovitch, Easy multiplications I. The Realm of Kleene's Theorem, *Information and Computation/* **74**, (1987), 173–197.

25. I. Simon, On semigroups of matrices over the tropical semiring, *Informatique Théorique et Applications* **28**, (1994), 277–294.

26. H. Straubing, The Burnside problem for semigroups of matrices, in *Combinatorics on Words, Progress and Perspectives, L.J. Cummings (ed.)*, Acad. Press, (1983), 279–295.

27. K. Sutner, Finite State Machines and Syntactic Semigroups, *The Mathematica Journal* **2**, (1991), 78–87.

28. R. E. Tarjan, Depth first search and linear graph algorithms, *SIAM Journal of Computing* **1** (1972) 146–160.

Extended Grzegorczyk Hierarchy in the BSS Model of Computability

Jean-Sylvestre Gakwaya*

Abstract. In this paper, we give an extension of the Grzegorczyk Hierarchy to the BSS theory of computability which is a generalization of the classical theory. We adapt some classical results ([3, 4]) related to the Grzegorczyk hierarchy in the new setting.

1 Introduction

The hierarchy of Grzegorczyk was originally defined in the 50's ([3]). Since then, it has been studied in order to quantify syntactically the complexity of some computable function classes. In Rose's book ([4]), an other definition of the hierarchy was introduced. This definition, equivalent to the original one, leads to simplifications of some Grzegorczyk's results ([3]).

The paper of Blum, Shub and Smale (BSS for short), published in 1989 ([1]), introduced a model of computability based upon an idealization of the representability of the real numbers. This setting of computability deals essentially with arithmetical manipulations instead of bit operations.

The aim of this paper is to define the *Grzegorczyk Hierarchy* in this model and adapt some related results from original ones ([2], [3] and [4]). In enlarging the discrete setting, we could hope to give the mathematical tools to quantify complexity of problems over any commutative totally ordered sub-ring R of the real numbers $\mathbb{R}$.

For the arguments of the functions defined in this paper, we will use two types of symbols. Indeed, we need to distinguish the variables of type 1 which take their values over R (called parameters) from those of type 0 over $\mathbb{N}$ (called counters). More generally, the type of a set of values is defined as the maximum type of all the elements of this set.

In the following, the functions are defined on space such as $R^{n_1} \times \mathbb{N}^{n_2}$. We give, at the end of this paper, the required functions to extend function domains to $R^\infty \times \mathbb{N}^\infty$ ($= \bigoplus_{n \in \mathbb{N}} R^n \times \bigoplus_{n \in \mathbb{N}} \mathbb{N}^n$).

---

* This work is supported by a FRIA fellowship and partially by NeuroCOLT ESPRIT (Working Group Nr.8556).

The paper is organized as follows[2]:

Section 2 introduces, in the BSS setting, the definition of initial functions and some effective operations. In this framework of computation, we also define the class RP_R of the *(extended) primitive recursive functions*.

After the definition of particular classes of computable functions, the Grzegorczyk classes $\mathcal{E}_{\mathbb{N}}^n$, the third section gives the definition of the extended Grzegorczyk classes $\mathcal{E}_R^n$. A sequence of functions $\mathbb{E}_R^n$ verifying some growth properties is introduced to construct the extension mentioned above. In order to use the usual floor function $\lfloor\ \rfloor$, we enlarge $\mathcal{E}_R^n$ to an other class of functions: $\overline{\mathcal{E}}_R^n$.

The growth lemma and their proofs are the topics of the section 4. We show that each class $\overline{\mathcal{E}}_R^n$ contains functions with a particular construction which bounds all of the functions of $\overline{\mathcal{E}}_R^n$.

The theorem of section 5 proves that the extended Grzegorczyk classes $\mathcal{E}_R^n$ constitute a hierarchy of classes of primitive recursive functions called the Extended Grzegorczyk Hierarchy. Moreover, we show that the countable union of the Extended Grzegorczyk Classes $\overline{\mathcal{E}}_R^n$ coincides with $\overline{RP}_R$.

Under a certain assumption, some results of the sections 4 and 5 hold for the classes $\mathcal{E}_R^n$.

Section 6 mainly shows that the classes $\mathcal{E}_R^n$ are closed for a certain effective operation and that some closure properties of the relation classes associated to the classes $\mathcal{E}_R^n$ are verified.

The last section proves that, for a particular class of functions, the scheme of limited primitive recursion can be reduced to an other effective operation. We achieve this section by an extension of the notion of the primitive recursive functions on domains such as $R^\infty \times \mathbb{N}^\infty$.

2   Definitions

First, we introduce some notations for more convenience. Afterwards, we define some functions and operations effectively computable (in the BSS setting) leading to the definition of particular classes of computable functions: *the Extended Grzegorczyk Classes*.

An essential idea in this paper is to distinguish, in the arguments of the functions introduced in the BSS setting, the different types of variable. We note, to further notice, the *parameters* (variable of type 1) by the letters $\overline{u}, \overline{u}', x, y, z, x_i,$

---

[2] We assume that the reader has some knowledge in the BSS theory ([1]).

the *counters* (variables of type 0) by i, j, m, n, t, t', t_i and the *constants* in R by c.

Let's define the following variable $\overline{\alpha}$ which notes a n–tuple of *parameters* and/or *counters*.
Let $\overline{u} \in R^{n_1}$ and $\overline{t} \in \mathbb{N}^{n_2}$,

$$\alpha_i = \begin{cases} u_i & \text{for } i \in \{1, \ldots, n_1\} \\ t_i & \text{for } i \in \{n_1 + 1, \ldots, n_1 + n_2\}. \end{cases}$$

If $n_1 + n_2 = 1$, we note α, a variable of type *parameter* or *counter*.

The following maps are called *initial functions* :

- the *null function* $Z : R \to R : x \to 0$
- the *functions* $(x, \alpha) \to \alpha \cdot X(x) + c$ with the *characteristic function* X defined as

$$X(x) = \begin{cases} +1 & \text{if } x > 0 \\ -1 & \text{if } x < 0 \\ 0 & \text{otherwise} \end{cases}$$

  and $c \in R$.
- the *projection functions* $p_i^{n_1 + n_2} : R^{n_1} \times \mathbb{N}^{n_2} \to R : \overline{\alpha} \to \alpha_i$ for each $i \in \{1, \ldots, n_1 + n_2\}$

The class of the *initial functions* is called B.

The definitions of the operations "effectively computable" are nearly the same as in classical theory with the particularity that the (primitive) recursion is made on a natural variable that we call the *counter* variable. Thus, for the *scheme of primitive recursion* $(O_{p.r.})$:
let two functions g and h, one says that f is defined by the *scheme of primitive recursion* $(O_{p.r.})$ if this function f verifies the following equalities:

$$(S) \begin{cases} f(\overline{t_1}, \overline{u}, 0) = g(\overline{t_1}, \overline{u}) \\ f(\overline{t_1}, \overline{u}, t + 1) = h(\overline{t_1}, \overline{u}, t, f(\overline{t_1}, \overline{u}, t)). \end{cases}$$

We can distinguish, through the scheme (S), the *active* counter t (i.e. the current counter on which the recursion acts) from the other counters $(\overline{t_1})$.

One says that f verifies the *scheme of primitive recursion limited* ($O_{l.p.r.}$) by a function f' if this function f is defined by $O_{p.r.}$ and verifies the following condition:

$$\forall \overline{u} \; \forall \overline{t} \; |f(\overline{u}, \overline{t})| \le |f'(\overline{u}, \overline{t})|$$

in which the *absolute value function* $| \; |$ is defined as $|u| = u \cdot X(u)$. Let's remark that $| \; |$ is an initial function.

The *limited minimum operation* $(O_{l.\mu})$ is defined as $\mu(t \le t')(F(\overline{t_1}, \overline{u}, t) = 0)$ which is a notation for the least natural number t bounded by the natural t' such as $F(\overline{t_1}, \overline{u}, t)$ vanishes if it exists or 0 otherwise.

In the same way, we introduce the *extended limited minimum operation* $(O_{e.l.\mu})$ if t is below (the absolute value of) a variable of type *parameter* in R.

The *operation of substitution* $(O_s.)$ consists in substituting for a variable, a function f whose type of range (i.e. the maximum type of the values $f(\overline{u}, \overline{t})$) is less or equal to the type of this variable.

We recall the following notation $\langle f_1, \ldots, f_{n_1} ; O_{oper.1}, \ldots, O_{oper.n_2} \rangle$ as the smallest class including the functions $f_1, \ldots, f_{n_1}$ and closed for the operations $O_{oper.1}, \ldots, O_{oper.n_2}$

Finally, we define RP_R, the class of the *(extended) primitive recursive functions*, as

$$RP_R = \langle B, x + y, x \cdot y ; O_s., O_{p.r.} \rangle$$

which clearly extends the *classical primitive recursive functions* ([2, 3, 4]).

3   Construction of the extended Grzegorczyk Hierarchy

First, we recall the construction of a particular classes of computable functions, the *Grzegorczyk classes* ([3, 4]). Secondly, we introduce, as in the discrete model, a sequence of functions which verifies some growth properties. This sequence is used to build the Extended Grzegorczyk Classes $\mathcal{E}_R^n$.

The original Grzegorczyk classes are based on a sequence of functions E_n defined on *natural variables* and introduced by Rose ([4]):
$$E_0(t_1, t_2) = t_1 + t_2$$
$$E_1(t) = t^2 + 2$$
$$\text{for } n \ge 1 \begin{cases} E_{n+1}(0) = 2 \\ E_{n+1}(t+1) = E_n(E_{n+1}(t)) . \end{cases}$$

These functions verify some growth properties:

$$\forall n \ge 1 \; \forall t$$
$$t < E_n(t)$$
$$E_n(t) < E_n(t+1)$$
$$E_n(t) < E_{n+1}(t)$$
$$\forall m \ge 0 \; E_n^m(t) \le E_{n+1}(t+m) .$$

Let's $B_{\mathbb{N}}$, the subset of the class $\langle B ; O_s. \rangle$ which contains functions depending exclusively on natural variables. For example, the usual *successor function*

$t \to t + 1$ is in B. We can introduce the *Grzegorczyk classes* $\mathcal{E}_{\mathbb{N}}^n$ as defined in Rose ([4][3]):

$$\mathcal{E}_{\mathbb{N}}^0 = \langle B_{\mathbb{N}} \; ; \; O_{s.}, O_{l.p.r.} \rangle$$
$$\text{for } n \geq 1 \;\; \mathcal{E}_{\mathbb{N}}^n = \langle B_{\mathbb{N}}, E_0, E_{n-1} \; ; \; O_{s.}, O_{l.p.r.} \rangle .$$

We extend the Rose's definition ([4]) of the functions E_n into the BSS model framework:

$$\mathbb{E}_2(x, t) = (1 + |x|)^t$$
$$\text{for } n \geq 3 \;\; \mathbb{E}_n(x, t) = \mathbb{E}_2(x, E_n(t)) .$$

As $\forall x \in R \; \forall n_1, n_2 \in \mathbb{N} \quad (1 + |x|)^{n_1} \leq (1 + |x|)^{n_2}$, we see that the growth properties transfer from the functions $E_n(t)$ into the functions $\mathbb{E}_n(x, t)$:

$$\forall x \; \forall t \; \forall n \geq 2$$
$$\mathbb{E}_2(x, t) \leq \mathbb{E}_{n+1}(x, t)$$
$$\mathbb{E}_n(x, t) \leq \mathbb{E}_{n+1}(x, t)$$
$$\mathbb{E}_n(x, t) \leq \mathbb{E}_n(x, t + 1) .$$

We have, clearly,

$$\forall x_1, x_2 \in R \quad |x_1| \leq |x_2| \Rightarrow \mathbb{E}_{n+2}(x_1, t) \leq \mathbb{E}_{n+2}(x_2, t) .$$

Remark: for $x \neq 0$, $\mathbb{E}_n(x, t)$ is an increasing function.

Now, we define the *Extended Grzegorczyk Classes* $\mathcal{E}_R^n$:

$$\mathcal{E}_R^0 = \langle B \; ; \; O_{s.}, O_{l.p.r.} \rangle$$
$$\mathcal{E}_R^1 = \langle B, c_1 \cdot x + c_2 \cdot y \text{ with } c_1, c_2 \in R \; ; \; O_{s.}, O_{l.p.r.} \rangle$$
$$\mathcal{E}_R^2 = \langle B, x + y, x \cdot y \; ; \; O_{s.}, O_{l.p.r.} \rangle$$
$$\text{for } n \geq 2 \;\; \mathcal{E}_R^{n+1} = \langle B, x + y, x \cdot y, \mathbb{E}_n \; ; \; O_{s.}, O_{l.p.r.} \rangle.$$

Let's remark that the functions $(x, y) \to x - y$ and $c_1 \cdot x + c_2 \cdot y$ with $c_1, c_2 \in R$ are in each class $\mathcal{E}_R^n$ from $n = 1$. Indeed, the constant functions c belong to each class $\mathcal{E}_R^n$, so the function $x \to c \cdot x$ is in each class (from $n = 1$).

The link between the Grzegorczyk Hierarchy originally developed ([3, 4]) and its extended version introduced in this paper is clear:

$$\forall n \geq 0 \;\; \mathcal{E}_{\mathbb{N}}^n \subseteq \mathcal{E}_R^n \quad \text{(from the definitions)}$$
$$\mathcal{E}_{\mathbb{N}}^n = \mathcal{E}_{R_{|\mathbb{N}}}^n \quad \text{(since } \mathbb{N} \subset R\text{)} .$$

In the sequel, we need to extend the classes to have the usual *floor function* $\lfloor \; \rfloor$ defined as the function $R \to \mathbb{N} : x \to n$ such as $n \leq x < n + 1$. Let's remark

---

[3] The definition of the Grzegorczyk classes given by Rose is equivalent to the original definition given by Grzegorczyk ([3]) but simplify his original results.

that, for x fixed, such a n exists and is unique since R is a sub-ring of $\mathbb{R}$. For $\overline{x}$, a n-tuple, we define $\lfloor \overline{x} \rfloor = (\lfloor x_1 \rfloor, \cdots, \lfloor x_n \rfloor)$.

We put a bar on any class X ($\overline{X}$) to mean the closure of X by $O_{e.l.\mu}$. For example, $\overline{RP}_R = \langle B, x+y, x \cdot y \; ; \; O_s., O_{p.r.}, O_{e.l.\mu} \rangle$. (A straightforward result is $X \subseteq \overline{X}$).

With this definition, the *floor function* is definable in each class $\overline{\mathcal{E}}_R^n$, indeed, $\lfloor u \rfloor = \mu(t \leq |u|)(u \dotminus t = 0)$ where the *extended modified subtraction*

$$u \dotminus t \triangleq \begin{cases} 0 & \text{if } |u| - t < 0 \\ |u| - t & \text{otherwise} \end{cases}$$

is defined in $\overline{\mathcal{E}}_R^n$, for any $u \in R$, by $O_{l.p.r.}$:

$$\begin{cases} u \dotminus 0 = |u| \\ u \dotminus (t+1) = ((|u| \dotminus t) - 1) \cdot \chi(\chi(|u| \dotminus t) + 1) \,. \\ u \dotminus t \leq |u| \end{cases}$$

The classes $\mathcal{E}_R^n$ and $\overline{\mathcal{E}}_R^n$ are not essentially different because of their definition. The results established for $\mathcal{E}_R^n$ sometimes hold when they are "extended" to $\overline{\mathcal{E}}_R^n$. So, we mark with a ($\diamond$) the results of this paper concerning the classes $\mathcal{E}_R^n$ (or $\mathcal{RE}_R^n$ [4]) still true (with some slight modifications) for the classes $\overline{\mathcal{E}}_R^n$ (or $\mathcal{R}\overline{\mathcal{E}}_R^n$).

4   The growth lemma

The growth functions E_n have been introduced in the previous section. In the discrete model, the growth lemma says that these functions E_n give bounds in $\mathcal{E}_{\mathbb{N}}^n$ of functions of $\mathcal{E}_{\mathbb{N}}^n$ (see [2, 4]). The main result of this section shows that, in the BSS setting, we have similar results for the extended Grzegorczyk classes $\overline{\mathcal{E}}_R^n$. The same results hold for the classes $\mathcal{E}_R^n$ with $n \geq 3$ under a certain condition.

In the following, we use the function max on natural variables with the obvious meaning. This function can be easily shown to be definable in all of the classical Grzegorczyk classes $\mathcal{E}_R^n$ (except for $n = 0$), indeed,

$$max(t_1, t_2) = t_1 \dotminus t_2 + t_2$$

with $\dotminus$ defined as usual in the discrete model (for example, see [4]). In order to generalize the function max to variables of parameter type, we extend this function such as max consists in comparing the maximum of the absolute value of its arguments. We can define it in any extended Grzegorczyk classes $\mathcal{E}_R^n$ (except for $n = 0$):

---

[4] Let X, a class of functions. We define $\mathcal{R}X$ as the class of the relations $f(\overline{\alpha}) = 0$ with f, a function of X.

$\max(x_1, x_2) = (x_1 \ominus x_2) + |x_2|$ with

$$x_1 \ominus x_2 \triangleq \begin{cases} |x_1| - |x_2| & \text{if } |x_1| \geq |x_2| \\ 0 & \text{otherwise} \end{cases}$$

which is definable (in $\mathcal{E}_R^n$) as $(|x_1| - |x_2|) \cdot \mathcal{X}(1 + \mathcal{X}(|x_1| - |x_2|))$
By definition, max is definable in $\mathcal{E}_R^1$ (let's recall that the addition and subtract
functions ($x_1 + x_2$ and $x_1 - x_2$) are in $\mathcal{E}_R^1$)
Let's note that

$$\max(x_1, x_2, x_3) = \max(\max(x_1, x_2), x_3)$$
$$= (\max(x_1, x_2) \ominus x_3) + |x_3|$$

extends max to any n–tuple.

Let X_R (resp. $X_{\mathbb{N}}$) a class of BSS-computable functions defined over R (resp.
$\mathbb{N}$). We introduce a condition such that if it is verified, then any function of X_R
whose range is included in $\mathbb{N}$ has a parameter–independent bound in $X_{\mathbb{N}}$.
We note $hyp(X_R, R)$ the following formula:

$$\forall p \in X_R \ (\ (\forall \overline{u}\ p(\overline{u}) \in \mathbb{N}) \Rightarrow \exists f \in X_{\mathbb{N}}\ \forall \overline{u}\ \forall \overline{t}\ p(\overline{u}, \overline{t}) \leq f(\overline{t})\ ).$$

For more convenience and in order to emphasize the analogy between the growth
lemma in the discrete model and in the BSS setting, we define the following
function:

$$\mathbb{E}_{n-1}^{[m]}(x, t) = \mathbb{E}_2(x, E_{n-1}^m(t)).$$

Let's recall the classical *growth lemma* ([2, 4]):

- The bound of $\mathcal{E}_{\mathbb{N}}^0$ [5] has a "weak" growth $(t + c)$:
  $\forall f \in \mathcal{E}_{\mathbb{N}}^0\ \exists i, c_f \in \mathbb{N}\ \forall \overline{t}$
  $$f(\overline{t}) \leq t_i + c_f.$$

- $\mathcal{E}_{\mathbb{N}}^1$ is bounded linearly:
  $\forall f \in \mathcal{E}_{\mathbb{N}}^1\ \exists c_0, \ldots, c_n \in \mathbb{N}\ \forall \overline{t}$
  $$f(\overline{t}) \leq c_0 + c_1 \cdot t_1 + \ldots + c_n \cdot t_n.$$

- For $n \geq 2$, the iterations of the growth function E_n bound $\mathcal{E}_{\mathbb{N}}^n$:
  $\forall n \geq 2\ \forall f \in \mathcal{E}_{\mathbb{N}}^n\ \exists m_f\ \forall \overline{t}$
  $$f(\overline{t}) \leq E_{n-1}^{m_f}(\max(\overline{t})).$$

In the BSS setting, the properties of the functions E_n and $\mathbb{E}_n$ (see previous
section) imply a similar version of the classical growth lemma:

---

[5] i.e. any function f of $\mathcal{E}_{\mathbb{N}}^0$ is bounded by a function f' whose constants depend on the
function f.

Proposition 4.1 (growth lemma in the BSS setting)

a) $\forall f \in \mathcal{E}_R^0 \; \exists c_f \in R^{>0} \; \exists i \in \mathbb{N} \; \forall \overline{\alpha} \in R^n$

$$|f(\overline{\alpha})| \leq |\alpha_i| + c_f \quad (\diamond).$$

Each function of $\mathcal{E}_R^n$ is limited by a function of "weak" growth $(|\alpha_i| + c)$.

b) $\forall f \in \mathcal{E}_R^1 \; \exists c_0, \ldots, c_{n_1+n_2} \in R^{>0} \; \forall \overline{\alpha} \in R^{n_1} \times \mathbb{N}^{n_2}$

$$|f(\overline{\alpha})| \leq c_0 + c_1 \cdot |\alpha_1| + \ldots + c_{n_1+n_2} \cdot |\alpha_{n_1+n_2}| \quad (\diamond).$$

In an equivalent way:
$\forall f \in \mathcal{E}_R^1 \; \exists c_f \in R^{>0} \; \forall \overline{\alpha} \in R^{n_1} \times \mathbb{N}^{n_2}$

$$|f(\overline{\alpha})| \leq c_f \cdot \mathbb{E}_2(\max(\overline{\alpha}), 1) \quad (\diamond).$$

In other terms, the bound is linear.

c) $\forall f \in \mathcal{E}_R^2 \; \exists c_f \in R^{>0} \; \exists n_f \in \mathbb{N} \; \forall \overline{\alpha} \in R^n$

$$|f(\overline{\alpha})| \leq (c_f \cdot \mathbb{E}_2(\max(\overline{\alpha}, n_f)) \quad (\diamond).$$

Hence, any function of $\mathcal{E}_R^2$ is bounded by a polynomial map.

d) $\forall n \geq 3 \; \forall f \in \overline{\mathcal{E}}_R^n \; \exists c_f \in R^{>0} \; \exists m_f \in \mathbb{N} \; \forall \overline{u} \in R^{n_1} \; \forall \overline{t} \in \mathbb{N}^{n_2}$

$$|f(\overline{u}, \overline{t})| \leq \mathbb{E}_2(\mathbb{E}_{n-1}^{[m_f]}(c_f + \max(\overline{u}), \max(\overline{t}, \lfloor \overline{u} \rfloor)), 2).$$

Let's remark the analogy with the classical growth lemma.

e) *For $n \geq 3$: if* $\mathrm{hyp}(\mathcal{E}_R^n, R)$ *is true, we have:*
$\forall f \in \mathcal{E}_R^n \; \exists c_f \in R^{>0} \; \exists m_f \in \mathbb{N} \; \forall \overline{u} \in R^{n_1} \; \forall \overline{t} \in \mathbb{N}^{n_2}$

$$|f(\overline{u}, \overline{t})| \leq \mathbb{E}_2(\mathbb{E}_{n-1}^{[m_f]}(c_f + \max(\overline{u}), \max(\overline{t})), 2).$$

Remarks:

- The function $|u|$ belongs to each class $\mathcal{E}_R^n$ ($|u| = u \cdot \mathcal{X}(u)$)
- The function $\lfloor u \rfloor$ has been defined in section 2. By definition, for any element u in R, $\lfloor u \rfloor$ is non-negative and vanishes for $u \leq 0$
  Let's recall that $\lfloor \overline{u} \rfloor$ notes $(\lfloor u_1 \rfloor, \cdots, \lfloor u_n \rfloor)$
- As the growth lemma in the discrete model, in each result, the bound belongs to the class $\mathcal{E}_R^n$ (or $\overline{\mathcal{E}}_R^n$) concerned (see lemma 5.1 in the next section).

First, we establish some technical results concerning the bounding of particular functions. Afterwards, we prove the growth lemma.

The following result shows that the composition of functions, whose shape is $(c \cdot (1 + x))^n$ and relatively to x, with themselves, is bounded by a function with the same shape (i.e. $(c \cdot (1 + x))^n$).

Lemma 4.1 $\forall c_1, c_2 > 1 \; \forall n_1, n_2 \in \mathbb{N} \; \forall x \in R^{>0} \; \exists c \in R \; \exists n \in \mathbb{N} :$

$$(c_1 \cdot (1 + (c_2 \cdot (1 + x))^{n_2}))^{n_1} \leq (c \cdot (1 + x))^n$$

Proof.

$$
\begin{aligned}
1 + (c_2 \cdot (1+x))^{n_2} \quad & \leq (1 + c_2 \cdot (1+x))^{n_2} \\
& \leq ((1 + c_2) \cdot (1+x))^{n_2} \\
& \leq (c_2' \cdot (1+x))^{n_2} \text{ with } c_2' = 1 + c_2 \\
c_1 \cdot (1 + (c_2 \cdot (1+x))^{n_2}) \quad & \leq c_1 \cdot (c_2' \cdot (1+x))^{n_2} \\
& \leq (c_1 \cdot c_2' \cdot (1+x))^{n_2} \\
(c_1 \cdot (1 + (c_2 \cdot (1+x))^{n_2}))^{n_1} & \leq ((c_1 \cdot c_2' \cdot (1+x))^{n_2})^{n_1} \\
& \leq ((c_1 \cdot c_2') \cdot (1+x))^{n_2 \cdot n_1} \\
& \leq (c \cdot (1+x))^n \\
& \quad \text{with } c = c_1 \cdot c_2' \text{ and } n = n_2 \cdot n_1
\end{aligned}
$$

$\square$

If a function f verifies some conditions, then $\max(\alpha_1, f(\alpha_2))$ can be bounded by $f(\max(\alpha_1, \alpha_2))$. (Let's recall that α notes a counter or a parameter.)

Lemma 4.2 *Let $f(\alpha)$, a function which is injective, increasing and such as $\forall \alpha \geq 0 \; f(\alpha) \geq \alpha$.*
Hence, for any α_1, α_2, we have $\max(\alpha_1, f(\alpha_2)) \leq f(\max(\alpha_1, \alpha_2))$

Proof. Straightforward

$\square$

Let's remark that we can apply this result to $f(x) = (c \cdot (1 + |x|))^n$ (in the sequel, we can assume that $c > 1$) which clearly verifies the hypothesis mentioned above.

In the same way, by noticing that

$$
\begin{aligned}
\max(\alpha_1, \alpha_2, f(\alpha_3)) &= \max(\alpha_1, \max(\alpha_2, f(\alpha_3))) \\
& \leq \max(\alpha_1, f(\max(\alpha_2, \alpha_3))) \\
& \leq f(\max(\alpha_1, \alpha_2, \alpha_3))
\end{aligned}
$$

we can generalize this last result to any n–tuples:

$$\max(\overline{\alpha_1}, f(\max(\overline{\alpha_2}))) \leq f(\max(\overline{\alpha_1}, \overline{\alpha_2})).$$

The next result gives, from an upper bound for the function $f(\overline{u}, \overline{t})$, two particular upper bounds for the function $\lfloor f(\overline{u}, \overline{t}) \rfloor$.

Lemma 4.3 *Let $f(\overline{u}, \overline{t}) \in \mathcal{E}_R^n$, with $n \geq 3$. If $(c_f \cdot (\max(\overline{u}) + 1))^{E_{n-1}^{m_f}(\max(\overline{t}))}$ is an upper bound of f, then the following functions:*

$$(c_f' \cdot (\max(\lfloor \overline{u} \rfloor) + 1))^{E_{n-1}^{m_f}(\max(\overline{t}))} \text{ with } c_f' = \lfloor 2 \cdot c_f \rfloor + 1 \text{ and } E_{n-1}^m(\max(\lfloor \overline{u} \rfloor, t))$$

are upper bounds of $\lfloor f(\overline{u}, \overline{t}) \rfloor$

Proof. The following computations are straightforward:

$\lfloor x \rfloor \leq |x| < \lfloor x \rfloor + 1$

$\lfloor x \rfloor + 1 \leq |x| + 1 < \lfloor x \rfloor + 2 < 2 \cdot (\lfloor x \rfloor + 1) .$

In the sequel, we can assume that $c > 1$

We have $\quad (1 <)c \cdot (\lfloor x \rfloor + 1) < c \cdot (|x| + 1) < 2 \cdot c \cdot (\lfloor x \rfloor + 1) .$
So $(c \cdot (\lfloor x \rfloor + 1))^n < (c \cdot (|x| + 1))^n < (2 \cdot c \cdot (\lfloor x \rfloor + 1))^n .$

Let $f(\overline{u}, \overline{t}) \in \mathcal{E}_R^n$ and $(c_f \cdot (\max(\overline{u}) + 1))^n$ (with $n = E_{n-1}^{m_f}(\max(\overline{t}))$) a bound of f.

Clearly, with $c'_f = \lfloor 2 \cdot c_f \rfloor + 1$ and $n = E_{n-1}^{m_f}(\max(\overline{t}))$,
we have $(c_f \cdot (\max(\overline{u}) + 1))^n < (c'_f \cdot (\max(\lfloor \overline{u} \rfloor) + 1))^n$ which is an upper bound of $f(\overline{u}, \overline{t})$.

As a consequence, the following function: $(\overline{n'}, t) \rightarrow (c' \cdot (\max(\overline{n'}) + 1))^t$ is a function of $\mathcal{E}_{R_{|\mathbb{N}}}^n (= \mathcal{E}_{\mathbb{N}}^n)$ for $n \geq 3$. So, by the growth lemma in the discrete setting (see the beginning of this section),
$\exists m$ such as $(c' \cdot (\max(\overline{n'}) + 1))^t \leq E_{n-1}^m(\max(\overline{n'}, t))$.

Hence, there is a m such as

$$\lfloor f(\overline{u}, \overline{t}) \rfloor \leq (c'_f \cdot (\max(\lfloor \overline{u} \rfloor) + 1))^{E_{n-1}^{m_f}(\max(\overline{t}))}$$
$$\leq E_{n-1}^m(\max(\lfloor \overline{u} \rfloor, \overline{t})) .$$

$\square$

The following result shows that any function in $\overline{\mathcal{E}}_R^n$ can be defined by functions included to $\mathcal{E}_R^n$ and by the floor function $\lfloor \ \rfloor$.
Let's recall that the *extended modified subtraction function* $\dot{-}$ is definable in any class $\mathcal{E}_R^n$.

Lemma 4.4 *Let $\lfloor \ \rfloor$, the usual floor function.*
$\forall n \geq 0 \quad f \in \overline{\mathcal{E}}_R^n$ *iff f is definable by the floor function $\lfloor \ \rfloor$, the functions and operations of $\mathcal{E}_R^n$.*

Proof.
$\Rightarrow$: any function $f(\overline{u}, x)$ defined as $\mu(t \leq |x|)(F(\overline{u}, t) = 0)$ can be defined as $\mu(t \leq \lfloor x \rfloor)(F(\overline{u}, t) = 0)$ $\quad$ (by def. of $\lfloor x \rfloor$)

$\Leftarrow$:
firstly, $\lfloor x \rfloor = \mu(t \leq |x|)(x \dot{-} t = 0)$
secondly, type of $\lfloor x \rfloor = 0 \leq$ type of y $(= 1)$
so, we can define $O_{e.l.\mu}$ by O_s. and $O_{l.\mu}$:

$$\mu(t \le \lfloor x \rfloor)(F(\overline{u},t) = 0)$$
$$= \mu(t \le \lfloor x \rfloor)(F(\overline{u},t) = 0)$$
$$= \mu(t \le (\mu(t \le |x|)(x \div t = 0)))(F(\overline{u},t) = 0)$$

□

Now, we prove the main result of this section

Proof of the proposition 4.1 (*growth lemma in the BSS setting*).

Idea: we show each point of the proposition by induction on the complexity of the function. As there are no difficulties for the initial functions (and the functions $\mathbb{E}_{n-1}$) and that it is straightforward for the functions defined by $O_{l.p.r.}$ (and $O_{l.\mu}$) , we will mainly be interested in the operation O_s. (and $O_{e.l.\mu}$).

a) and *b)* are straightforward, we take interest in *c)*, *d)* and *e)*

c) Let's assume that the result is verified for the functions $f_1(\overline{\alpha_1})$ and $f_2(\overline{\alpha_2})$:

$$|f_1(\ldots,\alpha_{1i},\ldots)| \le (c_{f_1} \cdot (1 + \max(\overline{\alpha_1})))^{n_{f_1}}$$
$$|f_2(\overline{\alpha_2})| \le (c_{f_2} \cdot (1 + \max(\overline{\alpha_2})))^{n_{f_2}}$$

We substitute $f_2(\overline{\alpha_2})$ to α_{1i}

Let $\tilde{f}_2(\alpha) = (c_{f_2} \cdot (1 + \alpha))^{n_{f_2}}$
(Clearly, $|f_2(\overline{\alpha})| \le |\tilde{f}_2(\max(\overline{\alpha}))|$)

Hence,
$$|f_1(\ldots,f_2(\overline{\alpha_2}),\ldots)| \le (c_{f_1} \cdot (1 + \max(\overline{\alpha_1}, f_2(\overline{\alpha_2}))))^{n_{f_1}}$$
$$\le (c_{f_1} \cdot (1 + \max(\overline{\alpha_1}, \tilde{f}_2(\max(\overline{\alpha_2})))))^{n_{f_1}}$$
$$\le (c_{f_1} \cdot (1 + \tilde{f}_2(\max(\overline{\alpha_1},\overline{\alpha_2}))))^{n_{f_1}} \text{ (by lemma 4.2)}$$
$$\le ((c_{f_1} \cdot c'_{f_2}) \cdot (1 + \max(\overline{\alpha_1},\overline{\alpha_2})))^{n_{f_1} \cdot n_{f_2}} \text{ (by lemma 4.1)}$$
$$\text{with } c'_{f_2} = c_{f_2} + 1$$
$$\le c \cdot (1 + \max(\overline{\alpha_1},\overline{\alpha_2}))^{n}$$
$$\text{with } c = (c_{f_1} \cdot c'_{f_2})^n \text{ and } n = n_{f_1} \cdot n_{f_2}$$
$$\le c \cdot \mathbb{E}_2(\max(\overline{\alpha_1},\overline{\alpha_2}), n)$$

□

In the sequel, we prove the following results:

d') $\forall n \ge 3 \forall f \in \overline{\mathcal{E}}_R^n \; \exists c_f$

$$|f(\overline{u},\overline{t})| \le (c_f \cdot (\max(\overline{u}) + 1))^{E_{n-1}^{m_f}(\max(\lfloor u \rfloor,\overline{t}))}$$

e') if, for $n \ge 3$, $hyp(\mathcal{E}_R^n, R)$ is true,

$$\forall f \in \mathcal{E}_R^n \; \exists c_f \; \exists m_f \; |f(\overline{u},\overline{t})| \le (c_f \cdot (\max(\overline{u}) + 1))^{E_{n-1}^{m_f}(\max(\overline{t}))}$$

The growth properties of the functions E_n and $\mathbb{E}_n$ (see section 3) imply the results $d)$ and $e)$ since the function E_{n-1} is in the class $\mathcal{E}_R^n$ (cf lemma 5.1 in the next section) and the following computation holds:

$$
\begin{aligned}
(c \cdot (\max(\overline{u}) + 1))^{E_{n-1}^m(\max(\overline{t}))} &= c^{E_{n-1}^m(\max(\overline{t}))} \cdot (\max(\overline{u}) + 1)^{E_{n-1}^m(\max(\overline{t}))} \\
&\leq \mathbb{E}_2(c, E_{n-1}^m(\max(\overline{t}))) \cdot \mathbb{E}_2(\max(\overline{u}), E_{n-1}^m(\max(\overline{t}))) \\
&\leq (\mathbb{E}_2(c + \max(\overline{u}), E_{n-1}^m(\max(\overline{t}))))^2 \\
&\leq \mathbb{E}_2(\mathbb{E}_{n-1}^{[m]}(c + \max(\overline{u}), \max(\overline{t})), 2)
\end{aligned}
$$

Proof of d')

Let the function $f(\overline{u}, \overline{t}) \in \overline{\mathcal{E}}_R^n$. We assume that $\lfloor \ \rfloor$ has been used at least once in the construction of f. So, there are two functions in $\overline{\mathcal{E}}_R^n$, $f_1(\overline{u_1}, \overline{t_1}, t)$ and $f_2(\overline{u_2}, \overline{t_2})$, such as $f(\overline{u}, \overline{t}) = f_1(\overline{u_1}, \overline{t_1}, \lfloor f_2(\overline{u_2}, \overline{t_2}) \rfloor)$ with $\overline{u_1}, \overline{u_2}$ (resp. $\overline{t_1}, \overline{t_2}$) containing components of $\overline{u}$ (resp. $\overline{t}$) (cf lemma 4.4)

We have, by hypothesis,

$$
|f_1(\overline{u_1}, \overline{t_1}, t)| \leq (c_{f_1} \cdot (\max(\overline{u_1}) + 1))^{E_{n-1}^{m_{f_1}}(\max(\overline{t_1}, t, \lfloor \overline{u_1} \rfloor))}
$$
$$
|f_2(\overline{u_2}, \overline{t_2})| \leq (c_{f_2} \cdot (\max(\overline{u_2}) + 1))^{E_{n-1}^{m_{f_2}}(\max(\overline{t_2}, \lfloor \overline{u_2} \rfloor))}
$$

By lemma 4.3,

$$
\begin{aligned}
\exists m, c'_{f_2} \quad \lfloor f_2(\overline{u_2}, \overline{t_2}) \rfloor &\leq (c'_{f_2} \cdot (\max(\lfloor \overline{u_2} \rfloor) + 1))^{E_{n-1}^{m_{f_2}}(\max(\overline{t_2}, \lfloor \overline{u_2} \rfloor))} \\
&\leq E_{n-1}^m(\max(\lfloor \overline{u_2} \rfloor, \overline{t_2})) \quad (\star)
\end{aligned}
$$

We compute a bound of $E_{n-1}^{m_{f_1}}(\max(\overline{t_1}, t, \lfloor \overline{u_1} \rfloor))$:

$$
\begin{aligned}
E_{n-1}^{m_{f_1}}(\max(\overline{t_1}, t, \lfloor \overline{u_1} \rfloor)) &= E_{n-1}^{m_{f_1}}(\max(\overline{t_1}, \lfloor \overline{u_1} \rfloor, \lfloor f_2(\overline{u_2}, \overline{t_2}) \rfloor)) \\
&\stackrel{(\star)}{\leq} E_{n-1}^{m_{f_1}}(\max(\overline{t_1}, \lfloor \overline{u_1} \rfloor, E_{n-1}^m(\max(\lfloor \overline{u_2} \rfloor, \overline{t_2})))) \\
&\leq E_{n-1}^{m+m_{f_1}}(\max(\overline{t_1}, \lfloor \overline{u_1} \rfloor, \lfloor \overline{u_2} \rfloor, \overline{t_2})) \quad (\star\star)
\end{aligned}
$$

So,

$$
\begin{aligned}
|f_1(\overline{u_1}, \overline{t_1}, t)| &= |f_1(\overline{u_1}, \overline{t_1}, \lfloor f_2(\overline{u_2}, \overline{t_2}) \rfloor)| \\
&\leq (c_{f_1} \cdot (\max(\overline{u_1}) + 1))^{E_{n-1}^{m_{f_1}}(\max(\overline{t_1}, t, \lfloor \overline{u_1} \rfloor))} \\
&\quad \text{with } t = \lfloor f_2(\overline{u_2}, \overline{t_2}) \rfloor \\
&\stackrel{(\star\star)}{\leq} (c_{f_1} \cdot (\max(\overline{u_1}) + 1))^{E_{n-1}^{m+m_{f_1}}(\max(\overline{t_1}, \lfloor \overline{u_1} \rfloor, \lfloor \overline{u_2} \rfloor, \overline{t_2}))}
\end{aligned}
$$

Hence

$$
|f(\overline{u}, \overline{t})| \leq (c_{f_1} \cdot (\max(\overline{u}) + 1)))^{E_{n-1}^{m+m_f}(\max(\overline{t}, \lfloor \overline{u} \rfloor))}
$$

$\square$

Let's remark that if $hyp(\mathcal{E}_R^n, R)$ is not verified, the previous result and the lemma 4.3 give some upper bounds which are, essentially, in $\mathcal{E}_{\mathbb{N}}^n$.

Proof of e'). We assume that $hyp(\mathcal{E}_R^n, R)$ is true.
Let the following functions $f(\overline{y}, \overline{t})$ and $g(\overline{x}, \overline{t'})$, both verifying the result:

$$|f(\overline{y}, \overline{t})| \le (c_f \cdot (\max(\overline{y}) + 1))^{E_{n-1}^{m_f}(\max(\overline{t}))}$$
$$|g(\overline{x}, \overline{t'})| \le (c_g \cdot (\max(\overline{x}) + 1))^{E_{n-1}^{m_g}(\max(\overline{t'}))}$$

Let's consider that we substitute $g(\overline{x}, \overline{t'})$ to one of the components of $\overline{y}$ or $\overline{t}$. Thus we have two cases according to the type of the substituted variable.

1st case: $t_i = g(\overline{x}, \overline{t'})$
As a consequence of the lemma growth in the discrete setting and by $hyp(\mathcal{E}_R^n, R)$:
$\exists m_g \ t_i = g(\overline{x}, \overline{t'}) \le E_{n-1}^{m_g}(\max(\overline{t'}))$
Hence $E_{n-1}^{m_f}(\max(\overline{t})) \le E_{n-1}^{m_f}(\max(t_1, \dots, E_{n-1}^{m_g}(\max(\overline{t'})), \dots, t_n))$
$$\le E_{n-1}^{m_f}(E_{n-1}^{m_g}(\max(\overline{t}, \overline{t'})))$$
$$\le E_{n-1}^{m_f + m_g}(\max(\overline{t}, \overline{t'}))$$
So, $|f(\overline{y}, \overline{t})| = |f(\overline{y}, t_1, \cdots, g(\overline{x}, \overline{t'}), \cdots, t_n)|$
$$\le (c_f \cdot (\max(\overline{y}) + 1))^{E_{n-1}^{m_f + m_g}(\max(\overline{t}, \overline{t'}))}$$

2nd case: $y_i = g(\overline{x}, \overline{t'})$, similar to the proof of c)

$\square$

5 The Extended Grzegorczyk Hierarchy

First, we establish a theorem showing that the extended Grzegorczyk classes $\mathcal{E}_R^n$ constitute a strict hierarchy called the Extended Grzegorczyk Hierarchy. Afterwards, we prove that the class $\overline{RP}_R$ of primitive recursive functions coincides with the countable union of the extended Grzegorczyk classes $\overline{\mathcal{E}}_R^n$.

The main results of the section are proved after some technical lemma.

The following result proves that the growth functions E_{n_1} used to build the classical Grzegorczyk classes are in the Extended Grzegorczyk Classes $\mathcal{E}_R^{n_2}$ with $n_1 < n_2$

Lemma 5.1 $\forall n \ge 1 \ \ i < n \Rightarrow E_i \in \mathcal{E}_R^n$ $(\diamond)$

Proof. We assume $n \ge 1$:

- for $i = 0$, $E_0(t_1, t_2) = t_1 + t_2 \in \mathcal{E}_R^n$ (in fact, $t_1 + t_2$ is in $\mathcal{E}_R^1$)

- for $i = 1$, $E_1(t) = t^2 + 2$ and the function $(x \cdot t_1) \cdot t_2$ is in $\mathcal{E}_R^n$ (by $O_{l.p.r.}$) hence $E_1 \in \mathcal{E}_R^n$ (in fact, in $\mathcal{E}_R^2$).
- for $i = 2$,

$$\begin{cases} E_2(0) &= 2 \\ E_2(t+1) &= E_1(E_2(t)) \end{cases}$$

$$E_2(t) \leq (1+1)^{E_2(t)}$$
$$= 2^{E_2(t)}$$
$$\leq 2^{E_{n-1}(t)}$$
$$= \mathbb{E}_{n-1}(1,t)$$

hence $E_2 \in \mathcal{E}_R^n$ (in fact, in $\mathcal{E}_R^3$).

- (by induction on i) let $E_i \in \mathcal{E}_R^n$ with $i < n-1$ and let's show that $E_{i+1} \in \mathcal{E}_R^n$. By definition,

$$\begin{cases} E_{i+1}(0) &= 2 \\ E_{i+1}(t+1) &= E_i(E_{i+1}(t)) \end{cases}$$

$$E_{i+1}(t) \leq (1+1)^{E_{i+1}(t)}$$
$$= 2^{E_{i+1}(t)}$$
$$\leq 2^{E_{n-1}(t)}$$
$$= \mathbb{E}_{n-1}(1,t)$$

hence, $E_{i+1} \in \mathcal{E}_R^n$

$\square$

We show that, with the previous lemma, for any $n \geq 3$, $\mathcal{E}_R^n$ contains the growth functions $\mathbb{E}_m$, for any m in $\{2, \ldots, n-1\}$

Lemma 5.2

$$\forall n \geq 3 \quad m \in \{2, \ldots, n-1\} \Rightarrow \mathbb{E}_m \in \mathcal{E}_R^n \quad (\diamond)$$

Proof. (We assume $n \geq 3$)

Firstly, by lemma 5.1, $\forall m \in \{2, \ldots, n-1\}$, the growth functions E_m are in $\mathcal{E}_R^n$.

Secondly, we can show that

$$\forall n \geq 3 \quad x^t \in \mathcal{E}_R^n$$

indeed, the function x^t is definable by $O_{l.p.r}$ in the following way:

$$\begin{cases} x^0 &= 1 \\ x^{t+1} &= x^t \cdot x \quad \text{(since mult.} \in \mathcal{E}_R^n \text{ for } n \geq 2) \end{cases}$$
$$|x^t| \leq (1+|x|)^t$$
$$= \mathbb{E}_2(x,t)$$
$$\leq \mathbb{E}_{n-1}(x,t) \quad \text{(by growth properties, see section 3)}$$

hence $x^t \in \mathcal{E}_R^n$ ($n \geq 3$)

So, for $m \geq 2$, $(1+|x|)^t = \mathbb{E}_2(x,t) \in \mathcal{E}_R^n$

and for $m \in \{3, \ldots, n-1\}$ $(1+|x|)^{E_m(t)} = \mathbb{E}_m(x,t) \in \mathcal{E}_R^n$ since both functions x^t and $E_m(t)$ are in $\mathcal{E}_R^n$ (lemma 5.1)

$\square$

The following result gives a particular bound in $\mathcal{E}_{\mathbb{N}}^3$ (thus, in $\mathcal{E}_R^3$) for the function $t_1^{t_2}$

Lemma 5.3 $\forall t_1, t_2$ $\quad t_1^{t_2} \leq E_2^2(\max(t_1, t_2))$

Proof.
The following computations are clear:

- $t_1^{t_2} = 2^{\log_2(t_1^{t_2})} = 2^{t_2 \cdot \log_2(t_1)} \leq 2^{t_2 \cdot t_1}$
- $2^{t_2 \cdot t_1} \leq 2^{(\max(t_1, t_2))^2 + 2} = 2^{E_1(\max(t_1, t_2))} \leq 2^{E_2(\max(t_1, t_2))}$
- $E_2(n) = E_1^n(2)$
  $$= (\underbrace{(\cdot^2 + 2) \circ \ldots \circ (\cdot^2 + 2)}_{n \text{ iterations}})(2)$$
  $$\geq (\underbrace{(2 \times \cdot) \circ \ldots \circ (2 \times \cdot)}_{n \text{ iterations}})(2) \quad \text{since } \cdot^2 + 2 \geq 2 \times \cdot$$
  $$\geq 2^n$$

These three inequalities imply the following computation:

$$t_1^{t_2} \leq 2^{t_2 \cdot t_1}$$
$$\leq 2^{E_1(\max(t_1, t_2))}$$
$$\leq E_2(E_1(\max(t_1, t_2)))$$
$$\leq E_2^2(\max(t_1, t_2))$$

$\square$

The next result implies that for $n \neq 1$, $\overline{\mathcal{E}}_R^{n+1}$ is the closure of the class $\overline{\mathcal{E}}_R^n$ by $O_{p.r.}$

Lemma 5.4 $\forall n \neq 1$, *if f is defined by the scheme of primitive recursion with functions of $\overline{\mathcal{E}}_R^n$ then f is in $\overline{\mathcal{E}}_R^{n+1}$. If $\text{hyp}(X_R, R)$ is true, the same result holds for the classes $\mathcal{E}_R^n$.*

Proof.

We prove the result for the classes $\overline{\mathcal{E}}_R^n$. For the classes $\mathcal{E}_R^n$, with $hyp(\mathcal{E}_R^n, R)$ true and by the lemma 4.3, the proof is analogous to the classical one ([2, 4]).

The proof for $n = 0$ and $n = 2$ is straightforward.

As we can show there is a function f defined by $O_{p.r.}$ from auxiliary functions of $\overline{\mathcal{E}}_R^1$ and belonging to $\overline{\mathcal{E}}_R^3 \setminus \overline{\mathcal{E}}_R^2$, we examine the case $n \geq 3$.

Let's remark that the general shape of a function f defined by primitive recursion on the variable t is $f(\overline{t_1}, \overline{u}, t)$ with t the *active* counter.

So, $f(\overline{t_1}, \overline{u}, t)$ verifies $\begin{cases} f(\overline{t_1}, \overline{u}, 0) = g(\overline{t_1}, \overline{u}) \\ f(\overline{t_1}, \overline{u}, t+1) = h(\overline{t_1}, \overline{u}, t, f(\overline{t_1}, \overline{u}, t)) \end{cases}$

with g and h, both functions in $\overline{\mathcal{E}}_R^n$ (hence verifying the growth lemma).

Our next step is to compute a bound of f in $\overline{\mathcal{E}}_R^{n+1}$ (so, f will be defined by $O_{l.p.r.}$ in $\overline{\mathcal{E}}_R^{n+1}$).

More precisely, in the sequel, we first prove, by induction on t, that:

$$|f(\overline{t_1}, \overline{u}, t)| \leq (\max(c_g, c_h) \cdot (\max(\overline{u})+1))^{E_{n-1}^{t \cdot (m_h+2)+(m_g+2)}(\max(c_g', \lfloor \overline{u} \rfloor, \overline{t_1}, c_h'))}$$

Afterwards, we show that this bound is in $\overline{\mathcal{E}}_R^{n+1}$

Firstly, we prove, by induction, the inequality:

- Let's bound $f(\overline{t_1}, \overline{u}, 0)$:
$$|f(\overline{t_1}, \overline{u}, 0)| = |g(\overline{t_1}, \overline{u})| \leq (c_g \cdot (\max(\overline{u})+1))^{E_{n-1}^{m_g}(\max(\overline{t_1}, \lfloor \overline{u} \rfloor))}$$
- Afterwards, we need, in the sequel, to bound $(c \cdot (\max(\lfloor \overline{u} \rfloor)+1))^{E_{n-1}^m(t)}$, so, we establish a bound for this function before pursuing the induction.

Clearly, we have, for $n \geq 2$:

$$c \leq c' \leq E_{n-1}(c') \text{ with } c' = \lfloor c \rfloor + 1$$

and

$$\max(\lfloor \overline{u} \rfloor)+1 \leq E_1(\max(\lfloor \overline{u} \rfloor)) \leq E_{n-1}(\max(\lfloor \overline{u} \rfloor))$$

Hence, by the growth properties of the classical growth function E_n (see section 3):

$$\begin{aligned} c \cdot (\max \lfloor \overline{u} \rfloor + 1) &\leq E_{n-1}(c') \cdot E_{n-1}(\max(\lfloor \overline{u} \rfloor)) \\ &\leq (E_{n-1}(\max(\lfloor \overline{u} \rfloor, c')))^2 + 2 \\ &\leq E_1(E_{n-1}(\max(c', \lfloor \overline{u} \rfloor))) \\ &\leq E_{n-1}(E_{n-1}(\max(c', \lfloor \overline{u} \rfloor))) \\ &\leq E_{n-1}^2(\max(c', \lfloor \overline{u} \rfloor)) \end{aligned}$$

So,

$$(1 \leq) \; c \cdot (\max \lfloor \overline{u} \rfloor + 1) \leq E_{n-1}^2(\max(c', \lfloor \overline{u} \rfloor)) \quad (n \geq 2)$$

Hence

$$\begin{aligned} &(c \cdot (\max(\lfloor \overline{u} \rfloor)+1))^{E_{n-1}^m(t)} \\ &\leq (E_{n-1}^2(\max(c', \lfloor \overline{u} \rfloor)))^{E_{n-1}^m(t)} \\ &\leq E_2^2(\max(E_{n-1}^2(\max(c', \lfloor \overline{u} \rfloor)), E_{n-1}^m(t))) \quad \text{by lemma 5.3} \\ &\leq E_2^2(E_{n-1}^m(\max(c', \lfloor \overline{u} \rfloor, t))) \quad \text{(we assume } m \geq 2\text{)} \\ &\leq E_{n-1}^2(E_{n-1}^m(\max(c', \lfloor \overline{u} \rfloor, t))) \\ &\leq E_{n-1}^{m+2}(\max(c', \lfloor \overline{u} \rfloor, t)) \end{aligned}$$

Finally,

$$(c \cdot (\max(\lfloor \overline{u} \rfloor) + 1))^{E_{n-1}^m(t)} \le E_{n-1}^{m+2}(\max(c', \lfloor \overline{u} \rfloor, t)) \quad (\star)$$

We come back to the induction on the active counter t (for $t \ge 1$):

$$|f(\overline{t_1}, \overline{u}, t)| \le (\max(c_g, c_h) \cdot (\max(\overline{u}) + 1))^{E_{n-1}^{t \cdot (m_h + 2) + (m_g + 2)}(\max(c'_g, \lfloor \overline{u} \rfloor, \overline{t_1}, c'_h))}$$

with $c_g, c'_g, c_h, c'_h, m_g$ and m_h some constants depending on the functions g and h (rem. m_g and m_h are constants in $\mathbb{N}$)

for $t = 1$

$$|f(\overline{t_1}, \overline{u}, 1)| = h(\overline{t_1}, \overline{u}, 0, f(\overline{t_1}, \overline{u}, 0))$$
$$\le (c_h \cdot (\max(\overline{u}) + 1))^{E_{n-1}^{m_h}(\max(\overline{t_1}, \lfloor \overline{u} \rfloor, \lfloor f(\overline{t_1}, \overline{u}, 0) \rfloor))}$$

By lemma 4.3,

$$\lfloor f(\overline{t_1}, \overline{u}, 0) \rfloor \le (c_g \cdot (\max(\lfloor \overline{u} \rfloor) + 1))^{E_{n-1}^{m_g}(\max(\overline{t_1}, \lfloor \overline{u} \rfloor))}$$
$$\le E_{n-1}^{m_g + 2}(\max(c'_g, \lfloor \overline{u} \rfloor, \overline{t_1})) \text{ (by } (\star))$$
$$\text{(with } c'_g = \lfloor c_g \rfloor + 1)$$
$$\text{hence } |f(\overline{t_1}, \overline{u}, 1)| \le (c_h \cdot (\max(\overline{u}) + 1))^{E_{n-1}^{m_h + (m_g + 2)}(\max(c'_g, \lfloor \overline{u} \rfloor, \overline{t_1}))}$$

for $t + 1 : |f(\overline{t_1}, \overline{u}, t + 1)| = |h(\overline{t_1}, \overline{u}, t, f(\overline{t_1}, \overline{u}, t))|$
$$\le \underbrace{(c_h \cdot (\max(\overline{u}) + 1))^{E_{n-1}^{m_h}(\max(\overline{t_1}, \lfloor \overline{u} \rfloor, \lfloor f(\overline{t_1}, \overline{u}, t) \rfloor))}}_{(\star\star)}$$

• By induction hypothesis:

$$|f(\overline{t_1}, \overline{u}, t)| \le (\max(c_g, c_h) \cdot (\max(\overline{u}) + 1))^{E_{n-1}^{t \cdot (m_h + 2) + (m_g + 2)}(\max(c'_g, \overline{t_1}, \lfloor \overline{u} \rfloor, c'_h))}$$

Hence,

$$\lfloor f(\overline{t_1}, \overline{u}, t) \rfloor \le (\max(c_g, c_h) \cdot (\max(\lfloor \overline{u} \rfloor) + 1))^{E_{n-1}^{t \cdot (m_h + 2) + (m_g + 2)}(\max(c'_g, \lfloor \overline{u} \rfloor, \overline{t_1}, c'_h))}$$
$$\le E_{n-1}^{2 + t \cdot (m_h + 2) + (m_g + 2)}(\max(c'_g, \lfloor \overline{u} \rfloor, \overline{t_1}, c'_h)) \quad (\text{ cf } (\star))$$

• Let's compute a bound of the exponent of $(\star\star)$:

$$E_{n-1}^{m_h}(\max(\overline{t_1}, \lfloor \overline{u} \rfloor, \lfloor f(\overline{t_1}, \overline{u}, t) \rfloor))$$
$$\le E_{n-1}^{m_h}(\max(\overline{t_1}, \lfloor \overline{u} \rfloor, E_{n-1}^{2 + t \cdot (m_h + 2) + (m_g + 2)}(\max(c'_g, \overline{t_1}, \lfloor \overline{u} \rfloor, c'_h))))$$
$$\le E_{n-1}^{m_h}(E_{n-1}^{2 + t \cdot (m_h + 2) + (m_g + 2)}(\max(c'_g, \overline{t_1}, \lfloor \overline{u} \rfloor, c'_h))) \quad (\text{cf lemma 4.2})$$
$$\le E_{n-1}^{(1 + t) \cdot (m_h + 2) + (m_g + 2)}(\max(c'_g, \overline{t_1}, \lfloor \overline{u} \rfloor, c'_h))$$

So,

$$
\begin{aligned}
&|f(\overline{t_1}, \overline{u}, t+1)| \\
&= |h(\overline{t_1}, \overline{u}, t, f(\overline{t_1}, \overline{u}, t))| \\
&\le (c_h \cdot (\max(\overline{u}) + 1))^{E_{n-1}^{m_h}(\max(\overline{t_1}, \lfloor \overline{u} \rfloor, \lfloor f(\overline{t_1}, \overline{u}, t) \rfloor))} \\
&\le (c_h \cdot (\max(\overline{u}) + 1))^{E_{n-1}^{(1+t) \cdot (m_h+2) + (m_g+2)}(\max(c_g', \overline{t_1}, \lfloor \overline{u} \rfloor, c_h'))}
\end{aligned}
$$

In conclusion,

$$
\forall t |f(\overline{t_1}, \overline{u}, t)| \le (\max(c_g, c_h) \cdot (\max(\overline{u}) + 1))^{E_{n-1}^{t \cdot (m_h+2) + (m_g+2)}(\max(c_g', \overline{t_1}, \lfloor \overline{u} \rfloor, c_h'))}
$$

Secondly, by one of the growth properties of the function E_n:
$(E_{n-1}^t(m) \le E_n(t+m)$, see section 3), we obtain the following bound of $|f(\overline{t_1}, \overline{u}, t)|$ in $\overline{\mathcal{E}}_R^{n+1}$:

$$
(\max(c_g, c_h) \cdot (\max(\overline{u}) + 1))^{E_n(t \cdot (m_h+2) + (m_g+2) + \max(c_g', \overline{t_1}, \lfloor \overline{u} \rfloor, c_h'))}
$$

$\square$

We prove the main results of this section

Proposition 5.1 *The classes $\mathcal{E}_R^n$ constitute a strict hierarchy called the Extended Grzegorczyk Hierarchy:*

$$
\mathcal{E}_R^0 \subsetneq \mathcal{E}_R^1 \subsetneq \mathcal{E}_R^2 \subsetneq \mathcal{E}_R^3 \subsetneq \ldots \subsetneq \mathcal{E}_R^n \subsetneq \ldots \qquad (\diamond)
$$

Proof.

$\forall n : \mathcal{E}_R^n \subseteq \mathcal{E}_R^{n+1}$:
We show that $\mathcal{E}_R^n$ contains the same initial functions and growth functions as $\mathcal{E}_R^{n+1}$ and is closed by the same operations.

- clearly, we have: $\mathcal{E}_R^0 \subseteq \mathcal{E}_R^1, \mathcal{E}_R^1 \subseteq \mathcal{E}_R^2$ and $\mathcal{E}_R^2 \subseteq \mathcal{E}_R^3$,
- for $n \ge 3$, we have: $\mathcal{E}_R^3 \subseteq \mathcal{E}_R^n$ because $\mathbb{E}_2 \in \mathcal{E}_R^n$ since it can be defined by $O_{l.p.r.}$:
$$
\begin{cases}
\mathbb{E}_2(x, 0) &= 1 \\
\mathbb{E}_2(x, t+1) &= (1 + |x|) \cdot \mathbb{E}_2(x, t)
\end{cases}
$$
$\mathbb{E}_2(x, t) \le \mathbb{E}_{n-1}(x, t)$
- by lemma 5.1, the following result holds:

$$
\forall n \ge 3 \quad \mathcal{E}_R^n \subseteq \mathcal{E}_R^{n+1}
$$

$\forall n \ge 0 : \mathcal{E}_R^n \ne \mathcal{E}_R^{n+1}$
By both a classical argument of diagonalization and the growth lemma, for each class $\mathcal{E}_R^n$, there are functions not belonging to the precedent class (see [2] for further details). In the following, we give one function and the lowest class of the hierarchy which contains it:

- $x_1 + x_2 \in \mathcal{E}_R^1$
- $(1+x)^2 \in \mathcal{E}_R^2$
- $(1+x)^t \in \mathcal{E}_R^3$
- $\forall\, n \geq 4\ :\ 10^{E_{n-1}(t)} \in \mathcal{E}_R^n$

□

Proposition 5.2 *The countable union of the Grzegorczyk classes coincides with the recursive primitive function classes:*

$$\bigcup_{n \in \mathbb{N}} \overline{\mathcal{E}}_R^n = \overline{RP}_R \ (\star)$$
$$\text{if } \mathrm{hyp}(X_R, R) \text{ is verified: } \bigcup_{n \in \mathbb{N}} \mathcal{E}_R^n = RP_R \ (\star\star)$$

The proof of $(\star\star)$ is less difficult technically (with $\mathrm{hyp}(X_R, R)$ verified) than $(\star)$, we give here the main ideas of the first result $(\star)$. Let's remark that the main difference consists in the using of the function $\lfloor\ \rfloor$ which belongs to each class $\overline{\mathcal{E}}_R^n$

Let's proof that $\underline{\bigcup_{n \in \mathbb{N}} \overline{\mathcal{E}}_R^n = \overline{RP}_R}$:

- any initial function in $\bigcup_{n \in \mathbb{N}} \overline{\mathcal{E}}_R^n$ is in $\overline{RP}_R$ and vice-versa,
- for any n, the functions $\mathbb{E}_n$ are in $\overline{RP}_R$ since the functions x^t and E_n belong to $\overline{RP}_R$,
- both classes are closed under the same operations: it is enough to verify for $O_{p.r.}$.
  Let f defined by recursion with two functions of $\bigcup_{n \in \mathbb{N}} \overline{\mathcal{E}}_R^n$: g and h. As the classes $\overline{\mathcal{E}}_R^n$ constitute a hierarchy, we can assume that g and h belong to the same class, say $\overline{\mathcal{E}}^{n_1}$. By lemma 5.4, f is a function of $\overline{\mathcal{E}}^{n_1+1}$ (we have assumed that $n \geq 3$), so, f is in $\bigcup_{n \in \mathbb{N}} \overline{\mathcal{E}}_R^n$.

□

6   Properties of the classes $\mathcal{RE}_R^n$

Firstly, we introduce some definitions about the class of relations associated to the Extended Grzegorczyk Classes $\mathcal{E}_R^n$. Secondly, we establish the closure of any classes $\mathcal{E}_R^n$ by the operation of limited minimum ($O_{l.\mu}$). Afterwards, we prove that the classes $\mathcal{RE}_R^n$ are closed for Boolean and limited quantifier operations.

Let's recall that for each class $\mathcal{E}_R^n$, we can define the following class of relations:

$$\mathcal{RE}_R^n \overset{\triangle}{=} \{\ \text{relations } f(\overline{t_1}, \overline{u}, t) = 0 \text{ with } f \in \mathcal{E}_R^n, \overline{u} \in R^{n_1}, \overline{t_1} \in \mathbb{N}^{n_2} \text{ and } t \in \mathbb{N}\}$$

A relation $T(\overline{t_1},\overline{u},t')$ (resp. $T(\overline{t_1},\overline{u},y)$) is defined from the relation $R(\overline{t_1},\overline{u},t)$ (resp: $R(\overline{t_1},\overline{u},y)$) by *limited quantifier* (resp. *extended limited quantifier operation*) if the following equivalence is true:

$$T(\overline{t_1},\overline{u},t') \Longleftrightarrow \exists t \le t'\ R(\overline{t_1},\overline{u},t)$$
$$(\text{resp. } T(\overline{t_1},\overline{u},y) \Longleftrightarrow \exists t \le |y|\ R(\overline{t_1},\overline{u},y)\ )$$

Let a function $F(\overline{t_1},\overline{u},t)$.
We define the following map: $F \to \sum_{i=0}^{t}(1-\chi^2(F(\overline{t_1},\overline{u},i)))$ called the *operation of natural zero counting* (from F). We will note $O_{z.c.}$ this operation. It consists in counting the number of times such as $F(\overline{t_1},\overline{u},i) = 0$ when i looks down $[0\ t]$ (rem.: $[0\ t]$ defines the following set: $\{n \in \mathbb{N} \mid 0 \le n \le t\}$)

Lemma 6.1 *If a function class X is closed for $O_{s.}$, $O_{z.c.}$, contains the initial functions and the extended modified subtraction $\dot{-}$ (see section 4 for the definition) then X is closed for $O_{l.\mu}$*

Proof. Let $F(\overline{t_1},\overline{u},i) \in X$ and $g(\overline{t_1},\overline{u},t) \stackrel{\Delta}{=} \mu(i \le t)(F(\overline{t_1},\overline{u},i) = 0)$.

We show that $g \in X$.

Let

$$\Sigma L\ F(\overline{t_1},\overline{u},t) \stackrel{\Delta}{=} \sum_{i=0}^{t}(1-\chi^2(F(\overline{t_1},\overline{u},i)))$$

and

$$f(\overline{t_1},\overline{u},t) \stackrel{\Delta}{=} \sum_{i=0}^{t}(1-\chi(\Sigma L\ F(\overline{t_1},\overline{u},i)))$$

We show that the function g is definable with the functions and operations of the class X:

$$g(\overline{t_1},\overline{u},t) = f(\overline{t_1},\overline{u},t) \cdot \chi(1-\chi^2(f(\overline{t_1},\overline{u},t) \dot{-} t))$$

Let's remark that the function $x \to \chi^2(x)$ is in X since $\chi^2(x) = \chi(x \cdot \chi(x))$

$f(\overline{t_1},\overline{u},t)$ is, by definition, the number of times (n_0) that $\Sigma L\ F(\overline{t_1},\overline{u},\cdot)$ vanishes on the interval $[0\ t]$

As $\forall j\ \Sigma LF(\overline{t_1},\overline{u},j+1) = 0$ implies $\Sigma LF(\overline{t_1},\overline{u},j) = 0$, n_0 is the smallest natural such as $\Sigma LF(\overline{t_1},\overline{u},n_0) \neq 0$. Consequently, n_0 is the first natural such as $F(\overline{t_1},\overline{u},\cdot)$ vanishes on $[0\ t]$.

- if $F(\overline{t_1},\overline{u},\cdot)$ vanishes on $[0\ t]$ then there is a smallest natural n_0 such as $n_0 \le t$ and $F(\overline{t_1},\overline{u},n_0) = 0$. Hence $f(\overline{t_1},\overline{u},t) = n_0$, so
$$f(\overline{t_1},\overline{u},t)\cdot\chi(1-\chi^2(f(\overline{t_1},\overline{u},t) \dot{-} t)) = n_0\cdot\chi(1-\chi^2(f(\overline{t_1},\overline{u},t) \dot{-} t))$$
$$= n_0\cdot\chi(1-\chi^2(n_0 \dot{-} t))$$
$$= n_0$$

– if $F(\overline{t_1}, \overline{u}, \cdot)$ does not vanish on $[0\ t]$ then $f(\overline{t_1}, \overline{u}, t) = t + 1 \ (= n_0)$, so
$$f(\overline{t_1}, \overline{u}, t) \cdot \mathcal{X}(1 - \mathcal{X}^2(f(\overline{t_1}, \overline{u}, t) \doteq t)) = (t + 1) \cdot \mathcal{X}(1 - \mathcal{X}^2(f(\overline{t_1}, \overline{u}, t) \doteq t))$$
$$= (t + 1) \cdot \mathcal{X}(1 - \mathcal{X}^2(n_0 \doteq t))$$
$$= (t + 1) \cdot \mathcal{X}(0)$$
$$= 0$$

$\square$

A consequence of the previous lemma is the following result:

Proposition 6.1

$\forall n \geq 0 \ \mathcal{E}_R^n$ *is closed for the limited minimum operation* $(\diamond)$

Proof. We show that $\mathcal{E}_R^0$ verifies the hypothesis of the lemma 6.1 and the result naturally transfers to upper class $\mathcal{E}_R^n$ since they constitute a hierarchy.

The following functions are in $\mathcal{E}_R^0$:

– $x \doteq t$ (see section 4 for the definition)
– we introduce the following function τ in order to define the operation $O_{z.c.}$ in $\mathcal{E}_R^0$: $\tau(x, t) \stackrel{\Delta}{=} x + 0^t$ (we define $0^0 = 1$)
τ is definable by $O_{l.p.r.}$
$$\begin{cases} \tau(x, 0) & = x + 1 \\ \tau(x, t + 1) = x \end{cases}$$
$|\tau(x, t)| \leq |x| + 1$
– if $F \in \mathcal{E}_R^0$, then $\Sigma L\ F(\overline{t_1}, \overline{u}, t) \stackrel{\Delta}{=} \sum_{i=0}^{t}(1 - \mathcal{X}^2(F(\overline{t_1}, \overline{u}, i)))$ is definable in $\mathcal{E}_R^0$ by $O_{l.p.r.}$,
indeed,
$$(S)\ \begin{cases} \Sigma L\ F(\overline{t_1}, \overline{u}, 0) & = 1 - \mathcal{X}^2(F(\overline{t_1}, \overline{u}, 0)) \\ \Sigma L\ F(\overline{t_1}, \overline{u}, t + 1) = \tau(\Sigma L\ F(\overline{t_1}, \overline{u}, t), \mathcal{X}^2(F(\overline{t_1}, \overline{u}, t + 1))) \end{cases}$$
$\Sigma L\ F(\overline{t_1}, \overline{u}, t) \leq t$
To justify (S), let's remark that

$$\Sigma L\ F(\overline{t_1}, \overline{u}, t + 1) = \sum_{i=1}^{t}(1 - \mathcal{X}^2(F(\overline{t_1}, \overline{u}, i))) + 0^{\mathcal{X}^2(F(\overline{t_1}, \overline{u}, t+1))}$$

We conclude by the lemma 6.1.

$\square$

The next result shows the closure of certain classes $\mathcal{R}X$ for connectives and limited quantifiers:

Proposition 6.2 *Let a class* X *of total functions[6], closed for the (extended) limited minimum and substitution operations then* $\mathcal{R}X$ *is closed for the (extended) limited quantifier. Moreover, if* $\mathcal{R}X$ *is closed for the negative connective then this class is closed for the (extended) limited universal operation*

---

[6] A total function is a function which is defined everywhere on $R^{n_1} \times \mathbb{N}^{n_2}$.

Proof. Let R $\in \mathcal{R}X$ and f, a function which represents it in X:

$$R(\overline{t_1}, \overline{u}, t) \Longleftrightarrow f(\overline{t_1}, \overline{u}, t) = 0 \quad (t \in \mathbb{N})$$

Let's define $h(\overline{t_1}, \overline{u}, y) = f(\overline{t_1}, \overline{u}, \mu(t \leq |y|)(f(\overline{t_1}, \overline{u}, t) = 0))$. The function h represents the relation $\exists t \leq |y|\ R(\overline{t_1}, \overline{u}, t)$ and hence is in $\mathcal{R}X$, indeed ,

- either, $\exists t_0 \leq |y|\ R(\overline{t_1}, \overline{u}, t_0)$ then $\mu(t \leq |y|)(f(\overline{t_1}, \overline{u}, t) = 0) = t_0$ and, hence, $h(\overline{t_1}, \overline{u}, y) = f(\overline{t_1}, \overline{u}, t_0) = 0$
- or, $\forall t_0 \leq |y|\ \neg R(\overline{t_1}, \overline{u}, t_0)$ then $\mu(t \leq |y|)(f(\overline{t_1}, \overline{u}, t) = 0) = 0$ and, particularly, $f(\overline{t_1}, \overline{u}, 0) \neq 0$, so $\mu(t \leq |y|)(f(\overline{t_1}, \overline{u}, t) = 0) = 0$ (by def. of μ) and $h(\overline{t_1}, \overline{u}, y) = f(\overline{t_1}, \overline{u}, 0) \neq 0$.

Moreover, if $\mathcal{R}X$ is closed for the negative Boolean operation, we have the closure of $\mathcal{R}X$ by the (extended) limited universal quantifier.

$\square$

We have the following result of the closure of $\mathcal{RE}_R^n$:

Proposition 6.3 $\forall n \geq 0\ \mathcal{RE}_R^n$ *is closed for Boolean and limited quantifiers operations* ($\diamond$)

Proof.
for Boolean operations:
  connective$\vee$: $f_1(\overline{x}) = 0 \vee f_2(\overline{x}) = 0 \Longleftrightarrow f_1(\overline{x}) \cdot \mathcal{X}(f_2(\overline{x})) = 0$
  connective$\neg$: $\neg(f(\overline{x}) = 0) \Longleftrightarrow \mathcal{X}^2(f(\overline{x})) = 1$

for limited quantifier operations: the classes $\mathcal{E}_R^n$ are closed for $O_{l.\mu}$ (cf proposition 6.1), we apply the proposition 6.2 to achieve the proof.

$\square$

7   The class of the extended elementary functions

We introduce a new class of computable functions in which $O_{l.p.r}$ is not more " powerful " than $O_{l.\mu}$. This class extends the well-known class of elementary functions in the classical setting ([2, 3, 4]).

The following result shows that if a class X of functions verifies some conditions, then the class of relations $\mathcal{R}X$ associated to X is closed for Boolean operations.

Lemma 7.1 *If a function class X is closed for O_s., contains the initial functions and exponential function (x^t) then $\mathcal{R}X$ is closed for Boolean operations.*

Proof. let's assume that $x = 0 \in \mathcal{R}X$.

So, $\qquad \neg(x = 0) \qquad \Leftrightarrow \mathcal{X}^2(x) = 1$

$\qquad (x = 0) \wedge (y = 0) \Leftrightarrow y^{0^{\mathcal{X}^2(x)}} = 0$

Let's remark that $x > 0 \Leftrightarrow \mathcal{X}(x) = 1$ and $x \leq 0 \Leftrightarrow (\mathcal{X}(x) = 0) \vee (\mathcal{X}(x) = -1)$

$\qquad\qquad\qquad\qquad\qquad\qquad\qquad\qquad\qquad\qquad\qquad\qquad\qquad\qquad$ □

Let $\mathcal{E}_R$, the class of the *extended elementary functions* which is the smallest class including the following functions: *initial functions*, $x + y$, $x \cdot y$, $\dot{-}$ *and closed for the limited sum, product operations (i.e.* $\Sigma_{t=0}^{t'} F(\overline{t_0}, \overline{u}, t)$ *and* $\Pi_{t=1}^{t'} F(\overline{t_1}, \overline{u}, t)$ *with F, a function of $\mathcal{E}_R$) and O_s.*.

In this class, we can define the function $x^t = \Pi_{i=0}^{t-1} x$. Moreover, by precedent results (lemma 6.1, 7.1 and propositions 6.2, 6.3), $\mathcal{R}\mathcal{E}_R$ is closed for Boolean and limited quantifier operations since $O_{z.c.}$ can be defined in this class.

If we define $\mathcal{E}_{R \to \mathbb{Z}}$ as the class of functions in $\mathcal{E}_R$ whose range is included in $\mathbb{Z}$, we have the following proposition:

Proposition 7.1

$$\mathcal{E}_{R \to \mathbb{Z}} \text{ is closed for } O_{l.p.r.} \quad (\diamond)$$

Idea:

We don't recall here the whole proof, we just point out the essential modifications to the original proof ([2, 3]) consisting in translating the scheme of primitive recursion into a first order logical formula ([2]).

Let's recall that an usual coding of a n–tuple of naturals $(n_0, \ldots, n_t)$ is $n = p_0^{n_0} \cdot \ldots \cdot p_t^{n_t}$ (p_i is the ith prime number). Here, we deal with integer values so, for each $m_i \in \mathbb{Z}$, let's define :

$$m^+ = \begin{cases} m & \text{if } m_i \geq 0 \\ 0 & \text{otherwise} \end{cases}$$

$$m^- = \begin{cases} -m & \text{if } m_i \leq 0 \\ 0 & \text{otherwise} \end{cases}$$

Of course, $m_i = m_i^+ - m_i^-$.

Hence $m = p_0^{m_0} \cdot \ldots \cdot p_t^{m_t} = p_0^{m_0^+} \cdot \ldots \cdot p_t^{m_t^+} \cdot p_0^{m_0^-} \cdot \ldots \cdot p_t^{m_t^-}$ with m_i^+ and m_i^- positive, for each $i \geq 0$.

Let a function f defined by $O_{p.r.}$ from g and h and limited by a function f'. Let's show we can translate it into a formula belonging to $\mathcal{R}\mathcal{E}_{R \to \mathbb{Z}}$. We have the following equivalences translating $O_{l.p.r.}$: (we have removed some parentheses for more readability)

$$y = f(\overline{t_1}, \overline{u}, t)$$
$$\Updownarrow$$
$$\exists m_0, \ldots, m_t \text{ such as}$$
$$m_0 = g(\overline{t_1}, \overline{u})$$
$$\wedge$$
$$(m_1 = h(\overline{t_1}, \overline{u}, 0, m_0)) \wedge \ldots \wedge (m_t = h(\overline{t_1}, \overline{u}, t-1, m_{t-1}))$$
$$\wedge$$
$$y = m_t$$
$$\Updownarrow$$
$$\exists m^+ \le p_t^{(t+1)\cdot\max_{t'\le t}|f'(\overline{t_1},\overline{u},t')|} \exists m^- \le p_t^{(t+1)\cdot\max_{t'\le t}|f'(\overline{t_1},\overline{u},t')|}$$
$$(\chi(g(\overline{t_1}, \overline{u})) = 1) \wedge (\exp(m^+, p_0) = g(\overline{t_1}, \overline{u})) \wedge (\exp(m^-, p_0) = 0)$$
$$\vee((\chi(g(\overline{t_1}, \overline{u})) \le 0) \wedge (\exp(m^-, p_0) = -g(\overline{t_1}, \overline{u})) \wedge (\exp(m^+, p_0) = 0)))$$
$$\wedge$$
$$\forall k < t \; \chi(h(\overline{t_1}, \overline{u}, k, \exp(m^+, p_k) - \exp(m^-, p_k))) = 1$$
$$\wedge$$
$$(\exp(m^+, p_{k+1}) = h(\overline{t_1}, \overline{u}, k, \exp(m^+, p_k) \dotminus \exp(m^-, p_k))) \wedge (\exp(m^-, p_{k+1}) =$$
$$\vee \qquad \chi(h(\overline{t_1}, \overline{u}, k, \exp(m^+, p_k) \dotminus \exp(m^-, p_k))) \le 0$$
$$\wedge$$
$$(\exp(m^-, p_{k+1}) = -h(\overline{t_1}, \overline{u}, k, \exp(m^+, p_k) \dotminus \exp(m^-, p_k))) \wedge (\exp(m^+, p_{k+1}) =$$
$$\wedge$$
$$((\exp(m^+, p_t) = y) \wedge (\chi(y) = 1) \wedge (\exp(m^-, p_t) = 0))$$
$$\vee((\exp(m^-, p_t) = -y) \wedge (\chi(y) \le 0) \wedge (\exp(m^+, p_t) = 0))$$

and we conclude by the proof of the result in the classical setting (see [2] and [3]).

□

To extend the function domains to $R^\infty \times \mathbb{N}^\infty$, we just add the following functions to the initial function class B:

- the *len* function which gives the length of $\overline{\alpha}$
- the *extended projection functions*:

$$p(f(i), \overline{\alpha}) = \begin{cases} \alpha_{f(i)} & \text{if } f(i) \le len(\overline{\alpha}) \\ 0 & \text{otherwise} \end{cases}$$

with f, a function in $RP_{\mathbb{N}}$ (i.e. the classical primitive recursive functions)

Acknowledgement

I want to thank Christian Michaux and Véronique Bruyère for their valuable advises during the preparation of this paper.

References

1. BLUM L., SHUB M., SMALE S., "On a theory of computation and complexity over the real numbers: NP-completeness, recursive functions and universal machines", Bull. Amer. Math. Soc., 21 (1989), pp. 1–46.
2. GAKWAYA J.S., "La Hiérarchie de Grzegorczyk", U. Mons-Hainaut, Mémoire de licence en mathématiques, juin 1993.
3. GRZEGORCZYK A., "Some classes of recursive functions", Warszawa 1953, Instytut Matematyczny Polskiej Akademii Nauk, pp.1–44.
4. ROSE H.E., "Subrecursion, Functions and Hierarchies", Clarendon Press, Oxford, 1984.

Affine-Invariant Symmetry Sets

Peter J. Giblin and Guillermo Sapiro

Abstract. Affine invariant symmetry sets of planar curves are introduced and studied in this paper. Two different approaches are investigated. The first one is based on affine invariant distances, and defines the symmetry set as the closure of the locus of points on (at least) two affine normals and affine-equidistant from the corresponding points on the curve. The second approach is based on affine bitangent conics. In this case the symmetry set is defined as the closure of the locus of centers of conics with (at least) three-point contact with two or more distinct points on the curve. Although the two analogous definitions for the classical Euclidean symmetry set are equivalent, this is not the case for the affine group. We present a number of properties of both affine symmetry sets, showing their similarities with and differences from the Euclidean case.

1   Introduction

Symmetry sets for planar shapes have received a great deal of attention from the mathematical [1, 2, 6, 7, 8, 10, 12, 25, 26], computational geometry [22], biological vision [16, 18, 17], and computer vision (see [21] and references therein) communities since original work by Blum [5]. The symmetry set of a planar curve is defined as the closure of the locus of points equidistant from at least two different points on the given curve, providing the distances are local extrema. The fact that the distances are local extrema means that the symmetry set point lies on the intersection of the (Euclidean) normals at the corresponding curve points. This leads to an equivalent definition of the symmetry set, as the closure of centers of bitangent circles. Blum actually defined the symmetry set in a different way: If fire is turned on at the boundary of the shape, and it travels with uniform speed, the symmetry set is the points were two or more fire fronts collide.

Symmetry sets, medial axes (centers of bitangent disks completely inside the shape), and the closely related Voronoi diagrams, [1] are based on the classical definition of the l_2 Euclidean distance between points, being therefore only Euclidean invariant (l_1 and l_∞ metrics were used to define Voronoi diagrams as

---

[1] A Voronoi diagram is the space partition obtained by classifying points according to their distance to a given set of points.

well). In this work we present and study symmetry sets which are affine invariant. Two alternatives to affine invariant symmetry sets are presented. The first one is based on a definition of affine invariant distances, obtaining the affine symmetry set as the closure of locus of points affine-equidistant from at least two points on the curve, providing that the distances are local extrema. In the Euclidean case, the analogous requirement is that the Euclidean distances from a point of the symmetry set to two points on the curve are equal and local extrema.

The second approach is based on affine bitangent conics. In this case, the symmetry set is defined as the closure of the locus of centers of conics having at least a three-point contact with two or more different points on the curve. This is equivalent to saying that the conic and the curve have the same affine tangent at those points. This also is analogous to the Euclidean case where 'three-point contact conic' is replaced by 'two-point contact circle.' As pointed out above, the analogous Euclidean definitions are equivalent. This is not true for the affine group.

In this paper, we present both definitions and give a number of basic properties. Proofs and details can be found in our extended report [13].

2 Basic planar affine differential geometry

Let $\gamma(t) : [0,1] \to I\!R^2$ be a planar curve, boundary of a planar shape, parametrized by t. The basic idea behind planar affine differential geometry [4] is to define a new parametrization, s, which is affine-invariant (we restrict our analysis to special, area-preserving, affine transformations). It can be shown that the simplest affine-invariant parametrization s is given by requiring at every curve point $\gamma(s)$ the relation $[\gamma_s, \gamma_{ss}] = 1$. Here subscripts denote derivatives [2] and $[\cdot, \cdot]$ stands for the determinant of the 2×2 matrix defined by the $I\!R^2$ vectors. The vector γ_s is the *affine tangent* and γ_{ss} the *affine normal*. The affine normal is the locus of centers of conics with at least four-point contact with the curve. It is clear that this relation can not hold at inflection points, and affine differential geometry is not defined at inflection points. Therefore, we assume from now on that γ is a strictly convex curve (this curve is also called an *oval*). [3] If v is the Euclidean arclength ($\langle \gamma_v, \gamma_v \rangle = 1$), then $ds = [\gamma_v, \gamma_{vv}]^{1/3} dv = [\mathbf{t}, \kappa \mathbf{n}]^{1/3} dv = \kappa^{1/3} dv$, where $\mathbf{t}$, $\mathbf{n}$, and κ are the Euclidean unit tangent, unit normal, and curvature respectively. This relation was introduced in [23, 24] for the study of affine invariant curve deformations. From it, the relation between the affine and Euclidean tangents is obtained, $\gamma_s = \kappa^{-1/3} \mathbf{t}$.

Differentiating the parametrization relation we obtain $[\gamma_s, \gamma_{sss}] = 0$. Therefore, $\gamma_{sss} + \mu \gamma_s = 0$, for some $\mu(s) \in I\!R$. This function μ is the *affine-invariant curvature*, and is the simplest non-trivial affine differential invariant, defining γ up to an affine transformation. It is easy to see that $\mu = [\gamma_{ss}, \gamma_{sss}]$. Curves have

---

[2] We will also denote derivatives by the prime $'$.

[3] Since inflection points are affine-invariant, segmenting the curve into convex portions does not bring a major problem for most shape understanding problems.

constant affine curvature if and only if they are conic sections. There are at least six points such that $\mu_s = 0$ in a closed convex curve [4, §19]. These points are called *sextactic* and are the affine vertices of a curve. Osculating conics passing through $\gamma(s)$ have five-point contact with the curve if $\gamma(s)$ is not sextactic, and at least six-point contact (generically exactly six-point) if it is.

3   Affine symmetry sets from invariant distances

In this Section we present and study the first of our affine invariant symmetry sets. Since it is based on distance functions, we begin with the presentation of an affine invariant distance [4, 15, 20] and its main properties.

3.1   Affine invariant distance

Definition 1 *Let* $\mathbf{X}$ *be a generic point on the plane,* [4] *and* $\gamma(s)$ *a strictly convex planar curve parametrized by affine arclength. The affine distance between* $\mathbf{X}$ *and a point* $\gamma(s)$ *on the curve is given by* [5]

$$d(\mathbf{X}, s) := [\mathbf{X} - \gamma(s), \gamma_s(s)]. \tag{1}$$

We present now some of the basic properties of $d(\mathbf{X}, s)$ (see also [15, 20] and [13] for proofs). Replacing the affine concepts by Euclidean ones, the same properties hold for the classical (squared) Euclidean invariant distance $\langle \mathbf{X} - \gamma, \mathbf{X} - \gamma \rangle$.

Proposition 2 ([13]) *The affine distance satisfies:*

1. *$d(\mathbf{X}, s)$ is an extremum (i.e., $d_s = 0$) if and only if $\mathbf{X} - \gamma(s)$ is parallel to γ_{ss}, i.e. $\mathbf{X}$ lies on the affine normal to the curve at $\gamma(s)$.*
2. *For non-parabolic points (where the affine curvature is non-zero), $d_s(\mathbf{X}, s) = d_{ss}(\mathbf{X}, s) = 0$ iff $\mathbf{X}$ is on the affine evolute.*
3. *γ is a conic and $\mathbf{X}$ its center if and only if $d(\mathbf{X}, s)$ is constant.*
4. *$\mathbf{X} - \gamma(s) = \alpha\gamma_s(s) - d(\mathbf{X}, s)\gamma_{ss}(s)$, for some real real number α.*

3.2   Affine invariant symmetry sets

Definition 3 *The affine distance symmetry set (ADSS) of a planar convex curve $\gamma(s)$, also called an oval, is the closure of the the locus of points $\mathbf{X}$ in $\mathbb{R}^2$ on (at least) two affine normals and affine-equidistant from the corresponding points on the curve. In other words, $\mathbf{X} \in \mathbb{R}^2$ is a point in the affine distance*

---

[4] We use capital bold letters to indicate points in $\mathbb{R}^2$.

[5] In order to be consistent with the Euclidean case and the geometric interpretation of the affine arclength, $d(\mathbf{X}, s)$ should be defined as the $1/3$ power of the determinant above. Since this does not imply any conceptual difference, we keep the definition above to avoid introducing further notation. In [15] the above function $d(\mathbf{X}, s)$ is called the *affine distance-cubed function*.

symmetry set of $\gamma(s)$ *if and only if there exist two different points* s_1, s_2 *such that* $d(\mathbf{X}, s_1) = d(\mathbf{X}, s_2)$, *and* $d_s(\mathbf{X}, s_1) = d_s(\mathbf{X}, s_2) = 0$, *or if* $\mathbf{X}$ *is a limit of such points.*

Proposition 4 ([13]) *Let* $\mathbf{X}(v) : \mathbb{R} \to \mathbb{R}^2$ *be a segment of the affine invariant symmetry set, and* $s_1(v), s_2(v)$ *the corresponding points on* γ, *where we use affine arclength* s *as parameter on* γ *(see §3.3 for smoothness conditions). Then*

1. *The tangent to* $\mathbf{X}(v)$ *is parallel to* $\gamma_s(s_1(v)) - \gamma_s(s_2(v))$.
2. *The ADSS has an inflection point iff* $(1 - [\gamma_s(s_2), \gamma_{ss}(s_1)])^2(1 - \mu_1 d(\mathbf{X})) = (1 - [\gamma_s(s_1), \gamma_{ss}(s_2)])^2(1 - \mu_2 d(\mathbf{X}))$, *where* $d(\mathbf{X})$ *is the common affine distance and* μ_1, μ_2 *the affine curvatures.*
3. *The tangent to the ADSS and the two tangents to the oval* γ *at the corresponding points* s_1 *and* s_2, *all intersect at one point.*
4. $[\gamma(s_1(v)) - \gamma(s_2(v)), \gamma_{ss}(s_1(v)) - \gamma_{ss}(s_2(v))] = 0$. *Conversely, if this holds, then the intersection* $\mathbf{X}$ *of the affine normals at* $\gamma(s_1)$ *and* $\gamma(s_2)$ *satisfies the conditions of Definition 3 above.*

The last result provides a method of drawing the affine distance symmetry set: Using arclength parameter on γ, find the solutions of the equation, in two variables s_1, s_2, $[\gamma(s_1) - \gamma(s_2), \gamma_{ss}(s_1) - \gamma_{ss}(s_2)] = 0$. This set in (s_1, s_2)-space can be called the *pre-ADSS*. Then for each solution, find the intersection $\mathbf{X}$ of the affine normals at $\gamma(s_1), \gamma(s_2)$.

In passing we note that the last result of the proposition makes it possible to draw also the affine invariant equivalent of the *midpoint locus* or *smoothed local symmetry* (see [12]), where the midpoint of the chord joining $\gamma(s_1)$ to $\gamma(s_2)$ is taken in preference to the point $\mathbf{X}$. Figure 1 shows an example of $ADSS$. These drawings were obtained using the relation in Proposition 4 to compute the pre-$ADSS$, and were implemented in Maple [11] and the Liverpool Surface Modeling Package [19].

Theorem 5 [13]. *Let* $\gamma(s)$ *be a strictly convex planar closed curve. For every non-sextactic point* $\gamma(s_0)$ *there exists a point* $\mathbf{X} \in \mathbb{R}^2$ *and at least a second point* $\gamma(s_1)$ *such that* $\mathbf{X}$ *is in the affine distance symmetry set (* $\mathbf{X} \in \gamma_{ss}(s_0)$, $\mathbf{X} \in \gamma_{ss}(s_1)$, *and* $d(\mathbf{X}, s_0) = d(\mathbf{X}, s_1)$ *). Exceptionally, the point* $\mathbf{X}$ *will be 'at infinity,' when the affine normals through* $\gamma(s_0)$ *and* $\gamma(s_1)$ *are parallel.*

This result means that (almost) all the points on the curve have an associated second point such that the pair defines a point on the affine distance symmetry set. Sextactic points will define the ends of the $ADSS$, where s_0 becomes equal to s_1 in the limit, as we see in the following Section. In fact with a little more trouble one can strengthen the result of Theorem 5 to show that, if $\gamma(s_0)$ is not a sextactic point, then there are at least *two* other points $\gamma(s_1), \gamma(s_2)$ giving points of the $ADSS$. We are grateful to S. Tabachnikov for pointing this out to us. (For relations to general questions see [27].)

As mentioned before, fronts motion can be used to obtain the Euclidean medial axis (a sub-set of the symmetry set such that the corresponding disks

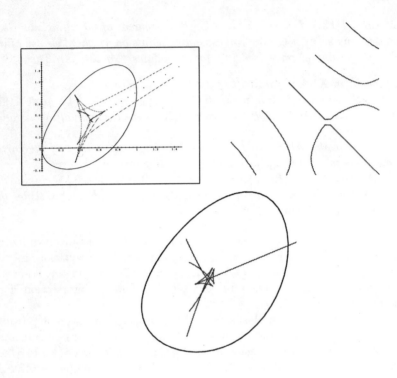

Fig. 1. *Example of $\mathcal{ADSS}$. The top-left figure shows the oval together with the affine evolute and points on the $\mathcal{ADSS}$. Note how the $\mathcal{ADSS}$ ends at evolute points, as shown in this paper. On the top-right, the pre-$\mathcal{ADSS}$ is shown, and a more complete picture of the $\mathcal{ADSS}$ on the bottom.*

are included in the shape). More specifically, being τ the evolution time, this set is obtained via the shocks (front collisions) of the flow $\frac{\partial \gamma}{\partial \tau} = \mathbf{n}$, where $\mathbf{n}$ is the Euclidean unit normal. The analogous flow in the affine case will be to (locally) move the front with constant affine distance in the direction of the affine normal, that is, $\frac{\partial \gamma}{\partial \tau} = \frac{\partial^2 \gamma}{\partial s^2}$, which is exactly the *affine heat flow* introduced and studied in [23]. Note than in contrast with the Euclidean constant motion above, this affine motion does not create singularities (the normal is not preserved), and smoothly deforms the shape to an elliptic point [23, 24].

If the Euclidean symmetry set is a straight line, then the curve γ consists of two portions which are symmetric by a reflexion in that line. It seems that there is no completely general analogy to this in the case of the $\mathcal{ADSS}$. We give a brief account of this matter now. As we shall see in §4, there is an alternative construction for affine invariant symmetry sets which does have a property completely analogous to the Euclidean case.

First of all, let's define what we mean by symmetry in the affine case.

Definition 6 *A planar shape is affine symmetric if it is the affine transformation of a shape which is symmetric by a reflexion in a straight line.*

It is straightforward to show that the $\mathcal{ADSS}$ of an affine symmetric curve contains a straight line. It is, however, possible to have an straight $\mathcal{ADSS}$ without the curve being affine symmetric. In [13], we derive the explicit relation between the two arcs of curve that define the straight $\mathcal{ADSS}$ and based on this present an example for a straight $\mathcal{ADSS}$ defined by a curve which is not affine symmetric. Note that when the $\mathcal{ADSS}$ is straight, from the condition on inflection points in Proposition 4, if the tangent to the $\mathcal{ADSS}$ does not pass trough the midpoint of the chord joining $\gamma(s_1)$ and $\gamma(s_2)$, then $(1 - [\gamma_s(s_2), \gamma_{ss}(s_1)])^2 \neq (1 - [\gamma_s(s_1), \gamma_{ss}(s_2)])^2$, and since the equality in Proposition 4 holds (the $\mathcal{ADSS}$ is a straight line), $\mu_1 \neq \mu_2$ and the shape is not affine symmetric. We then obtain that for the shape to be affine symmetric, in addition to the $\mathcal{ADSS}$ to be a straight line, the midpoint of the chord joining two corresponding points in the curve γ has to lie on the $\mathcal{ADSS}$ as well. This second property will automatically hold in the alternative affine symmetry set that we present in §4, and the fact that the symmetry set is a straight line will be enough to conclude that the shape is affine symmetric.

3.3   Singularity theory and the affine distance symmetry set

It is shown in Proposition 2 and in [15] that the function d of Definition 1, regarded as a family of functions on γ parametrized by the points of the plane, has bifurcation set which is precisely the classical affine evolute [4] of the curve γ. The bifurcation set is by definition the set of points $\mathbf{X}$ in the plane (or, exceptionally, at infinity) where there exists a value of s for which $d_s(\mathbf{X}, s) = d_{ss}(\mathbf{X}, s) = 0$, s being, as usual, affine arclength. This makes the function d completely analogous to the well-known *distance-squared function* of Euclidean differential geometry, whose bifurcation set is the Euclidean evolute of γ (for an exposition of this see for example [7, §§2.27, §§7.1]).

We can include the affine distance symmetry set in the bifurcation set of the family of functions d: it then becomes the *full bifurcation set* or *levels bifurcation set* (see e.g. [6]) which includes all non-Morse functions in the family. A function is non-Morse if it has a degenerate singularity (first two derivatives zero, as above) *or* if it has two singularities (where $d_s = 0$) at which the function has the same value. The latter case is precisely the affine distance symmetry set as defined in Definition 3. The limit points will be points which belong to the affine evolute, in exactly the same way as the Euclidean symmetry set [6, 8, 12] has end-points in the cusps of the Euclidean evolute.

Geometrically, Definition 3 amounts to saying that $\mathbf{X}$ is the center of two conics, having (at least) 4-point contact with γ at $\gamma(s_1), \gamma(s_2)$, and having the same affine distance from γ (see Proposition 2 and Theorem 5). In terms of these, and using the full bifurcation set structure as in [6], we find the local structure of the affine distance symmetry set.

Theorem 7 *Locally, the affine distance symmetry set is:*

- *smooth when both conics have exactly 4-point contact with γ,*
- *an ordinary cusp when one of the conics has 5-point contact with γ (**X** is then on the affine evolute too but at a smooth point of it),*
- *an endpoint when **X** is the center of a 6-point contact conic (tangent to γ at a sextactic point): the endpoint is then in a cusp of the affine evolute,*
- *a triple crossing when there are three conics centered at **X** having equal affine distance and 4-point contact with γ; this is a stable phenomenon.*

Note that the pre-$\mathcal{ADSS}$, defined in the parameter space of pairs of points of γ, can be used, as in [12], to analyze the simpler structures of the $\mathcal{ADSS}$. For example, the pre-$\mathcal{ADSS}$ will have a tangent parallel to the s_1-axis when the $\mathcal{ADSS}$ has a cusp on account of the conic tangent at $\gamma(s_1)$ having 5-point contact there. The pre-$\mathcal{ADSS}$ is smooth unless *both* conics have 5-point contact. One can also use the techniques of [6] to analyze the events in the evolution of affine distance symmetry sets when γ undergoes a generic change in shape. For example, we can analyze the behavior of the $\mathcal{ADSS}$ for a curve deforming according to the affine invariant heat flow studied in [23]. This is the subject of further work and will be reported elsewhere; see §5.

4   Affine symmetry sets from envelopes of conics or lines

We will now consider an alternative definition of affine invariant symmetry sets. Let γ be a simple closed smooth curve in the plane. Strictly we do not any longer need γ to be an oval, though we shall explore the non-oval case in more detail elsewhere. The alternative definition is based on conics having 3-point contact with the oval in at least two places. In the Euclidean case there is only one reasonable definition of symmetry set, based on bitangent circles, and indeed it can be argued that in the Euclidean case the distance based and contact based definitions reduce to the same thing. But for the affine group there do seem to be two alternative constructions.

We shall see that the new definition has natural relations with the theory of envelopes and for that reason we call it the *affine envelope symmetry set* ($\mathcal{AESS}$).

Definition 8 *Given a simple closed smooth curve γ (which we generally take to be an oval below), the affine envelope symmetry set ($\mathcal{AESS}$) is the closure of locus of centers of conics with at least 3-point contact with the curve in two or more different points. This is equivalent to saying that the curve and the conic have the same affine tangent in at least two points.*

The properties of the $\mathcal{AESS}$ can be obtained by considering either envelopes of *conics* or envelopes of *lines*. We shall give details of each approach here since they give different insights into the $\mathcal{AESS}$. The 'envelope of lines' approach is analogous to that adopted in [14] for the Euclidean case, where the lines are perpendicular bisectors of segments joining points on the curve γ and these lines are shown to envelope the symmetry set. There does not seem to be an analogue of the 'envelope of conics' approach in the Euclidean case.

4.1   Envelopes of conics

Consider an oval γ and a fixed point of the oval, which we take to be the origin. Suppose that γ is tangent to the x-axis at the origin. We seek a conic which has 3-point contact with γ at the origin and at another point. Initially we consider conics which have 3-point contact with γ at the origin and pass through a second point. We shall in particular be interested in the *centers* of such conics. Note again that from §2, 3-point contact is equivalent to having the same affine tangent (γ_s). We shall see that the $\mathcal{AESS}$ can be described as the set of *points of regression* on a certain envelope of conics.

Consider a general conic $\mathcal{C}$ with equation $ax^2 + by^2 + 2hxy + 2gx + 2fy + c = 0$. Using the definition of the center as the pole of the line at infinity with respect to the conic, the coordinates (p, q) of the center satisfy $ap + hq + g = 0$, $\quad hp + bq + f = 0$. If $\mathcal{C}$ is to touch γ at the origin then $g = c = 0$. Suppose γ has local parametrization of the form

$$y = \alpha x^2 + \beta x^3 + \dots \tag{2}$$

near the origin. We may suppose $\alpha > 0$. The (Euclidean) curvature κ at the origin is 2α, and substituting the parametrization into the equation of the conic we find that the condition for 3-point contact between $\mathcal{C}$ and γ is $a + 2f\alpha = 0$. If $a = 0$ this means that $f = 0$, and $\mathcal{C}$ is the degenerate conic consisting of the tangent to γ at the origin together with the chord joining the origin to $\gamma(t)$. If $a \neq 0$ then we can take $a = 1$ and f is determined by the 3-point contact condition: $f = -1/2\alpha = -1/\kappa(0)$. Note that this makes $f < 0$, assuming $\alpha > 0$. Solving the equations for the center of the conic γ we have $p = -\frac{fh}{h^2 - ab}$, $\quad q = \frac{af}{h^2 - ab}$. In fact we are more interested in the reverse process, expressing the coefficients in $\mathcal{C}$ in terms of p and q. We suppose from now on that $a \neq 0$, and divide through to make $a = 1$. Then $h = -\frac{p}{q}$, $b = \frac{qh^2 - f}{q} = \frac{p^2 - fq}{q^2}$. The conic $\mathcal{C}$ therefore has the form $q^2 x^2 + (p^2 - fq)y^2 - 2pqxy + 2fyq^2 = 0$. Now suppose that $\mathcal{C}$ passes through a second point $(X(t), Y(t))$ of γ, which we take to be other than the origin. This amounts to replacing x, y by $X(t), Y(t)$ in the above equation. Omitting t we get

$$q^2 X^2 + (p^2 - fq)Y^2 - 2pqXY + 2fYq^2 = 0. \tag{3}$$

Important things to notice for the moment are:

1. For a fixed t, (3) now gives an equation of degree 2 in the variables p, q. This shows that *the locus of centers (p, q) of conics C having 3-point contact with γ at the origin and passing through $\gamma(t)$ is itself a conic.* We denote it by $\mathcal{D}(t)$, or later by $\mathcal{D}(t_0, t)$ where t_0 is the parameter value of the origin on γ. Note that the conic (3) passes through the origin $p = q = 0$. This corresponds to the degenerate conic C consisting of the tangent to γ at the origin together with the chord joining the origin to $\gamma(t)$: the 'center' of this conic is at the origin (the case $a = 0$ above). The discriminant of the conic (3) is $-2fY^3$. With $f < 0$ and the oval γ in the region $y \geq 0$ this discriminant is positive, so the conic $\mathcal{D}(t)$ is a *hyperbola*.

2. Equation (3) can be written in the form $(pY - qX)^2 = fqY(Y - 2q)$. Note that, in p, q coordinates, $pY - qX = 0$ is the equation of the line $\mathcal{L}$ joining the origin to (X, Y), $q = 0$ is the equation of the tangent $\mathcal{T}$ at the origin, and $2q = Y$ is the equation of the line $\mathcal{M}$ parallel to $\mathcal{T}$ passing through the midpoint of the segment from the origin to (X, Y). Thus with a suitable choice of equations for these lines (multiplying by constants), the conic $\mathcal{D}(t)$ can be written $\mathcal{L}^2 = \mathcal{M}\mathcal{T}$.

3. One consequence of the last fact is that $\mathcal{D}(t)$ passes through the midpoint of the segment from the origin to (X, Y), and is tangent to the line $\mathcal{M}$ there. Thus it is a hyperbola with one branch having 3-point contact with the oval γ at the origin, and the other branch tangent to $\mathcal{M}$ at the point $\mathcal{L} \cap \mathcal{M}$.

Equation (3) is a 1-parameter family of conics, in (p, q) coordinates, parametrized by t. The envelope of this family is obtained by differentiating (3) with respect to t:

$$q^2 XX' + (p^2 - fq)YY' - pq(XY' + X'Y) + fq^2Y' = 0. \tag{4}$$

Equation (4) is the same as the additional condition required for the conic C to be *tangent* to γ at $(X(t), Y(t))$. Thus *the envelope of the conics (3) is the locus of centers of conics C having 3-point contact with γ at the origin and tangent to γ somewhere else.* For a given t we expect (3) and (4) to have a unique solution (p, q), since there should be a unique conic having 3-point contact with γ at the origin $\gamma(t_0)$ and 2-point contact with γ elsewhere – at $\gamma(t)$. (If $t = t_0$ then we might still expect a unique solution, namely the osculating conic for γ at $\gamma(t_0)$.)

Taking $Y' \times (3) - Y \times (4)$ gives

$$-pY(XY' - X'Y) + q(fYY' + X(XY' - X'Y)) = 0. \tag{5}$$

Note that assuming $t \neq t_0$, so that $Y \neq 0$, we can recover (4) from (3) and (5). It follows that the envelope point of the family of conics (3) is the unique intersection of (3) for a particular t with the line (5), disregarding the intersection at the origin $\gamma(t_0)$.

A short calculation shows that the limiting line (5) as $t \to t_0$ is the affine normal $2p\alpha^2 + q\beta = 0$, using the local parametrization of γ is (2). The limiting conic (3) is the affine normal together with the tangent to γ at the origin.

Double 3-point contact conics We now derive the condition for the conic $\mathcal{C}$ to have 3-point contact with γ at $\gamma(t)$ as well as at the origin. This amounts to differentiating (4) again, so corresponds to *points of regression* on the envelope of the conics (3): *The point* (p, q) *is on the $\mathcal{AESS}$, i.e. is the center of a conic $\mathcal{C}$ having 3-point contact with γ at $\gamma(t_0)$ and $\gamma(t)$, precisely when $\gamma(t)$ is a point of regression [7, §5.26] – generically a cusp – on the envelope of the conics (3).*

There are various ways of calculating the condition on t (and t_0) for this to hold. One way is to differentiate (3) again, and to derive a second line through the origin which meets $\mathcal{D}(t)$ in the intersection of $\mathcal{D}(t)$ with this second derivative. We can then calculate the condition for this line to coincide with (5). This is then extended to allow $\gamma(t_0)$ to be a general point in the curve, obtaining, following the geometric interpretation of the result (see our extended report [13] for the complete computations):

Theorem 9 [13]. *The (equivalent) conditions for a conic to have 3-point contact with γ at tow points $\gamma(t_0)$ and $\gamma(t)$ are*

1. *Euclidean: Being κ the Euclidean curvature, then up to a sign,*

$$\left(\frac{\kappa(t)}{\kappa(t_0)} \right)^{1/3} = \frac{\text{distance from } \gamma(t_0) \text{ to tangent at } \gamma(t)}{\text{distance from } \gamma(t) \text{ to tangent at } \gamma(t_0)}.$$

2. *Affine: Having the curve parametrized by affine arc-length s ($\gamma(t_0) = \gamma(s_0)$, $\gamma(t_1) = \gamma(s_1)$), then*

$$[\gamma(s) - \gamma(s_0), \gamma'(s) + \gamma'(s_0)] = 0. \tag{6}$$

The last relation defines the *pre-$\mathcal{AESS}$* in parameter space (s_0, s). (In [13] we also derive the explicit formulas to compute the center of the conic from the pre-$\mathcal{AESS}$.) Note that this is exactly analogous to the Euclidean case, where the condition for a bitangent circle can be written as the above when Euclidean arclength t is used, so that $\gamma'(t) + \gamma'(t_0)$ is along the bisector of the angle between the two oriented tangents to the curve γ at $\gamma(t)$ and $\gamma(t_0)$. In spite of this, Equation (6) might still be a bit of a surprise. The following Theorem gives the geometric interpretation of this relation.

Theorem 10 [13]. *If γ is a conic, then (6) holds for every two points on γ.*

An important consequence of the condition for a double 3-contact point is that we shall always find a conic having 3-point contact at $\gamma(t_0)$ and *somewhere else*:

Theorem 11 [13]. *Provided γ does not have a sextactic point at $\gamma(t_0)$, there is always a conic with 3-point contact at $\gamma(t_0)$ and at some other point $\gamma(t), t \neq t_0$. If there is a sextactic point at $\gamma(t_0)$ then of course there is a six-point contact conic at $\gamma(t_0)$, which can be regarded as a limiting case of the double-3-point contact conic.*

Examples of the $\mathcal{AESS}$ are given in Figure 2. The top row shows the $\mathcal{AESS}$ and *midlocus* for an oval. This is repeated for a non-oval in the bottom row. Basically, the inflection points are 'ignored' where searching for the pre-$\mathcal{AESS}$. The analysis of the $\mathcal{AESS}$ for non-convex curves will be reported elsewhere.

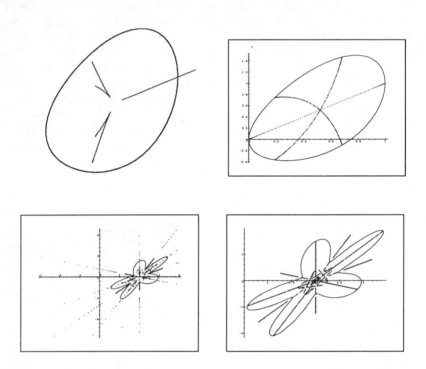

Fig. 2. *Examples of the $\mathcal{AESS}$. The top figure shows (left) points on the $\mathcal{AESS}$ for the oval in Figure 1, and its midlocus (right). On the bottom, the $\mathcal{AESS}$ and midlocus of a non-oval shape are given. Note how both the $\mathcal{AESS}$ and midlocus capture the affine symmetry of the different parts of the shape.*

4.2   Envelopes of lines

There is an attractive way to describe the $\mathcal{AESS}$ in terms of envelopes of lines instead of envelopes of conics and since this gives certain information which the other approach does not (at any rate not so easily), we give some details here. In fact this approach makes it clear that the $\mathcal{AESS}$ is from the point of view of singularity theory *the dual of a discriminant* as opposed to the $\mathcal{ADSS}$ which is a *full bifurcation set*. The Euclidean symmetry set is *both* of these things [14].

Consider an oval γ and two points on it, $\gamma(t_1), \gamma(t_2)$. We consider initially conics $\mathcal{C}$ which are tangent to γ at these two points. It is known from the general theory of conics that these conics form a pencil; in fact if T_1, T_2 are the tangent lines to the oval at the two points and $\mathcal{L}$ is the line joining the two points, then the general conic of the pencil is $T_1 T_2 + \lambda \mathcal{L}^2 = 0$. Also the locus of centers of these conics is the line $\mathcal{M}$ passing through the intersection of T_1 and T_2, and the midpoint of the chord from $\gamma(t_1)$ to $\gamma(t_2)$. (There is another degenerate component of the locus of centers, namely the line $\mathcal{L}$, which corresponds to the double line $\mathcal{L}^2$ of the pencil, whose center is indeterminate on $\mathcal{L}$.)

We examine this situation in more detail as follows. The conic $\mathcal{C}$ will have six homogeneous coefficients, being the general equation of the form $ax^2 + by^2 + 2hxy + 2gx + 2fy + c = 0$. Writing $\gamma(t) = (X(t), Y(t))$, we can substitute $x = X(t)$, $y = Y(t)$ into this conic equation to measure the contact of $\mathcal{C}$ with the oval at $\gamma(t)$. Calling the resulting function $\mathcal{C}(t)$ we say that there is k-point contact when $\mathcal{C}$ and its first $k - 1$ derivatives vanish at t while the k-th derivative is non-zero there. Furthermore it is well known that the coordinates of the center (p, q) of $\mathcal{C}$ satisfy the two linear equations $\mathcal{C}_x(p, q) = 0, \mathcal{C}_y(p, q) = 0$, where here and later subscripts x, y, t denote again partial derivatives. Assuming $\mathcal{C}$ is tangent to γ at both points $\gamma(t_1), \gamma(t_2)$ we can eliminate a, b, h, g, f, c from six linear equations, two coming from each point of contact and two from the center of the conic. In order to write down the resulting determinant condition, we introduce some abbreviations. Thus $\mathcal{C}(t_1)$ stands for the vector $(X^2, Y^2, 2XY, 2X, 2Y, 1)$, evaluated at $t = t_1$. This is the vector in $I\!\!R^6$ such that $\mathcal{C}(t_1)(a, b, h, g, f, c)^{\top} = 0$ gives the condition for the conic $\mathcal{C}$ to pass through $(X(t_1), Y(t_1))$. The vector $\mathcal{C}(t_2)$ is similarly defined. Furthermore, by $\mathcal{C}_t(t_1)$ we mean the vector $(2XX', 2YY', 2(XY' + X'Y), 2X', 2Y', 0)$, evaluated at $t = t_1$. Thus $\mathcal{C}_t(t_1)(a, b, h, g, f, c)^{\top} = 0$ is the additional condition for $\mathcal{C}$ to be tangent to the $\gamma = (X, Y)$ curve at $(X(t_1), Y(t_1))$. Likewise $\mathcal{C}_x(p, q)$ denotes the vector obtained by differentiating the equation for the general conic with respect to x, substituting $x = p, y = q$ and writing down the resulting coefficients of a, b, h, g, f, c. The same goes for $\mathcal{C}_y(p, q)$.

The determinant condition that there should exist a conic, with not all coefficients zero, tangent to γ at $\gamma(t_1), \gamma(t_2)$ and having center (p, q) is then

$$\mathcal{G}(t_1, t_2, p, q) := \begin{vmatrix} \mathcal{C}(t_1) \\ \mathcal{C}_t(t_1) \\ \mathcal{C}(t_2) \\ \mathcal{C}_t(t_2) \\ \mathcal{C}_x(p, q) \\ \mathcal{C}_y(p, q) \end{vmatrix} = 0. \tag{7}$$

Of course, (7) is nothing other than the equation, in (p, q) coordinates, of the locus of centers of conics $\mathcal{C}$. Hence, writing $\mathcal{M}(t_1, t_2, p, q) = 0$ for an equation of the line $\mathcal{M}$ above (using again (p, q) as current coordinates), and $\mathcal{L}(t_1, t_2, p, q) = 0$ for an equation of the chord, it must happen that

$$\mathcal{G} = \mathcal{M}\mathcal{L}.$$

Let $\mathcal{G}(t_1, t_2, p, q) = 0$, that is, (p, q) is the center of a conic tangent to γ at $\gamma(t_1)$ and $\gamma(t_2)$. So long as (p, q) is not the center of the chord, where $\mathcal{M}$ meets

$\mathcal{L}$, we can deduce that $\mathcal{L}(t_1, t_2, p, q) \neq 0$. Since tangents to a conic at opposite ends of a diameter are parallel, the only case that gives a problem here is where the tangents to γ at $\gamma(t_1)$ and $\gamma(t_2)$ are parallel. For an oval γ this determines t_2 uniquely from t_1. Apart from that, the zeroes of $\mathcal{G}$ coincide with those of $\mathcal{M}$. The same holds for derivatives; for example

$$\mathcal{G}_t = \mathcal{M}_t \mathcal{L} + \mathcal{M} \mathcal{L}_t,$$

so that $\mathcal{G} = \mathcal{G}_t = 0$ is equivalent, away from $\mathcal{L} = 0$, to $\mathcal{M} = \mathcal{M}_t = 0$.

The function $\mathcal{M}$ just defined gives us a 2-parameter family of lines in the plane, parametrized by the pairs of distinct points of γ. It is natural to ask whether $\mathcal{M}$ can be extended smoothly to all pairs of points. The answer is yes, provided we define $\mathcal{M}(t, t, p, q)$ to be an equation of the affine normal to γ at $\gamma(t)$. [6] In fact $\mathcal{M}$ gives rise to a map (compare with [14])

$$\mathcal{B}: \quad \gamma \times \gamma \to \mathcal{DL}, \tag{8}$$

where $\mathcal{DL}$ is the dual plane, or the set of lines in the ordinary plane $\mathbb{R}^2$, and $\mathcal{B}(t_1, t_2)$ is the line whose equation in current coordinates (p, q) is $\mathcal{M}(t_1, t_2, p, q) = 0$.

We can consider $\mathcal{M}$ as defining two envelopes of lines, one by fixing t_1 and the other by fixing t_2. We can also use the function $\mathcal{G}$ to measure contact between the conic $\mathcal{C}$ and γ and use the relation $\mathcal{G} = \mathcal{M}\mathcal{L}$ to compare this contact with the properties of the two envelopes. We collect the main results below. The proof, which amounts merely to differentiating the functions $\mathcal{G}$ and $\mathcal{M}$ sufficiently many times, can be found in our extended report [13]. In what follows, $\mathcal{C}$ will be the conic tangent to γ at $\gamma(t_1)$ and $\gamma(t_2)$, and having center (p, q). As before, the $\mathcal{AESS}$ of γ is the locus of centers of conics which have 3-point contact with γ at two different places (or, in the limit, the centers of sextactic conics). We are now ready to derive the $\mathcal{AESS}$ as envelope of lines.

Theorem 12 [13]. *For $\mathcal{M}$, $\mathcal{G}$, and $\mathcal{B}$ as defined above, the following holds:*

1. *Consider the 1-parameter family of lines $\mathcal{M}(t_1, t_2, p, q) = 0$ where t_2 is fixed. A point (p, q) belongs to the envelope of this family, given by $\mathcal{M} = \partial \mathcal{M}/\partial t_1 = 0$, if and only if the conic $\mathcal{C}$ has 3-point contact (at least) with γ at $\gamma(t_1)$.*
2. *Fixing t_2, p, q, the function $\mathcal{G}$ is zero and has a singularity of type A_k at t_1 (i.e. the first k derivatives vanish there but the $(k+1)$-st derivative is nonzero), if and only if the conic $\mathcal{C}$ has $(k+2)$-point contact with γ at $\gamma(t_1)$. The previous result is the case $k = 1$ of this. The functions $\mathcal{G}$ and $\mathcal{M}$ have the same singularity type so long as (p, q) is not on the line $\mathcal{L}$.*

---

[6] A geometric reason for this is the following: For fixed t_1, t_2, the line $\mathcal{M}$ is the locus of centers of conics having (at least) 2-point contact with γ at $\gamma(t_1)$ and $\gamma(t_2)$. When $t_2 \to t_1$ we shall obtain the locus of centers of conics having at least 4-point contact with γ at $\gamma(t_1)$, which is the affine normal at this point.

3. (t_1, t_2) is a critical point of $\mathcal{B}$ if and only if there is a conic with 3-point contact at both $\gamma(t_1)$ and $\gamma(t_2)$. To say that (t_1, t_2) is a critical point of $\mathcal{B}$ amounts to saying that (p, q) is a point of the envelope of both families of lines given by $\mathcal{M}$, keeping t_1 constant and keeping t_2 constant, so this follows from Part 1 above.

4. At points where $\mathcal{M} = \partial \mathcal{M}/\partial t_1 = \partial \mathcal{M}/\partial t_2 = 0$, we have $\partial^2 \mathcal{M}/\partial t_1 \partial t_2 = 0$. (It is not obvious to us what this condition means geometrically.) We note here that taking the Euclidean case where $\mathcal{M}$ is replaced by the perpendicular bisector of the segment joining $\gamma(t_1)$ and $\gamma(t_2)$ (compare with [14]), the formula $\partial^2 \mathcal{M}/\partial t_1 \partial t_2 = 0$ holds identically since it is easy to see that $\mathcal{M}$ is a sum of functions of t_1 and t_2. But in the present case this is not so: we need $\mathcal{M} = \partial \mathcal{M}/\partial t_1 = \partial \mathcal{M}/\partial t_2 = 0$.

5. The line $\mathcal{M}$ is tangent to the $\mathcal{AESS}$ when there is a conic $\mathcal{C}$ having 3-point contact with γ at $\gamma(t_1), \gamma(t_2)$. This says that the tangent to the $\mathcal{AESS}$ passes through the midpoint of the chord joining the two points of γ and through the intersection of the tangents to γ at these two points.

   Putting this in a more formal way: the critical locus of $\mathcal{B}$ (the image of the critical set) is the set of tangent lines to the $\mathcal{AESS}$ of γ, that is the dual of the $\mathcal{AESS}$ of γ. This is completely analogous to the situation studied in [14].

6. If the conic $\mathcal{C}$ has k-point contact $(k \geq 4)$ with γ at $\gamma(t_1)$, then the dual of the $\mathcal{AESS}$ of γ has an inflexion of order $k - 3$. An inflexion of order 1 – an ordinary inflexion – corresponds to an ordinary cusp on the $\mathcal{AESS}$, and arises when the conic $\mathcal{C}$ has 4-point contact.

Straight affine envelope symmetry sets We have seen that a straight $\mathcal{ADSS}$ does not necessarily mean that the curve giving rise to it is affine symmetric. For the $\mathcal{AESS}$ the situation is different:

Theorem 13 [13]. *Suppose that the $\mathcal{AESS}$ for two arcs of an oval γ is a straight line. Then either of these arcs is obtained from the other by an affine transformation of determinant -1.*

It is interesting to note that when the $\mathcal{AESS}$ has an inflection point, then the pre-$\mathcal{ADSS}$ and pre-$\mathcal{AESS}$ coincide. Then, for affine symmetric shapes, both the $\mathcal{ADSS}$ and $\mathcal{AESS}$ and their corresponding pre-images have common segments. In figures 1 and 2 we observe the relation between affine symmetric shapes and straight symmetry sets. Note that both the $\mathcal{ADSS}$ and the $\mathcal{AESS}$ contain straight segments, since the shapes are affine symmetric. Additional segments appear, since the conditions for the pre-images, both for the $\mathcal{ADSS}$ and $\mathcal{AESS}$, do not have to hold just for pair of points belonging to the affine related segments of the curve.

5   Discussion and further research

We have presented two different definitions for affine invariant symmetry sets, one based on distance functions and the second one on affine bitangent conics.

Proofs and details can be found in [13]. A number of interesting issues remain open and will be the subject of subsequent reports. We want to conclude this paper by mentioning some of these issues.

The relation between the $\mathcal{ADSS}$ and the $\mathcal{AESS}$ should be further investigated. We already observed that when the shape is affine symmetric, both the pre-images and the symmetry sets have common segments. This raises questions like if the 'differences' between both symmetry sets can give a measurement of the deviation from affine symmetry of the given shape.

Another open question is the reconstruction from the $\mathcal{ADSS}$ and $\mathcal{AESS}$, being this of extreme significance from the point of view of shape representation.

As pointed out before, following [6], the deformation of the $\mathcal{ADSS}$ and the $\mathcal{AESS}$ can be studied. Actually, most of the research done for the classical Euclidean symmetry set can be now carried on for the affine ones described in this paper.

The study of affine invariant Voronoi diagrams is an interesting topic as well. Assume we have a finite set of points $\{\mathcal{P}_i\}_{i=0}^{i=N-1} \in I\!R^2$, defining a polygon $\gamma(\mathcal{P}_i)$, and we want to extend the classical Euclidean Voronoi diagram to the affine group. Following the results for the $\mathcal{ADSS}$, the basic idea is to define an affine distance between a given point $\mathbf{X} \in I\!R^2$ and the points $\mathcal{P}_i$. For this, we have to define the affine tangent to the polygon at $\gamma(\mathcal{P}_i)$. One possible way of doing this is to consider the vector $\mathcal{P}_{i+1} - \mathcal{P}_i$ (points are counted modulo N). (Note that all points on a parallel to a line segment have the same affine distance with the segment according to this definition.) In order to avoid bias towards one direction on the polygon, both vertices around $\mathcal{P}_i$ can be used to define the discrete tangent. Another possibility is to follow the approach approach in [9], that is, construct the conic defined by $(\mathcal{P}_{i-2}, \mathcal{P}_{i-1}, \mathcal{P}_i, \mathcal{P}_{i+1}, \mathcal{P}_{i+2})$, and compute the affine tangent to this conic at $\mathcal{P}_i$. After the discrete affine tangent is defined, the affine distance follows as before, and the affine Voronoi diagram is defined as in the Euclidean case. Note that the affine Voronoi diagram can be defined for non-convex polygons as well.

From the $\mathcal{AESS}$, symmetry sets for polygons can be defined extending the definitions in [3] to 3-point contact of conics with polygons.

Probably more interesting is to extend the definitions and analysis of the $\mathcal{ADSS}$ and $\mathcal{AESS}$ to other groups and dimensions. Regarding the extension to higher dimensions, the $\mathcal{ADSS}$ can be defined for any dimension since the the affine distance exists for higher dimensions as well [20]. Some of the properties in Proposition 2 have their analogue in higher dimensions as well. In the same way, the $\mathcal{AESS}$ can be extended to higher dimensions, requesting for the right order of contact with the right high dimensional shapes. Extending the definition of the $\mathcal{AESS}$ to other groups, as the projective one, requires the use of special projective curves instead of conics, and of course requesting the right degree of contact. The extension of the $\mathcal{ADSS}$ means that a projective invariant distance has to be defined first.

To conclude, we believe that beyond the results here presented, this paper opens the door to some new research in the area of symmetry sets.

Acknowledgements

This work was initiated when the authors were visiting the *Isaac Newton Institute for Mathematical Sciences*, Cambridge University, participating at the *Workshop on New Connections between Mathematics and Computer Science*, November 1995. We thank Prof. Jeremy Gunawardena from BRIMS-HP for inviting us to the workshop, the Newton Institute for its hospitality, and the corresponding funding agencies for supporting our visit. GS thanks Prof. Peter Olver and Prof. Allen Tannenbaum for early conversations on symmetry sets.

References

1. V. I. Arnold, *Singularity Theory - Selected Papers*, London Math. Soc. Lecture Notes 53, Cambridge University Press, Cambridge, 1981.
2. V. I. Arnold, *Singularities of Caustics and Wavefronts*, Kluwer, 1990.
3. T. F. Banchoff and P. J. Giblin, "Global theorems for symmetry sets of smooth curves and polygons in the plane," *Proc. Royal Soc. Edinburgh* **106A**, pp. 221-231, 1987.
4. W. Blashke, *Vorlesungen uber Differentialgeometrie II: Affine Differentialgeometrie*, Springer, 1923.
5. H. Blum, "Biological shape and visual science", *J. Theor. Biology* **38**, pp. 205-287, 1973.
6. J. W. Bruce and P. J. Giblin, "Growth, motion and 1-parameter families of symmetry sets," *Proc. Royal Soc. Edinburgh* **104A**, pp. 179-204, 1986.
7. J. W. Bruce and P. J. Giblin, *Curves and Singularities*, Cambridge University Press, Cambridge, 1984; second edition 1992.
8. J. W. Bruce, P .J. Giblin and C. G. Gibson, "Symmetry sets," *Proc. Royal Soc. Edinburgh* **101A** pp. 163-186, 1985.
9. E. Calabi, P. J. Olver, and A. Tannenbaum, "Affine geometry, curve flows, and invariant numerical approximations," *Advances in Mathematics*, to appear.
10. L. Calabi and W. E. Hernett, "Shape recognition, prairie fires, convex deficiencies and skeletons," *Amer. Math. Monthly* **75**, pp. 335-342, 1968.
11. B. W. Char, K. O. Geddes, G. H. Gonnet, B. L. Leong, M. B. Monogan, and S. W. Watt, *Maple V*, Springer-Verlag, New York, 1991.
12. P. J. Giblin and S. A. Brassett, "Local symmetry of plane curves," *Amer. Math. Monthly* **92**, pp. 698-706, 1985.
13. P. J. Giblin and G. Sapiro, "Affine-invariant distances, envelopes, and symmetry sets," *Hewlett-Packard Laboratories Technical Report* **96:93**, June 1996.
14. P. J. Giblin and F. Tari, "Perpendicular bisectors, duality and local symmetry of plane curves," *Proc. Royal Soc. Edinburgh* **125A**, pp. 181-194, 1995.
15. S. Izumiya and T. Sano, "Generic affine differential geometry of plane curves," *Technical Report* **310**, Department of Mathematics, Hokkaido University, Sapporo, Japan, October 1995.
16. I. Kovács and B. Julesz, "Perceptual sensitivity maps within globally defined visual shapes," *Nature* **370** pp. 644-646, 1994.
17. T. S. Lee, D. Mumford, K. Zipser, and P. Schiller, "Neural correlates of boundary and medial axis representation in primate striate cortex," *Investigative Opthalmology and Visual Science Supplement (ARVO Abstract)*, 1995.
18. M. Leyton, *Symmetry, Causality, Mind*, MIT-Press, Cambridge, 1992.

19. R. Morris, *Liverpool Surface Modeling Package for Silicon Graphics and X windows*, 1996.
20. K. Nomizu and T. Sasaki, *Affine Differential Geometry*, Cambridge University Press, Cambridge, 1993.
21. R. L. Ogniewicz, *Discrete Voronoi skeletons*, Hartung-Gorre Verlag Konstanz, Zurich, 1993.
22. F. P. Preparata and M. I. Shamos, *Computational Geometry*, Texts and Monographs in Computer Science, Springer-Verlag, New York, 1990.
23. G. Sapiro and A. Tannenbaum, "On affine plane curve evolution," *Journal of Functional Analysis* **119:1**, pp. 79-120, 1994.
24. G. Sapiro and A. Tannenbaum, "On invariant curve evolution and image analysis," *Indiana University Mathematics Journal* **42:3**, 1993.
25. J. Serra, *Image Analysis and Mathematical Morphology*, Vol. 1, Academic Press, New York, 1982.
26. J. Serra, Editor, *Image Analysis and Mathematical Morphology*, Vol. 2, Academic Press, New York, 1988.
27. S. Tabachnikov, "Parametrized plane curves, Minkowski caustics, Minkovski vertices and conservative line fields," *Max Planck Institute Technical Report*, Bonn, Germany, 1996.

On the Qualitative Properties of Modified Equations

O. Gonzalez[1,2] and A.M. Stuart[3,4]

Abstract. An arbitrary consistent one-step approximation of an ordinary differential equation is studied. If the scheme is assumed to be accurate of $\mathcal{O}(\Delta t^r)$, then it may be shown to be $\mathcal{O}(\Delta t^{r+m})$ accurate as an approximation to a modified equation for any integer $m \geq 1$. A technique is introduced for proving that the modified equations inherit qualitative properties from the numerical method. For Hamiltonian problems the modified equation is shown to be Hamiltonian if the numerical method is symplectic, and for general problems the modified equation is shown to possess integrals shared between the numerical method and the underlying system. The technique for proving these results does not use the well-known theory for power series expansions of particular methods such as Runge-Kutta schemes, but instead uses a simple contradiction argument based on the approximation properties of the numerical method. Although the results presented are known, the method and generality of the proofs are new and may be of independent interest.

1 Introduction

In this note we consider the relationship between solutions to a given system of ordinary differential equations, numerical approximations to them, and solutions to associated modified equations. Our goal is to show that if an underlying system and an approximation scheme possess solutions sharing certain qualitative properties, then there is a family of associated modified equations possessing solutions that also share these properties.

The theoretical developments of this paper begin in Section 2 where we establish the notation for the semigroups generated by the underlying differential equation and the numerical method. We work within a general class of one-step methods which satisfy a certain local approximation property: the expansions of the true and approximate semigroups in powers of the time-step Δt agree up

---

1. Graduate fellow supported by the National Science Foundation

2. Current Address: University of Maryland, Institute for Physical Science and Technology, CSS Bldg, College Park, MD 20742

3. Work supported by the National Science Foundation under grant DMS-9201727 and by the Office of Naval Research under grant N00014-92-J-1876

4. Program in Scientific Computing and Computational Mathematics

to order r. Standard error estimates for such methods show that the error over a finite time interval is of $\mathcal{O}(\Delta t^r)$. Runge-Kutta methods are included in our framework, together with a variety of non-standard one-step methods used in practice, such as those used to ensure conservation of invariants for Hamiltonian problems.

In Section 3 we discuss modified equations; in particular, for any integer $m \geq 1$, the idea is to find an $\mathcal{O}(\Delta t^r)$ modification of the original ordinary differential equation with the property that the numerical method is $\mathcal{O}(\Delta t^{r+m})$ accurate as an approximation of this *modified equation*. We prove a general result concerning the existence and approximation properties of modified equations for the general class of one-step methods introduced in Section 2. Note that the idea of modified equations is well-known and was first studied in detail in the paper Warming & Hyett [1974] within the context of partial differential equations. Therein the modified equation approach was found to be useful in interpreting the qualitative properties of errors introduced by numerical approximation; for example, numerical dissipation or dispersion for wave propagation problems can often be clearly understood by studying associated modified equations. For further results on the usefulness and applicability of the modified equation approach see Griffiths & Sanz-Serna [1986].

The main contribution of this paper is contained in Section 4 where the qualitative properties of modified equations are studied. By use of a straightforward contradiction argument we show that if the numerical method inherits a certain structural property from the underlying ordinary differential equation, then the family of associated modified equations also inherits this property. The specific structural properties that we consider are conservation of a scalar function for general problems, and conservation of the canonical symplectic two-form for Hamiltonian problems.

Results similar to those in Section 4 are already contained in the literature. However, the technique of proof that we employ is new and, since it is very straightforward to apply, may have some merit when compared to existing proofs which, although somewhat shorter and more elegant, require more sophisticated mathematical machinery.

A result, similar to our Theorem 4.1 concerning conservation of a scalar function, is proved in Reich [1996]. The result of Theorem 4.2, showing that a symplectic numerical method has a family of modified equations which are Hamiltonian, is known in a wide variety of cases – see Auerbach & Friedman [1991] and Yoshida [1993] for specific examples, and see Mackay [1992], Sanz-Serna [1992] and Sanz-Serna & Calvo [1994] for general discussions concerning the backward error interpretation of symplectic schemes. The first general result concerning symplectic numerical methods and their associated modified equations is due to Hairer [1994] (see also Hairer & Lubich [1995]), who proves that all symplectic partitioned Runge-Kutta methods have Hamiltonian modified equations. Despite the great generality of this result, the proof is tied in a fundamental way to the specific form of Runge-Kutta methods. In Benettin & Giorgilli [1994], it is proved that the modified equations of symplectic methods are Hamiltonian; their analysis is not restricted to Runge-Kutta methods either and, in addition, is considerably more elegant than the one we present here. However, it is tied in a very specific way to the use of Poisson brackets to elucidate the symplectic structure. In contrast, our proof is a simple example of a more general contradiction approach which will apply to a wide variety of other structure-preserving numerical methods, including the integral-conserving methods studied here.

In summary, although the results proved here are for the most part not new, the techniques used, and the generality of the framework used, may be of independent interest. We present full proofs only for selected results. Complete details may be found in Gonzalez & Stuart [1995].

2   Background

Consider a system of ordinary differential equations in $\mathbf{R}^p$ of the form

$$\frac{du}{dt} = f(u), \tag{2.1}$$

where the vector field $f : \mathbf{R}^p \to \mathbf{R}^p$ is assumed to be of class C^∞. For any $u_0 \in \mathbf{R}^p$ we denote by $S : B \times [0,T] \to \mathbf{R}^p$ the local evolution semigroup generated by (2.1) where B is a closed ball at u_0 and $T > 0$. In particular, for any $U \in B$ the curve

$$u(t) = S_t(U) = S(U,t) \tag{2.2}$$

is a solution to (2.1) with initial condition $u(0) = U$ and is defined for all $t \in [0,T]$. Furthermore, for each $t \in [0,T]$ the mapping $S_t : B \to \mathbf{R}^p$ is a C^∞ diffeomorphism onto its image, and we denote its derivative at a point $U \in B$ by $dS_t(U) \in \mathbf{R}^{p \times p}$. We will use the fact that the mapping $B \times [0,T] \ni (U,t) \mapsto dS_t(U) \in \mathbf{R}^{p \times p}$ is continuous in U and continuously differentiable in t, and we note that $dS_t(U)$ is invertible for each $U \in B$ and $t \in [0,T]$. Hence, by compactness, there exists real numbers $C_i > 0$ $(i = 1, \ldots, 4)$ such that

$$C_1 \leq |||dS_t(U)||| \leq C_2 \quad \text{and} \quad C_3 \leq |||dS_t(U)^{-1}||| \leq C_4, \tag{2.3}$$

for all $U \in B$ and $t \in [0,T]$, where $||| \cdot |||$ denotes the Frobenius norm on $\mathbf{R}^{p \times p}$.

We will consider one-step numerical methods for (2.1) of the form

$$\mathcal{G}_{\Delta t}(U_n, U_{n+1}) = 0, \tag{2.4}$$

where $\mathcal{G}_{\Delta t} : \mathbf{R}^p \times \mathbf{R}^p \to \mathbf{R}^p$ is a given C^∞ map which depends smoothly on the parameter Δt. For any $u_0 \in \mathbf{R}^p$ we assume the numerical scheme generates an evolution semigroup in the sense that there is a closed ball $\mathcal{B}$ at u_0, real numbers $h, \mathcal{T} > 0$, and a mapping $\bar{S} : \mathcal{B} \times [0,h] \to \mathbf{R}^p$ such that for any $U \in \mathcal{B}$ and $\Delta t \in [0,h]$ the sequence (U_n) generated by

$$U_n = \bar{S}_{\Delta t}^n(U) = \bar{S}^n(U, \Delta t) \tag{2.5}$$

satisfies (2.4) for all $n\Delta t \in [0,\mathcal{T}]$. Here $\bar{S}^n(U, \Delta t)$ denotes the n-fold composition of the map $\bar{S}_{\Delta t} : \mathcal{B} \to \mathbf{R}^p$.

Given any $u_0 \in \mathbf{R}^p$ we assume without loss of generality that $\mathcal{B} = B$ and $\mathcal{T} = T$. Furthermore, we assume the numerical scheme is consistent of order r as an approximation to (2.1); that is, for any $U \in B$ we have

$$\left.\frac{\partial^i}{\partial \tau^i}\right|_{\tau=0} \bar{S}(U, \tau) = \left.\frac{\partial^i}{\partial \tau^i}\right|_{\tau=0} S(U, \tau), \qquad i = 1, \dots, r \qquad (2.6)$$

where $r \geq 1$ by consistency.

For any $n\Delta t \in [0, T]$ with $\Delta t \in [0, h]$ let $d\bar{S}_{\Delta t}^n(U) \in \mathbf{R}^{p \times p}$ denote the derivative of $\bar{S}_{\Delta t}^n : B \to \mathbf{R}^p$ at a point $U \in B$, and let $\| \cdot \|$ denote the standard Euclidean norm on $\mathbf{R}^p$. Then, by standard results from the numerical analysis of ordinary differential equations (see e.g. Stuart & Humphries [1996, Theorem 6.2.1]) there exist real numbers $C_5 > 0$ and $C_6 > 0$ depending on $U \in B$ and T such that

$$\|S_t(U) - \bar{S}_{\Delta t}^n(U)\| \leq C_5 \Delta t^r \qquad (2.7)$$

and

$$\||dS_t(U) - d\bar{S}_{\Delta t}^n(U)\|| \leq C_6 \Delta t^r \qquad (2.8)$$

for any $t = n\Delta t \in [0, T]$ with $\Delta t \in [0, h]$. Additionally, in view of (2.3) and (2.8), there is a real number $C_7 > 0$ depending on $U \in B$ and T such that, for any $t = n\Delta t \in [0, T]$ with $\Delta t \in [0, h]$, the derivative of the mapping $\bar{S}_{\Delta t}^n : B \to \mathbf{R}^p$ satisfies

$$\||d\bar{S}_{\Delta t}^n(U)\|| \leq C_7. \qquad (2.9)$$

3   Associated Modified Equations

To any ordinary differential equation of the form (2.1), and numerical approximation scheme (2.4) of order r, we can associate a *modified equation* of index N of the form

$$\frac{dv}{dt} = \tilde{f}_{\Delta t}^{(N)}(v), \qquad (3.1)$$

where $N \geq 1$ is an integer and the *modified vector field* $\tilde{f}_{\Delta t}^{(N)} : \mathbf{R}^p \to \mathbf{R}^p$ is defined as

$$\tilde{f}_{\Delta t}^{(N)}(v) = f(v) + \sum_{i=1}^{N} \Delta t^{r+i-1} q_i(v) \qquad (3.2)$$

for some functions $q_i : \mathbf{R}^p \to \mathbf{R}^p$ $(i = 1, \dots, N)$. With the appropriate choice of the functions q_i $(i = 1, \dots, N)$ the numerical scheme (2.4) is an order $r + N$ approximation to (3.1) as we now show.

For any $v_0 \in \mathbf{R}^p$ denote by $\tilde{S}^{(N)} : \tilde{B} \times [0, \tilde{T}] \times [0, \tilde{h}] \to \mathbf{R}^p$ the local evolution semigroup generated by (3.1) where $\tilde{B}$ is a closed ball at v_0 and $\tilde{h}, \tilde{T} > 0$. In particular, for any $V \in \tilde{B}$ and $\Delta t \in [0, \tilde{h}]$ the curve defined by

$$v_{\Delta t}(t) = \tilde{S}^{(N)}(V, t, \Delta t) = \tilde{S}_{\Delta t}^{(N)}(V, t) = \tilde{S}_{t,\Delta t}^{(N)}(V) \qquad (3.3)$$

is a solution to (3.1) with initial condition $v_{\Delta t}(0) = V$ and defined for all $t \in [0, \tilde{T}]$. For any $t \in [0, \tilde{T}]$ and $\Delta t \in [0, \tilde{h}]$ we denote by $d\tilde{S}_{t,\Delta t}^{(N)}(V) \in \mathbf{R}^{p \times p}$ the derivative of the mapping $\tilde{S}_{t,\Delta t}^{(N)} : \tilde{B} \to \mathbf{R}^p$ at a point $V \in \tilde{B}$. As for the underlying system, we will use the fact that, for any $\Delta t \in [0, \tilde{h}]$, the mapping $\tilde{B} \times [0, \tilde{T}] \ni (V, t) \mapsto d\tilde{S}_{t,\Delta t}^{(N)}(V) \in \mathbf{R}^{p \times p}$ is continuous in V and continuously differentiable in t, and we note that $d\tilde{S}_{t,\Delta t}^{(N)}(V)$ is invertible for each $V \in \tilde{B}$ and $t \in [0, \tilde{T}]$.

Consider the local evolution semigroups at v_0 generated by (2.1) and (2.4), and without loss of generality assume $\tilde{B} = B$, $\tilde{T} = T$, and $\tilde{h} = h$. For any $U \in B$, and for $t \in [0, T]$ and $\Delta t \in [0, h]$ sufficiently small, we may expand $S(U, t)$, $\bar{S}(U, \Delta t)$ and $\tilde{S}^{(N)}(U, t, \Delta t)$ in Taylor series about $t = 0$ and $\Delta t = 0$ as

$$S(U, t) = \sum_{j=0}^{k} \frac{t^j}{j!} \alpha_j(U) + \mathcal{O}\left(t^{k+1}\right) \tag{3.4}$$

$$\bar{S}(U, \Delta t) = \sum_{j=0}^{k} \frac{\Delta t^j}{j!} \beta_j(U) + \mathcal{O}\left(\Delta t^{k+1}\right) \tag{3.5}$$

$$\tilde{S}^{(N)}(U, t, \Delta t) = \sum_{j=0}^{k} \sum_{\ell=0}^{j} \binom{j}{\ell} \frac{t^{j-\ell} \Delta t^\ell}{j!} \tilde{\alpha}_{j\ell}(U) + \mathcal{O}\left((t + \Delta t)^{k+1}\right) \tag{3.6}$$

where $k \geq (r + N)$ is an integer and the coefficients are defined as

$$\alpha_j(U) = \left.\frac{\partial^j}{\partial \tau^j}\right|_{\tau=0} S(U, \tau) \tag{3.7}$$

$$\beta_j(U) = \left.\frac{\partial^j}{\partial \tau^j}\right|_{\tau=0} \bar{S}(U, \tau) \tag{3.8}$$

$$\tilde{\alpha}_{j\ell}(U) = \left.\frac{\partial^j}{\partial \tau^{j-\ell} \partial s^\ell}\right|_{\tau=0, s=0} \tilde{S}^{(N)}(U, \tau, s). \tag{3.9}$$

By definition of the semigroups we have

$$\alpha_0(U) = \beta_0(U) = \tilde{\alpha}_{00}(U) = U \tag{3.10}$$

and, since $\tilde{S}^{(N)}(U, 0, \Delta t) = U$ for all $\Delta t \in [0, h]$, we have

$$\tilde{\alpha}_{jj}(U) = 0, \quad 1 \leq j \leq k. \tag{3.11}$$

Now, since (2.4) is an order r approximation to (2.1), we have

$$\alpha_j(U) = \beta_j(U), \quad 1 \leq j \leq r \tag{3.12}$$

for all $U \in B$. Furthermore, from (3.1), (3.2) and its relation with (2.1) we deduce that

$$\tilde{\alpha}_{j0}(U) = \alpha_j(U), \quad 1 \leq j \leq k \tag{3.13}$$

and that

$$\tilde{\alpha}_{j\ell}(U) = 0, \quad 1 \leq j \leq k, \quad 1 \leq \ell < \min\{r, j+1\}. \tag{3.14}$$

Hence the coefficients $\tilde{\alpha}_{j\ell}(U)$ for $0 \leq j \leq r$ and $0 \leq \ell \leq j$ are fully determined by properties of the underlying evolution semigroups. Our task now is to examine the remaining coefficients $\tilde{\alpha}_{j\ell}(U)$ for $r < j \leq k$ and $r \leq \ell \leq j$, and choose the functions q_i such that (2.4) is an order $r + N$ approximation to (3.1). From (3.1) and the definition of its local evolution semigroup we deduce that the functions q_i appear in the coefficients $\tilde{\alpha}_{j\ell}$ for $r+1 \leq j \leq r+N$ and $\ell = j-1$; in particular,

$$q_i(U) = \tilde{\alpha}_{(i+r)\,(i+r-1)}(U)/(i+r-1)!, \quad i = 1, \ldots, N. \tag{3.15}$$

This follows from (3.9) using the fact that

$$q_i(U) = \frac{1}{(i+r-1)!} \left. \frac{\partial^{i+r-1}}{\partial s^{i+r-1}} \right|_{s=0} \tilde{f}_s^{(N)}(U) \tag{3.16}$$

and

$$\tilde{f}_s^{(N)}(U) = \left. \frac{\partial}{\partial \tau} \right|_{\tau=0} \tilde{S}^{(N)}(U, \tau, s). \tag{3.17}$$

We now determine the functions q_i such that (2.4) approximates (3.1) to order $r + N$. Note that this order is achieved if over one time step, i.e. $t = \Delta t$, the Taylor expansions of $\tilde{S}^{(N)}(U, \Delta t, \Delta t)$ and $\bar{S}(U, \Delta t)$ agree through order $r+N$; that is, if

$$\sum_{\ell=0}^{j} \binom{j}{\ell} \tilde{\alpha}_{j\ell}(U) = \beta_j(U), \quad 0 \leq j \leq r + N. \tag{3.18}$$

In view of (3.10) through (3.14) we have this equality for $0 \leq j \leq r$, and we now choose the functions q_i $(i = 1, \ldots, N)$ so that this equality holds for $r + 1 \leq j \leq r + N$. Replacing j in (3.18) by $i + r$ we get

$$\sum_{\ell=0}^{i+r} \binom{i+r}{\ell} \tilde{\alpha}_{i+r\,\ell}(U) = \beta_{i+r}(U), \quad 1 \leq i \leq N \tag{3.19}$$

and, since $i + r \geq 2$ for $1 \leq i \leq N$, we may write

$$\sum_{\ell=0}^{i+r-2} \binom{i+r}{\ell} \tilde{\alpha}_{i+r\,\ell}(U) + \binom{i+r}{i+r-1} \tilde{\alpha}_{(i+r)\,(i+r-1)}(U)$$

$$+ \binom{i+r}{i+r} \tilde{\alpha}_{(i+r)\,(i+r)}(U) = \beta_{i+r}(U), \quad 1 \leq i \leq N. \tag{3.20}$$

In view of (3.11) and (3.15) we obtain

$$q_i(U) = \frac{1}{(i+r)!} \left(\beta_{i+r}(U) - \sum_{\ell=0}^{i+r-2} \binom{i+r}{\ell} \tilde{\alpha}_{i+r\,\ell}(U) \right)$$

and hence, by (3.13) and (3.14), we find that

$$q_i(U) = \frac{1}{(i+r)!} \left(\beta_{i+r}(U) - \alpha_{i+r}(U) - \sum_{\ell=r}^{i+r-2} \binom{i+r}{\ell} \tilde{\alpha}_{i+r\,\ell}(U) \right). \tag{3.21}$$

Here we use the convention that a sum from $\ell = a$ to $\ell = b$ with $b < a$ is zero.

By direct calculation one can show that the coefficients $\{\tilde{\alpha}_{(i+r)\ell}\}_{\ell=r}^{i+r-2}$ only depend upon the functions $\{q_j\}_{j=1}^{i-1}$, and thus (3.21) provides a recursive definition for the functions q_i (see Gonzalez & Stuart [1995] for details). Also, since the evolution semigroups are by assumption C^∞ smooth, we note that the functions q_i are C^∞ smooth.

Remarks

1) The foregoing developments provide only *local* definitions of the functions q_i in a ball B about an arbitrary point $v_0 \in \mathbf{R}^p$. However, since the above constructions can be performed at any point, we can use these local definitions to construct mappings on all of $\mathbf{R}^p$. Note that these global mappings are well-defined since the local ones are equal on the intersection of their domains; in particular, this follows from the fact that the underlying evolution semigroups are equal on the intersection of their domains, i.e. uniqueness of solutions to the systems in (2.1), (2.4) and (3.1).

2) Arguments similar to those just given may also be found in Section 3 of Benettin and Giorgilli [1994].　　■

Since the semigroup for the modified equation (3.1) over a time interval of length Δt agrees with that of the numerical method over one time step to order $\mathcal{O}(\Delta t^{r+N})$, it follows by standard techniques (see e.g. Stuart & Humphries [1996, Theorem 6.2.1]) that the solution operator and its derivative with respect to initial data converge with order $r + N$. Furthermore, because the modified vector field is $\mathcal{O}(\Delta t^r)$ close to the original vector field in the C^1 sense, it follows that any local evolution semigroup of the modified equation is $\mathcal{O}(\Delta t^r)$ close to the corresponding local evolution semigroup of the original equation in the C^1 sense. These observations are combined with the foregoing developments in the following

Theorem 3.1 *Given any $u_0 \in \mathbf{R}^p$ and any integer $N \geq 1$ there exists a ball B at u_0, real numbers $h, T > 0$, and smooth functions q_i $(i = 1, \ldots, N)$ such that the local evolution semigroups $S_t, \bar{S}_{\Delta t}, \tilde{S}_{t,\Delta t}^{(N)} : B \to \mathbf{R}^p$ for (2.1), (2.4) and (3.1), respectively, are defined for all $t = n\Delta t \in [0, T]$ with $\Delta t \in [0, h]$. Furthermore, for each $U \in B$, there is a constant $C_8 = C_8(T, N, U) > 0$ such that*

$$|||d\tilde{S}_{t,\Delta t}^{(N)}(U) - d\bar{S}_{\Delta t}^n(U)||| + ||\tilde{S}_{t,\Delta t}^{(N)}(U) - \bar{S}_{\Delta t}^n(U)|| \leq C_8 \Delta t^{r+N} \qquad (3.22)$$

and

$$|||d\tilde{S}_{t,\Delta t}^{(N)}(U) - dS_t(U)||| + ||\tilde{S}_{t,\Delta t}^{(N)}(U) - S_t(U)|| \leq C_8 \Delta t^r \qquad (3.23)$$

for all $t = n\Delta t \in [0, T]$ with $\Delta t \in [0, h]$.

4   Qualitative Properties of the Modified Equations

In this section we will employ an induction on N to prove various properties of the modified equation (3.1). In view of Theorem 3.1 we see that, given any $u_0 \in \mathbf{R}^p$, the ball B at u_0 and the numbers $h, T > 0$ will in general depend upon N. In the following induction arguments we will choose a ball B and numbers $h, T > 0$ such that all the local evolution semigroups $S_t, \bar{S}_{\Delta t}, \tilde{S}_{t,\Delta t}^{(m)} : B \to \mathbf{R}^p$ ($m = 1, \ldots, N + 1$) are defined for any $t = n\Delta t \in [0, T]$ with $\Delta t \in [0, h]$. Note that h may shrink to zero as $N \to \infty$, but will be finite for every fixed integer $N \geq 1$.

By virtue of Theorem 3.1 we may assume, without loss of generality, that the same constants C_1, C_2, C_3 and C_4 which appear in (2.3) may be used to bound the derivatives of the semigroups for the modified equations up to order $N + 1$. Thus, for $1 \leq m \leq N + 1$,

$$C_1 \leq |||d\tilde{S}_{t,\Delta t}^{(m)}(U)||| \leq C_2 \quad \text{and} \quad C_3 \leq |||d\tilde{S}_{t,\Delta t}^{(m)}(U)^{-1}||| \leq C_4. \qquad (4.1)$$

For simplicity we define the modified equation of order $N = 0$ to be the original unperturbed equation (2.1) itself. Thus

$$\tilde{S}_{\Delta t}^{(0)}(u, t) = S(u, t) \quad \text{and} \quad \tilde{f}_{\Delta t}^{(0)}(u) = f(u), \qquad \forall u \in \mathbf{R}^p. \qquad (4.2)$$

4.1 Integrals for the Modified Semigroup

Suppose that the underlying system (2.1) and the approximation scheme (2.4) share an *integral* $\mathcal{F} \in C^1(\mathbf{R}^p, \mathbf{R})$. That is, for any $u_0 \in \mathbf{R}^p$ the function $\mathcal{F}$ is invariant under the local semigroups S and $\bar{S}$ in the sense that, for any $U \in B$ and $\Delta t \in [0, h]$, we have $\mathcal{F}(S_t(U)) = \mathcal{F}(U)$ and $\mathcal{F}(\bar{S}_{\Delta t}^n(U)) = \mathcal{F}(U)$ for all $t \in [0, T]$ and $n\Delta t \in [0, T]$. Given a modified equation for (2.1) and (2.4) the question arises as to whether or not $\mathcal{F}$ is an integral for the modified system. In this section we show that $\mathcal{F}$ is indeed an integral for the associated modified equation of index N for any integer $N \geq 1$. Precisely, we have the following

Theorem 4.1 *Suppose the underlying system* (2.1) *and the approximation scheme* (2.4) *share an integral* $\mathcal{F} \in C^1(\mathbf{R}^p, \mathbf{R})$. *Then* $\mathcal{F}$ *is an integral for the associated modified equation* (3.1) *of index* N *for any integer* $N \geq 1$. *In particular, the modified equation* (3.1) *has the form*

$$\frac{dv}{dt} = f(v) + \Delta t^r \sum_{i=1}^{N} \Delta t^{i-1} q_i(v)$$

where

$$\nabla \mathcal{F}(v) \cdot q_i(v) = 0, \quad \forall v \in \mathbf{R}^p, \ i = 1, \ldots, N.$$

Proof F. or induction assume the modified equation of index N, with local semigroup denoted by $\tilde{S}^{(N)}$, has $\mathcal{F} : \mathbf{R}^p \to \mathbf{R}$ as an integral. Note that this is true for $N = 0$ since the modified equation of order 0 is the original equation (2.1) itself.

Consider any $u_0 \in \mathbf{R}^p$. Then, for any $U \in B$ and $\Delta t \in [0, h]$ we have

$$\mathcal{F}(\tilde{S}_{\Delta t}^{(N)}(U, t)) = \mathcal{F}(U), \tag{4.3}$$

for all $t \in [0, T]$. Equivalently, for any $\Delta t \in [0, h]$, we have

$$\nabla \mathcal{F}(u) \cdot \tilde{f}_{\Delta t}^{(N)}(u) = 0, \qquad \forall u \in \mathcal{I}m(\tilde{S}_{\Delta t}^{(N)}) \tag{4.4}$$

where

$$\mathcal{I}m(\tilde{S}_{\Delta t}^{(N)}) = \{ u \in \mathbf{R}^p \mid u = \tilde{S}_{\Delta t}^{(N)}(U, t), \quad U \in B, \quad t \in [0, T] \}. \tag{4.5}$$

Now assume, for contradiction, that $\mathcal{F}$ is not an integral for the modified equation of index $N + 1$, which is of the form

$$\frac{dv}{dt} = \tilde{f}_{\Delta t}^{(N+1)}(v) = \tilde{f}_{\Delta t}^{(N)}(v) + \Delta t^{r+N} q_{N+1}(v). \tag{4.6}$$

Then there exists $u_0 \in \mathbf{R}^p$ such that

$$\nabla \mathcal{F}(u_0) \cdot q_{N+1}(u_0) \neq 0. \tag{4.7}$$

Otherwise, $\nabla \mathcal{F}(u) \cdot \tilde{f}_{\Delta t}^{(N+1)}(u) = 0$ for all $u \in \mathbf{R}^p$ and $\mathcal{F}$ would be an integral.

Let $C_9(u_0) = \nabla \mathcal{F}(u_0) \cdot q_{N+1}(u_0)/2 \neq 0$ and assume, without loss of generality, that $C_9(u_0) > 0$; otherwise, if $C_9 < 0$, then one can redefine $\mathcal{F}$ by changing sign. By continuity there is a closed ball D at u_0 such that

$$\nabla \mathcal{F}(U) \cdot q_{N+1}(U) \geq C_9 > 0, \qquad \forall U \in D. \tag{4.8}$$

Consider a point $U \in D \cap B$ and let $h, T > 0$ be such that, for any $\Delta t \in [0, h]$, the evolution semigroups satisfy $\tilde{S}_{\Delta t}^{(N+1)}(U, t) \in D$ for all $t \in [0, T]$ and $U_n = \bar{S}_{\Delta t}^n(U) \in D$ for all $n\Delta t \in [0, T]$. Then, for any $\Delta t \in [0, h]$ and $t \in [0, T]$, we have by (4.4) and (4.6)

$$\left. \frac{\partial}{\partial \tau} \right|_{\tau=t} \mathcal{F}(\tilde{S}_{\Delta t}^{(N+1)}(U, \tau)) = \nabla \mathcal{F}(\tilde{S}_{\Delta t}^{(N+1)}(U, t)) \cdot \tilde{f}_{\Delta t}^{(N+1)}(\tilde{S}_{\Delta t}^{(N+1)}(U, t))$$

$$= \Delta t^{r+N} \nabla \mathcal{F}(\tilde{S}_{\Delta t}^{(N+1)}(U, t)) \cdot q_{N+1}(\tilde{S}_{\Delta t}^{(N+1)}(U, t))$$

$$\geq C_9 \Delta t^{r+N}, \tag{4.9}$$

which implies

$$|\mathcal{F}(\tilde{S}_{\Delta t}^{(N+1)}(U, T)) - \mathcal{F}(U)| \geq C_9 T \Delta t^{r+N}, \tag{4.10}$$

for all $\Delta t \in [0, h]$.

By compactness of the closed ball D, since $\mathcal{F} \in C^1(\mathbf{R}^p, \mathbf{R})$, there is a real number $C_{10} > 0$ such that

$$|\mathcal{F}(U) - \mathcal{F}(V)| \leq C_{10} \|U - V\|, \qquad \forall U, V \in D. \tag{4.11}$$

Furthermore, in view of (3.22), the modified equation of index $N + 1$ and the numerical scheme (2.4) have solutions satisfying

$$\|\tilde{S}_{\Delta t}^{(N+1)}(U,T) - \bar{S}_{\Delta t}^n(U)\| \leq C_8 \Delta t^{r+N+1}, \tag{4.12}$$

for all $\Delta t = T/n$ and $n \geq n^*$, where n^* is any positive integer such that $T/n^* \in [0,h]$. Since by hypothesis $\mathcal{F}$ is an integral for the local numerical semigroup $\bar{S}$ we use (4.11) and (4.12) to write

$$
\begin{aligned}
|\mathcal{F}(\tilde{S}_{\Delta t}^{(N+1)}(U,T)) - \mathcal{F}(U)| &= |\mathcal{F}(\tilde{S}_{\Delta t}^{(N+1)}(U,T)) - \mathcal{F}(\bar{S}_{\Delta t}^n(U))| \\
&\leq C_{10}\|\tilde{S}_{\Delta t}^{(N+1)}(U,T) - \bar{S}_{\Delta t}^n(U)\| \\
&\leq C_8 C_{10} \Delta t^{r+N+1},
\end{aligned}
\tag{4.13}
$$

for all $\Delta t = T/n$ and $n \geq n^*$. This yields a contradiction, since for $\Delta t < TC_9/C_8C_{10}$ both (4.10) and (4.13) cannot hold. Hence $\mathcal{F}$ must be an integral for the modified equation of index $N+1$. Since $N = 0$ gives the original equation (2.1), the result follows by induction. ∎

Remark A more general result which includes the above as a special case has recently appeared in Reich [1996].

4.2 Symplecticity of the Modified Semigroup

Suppose now that the underlying system (2.1) is Hamiltonian on $\mathbf{R}^p$ with the canonical symplectic structure (assuming p is even, say $p = 2m$) and local semigroup denoted by S. Furthermore, assume the numerical approximation scheme (2.4) generates a semigroup $\bar{S}$ which is symplectic. Given an associated modified equation for (2.1) and (2.4) the question now arises as to whether or not the semigroup for the modified system defines a symplectic map. In this section we show that the semigroup for the associated modified equation of index N is indeed symplectic for any integer $N \geq 1$. Before doing this we introduce some notation.

For simplicity, we assume that the vector field $f : \mathbf{R}^p \to \mathbf{R}^p$ is Hamiltonian with respect to the canonical symplectic structure, that is

$$f(u) = J\nabla H(u) \tag{4.14}$$

for some smooth function $H : \mathbf{R}^p \to \mathbf{R}$ where $J \in \mathbf{R}^{p\times p}$ is of the form

$$J = \begin{pmatrix} O_m & I_m \\ -I_m & O_m \end{pmatrix}. \tag{4.15}$$

Here O_m and I_m denote the zero and identity matrices in $\mathbf{R}^{m\times m}$, respectively.

Theorem 4.2 *Suppose the underlying system* (2.1) *and the approximation scheme* (2.4) *both generate symplectic semigroups. Then the associated modified equation* (3.1) *of index N generates a symplectic semigroup for any integer $N \geq 1$. Thus the modified equation* (3.1), (3.2) *has the form*

$$\frac{dv}{dt} = J\nabla[H(v) + \Delta t^r Q^{(N)}(v; \Delta t)]. \tag{4.16}$$

Proof T. he result follows by a contradiction argument similar to that used in the proof of Theorem 4.1. The main idea is to show that $\tilde{f}_{\Delta t}^{(N+1)}$ is *infinitesimally symplectic* given that $\tilde{f}_{\Delta t}^{(N)}$ is. One then uses induction and the results of Dragt & Finn [1976] to establish the result. See Gonzalez & Stuart [1995] for details.

∎

5 References

S.P Auerbach & A. Friedman (1991) " Long-time Behaviour of Numerically Computed Orbits: Small and Intermediate Time-Step Analysis of One-Dimensional Systems," *J. Computational Physics,* **93**, 189–223.

G. Benettin & A. Giorgilli (1994) "On the Hamiltonian Interpolation of Near-to-the-Identity Symplectic Mappings with Application to Symplectic Integrators", *J. Stat. Phys.,* **74**, 1117–1143.

A.J. Dragt & J.M. Finn (1976) "Lie Series and Invariant Functions for Analytic Symplectic Maps," *J. Math. Phys.,* **17**, 2215–2227.

O. Gonzalez & A.M. Stuart (1995) "Remarks on the Qualitative Properties of Modified Equations," SC/CM Technical Report, Scientific Computing and Computational Mathematics Program, Stanford University.

D.F. Griffiths & J.M. Sanz-Serna (1986) "On the Scope of the Method of Modified Equations" *SIAM J. Sci. Stat. Comp.,* **7**, 994–1008.

R.S. Mackay (1992) " Some Aspects of the Dynamics and Numerics of Hamiltonian Systems," *Proceedings of the IMA Conference on The Dynamics of Numerics and the Numerics of Dynamics, 1990,* editors D. Broomhead and A. Iserles, Cambridge University Press.

E. Hairer (1994) "Backward Analysis of Numerical Integrators and Symplectic Methods," *Annals of Numerical Mathematics,* **1**, 107–132.

E. Hairer and Ch. Lubich (1995) "The Life-Span of Backward Error Analysis for Numerical Integrators," preprint.

S. Reich (1996) "Backward Error Analysis for Numerical Integrators," preprint.

J.M. Sanz-Serna (1992) "Symplectic Integrators for Hamiltonian Problems: An Overview" *Acta Numerica 1992,* 243–286.

J.M. Sanz-Serna & M.P. Calvo (1994) *Numerical Hamiltonian Problems,* Chapman and Hall.

A.M. Stuart & A.R. Humphries (1996) "Dynamical Systems and Numerical Analysis," Cambridge University Press.

R.F. Warming & B.J. Hyett (1974) "The Modified Equation Approach to the Stability and Accuracy Analysis of Finite Difference Methods," *J. Computational Physics,* **14**, 159–179.

H. Yoshida (1993) "Recent Progress in the Theory and Application of Symplectic Integrators," *Celestial Mechanics and Dynamic Astronomy,* **56**, 27–43.

Numerical Methods on (and off) Manifolds

Arieh Iserles

Abstract. It is often known from theoretical analysis that the exact solution of an ordinary differential system lies on a specific differentiable manifold and there are important advantages in retaining this feature under discretization. In this paper we examine whether the correct manifold is retained by a class of discretization methods that includes explicit multistep and multiderivative schemes. We obtain a necessary and sufficient condition for the retention of invariance under discretization. In particular, we prove that no such method can be expected to stay on a quadratic manifold. More specifically, given such a method and an arbitrary quadratic manifold $\mathcal{N}$, there exists an ordinary differential equation whose exact solution lies on $\mathcal{N}$, yet its discretization by the underlying method departs from the manifold.

1 Numerical Methods and Differentiable Manifolds

This paper addresses itself to a topic of an increasing interest in the numerical community, the discretization of ordinary differential equations (ODEs) on differentiable manifolds. Suppose that it is known that the solution of the ODE system

$$y' = f(t, y) \quad t \geq 0, \qquad y(0) = y_0 \in \mathcal{U} \subseteq \mathbb{R}^d, \tag{1}$$

where $f \in C[\mathbb{R} \times \mathbb{R}^d \to \mathbb{R}^d]$, lies for all $t \geq 0$ on the differentiable manifold $\mathcal{M}(y_0)$. In other words, the equation (1) defines a foliation of $\mathcal{U}$ into invariant manifolds. This is often a crucially important information, since invariance corresponds to the satisfaction of conservation laws by (1).

To establish common terminology, we let

$$\mathcal{M} = \bigcup_{x \in \mathcal{U}} \mathcal{M}(x)$$

and say that a numerical method is $\mathcal{M}$-*invariant* if, whenever the ODE (1) possesses a solution on $\mathcal{M}(y_0)$, so does the method.[1]

Traditionally, there has been little interest in the numerical community in investigating how well is invariance retained under discretization. The prevailing attitude was (and often still is) that the ultimate goal of numerical analysis

---

[1] There should be no confusion with the entirely different concept of G-invariance of the action of a Lie group G on a manifold.

is to devise, programme and analyse general algorithms, that can cater to the widest-possible range of ODEs, rather than affording special attention to specific equations or classes of equations [7]. This paradigm has been increasingly challenged in the last few years. A new school of thought, sometimes termed *geometric integration* or *qualitative numerical analysis* [7], aims to blend analytic and numerical means: the goal is not computation *per se* but the understanding of the ODE (1) and its behaviour. With growing attention being paid to the question of invariance, several general techniques have been proposed for invariant integration:

1. *Serendipitous methods:* Existing numerical methods which can be shown to respect specific manifolds. In particular, Runge–Kutta methods of certain type always stay on the symplectic group $Sp(2d)$ [11], the orthogonal group $O(d, \mathbb{R})$ and the unitary group $U(d, \mathbb{C})$ [2, 7].

2. *Projective methods:* An obvious idea (yet, with nontrivial ramifications) is to solve (1) with an arbitrary numerical method, subsequently projecting the result on $\mathcal{M}(\boldsymbol{y_0})$. Although this approach is sometimes infeasible or inordinately expensive – symplectic [11] and isospectral [2] flows are two important examples – it often offers a cheap and affordable means of retaining invariance. There exists an intimate connection between projective methods and differential-algebraic equations [1, 6].

3. *Rigid frame methods:* In a pioneering paper, Crouch and Grossman generalized conventional Runge–Kutta methods from Euclidean space to the setting of differentiable manifolds by correcting in a local rigid frame of $\mathcal{M}(\boldsymbol{y_0})$ [4]. This approach has been recently extended by Marthinsen and Owren [8] but its main shortcoming is that the complexity of order analysis grows very fast, limiting its present scope. Moreover, the method requires the explicit knowledge of the tangent space $T\mathcal{M}(\boldsymbol{y_0})$ at each solution point – an easy commodity in a Lie group but much less so in a general manifold.

4. *Tangent space methods:* Unlike the former class, these methods operate exclusively in the tangent space and restrict themselves to linear space operations and to commutation, subsequently returning to $\mathcal{M}(\boldsymbol{y_0})$ by means of the exponential map. Runge–Kutta methods of this kind have been recently introduced by Munthe-Kaas [9, 10] and they are particularly effective in a Lie group setting. A very interesting class of explicit algorithms that always respect a Lie group by correcting in a Lie algebra with a numerical equivalent of the Lie reduction method has been recently introduced by Zanna [12, 13].

5. *Diffeomorphism methods:* A simple idea has been presented in [3]. The manifold $\mathcal{M}(\boldsymbol{y_0})$ might sometimes have complicated geometry, yet be of a simple diffeotype. In particular, it might be diffeomorphic to a manifold that can be respected under discretization. In that case [3] advocates subjecting (1) to the same diffeomorphism, solving in the 'simpler' manifold and mapping the numerical solution back to $\mathcal{M}(\boldsymbol{y_0})$.

The purpose of this paper is to explore the potential and the limitations of a 'serendipitous' method. In other words, we consider the scope for the conservation of manifolds by conventional numerical schemes. We have already proved, together with Mari Paz Calvo and Antonella Zanna, that no noncon-fluent Runge–Kutta methods are assured of staying on manifolds that are level sets of r-degree tensors with $r \geq 3$ [3]. Here we address ourselves to explicit multistep-multiderivative methods, which fall outside the Runge–Kutta formal-ism. We prove in the sequel that such methods are, if at all, even less suitable in the integration of ODEs on manifolds and no such method can be invariant on a quadratic manifold.

By showing the limitations of 'serendipity', this paper, together with [3], provides motivation and philosophical underpinning to the ongoing search for new types of numerical ODE methods.

2   Taylor-type Discretization Methods

We assume in the sequel that the function f is analytic, hence so is the exact solution of (1). Analyticity of f implies that we can repeatedly differentiate (1), thereby explicitly expressing arbitrarily high derivatives of $y(t)$,

$$y^{(\ell)} = g_\ell(t, y), \qquad t \geq 0, \quad \ell \in \mathbb{Z}^+,$$

where

$$g_0(t, y) = y, \quad g_1(t, y) = f(t, y), \quad g_2(t, y) = \frac{\partial f(t, y)}{\partial t} + \frac{\partial f(t, y)}{\partial y} f(t, y), \quad \ldots$$

We say that a discretization method at step $n \in \mathbb{Z}^+$ is of a *Taylor type* if, perhaps after some algebraic manipulation, it can be written in the form

$$y_{n+1} = \sum_{\ell=0}^{\infty} c_{n,\ell} h^\ell g_\ell(nh, y_n), \tag{2}$$

where $y_n \approx y(nh)$, $h > 0$, and the coefficients $\{c_{n,\ell}\}$ are independent of the ODE (1). In order to make (2) consistent with the differential equation, we stipulate that $c_{n,0} = c_{n,1} = 1$.

An obvious example of a Taylor-type method (2) is the truncated Taylor expansion

$$y_{n+1} = \sum_{\ell=0}^{q} \frac{1}{\ell!} h^\ell g_\ell(nh, y_n), \qquad n \in \mathbb{Z}^+.$$

For a more interesting example let us consider the *explicit multistep* scheme

$$y_{n+1} + \sum_{k=0}^{m} a_k y_{n-m+k} = h \sum_{k=0}^{m} b_k f((n-m+k)h, y_{n-m+k}), \qquad n \geq m. \tag{3}$$

As usual with multistep methods, we need to provide *starting values* y_k, $k = 1, 2, \ldots, m$, in addition to the initial value y_0. Let us assume that the starting values are exact, hence $y_k = y(kh)$, $k = 0, 1, \ldots, m$, and consider (3) for $n = m$. Therefore, expanding in Taylor series,

$$
\begin{aligned}
y_{m+1} &= - \sum_{k=0}^{m} a_k y(kh) + h \sum_{k=0}^{m} b_k y'(kh) \\
&= - \sum_{\ell=0}^{\infty} \frac{(-1)^\ell}{\ell!} \left[\sum_{k=0}^{m} a_k (m-k)^\ell \right] g_\ell(mh, y(mh)) h^\ell \\
&\quad - \sum_{\ell=1}^{\infty} \frac{(-1)^\ell}{(\ell-1)!} \left[\sum_{k=0}^{m} (m-k)^{\ell-1} b_k \right] g_\ell(mh, y(mh)) h^\ell .
\end{aligned}
$$

Similar expansion applies to the explicit multistep-multiderivative scheme

$$
y_{n+1} = \sum_{k=0}^{m} \sum_{\ell=0}^{q} \alpha_{k,\ell} h^\ell g((n-m+k)h, y_{n-m+k}), \qquad n \geq m, \qquad (4)
$$

whereby

$$
c_{m,\ell} = \sum_{j=(\ell-q)_+}^{\ell} \frac{(-1)^j}{j!} \sum_{k=0}^{m} (m-k)^j \alpha_{k,\ell-j}, \qquad \ell \in \mathbb{Z}^+ .
$$

There is nothing special about a constant step-size h and our formalism can easily accomodate variable step sequences.

Note that a scheme may well be of a Taylor type for just some values of n: both (3) and (4) are of Taylor type just for $n = m$. Yet, this is sufficient for the purpose of the present paper.

3   A Necessary Condition for $\mathcal{M}$-invariance

Given a differentiable manifold $\mathcal{M}(x)$, embedded in $\mathbb{R}^d$, it is well known that there exists a smooth function $\rho = \rho(\,\cdot\,, \omega) : \mathbb{R}^m \to \mathbb{R}$ such that the manifold coincides with the variety

$$
\mathcal{M}_\omega = \{ x \in \mathbb{R}^d : \rho(x) = 0 \}
$$

[5]. Therefore we can rewrite the foliation from Sect. 1 in the form

$$
\mathcal{M} = \bigcup_{\omega \in \Omega} \mathcal{M}_\omega
$$

for some parameter set Ω. We henceforth stipulate that each ρ is an analytic function of x within the domain of interest.

$\mathcal{M}$-invariance of a numerical method is equivalent to the statement that

$$
\rho(y_n, \omega) = 0, \qquad n \in \mathbb{Z}^+, \quad \omega \in \Omega,
$$

where ω is such that $\rho(\boldsymbol{y}_0, \omega) = 0$. Substituting (2) into this identity (letting $c_\ell = c_{n,\ell}$ and suppressing the dependence of ρ on ω for the sake of clarity) we have

$$\rho(\boldsymbol{y}_{n+1}) = \rho\left(\boldsymbol{y}_n + \sum_{\ell=1}^{\infty} c_\ell \boldsymbol{g}_\ell\right)$$

$$= \rho(\boldsymbol{y}_n) + \sum_{m=1}^{\infty} \frac{1}{m!} \sum_{\ell_1,\ldots,\ell_m=1}^{\infty} c_{\ell_1} \cdots c_{\ell_m} h^{\ell_1 + \cdots + \ell_m}$$

$$\times \sum_{\alpha_1,\ldots,\alpha_m=1}^{d} \rho_{\boldsymbol{\alpha}} g_{\ell_1,\alpha_1} \cdots g_{\ell_m,\alpha_m},$$

where $\boldsymbol{g}_\ell = \boldsymbol{g}_\ell(nh, \boldsymbol{y}_n)$, the scalar $g_{\ell,\beta}$ is the βth coordinate of $\boldsymbol{g}_\ell$, $1 \le \beta \le d$, and

$$\rho_{\boldsymbol{\alpha}} = \frac{\partial^{\alpha_1 + \cdots + \alpha_m}}{\partial x_{\ell_1}^{\alpha_1} \cdots \partial x_{\ell_m}^{\alpha_m}} \rho(\boldsymbol{y}_n).$$

Therefore, changing the order of summation,

$$\rho(\boldsymbol{y}_{n+1}) - \rho(\boldsymbol{y}_n)$$

$$= \sum_{m=1}^{\infty} \frac{1}{m!} \sum_{r=m}^{\infty} h^r \sum_{\substack{\ell_1+\cdots+\ell_m=r \\ \ell_1,\ldots,\ell_m\ge 1}} \left(\prod_{k=1}^{m} c_{\ell_k}\right) \sum_{\alpha_1,\ldots,\alpha_m=1}^{d} \rho_{\boldsymbol{\alpha}} \prod_{j=1}^{m} g_{\ell_j,\alpha_j}$$

$$= \sum_{r=1}^{\infty} h^r \sum_{m=1}^{r} \frac{1}{m!} \sum_{\substack{\ell_1+\cdots+\ell_m=r \\ \ell_1,\ldots,\ell_m\ge 1}} \left(\prod_{k=1}^{m} c_{\ell_k}\right) \sum_{\alpha_1,\ldots,\alpha_m=1}^{d} \rho_{\boldsymbol{\alpha}} \prod_{j=1}^{m} g_{\ell_j,\alpha_j}. \tag{5}$$

Next, we examine the derivatives of ρ. We have

$$\frac{\mathrm{d}}{\mathrm{d}t}\rho(\boldsymbol{y}_n) = \sum_\alpha \rho_\alpha y'_\alpha,$$

$$\frac{\mathrm{d}^2}{\mathrm{d}t^2}\rho(\boldsymbol{y}_n) = \sum_\alpha \rho_\alpha y''_\alpha + \sum_{\alpha,\beta} \rho_{\alpha\beta} y'_\alpha y'_\beta,$$

$$\frac{\mathrm{d}^3}{\mathrm{d}t^3}\rho(\boldsymbol{y}_n) = \sum_\alpha \rho_\alpha y'''_\alpha + 3\sum_{\alpha,\beta} \rho_{\alpha\beta} y'_\alpha y''_\beta + \sum_{\alpha,\beta,\gamma} \rho_{\alpha\beta\gamma} y'_\alpha y'_\beta y'_\gamma,$$

$$\frac{\mathrm{d}^4}{\mathrm{d}t^4}\rho(\boldsymbol{y}_n) = \sum_\alpha \rho_\alpha y_\alpha^{(iv)} + 4\sum_{\alpha,\beta} \rho_{\alpha\beta} y'_\alpha y'''_\beta + 3\sum_{\alpha,\beta} \rho_{\alpha\beta} y''_\alpha y''_\beta + 6\sum_{\alpha,\beta,\gamma} \rho_{\alpha\beta\gamma} y'_\alpha y'_\beta y''_\gamma$$

$$+ \sum_{\alpha,\beta,\gamma,\delta} \rho_{\alpha\beta\gamma\delta} y'_\alpha y'_\beta y'_\gamma y'_\delta,$$

where $y_\alpha^{(\ell)}$ is the αth component of $\boldsymbol{y}^{(\ell)}(nh)$. With greater generality, it is easy to prove by induction that

$$\frac{\mathrm{d}^r}{\mathrm{d}t^r}\rho(\boldsymbol{y}_n) = \sum_{m=1}^{r} \sum_{\substack{\ell_1+\cdots+\ell_m=r \\ \ell_1,\ldots,\ell_m\geq 1}} s_\ell \sum_{\alpha_1,\ldots,\alpha_m=1}^{d} \rho_\alpha y_{\alpha_1}^{(\ell_1)}\cdots y_{\alpha_m}^{(\ell_m)}, \qquad (6)$$

where $s_r = 1$ and the only occurrence of $\boldsymbol{y}^{(r)}$ is for $m = 1$ and $\ell_1 = r$.

Let us suppose that (2) is of order $p \geq 1$, therefore $c_k = 1/k!$, $k = 1, 2, \ldots, p$. We compare (5) with (6), recalling that $g_\ell(nh, \boldsymbol{y}_n) = \boldsymbol{y}^{(\ell)}(nh)$, $\ell \in \mathbb{Z}^+$. Our first observation is that, except for $m = 1$ and $\ell_1 = p + 1$, all the h^{p+1} coefficients in (5) depend solely upon $c_1, \ldots, c_p$ and are independent of c_{p+1}. We consider c_{p+1} as a parameter: had it equalled $1/(p+1)!$, all the h^{p+1} terms in (5) and (6) would have matched perfectly, since (2) would have been of order $p + 1$. Therefore all the terms in the two expansions, except for $m = 1$, $\ell_1 = p + 1$, are identical and we derive explicitly an expression of the leading error term.

Theorem 1 *The departure of a p-order Taylor-like method (2) from the manifold $\mathcal{M}_\omega$ is*

$$\rho(\boldsymbol{y}_{n+1}, \omega) - \rho(\boldsymbol{y}_n, \omega) = \left(c_{p+1} - \frac{1}{(p+1)!}\right) h^{p+1} \sum_{\alpha=1}^{d} \rho_\alpha y_\alpha^{(p+1)} + \mathcal{O}(h^{p+2}). \quad (7)$$

Corollary 2 *The Taylor-like method (2) is $\mathcal{M}$-invariant if and only if for every $\mathcal{M}$-invariant ODE (1) and every $r \in \mathbb{N}$ it is true either that $c_r = 1/r!$ or that*

$$\sum_{\alpha=1}^{d} \rho_\alpha y_\alpha^{(r)} = 0. \qquad (8)$$

Corollary 3 *Suppose that there exists an ODE (1) such that $\boldsymbol{y}(t) \in \mathcal{M}_\omega$, $t \geq 0$, and such that*

$$\sum_{\alpha=1}^{d} \rho_\alpha y_\alpha^{(r)} \neq 0, \qquad r \geq 2. \qquad (9)$$

Then no Taylor-like method (2) may be $\mathcal{M}$-invariant, except for the exact solution of the ODE.

4 Nonexistence of Quadratically-Invariant Taylor-type Methods

We restrict our attention to nontrivial quadratic manifolds embedded in $\mathbb{R}^2$. Each such manifold can be linearly mapped onto the unit circle $y_1^2 + y_2^2 = 1$ hence, without loss of generality, we consider ODEs in $\mathbb{R}^2$ whose solution obeys $\|\boldsymbol{y}(t)\|_2 \equiv 1$, $t \geq 0$. Quadratic manifolds are significant as the simplest example of a genuinely nonlinear manifold but they are important also for their own sake since they include as special cases orthogonal, unitary, Stiefel and Grassmann manifolds and their retention by physical systems can be often interpreted as conservation of mechanical energy.

The restriction to two dimensions does not lead to any loss of generality: Our purpose is to show that no Taylor-like method (2) can be invariant on a Euclidean sphere in $\mathbb{R}^d$, $d \geq 2$. It is enough to prove this result by singling out an ODE that escapes from the sphere through a two-dimensional slice. (In the context of Runge–Kutta methods a similar, albeit more complicated, technique of cutting a manifold into two-dimensional slices has led to the descriptivelly named Cheesecutter Theorem [6].)

We consider the linear system

$$\boldsymbol{y}' = \psi(t)J\boldsymbol{y}, \qquad \text{where} \qquad J = \begin{bmatrix} 0 & 1 \\ -1 & 0 \end{bmatrix} \tag{10}$$

where the analytic function ψ is yet to be determined. We assume that $\|\boldsymbol{y}_0\|_2 = 1$. Since $\boldsymbol{y}^T \boldsymbol{f}(t, \boldsymbol{y}) \equiv 0$ for all $t \geq 0$, $\boldsymbol{y} \in \mathbb{R}^2$, it trivially follows that $\boldsymbol{y}(t)$ indeed lives on the unit circle.[2] Moreover, repeatedly differentiating (10) and using $J^2 = -I$, it is easy to prove by induction that for every $k \in \mathbb{Z}^+$ there exist functions $\mathfrak{p}_k$ and $\mathfrak{q}_k$ such that the kth derivative of $\boldsymbol{y}$ has the form

$$\boldsymbol{g}_k(t, \boldsymbol{y}) = [\mathfrak{p}_k(t)I + \mathfrak{q}_k(t)J]\boldsymbol{y}. \tag{11}$$

These functions obey the differential recurrence relations

$$\mathfrak{p}_0 \equiv 1, \qquad\qquad \mathfrak{q}_0 \equiv 0,$$
$$\mathfrak{p}_{k+1} = \mathfrak{p}_k' - \psi\mathfrak{q}_k, \qquad \mathfrak{q}_{k+1} = \psi\mathfrak{p}_k + \mathfrak{q}_k', \qquad k \in \mathbb{Z}^+. \tag{12}$$

Since $\rho(\boldsymbol{x}) = \frac{1}{2}(x_1^2 + x_2^2 - 1)$ and $\boldsymbol{x}^T J \boldsymbol{x} = 0$ for all $\boldsymbol{x} \in \mathbb{R}^2$, we deduce that

$$\sum_{\alpha=1}^{s} \rho_\alpha y_\alpha^{(r)} = \boldsymbol{y}^T \boldsymbol{g}_r(t, \boldsymbol{y}) = \mathfrak{p}_r(t)\|\boldsymbol{y}\|_2^2, \qquad r \in \mathbb{Z}^+.$$

Therefore, *pace* Corollary 3, it is enough to prove that there exists a continuous function ψ such that $\mathfrak{p}_r(0) \neq 0$ for all $r \geq 2$. As a matter of fact, even less is required to prove that no Taylor-type method can be invariant on the unit circle, namely that, given any integer $m \geq 2$, there exists $\psi = \psi_m$ such that $\mathfrak{p}_r(0) \neq 0$ for all $r = 2, 3, \ldots, m$.

Let us examine the first few iterates of (12):

$$k = 1: \quad \mathfrak{p}_2 = -\psi^2,$$
$$\mathfrak{q}_2 = \psi';$$
$$k = 2: \quad \mathfrak{p}_3 = -3\psi\psi',$$
$$\mathfrak{q}_3 = \psi'' - \psi^3;$$

---

[2] For another proof note that ψJ, being skew-symmetric, lives in the Lie algebra of $O(2, \mathbb{R})$.

$$k = 3: \quad \mathfrak{p}_4 = -4\psi\psi'' - 3\psi'^2 + \psi^4,$$
$$\mathfrak{q}_4 = \psi''' - 6\psi^2\psi';$$
$$k = 4: \quad \mathfrak{p}_5 = -5\psi\psi''' - 10\psi'\psi'' + 10\psi^3\psi',$$
$$\mathfrak{q}_5 = \psi^{(\mathrm{iv})} - 10\psi^2\psi'' - 15\psi\psi'^2 + \psi^5;$$
$$k = 5: \quad \mathfrak{p}_6 = -6\psi\psi^{(\mathrm{iv})} - 10\psi''^2 - 15\psi'\psi''' + 20\psi^3\psi'' + 45\psi^2\psi'^2 - \psi^6,$$
$$\mathfrak{q}_6 = \psi^{(\mathrm{v})} - 15\psi^2\psi''' - 60\psi\psi'\psi'' - 15\psi'^3 + 15\psi^4\psi'.$$

In general, our claim is that for every $k \geq 3$

$$\mathfrak{p}_k = -k\psi\psi^{(k-2)} + \cdots, \tag{13}$$

$$\mathfrak{q}_k = \psi^{(k-1)} - \binom{k}{2}\psi^2\psi^{(k-3)} + \cdots, \tag{14}$$

where the remaining terms in (13) do not include $\psi^{(j)}$ for $j \geq k - 2$, while the remaining terms in (14) do not include $\psi^{(j)}$ for $j \geq k - 3$. The proof is by induction: We directly verify (13–14) when $k = 3$, while for every $k \geq 3$ the recurrences (12) yield

$$\mathfrak{p}_{k+1} = [-k\psi\psi^{(k-2)} + \cdots]\psi[\psi^{(k-1)} + \cdots] = -(k+1)\psi\psi^{(k-1)} + \cdots$$

and

$$\mathfrak{q}_{k+1} = \psi[-k\psi\psi^{(k-2)} + \cdots] + \left[\psi^{(k)} - \binom{k}{2}\psi^2\psi^{(k-2)} + \cdots\right]$$
$$= \psi^{(k)} - \binom{k+1}{2}\psi^2\psi^{(k-2)} + \cdots.$$

Let an integer $m \geq 2$ be given. We choose a function ψ such that $\mathfrak{p}_k(0) \neq 0$, $k = 2, 3, \ldots, m$, by specifying its derivatives consecutively. We commence by letting $\psi(0) = 1$. Let us suppose that we have already chosen $\psi(0), \psi'(0), \ldots, \psi^{(k-3)}(0)$ so that $\mathfrak{p}_j(0) \neq 0$, $j = 2, 3, \ldots, k - 1$. According to (13), it is thus true that $\mathfrak{p}_k(0) = -k\psi(0)\psi^{(k-2)}(0) + \xi_k$, where ξ_k depends (in a nonlinear fashion) exclusively upon $\psi^{(j)}(0)$, $j = 0, 1, \ldots, k - 3$. We choose an arbitrary $|\psi^{(k-2)}(0)| > |\xi_k|/k$, whence $\mathfrak{p}_k(0) \neq 0$. Finally, we let $\psi^{(k)}(0) = 0$, $k \geq m - 1$, to ensure analyticity.

Theorem 4 *No Taylor-type method (2) may be invariant on the unit circle in* IR^2, *except for the exact solution of the underlying ODE.*

Proof.. Let us suppose that a p-order method (2) is invariant on the unit circle. Note that $c_k = 1/k!$, $k = 0, 1, \ldots, p$, and $c_{p+1} \neq 1/(p+1)!$. We consider the linear system (10) where the function ψ is chosen as above so that $\mathfrak{p}_2(0), \mathfrak{p}_3(0), \ldots, \mathfrak{p}_{p+1}(0) \neq 0$. By continuity, there exists $t^* > 0$ such that $\mathfrak{p}_j(t) \neq 0$ for all $j = 2, 3, \ldots, p + 1$ and $t \in (0, t^*]$. Integrating from the origin with step size $0 < h \leq t^*$, we reach a contradiction with Corollary 2. Hence no such method exists. $\qquad\square$

5   Implicit Methods

The analysis of this paper does not apply to implicit multistep methods. Yet, there are good reasons to believe that implicitness does not add much insofar as invariance is concerned. Herewith we present a direct, brute-force approach to the retention of quadratic invariance by the *theta method*

$$\boldsymbol{y}_{n+1} = \boldsymbol{y}_n + h[\theta \boldsymbol{f}(nh, \boldsymbol{y}_n) + (1 - \theta)\boldsymbol{f}((n + 1)h, \boldsymbol{y}_{n+1})], \qquad n \in \mathbb{Z}^+, \qquad (15)$$

where $\theta \in [0, 1]$. We note in passing that it is possible to represent (15) as a Runge–Kutta method and apply the analysis from [2]. This shows that, regardless of the choice of θ, the method cannot be invariant on quadratic manifolds. However, the direct proof adds insight. The method of proof in [2], like that in [3] and, indeed, in Sect. 4, rests upon the presence of a counterexample, a specific ODE (1) which is invariant on a manifold yet which cannot be discretized invariantly by the underlying method. An obvious rejoinder to this approach is that the example might be non-generic and it is perhaps possible that, in a well-defined sense, 'most' invariant ODEs can be discretized invariantly. The following proof demonstrates that this is not true.

By following the argument from Sect. 4, we consider invariance on the Euclidean sphere $\|\boldsymbol{y}\|_2 = 1$. We write (15) in the form

$$\boldsymbol{y}_{n+1} - (1 - \theta)h\boldsymbol{f}((n + 1)h, \boldsymbol{y}_{n+1}) = \boldsymbol{y}_n + \theta h \boldsymbol{f}(nh, \boldsymbol{y}_n)$$

and take the Euclidean norm. Since invariance and (1) imply that $\boldsymbol{y}^T \boldsymbol{f}(t, \boldsymbol{y}) \equiv 0$, the outcome is

$$\|\boldsymbol{y}_{n+1}\|_2^2 + (1 - \theta)^2 h^2 \|\boldsymbol{f}((n + 1)h, \boldsymbol{y}_{n+1})\|_2^2 = \|\boldsymbol{y}_n\|_2^2 + \theta^2 h^2 \|\boldsymbol{f}(nh, \boldsymbol{y}_n)\|_2^2.$$

Therefore, assuming invariance and considering small $h > 0$ (hence $\boldsymbol{y}_{n+1}$ arbitrarily close to $\boldsymbol{y}_n$) we deduce that $\theta = \frac{1}{2}$ (the *trapezoidal rule*) and

$$\|\boldsymbol{f}((n + 1)h, \boldsymbol{y}_{n+1})\|_2 = \|\boldsymbol{f}(nh, \boldsymbol{y}_n)\|_2. \qquad (16)$$

This, however, is impossible for all ODEs invariant on the unit sphere, as can be easily seen in $\mathbb{R}^2$. In that case the most general sphere-invariant system (1) has the form

$$\boldsymbol{f}(t, \boldsymbol{y}) = g(t, \boldsymbol{y}) \begin{bmatrix} y_2 \\ -y_1 \end{bmatrix},$$

where the function $g : \mathbb{R} \times \mathbb{R}^2 \to \mathbb{R}$ is sufficiently smooth but otherwise arbitrary. Since

$$\|\boldsymbol{f}(t, \boldsymbol{y})\|_2 = |g(t, \boldsymbol{y})| \|\boldsymbol{y}\|_2,$$

we deduce from (16) that necessarily

$$|g((n + 1)h, \boldsymbol{y}_{n+1})| = |g(nh, \boldsymbol{y}_n)|,$$

in variance with the arbitrariness of the function g. Hence, not even the trapezoidal rule may be invariant on the unit sphere and, by translation, on an arbitrary nontrivial quadratic manifold.

Acknowledgements

The author wishes to thank Antonella Zanna who has read an early version of the manuscript and offered a number of helpful and perceptive comments.

References

1. Ascher, U.M.: Stabilization of invariants of discretized differential systems. Univ. of British Columbia Tech. Rep. (1996).
2. Calvo, M.P., Iserles, A., Zanna, A.: Numerical solution of isospectral flows. Cambridge University Tech. Rep. 1995/NA3 (1995).
3. Calvo, M.P., Iserles, A., Zanna, A.: Runge–Kutta methods on manifolds. In Numerical Analysis: A.R. Mitchell's 75th Birthday Volume (G.A. Watson & D.F. Griffiths, eds), World Scientific, Singapore, 57–70.
4. Crouch, P., Grossman, R.: Numerical integration of ordinary differential equations on manifolds. J. Nonlinear Sci. **3** (1993) 1–33.
5. Guillemin, V., Pollack, A.: Differential Topology. Prentice-Hall, Englewood Cliffs NJ (1974).
6. Iserles, A.: Beyond the classical theory of computational ordinary differential equations. In State of the Art in Numerical Analysis (I.S. Duff & G.A. Watson, eds), Oxford University Press (to appear).
7. Iserles, A., Zanna, A.: Qualitative numerical analysis of ordinary differential equations. In Lectures in Applied Mathematics (J. Renegar, M. Shub & S. Smale, eds), American Mathematical Society, Providence RI (to appear).
8. Marthinsen, A., Owren, B.: Order conditions for integration methods based on rigid frames (to appear).
9. Munthe-Kaas, H.: Lie–Butcher theory for Runge–Kutta methods. BIT **35** (1995) 572–587.
10. Munthe-Kaas, H.: Runge–Kutta methods on Lie groups (to appear).
11. Sanz-Serna, J.M., Calvo, M.P.: Numerical Hamiltonian Problems. Chapman & Hall, London (1994).
12. Zanna, A: The method of iterated commutators for ordinary differential equations on Lie groups. Cambridge University Tech. Rep. DAMTP 1996/NA12 (1996).
13. Zanna, A., Munthe-Kaas, H.: Iterated commutators, Lie's reduction method and ordinary differential equations on matrix Lie groups. This volume.

On One Computational Scheme Solving the Nonstationary Schrödinger Equation with Polynomial Nonlinearity

Kh.T.Kholmurodov, I.V.Puzynin, Yu.S.Smirnov

In the present paper we propose a new technique for the numerical solution of
the Schrödinger equation with the commonly arising nonlinearity

$$i\psi_t + \psi_{xx} + f(|\psi|^2)\psi = 0, \tag{1}$$

where in the term of f we take the polynomial of the l-th degree:

$$f \equiv f(|\psi|^2) = \sum_{l=0}^{N} \alpha_l |\psi|^{2l} = \alpha_0 + \alpha_1|\psi|^2 + \alpha_2|\psi|^4 + \cdots + \alpha_N|\psi|^{2N}. \tag{2}$$

Depending on the number of terms in the series and also on the coefficients of
(2) (we suppose that $\alpha_0, \alpha_1, \cdots$ are arbitrary in general), eq.(1) arises in various
branches of physics. For example, in the case of $N = 0$ and $\alpha_0 = U \equiv U(x)$,
eq.(1) describes the motion of quantum particles (electrons, protons and so on)
in the external field U (1926, E. Schrödinger). The eq.(1) for $N = 1$ is the
Schrödinger equation with the cubic nonlinearity (the so called $\psi^3 - NLSE$).
Corresponding to the sign of $\alpha_1 = const$, the $\psi^3 - NLSE$ has a wide application
in physics. For $\alpha_1 > 0$ it describes the nonideal Bose gas of the attracting
particles [1], the propagation of light beams in a nonlinear dispersive media [2]
(the propagation of "bright solitons"), and it also arises in the study of some
problems of the theory of magnetism and molecular crystals. The $\psi^3 - NLSE$
with the $\alpha_1 = const < 0$ serves as a phenomenogical model of superfluidity for
the inhomogeneous and nonstationary order parameter ψ [3], it describes the
propagation of "dark solitons" in nonlinear optics [4], the nonideal Bose gas of
the repulsive particles [5] etc.

The addition of the following members in (2) leads to the description of the
more complicated phenomena apart from the pure mathematical variety. For
$N = 2$ one has the $\psi^3 - \psi^5 - NLSE$, which were applied to the nuclear hydro-
dynamics with Skyrma forces [6], to the Bose gas with two-particle attractive
and three-particle repulsive δ-function-like interaction potential [7] and so on.
In view of such a rich application in physics and other sciences, the development
of numerical methods for solving the NLSE is still a challenging task. A good
library of the various finite difference, spectral, iterative and other numerical

methods can be found in the remarkable work of Taha and Ablowitz [8]. Recently [9], a modification of the multigrid methods for the parallel numerical simulation of the $\psi^3 - NLSE$ was proposed.

Nevertheless, many methods applied to numerical analysis of the NLSE will work when the initial condition or some information about the initial configuration is known. In the case of the absence of such an information we propose the following scheme to solve the NLSE. Using the idea of "continuation on parameter" [10], we introduce a parametric dependence on T for $f(|\psi|^2)\psi$ the following way:

$$f(T) = Tf(|\psi|^2)\psi, \tag{3}.$$

Further, by carrying out the discretization of (1), taking into account the (3), we will have:

$$i\frac{\psi_m^{n+1} - \psi_m^n}{\Delta t} + \frac{(\delta^2\psi)_m^n + (\delta^2\psi)_m^{n+1}}{2(\Delta x)^2} + T\frac{\Phi(\psi_m^n) + \Phi(\psi_m^{n+1})}{2} = 0. \tag{4}$$

Here in (4) we introduce the following notation for the space differences and nonlinear term approximation:

$$(\delta^2\psi)_m^n = \psi_{m+1}^n - 2\psi_m^n + \psi_{m-1}^n, \quad \Phi(\psi_m^n) = f(|\psi_m^n|^2)\psi_m^n, \tag{5}.$$

The main idea of the developed scheme is that the numerical solution of the all $\psi^{2l+1} - NLSE$ can be carried out on the basis of the solution of the linear problem. Starting from the same initial condition at $T = 0$, we will come to the various results at $T = 1$ depending on the type of nonlinearity in (1). Thus, the solution of $\psi_m^n = \psi(m\Delta x, n\Delta t) \equiv \psi(x,t)$ will be a function of T for every value of the parameter $T \in [0,1]$. Obviously, for $T = 1$ we have the original system of equations (1)-(2), and for $T = 0$ we obtain the simpler linear problem.

We introduce a discretization on $T : T_j; j = 0, 1, \cdots, M(T_0 = 0, T_M = 1)$. In order to find a solution of the problem (1)-(2) in the point T_{j+1}, having the $\psi_m^n(T_j)$, one can use Newton's method. We suppose that the difference $|T_{j+1}-T_j|$ is enough small and we have the good initial approximation for Newton's method:

$$F'(\vec{p}, z_k)v_k = -F(\vec{p}, z_k),$$

$$\vec{p} = \{T, \psi_m^n, \alpha_0, \alpha_1, \cdots\}, \quad z_k = \{\psi_m^{n+1,k}\},$$

and

$$z_{k+1} = z_k + v_k, \quad k = 0, 1, \cdots.$$

It is important that the proposed approach can be easily generalized for investigation of multidimensional nonlinear Schrödinger systems. The more detailed practical analysis of the described above numerical scheme is in progress. Another realization of the proposed method has been done in [11] to carry out the numerical analysis of three-dimensional polaron equations.

References

1. (a) Eilbeck J.C. Numerical studies of solitons, in: *Solitons and Condensed Matter Physics*, Springer-Verlag, Berlin, 1978; (b) Zakharov V.E., Shabat A.B. *Sov. Phys. JETP, 34 (1972) 62.*
2. Yajima N., Oikawa M., Satsuma J. and Namba C. *Rep. Res. Inst. Appl. Mech.* XXII, No 70, 1975.
3. Ginzburg V.L., Sobyanin A.A. *Sov. J. Usp. Fiz. Nauk,* 120, 1976, 153.
4. Pitaevskii L.P. *Sov. Phys. JETP,* 13, 1961, 451.
5. (a) Hasegawa A., Tappet F. *Appl. Phys. Lett.,* 23, 171, 1973; (b) Gredeskul, S.A., Kivshar Yu.S. *Phys. Rev. Lett.,* 62, 977, 1989.
6. Kartavenko V.G. *Sov. J. Nucl. Phys.,* 1984, 40, 2, p.337.
7. (a) Kovalev A.S., Kosevich A.M. *Sov. J. Low Temp. Phys.,* 1976, 2, 7, p.913; (b) Barashenkov I.V., Kholmurodov Kh.T. *Proc. Int. Symp. Selected Prob. of Stat. Mech.,* Dubna, JINR, 1987; (c) Makhankov V.G., Kholmurodov Kh.T. *Sov. J. Low Temp. Phys.,* 1989, 15, 1, p.72.
8. Taha T.R., Ablowitz M. *J. Comp. Phys.,* 1984, 55, pp.203-220.
9. Martin I., Tirado F., Vazquez L. and Fabero J.C. *Proc. Int. Conf. PC 94,* Lugano, Switzerland, 1994.
10. Davidenko D.F. *Math. J. of Ukraine,* 1955, v.7, 1, p.18.
11. Akishin P.G., Puzynin I.V., Smirnov Yu.S. On numerical solving of nonlinear polaron equations, in: *Lectures in Applied Mathematics*, M.Shub, J.Renegar, S.Smale (editors), AMS, 1996, (to be published).

Newton Iteration Towards a Cluster of Polynomial Zeros

Peter Kirrinnis

Abstract. Let $P \in \mathbf{C}[z]$ be a polynomial which has a cluster of k zeros near $v_0 \in \mathbf{C}$. The coefficients of the factor F of P corresponding to this root cluster can be computed from the coefficients of P by Newton iteration in $\mathbf{C}^k$, applied to the mapping $F \mapsto P \bmod F$.

This paper presents and analyzes a numerical algorithm based on this idea. The algorithm uses only polynomial multiplication and division. In particular, there is no need to solve linear systems.

The analysis results in an explicit quantitative condition for the coefficients of P which guarantees the algorithm to converge when started with $F_0 = (z - v_0)^k$. Efficient test procedures for this condition are described. Finally, another starting value condition in terms of root sizes is proved and illustrated with some examples.

1 Introduction

1.1 Newton Iteration and Polynomial Zero Finding

Newton's method is one of the most important tools for solving nonlinear equations. Algorithms for computing zeros of univariate complex polynomials involve Newton iteration in several ways. There are also several approaches to measure the efficiency of such algorithms:

A frequently used method to compute a single root of a polynomial P is to choose an initial value v_0 at random, try some steps of the iteration $v_{j+1} = v_j - P(v_j)/P'(v_j)$ and repeat with another random v_0 if this does not work. Practical experience shows this approach to be successful and efficient in most cases. This experience is founded theoretically by the work of Smale [27, 29, 30] and Shub and Smale [25, 26] who have analyzed the probability of success and the average running time (in terms of the number of iterations resp. of arithmetic operations) of such algorithms, using suitably parameterized Newton and Euler type iterations. Friedman [2] has shown that the algorithm as described above succeeds with positive probability. Renegar [18] has discussed the efficiency of a homotopy method for computing all (instead of one) roots of p in the above probabilistic framework.

In the last years there has also been great progress in the development of asymptotically fast deterministic root finding algorithms. They involve Newton iteration essentially in two ways: One way is to make Newton's method for one zero as described above deterministic by computing suitable starting values v_0 with a deterministic (and usually slow) search method. Example for such algorithms are Renegar's [19], Pan's [16], and Schönhage's [23], see Sect. 1.2 for details. The other approach is to split a polynomial P into two factors F and G, preferably of roughly the same degree, and proceed recursively. These algorithms use multidimensional Newton iteration to refine the coefficients of the factors F and G as a zero of the mapping $(F, G) \to P - F \cdot G$. Examples are the algorithms proposed by Schönhage [21, 22], Pan [15, 17], Neff and Reif [14], and Malajovich and Zubelli [13]. For both approaches, the crucial (and for moderate accuracy requirements the most expensive) part of the algorithm is to compute initial values v_0 resp. (F_0, G_0) with which Newton iteration is guaranteed to work.

1.2 Search Algorithms and Newton Iteration

While most asymptotically fast algorithms use the splitting strategy described above, for polynomials of moderate degree other methods may be much faster in practice. A promising approach is to compute a crude approximation v_0 for a root with a classical search algorithm and then test whether v_0 can be used as a starting value for Newton iteration. These search algorithms combine some sort of algorithm to detect (not to locate!) or count zeros of a polynomial in a given region in the complex plane (say, a disk or a rectangle) with geometric constructions. Classical root detection algorithms (like Schur-Cohn tests) are discussed in chapter 6 of Henrici's monograph [6]. Prominent examples for the use of geometric constructions are Weyl's [31] and Lehmer's [12] algorithms. The combination with Newton iteration is studied, e.g., by Renegar [19], who uses Schur-Cohn tests to find starting values, Pan [16], who uses Weyl's classical algorithm and Turan's proximity test, and Schönhage [23], who combines Graeffe's root squaring method with a geomtric construction similar to Lehmer's. The latter algorithm is implemented.

The classical criteria for a point to be a starting value for Newton iteration are Kantorovich's [8]. Smale [28] and Kim [9] have derived criteria which use only the Taylor coefficients of P in v_0.

The present paper deals with the problem that v_0 may be close to a multiple root or a cluster of roots. In this case Kantorovich's resp. Smale's or Kim's criteria are never fulfilled, and so the "slow" search process goes on, usually with linear convergence. If the cluster is in fact a multiple root which is well isolated from the others, then the usual Newton iteration $v_{j+1} = v_j - P(v_j)/P'(v_j)$ still converges, but only linearly. In addition, the iteration becomes unstable, because $P'(v_j)$ tends to zero

Quadratic convergence for multiple roots can be achieved with other iterations. The most common solutions are Newton's method for P/P' and the modified Newton iteration $v_{j+1} = v_j - k \cdot P(v_j)/P'(v_j)$ for a k-fold zero v. Pan [16] has described an efficient root finding algorithm which combines the latter iteration with a modification of Weyl's algorithm. Renegar [19] computes the root of the $(k-1)$-th derivative $P^{(k-1)}$ corresponding to the cluster and maintains geometric information about the corresponding roots of P.

1.3 Newton Iteration towards a Cluster

The present paper deals with clusters of zeros in the following way: Assume that a polynomial $P \in \mathbf{C}[z]$ and a complex number v_0 are given such that v_0 is close to a cluster of k zeros of P, i.e., $|P(v_0)|$, $|P'(v_0)|$, $|P''(v_0)|$, $\ldots$, $|P^{(k-1)}(v_0)|$ are small, while $|P^{(k)}(v_0)|$ is large. Making terms like "close", "small" and "large" precise is one of the subjects of this paper.

Further information about this cluster of roots can be obtained as follows: The normalizing simplification $v_0 = 0$ can be assumed w.l.o.g. It is achieved by replacing $P(z)$ with $P(z + v_0)$. Then P has a cluster of k roots $u_1, \ldots, u_k$ near the origin. Now let F denote the factor of P whose roots are exactly those belonging to the cluster,

$$F(z) \;=\; (z - u_1) \cdots (z - u_k) \; .$$

Once the coefficients of F are known, the root cluster can be further investigated after "blowing up" the problem by scaling with a suitable factor $A > 0$, i.e., replacing $F(z)$ with $A^{-k} \cdot F(A \cdot z)$. The other factor $G = P/F$ is investigated recursively.

This paper deals with the problem of computing numerical approximations for F and G by multidimensional Newton iteration. Let $F(z) = z^k - E(z)$. As the u_j are small, E has small coefficients. Hence $F_0(z) = z^k$ is a first approximation to F. If F_0 approximates F sufficiently well (in other words, if v_0 is sufficiently close to the u_j or if the coefficients of E are small enough), then better approximations for F can be computed by multidimensional Newton iteration, applied to the function from $\mathbf{C}^k$ to $\mathbf{C}^k$ which maps the vector of coefficients of $E(z)$ to the coefficient vector of $P(z)$ mod $(z^k - E(z))$, taking, of course, the unique representative of degree less than k. The complementary factor $G = P/F$ is obtained by polynomial division.

An algorithm based on this idea is described and analyzed in detail. The analysis of the algorithm yields an explicit quantitative condition in terms of the coefficients of P which guarantees convergence when the iteration is started with $F_0(z) = z^k$. Then the problem of computing a starting value is trivially solved. It is discussed how this convergence criterion can be checked easily (and fast) from the coefficients of P. Having a fast test for root clusters is important because in most applications they do not occur frequently, and so the test will fail. Finally, a starting value condition in terms of the size of the roots of P is proved and illustrated with some examples.

The special structure of the problem is exploited in three aspects: First, the convergence criterion is derived directly from the analysis of the algorithm. Thus the general theory of convergence criteria for multidimensional Newton iteration by Kantorovich [8] or Smale [28] is not used.

Second, it is not necessary to solve the linear system implicit in the Newton step explicitly. This is explained in the general framework of multidimensional Newton iteration: The Newton step for a mapping $\Phi : \mathbf{C}^k \to \mathbf{C}^k$ is $x_{j+1} = x_j - (D\Phi(x_j))^{-1} \Phi(x_j)$, where $D\Phi(x)$ denotes the Jacobian of Φ in x. A straightforward implementation is to compute the "Newton correction" ξ_j from the linear system $D\Phi(x_j)\xi_j = -\Phi(x_j)$ and then set $x_{j+1} = x_j + \xi_j$. Here the subproblem of computing ξ_j is solved by multiplying $\Phi(x_j)$ with a sufficiently precise approximation C_j for $(D\Phi(x_j))^{-1}$. The point is that C_j is computed from C_{j-1} with a simple quadratically convergent iteration, and even more, one "correction step" per Newton step is sufficient.

Third, because the Jacobians (and their inverses) are Toeplitz matrices, all computations can be expressed in terms of polynomial arithmetic. The algorithm involves only basic operations with complex polynomials, namely addition, multiplication, and computation of remainders. Most of the operations involve only polynomials of degree less than $2k$, which is particularly favourable when k is small with respect to n. This is likely to be the case in most applications.

1.4   Implementation of Fast Algorithms

The interest in bounds for the complexity of root finding is complemented by efforts on efficient implementations. Besides the widely used Jenkins-Traub algorithm [7] there have been recent efforts in implementing Schönhage's "splitting circle method" [21]. Gourdon has implemented this algorithm in MAPLE [3] and on the basis of the number theory package PARI [4].

Schönhage has started a project of implementing reliable numerical algorithms (in particular polynomial arithmetic and root finding) with a system based on fast (FFT based) multiprecision arithmetic. This project is motivated by the complexity studies [20] and [21]. The algorithms implemented so far comprise integer and arbitrary complex number and polynomial arithmetic and solving quadratic and cubic equations numerically with radicals. For further details see [24].

As a first root finding algorithm, particularly for polynomials of moderate degree, Schönhage has implemented an "ordinary" method [23] which roughly works as follows: Graeffe's root squaring method is used to determine thin annuli in which P has a zero. The intersection of such annuli eventually produces a moderate precision approximation v_0 for a root, and as soon as Kantorovich's criterion is fulfilled, Newton iteration is used to approximate the root more precisely. If v_0 is close to a root cluster, then the performance slows down from quadratic to linear convergence. The "cluster splitting" method described here shall complement the algorithm in this case.

One aim of the present paper is to serve as a theoretical basis for an efficient implementation. This has motivated to make error estimates explicit and to try to achieve small (instead of O) error bounds. On the other hand, a little increase in the constants is accepted, because it allows a more simple analysis. A full discussion of rounding errors and bit complexity issues would only introduce technical complications without giving new insight. Therefore, polynomial operations are assumed to be performed exactly. In practice, an algorithm for polynomial multiplication would, e.g., be specified as follows: Given polynomials U and V and $\epsilon > 0$, compute a polynomial P such that $\|U \cdot V - P\| < \epsilon$, where $\|\cdot\|$ denotes some norm on the space of polynomials; for details see, e.g., [20] or [24, Chap. 9].

1.5   Related Research

The idea of using multidimensional Newton iteration for splitting a polynomial into factors and the technique of using an auxiliary polynomial (as an encoding of the inverse of the Jacobian) and the quadratic iteration for it have already been used by Schönhage [21, §§ 10,11], [22, § 3]. The new contributions of the present paper are the following: First, Newton's method is applied to the mapping $F \mapsto P \bmod F$ instead of $(F, G) \mapsto P - F \cdot G$. This allows to do most of the computations with polynomials of degree k instead of n. Second, explicit error bounds and starting value conditions for the particular case $F_0(z) = z^k$ are given, and third, simple and efficient tests for root clusters are provided.

Malajovich and Zubelli [13] have proposed an algorithm which is closely related to the one described here. They also describe an algorithm for splitting a polynomial P with k small and $n - k$ large zeros into the corresponding factors F and G by multidimensional Newton iteration and discuss conditions for $F_0(z) = z^k$ to be a suitable starting value. Their approach and the present one differ in the following aspects: First, in [13], Newton iteration is applied to the mapping $(F, G) \mapsto P - F \cdot G$ instead of $F \mapsto P \bmod F$, i.e., n coefficients are involved throughout. Second, the linear system involved in the Newton iteration is solved explicitly (in an efficient way), cf. the end of Sect. 1.3. Third, the algorithm of [13] is described within the usual framework of multidimensional Newton iteration, i.e., in terms of linear systems (with Toeplitz like matrices) rather than polynomial arithmetic. The convergence of the algorithm is proved by applying general results from Smale [28]. Finally, the resulting starting value condition is expressed in terms of the roots of P rather than its coefficients. This is motivated by the intended application of the algorithm which differs from the one intended here, see the introduction of [13] for details. Theorem 8.1 gives a similar starting value condition which is weaker than the one stated by Malajovich and Zubelli, see Chap. 8 for further discussion.

A related topic is the use of multidimensional Newton iteration for the simultaneous computation of all roots of a polynomial. This has been studied, e.g., by Grau [5], whose technique has been extended and analyzed with respect to bit complexity by Kirrinnis [10, 11], also using explicit error bounds.

2   Definitions and Notation

2.1   General Definitions

The leading coefficient and the exact degree of a polynomial $P \in \mathbf{C}[z]$ are denoted by $\mathrm{lc}(P)$ and $\deg P$, respectively. Besides congruences, the symbol "mod" also denotes the corresponding operator: For $F, P \in \mathbf{C}[z]$, $R = F \bmod P$ means R to be the polynomial determined uniquely by $R \equiv F \bmod P$ and $\deg R < \deg P$. A "div" operator denoting the quotient is defined correspondingly.

Π denotes the algebra of univariate complex polynomials, and $\| \cdot \|$ denotes the ℓ_1-norm, $\|P\| := |a_0| + |a_1| + \cdots + |a_n|$ for $P(z) = a_0 + a_1 z + \cdots + a_n z^n$. This norm is submultiplicative, $\|P_1 \cdot P_2\| \leq \|P_1\| \cdot \|P_2\|$.

For $n \in \mathbf{N}$, Π_n denotes the subspace of Π of polynomials of degree $\leq n$ and is identified with the space $\mathbf{C}^{n+1}$ of the coefficient vectors in the natural way. Π_n^0 denotes the set of monic polynomials of degree n:

$$\Pi_n^0 \quad := \quad \{z^n + U(z) : U \in \Pi_{n-1}\} \ .$$

This is an n-dimensional affine subspace of Π_n; it is identified with the space $\mathbf{C}^n$ of the coefficient vectors of the Us in the natural way. For $\tau \geq 0$, let

$$\Pi_n^0(\tau) \quad := \quad \{z^n + U(z) : U \in \Pi_{n-1}, \|U\| \leq \tau\} \ .$$

2.2   Special Assumptions

Now it is time to fix some notation used throughout this paper: The polynomial P to be factored has degree n and is w.l.o.g. assumed to have norm 1:

$$P \in \Pi_n, \quad \|P\| = 1 \ .$$

The size of the root cluster in question is k, whence it is reasonable to assume

$$2 \leq k < n \ .$$

For notational convenience, define

$$m := n - k \ .$$

3   Algorithmic Ideas

The problem of computing a factor F of P can be expressed as a multidimensional zero computation problem in two "natural" ways: F divides P iff there is a complementary factor G such that $P - F{\cdot}G = 0$. On the other hand, a factor F of P annihilates the remainder of polynomial division, i.e., $P \bmod F = 0$. The Newton steps implied by both approaches are derived in Sects. 3.1 and 3.2. Sect. 3.3 discusses how these Newton steps can be implemented.

3.1   Newton Iteration for Both Factors

The first approach is to compute a zero of the mapping $(F, G) \mapsto P - F{\cdot}G$, which is more precisely defined as follows:

Definition 3.1 *Let $a_n := \mathrm{lc}(P)$ and define*

$$\mathcal{E}(U, V) \quad := \quad P - (z^k + U) \cdot (a_n \cdot z^m + V)$$

for $U \in \Pi_{k-1}$ and $V \in \Pi_{m-1}$. Then $\mathcal{E}(U, V) \in \Pi_{n-1}$, and

$$\mathcal{E} \quad : \quad \mathbf{C}^k \oplus \mathbf{C}^m = \mathbf{C}^n \quad \rightarrow \quad \mathbf{C}^n$$

in the canonical way, identifying U, V, and $\mathcal{E}(U, V)$ with their coefficient vectors. It is convenient to abuse the notation as follows: For $F(z) = z^k + U(z)$ and $G(z) = a_n z^m + V(z)$ write $\mathcal{E}(F, G) = P - F \cdot G$ instead of $\mathcal{E}(U, V)$.

The mapping $\mathcal{E}$ is given by polynomials of degree at most two and is therefore easy to analyze: Let $f \in \Pi_{k-1}$ and $g \in \Pi_{m-1}$. Then

$$\mathcal{E}(F+f, G+g) \;=\; P - (F+f) \cdot (G+g) \;=\; \mathcal{E}(F,G) - (f \cdot G + g \cdot F) - f \cdot g \quad (3.1)$$

shows that the derivative $D\mathcal{E}(F,G)$ of $\mathcal{E}$ at (F,G) is the linear operator from $\Pi_{k-1} \oplus \Pi_{m-1}$ to Π_{n-1} defined by

$$D\mathcal{E}(F,G)(f,g) \;=\; -(f \cdot G + g \cdot F) \ .$$

This operator is nonsingular iff F and G are relatively prime which is reflected by the fact that the Jacobian of $\mathcal{E}$ in (F,G), i.e., the matrix representation of $D\mathcal{E}(F,G)$, is the Sylvester matrix of F and G, whence its determinant is the resultant of F and G. Now let F and G be relatively prime and let f and g be the *Newton corrections* for F and G,

$$(f,g) \;=\; -\left(D\mathcal{E}(F,G)\right)^{-1} \mathcal{E}(F,G) \ .$$

This means that f and g are chosen such that in (3.1) only the second order term is left, i.e., f and g are defined by

$$\mathcal{E}(F,G) \;=\; P - F \cdot G \;=\; f \cdot G + g \cdot F \ .$$

This simple approach involves n coefficients throughout. This is acceptable for applications where usually $k, m = O(n)$, cf. [21, 17, 13] For the applications intended here, k will in most cases be small compared with n, whence computations involving only k coefficients should perform better. The idea is therefore to iterate only for F and to compute a corresponding G by division, which leads to the second approach.

3.2   Newton Iteration for Annihilating the Remainder

The second approach is to compute a zero of the mapping $F \mapsto P \bmod F$.

Definition 3.2 *For $F \in \Pi_k^0$, define $\mathcal{R}(F) \in \Pi_{k-1}$ and $\mathcal{Q}(F) \in \Pi_m$ as the unique polynomials with*

$$P \;=\; F \cdot \mathcal{Q}(F) + \mathcal{R}(F) \ .$$

Then $\mathcal{R} : \mathbf{C}^k \to \mathbf{C}^k$ and $\mathcal{Q} : \mathbf{C}^k \to \mathbf{C}^m$ in the canonical way, mapping the coefficients of $F(z) - z^k$ to the coefficients of $\mathcal{R}(F)$ or $\mathcal{Q}(F)$, respectively.

As F is assumed to be monic, the coefficients of $\mathcal{R}(F)$ and $\mathcal{Q}(F)$ are polynomials in the coefficients of P and F. To determine the derivative of $\mathcal{R}$ and $\mathcal{Q}$, let $f \in \Pi_{k-1}$ be a perturbation of F and denote

$$Q \;=\; \mathcal{Q}(F), \quad R \;=\; \mathcal{R}(F), \quad q_1 \;=\; D\mathcal{Q}(F)f, \quad r_1 \;=\; D\mathcal{R}(F)f \ .$$

Then $\mathcal{Q}(F+f) = Q + q_1 + O(\|f\|^2)$ and $\mathcal{R}(F+f) = R + r_1 + O(\|f\|^2)$. The equation

$$P \;=\; (F+f) \cdot \mathcal{Q}(F+f) + \mathcal{R}(F+f) \;=\; F \cdot Q + R + (f \cdot Q + q_1 \cdot F + r_1) + O(\|f\|^2)$$

shows that the first order terms q_1 and r_1 are given by

$$f \cdot Q + q_1 \cdot F + r_1 \;=\; 0 \; .$$

In particular,

$$D\mathcal{R}(F)f \;=\; -f \cdot \mathcal{Q}(F) \bmod F \; .$$

Hence $D\mathcal{R}(F)$ is nonsingular iff $\mathcal{Q}(F)$ has an inverse modulo F, i.e., is a unit in the ring $\mathbf{C}[z]/(F)$. This is true iff F and $\mathcal{Q}(F)$ are relatively prime. In this case the Newton correction $f = (D\mathcal{R}(F))^{-1}\mathcal{R}(F)$ for F is given by the condition

$$f \cdot \mathcal{Q}(F) \;\equiv\; \mathcal{R}(F) \;\equiv\; P \bmod F \; . \tag{3.2}$$

This is the same condition as in the first approach, because $\mathcal{R}(F) = \mathcal{E}(F, \mathcal{Q}(F))$.

3.3   Implementation of the Newton Step

From now on, the discussion is restricted to the second approach, computing a zero of $\mathcal{R}$. This section discusses how to *compute* the Newton correction f defined by (3.2).

Definition 3.3 *Let $F \in \Pi_k^0$ be such that F and $Q := \mathcal{Q}(F)$ are relatively prime. Then Q has a (unique) inverse $H \in \Pi_{k-1}$ modulo F, i.e., $H \cdot Q \equiv 1 \bmod F$, which is denoted as $H = \mathcal{H}(F)$. With the canonical identification, $\mathcal{H}$ is a partial function from $\mathbf{C}^k$ to $\mathbf{C}^k$.*

Now let F, Q, and H as in Def. 3.3 and let $R = \mathcal{R}(F)$. $H = \mathcal{H}(F)$ encodes the inverse $(D\mathcal{R}(F))^{-1}$ of the Jacobian of $\mathcal{R}(F)$ in the following sense: The Newton correction $f \in \Pi_{k-1}$ for F with respect to $\mathcal{R}$ as defined by (3.2) is given by

$$R \cdot H \;\equiv\; f \cdot Q \cdot H \;\equiv\; f \bmod F \; .$$

Hence f can be computed from R and H by a multiplication of two polynomials of degree less than k and a division of a polynomial of degree at most $2k - 2$ by a polynomial of degree k with small roots.

The approximate factor F and $\mathcal{H}(F)$ are maintained simultaneously. More precisely, a sufficiently precise approximation H for $\mathcal{H}(F)$ is used. Let $D \in \Pi_{k-1}$ be such that $H \cdot Q \equiv 1 - D \bmod F$. A better approximation for $\mathcal{H}(F)$ can be computed from H with the *"defect squaring step"* $\tilde{H} \equiv H \cdot (1 + D) \bmod F$. Then $\tilde{H} \cdot Q \equiv 1 - D^2 \equiv 1 - \tilde{D} \bmod F$, whence the new defect $\tilde{D} \in \Pi_{k-1}$ fulfils $\tilde{D} \equiv D^2 \bmod F$, which implies $\|\tilde{D}\| \leq C \cdot \|D\|^2$ with some constant $C > 0$. If $F \in \Pi_k^0$, then $C = 1$ is sufficient. Applying this repeatedly gives a quadratic iteration which computes approximations for $\mathcal{H}(F)$ with any desired precision.

The Newton step for approximating a factor of P as a root of $\mathcal{R}$ is now organized as follows: Given F, R, and H with sufficiently small defect D, the Newton correction $f = R \cdot H \bmod F$ yields a (hopefully better) approximate factor $\hat{F} = F + f$. The inaccuracy of H must be accounted for in the estimate of the new error $\mathcal{R}(\hat{F})$. Now let $\hat{Q} = \mathcal{Q}(\hat{F})$ and $\tilde{D} \in \Pi_{k-1}$ with $H \cdot \hat{Q} \equiv 1 - \tilde{D} \bmod \hat{F}$. The norm of the new defect $\tilde{D}$ will usually be too large for the next Newton step. Therefore the above quadratic iteration is used to replace H with $\hat{H}$ such that the defect is sufficiently small again. It turns out that *one* "defect squaring step" is sufficient.

4   Auxiliaries

Lemma 4.1 *Let $n, k \in \mathbf{N}$ with $1 \le k \le n+1$, $G \in \Pi_n$ and $F(z) = z^k - E(z)$, where $E \in \Pi_{k-1}$ with $\|E\| \le 1$. Furthermore let $Q \in \Pi_{n-k}$ and $R \in \Pi_{k-1}$ with $G = Q \cdot F + R$. Then $\|R\| \le \|G\|$ and $\|Q\| \le (n-k+1) \cdot \|G\|$. Both estimates are sharp.*

Proof.. The proof is by induction in n. If $n = k-1$, then $Q = 0$ and $R = G$, hence both estimates hold. For $n \ge k$, let $G(z) = G_0(z) + z^k \cdot G_1(z)$ with (unique) $G_0 \in \Pi_{k-1}$ and $G_1 \in \Pi_{n-k}$. Let $\tilde{G}(z) = G_0(z) + E(z) \cdot G_1(z)$. Then $\deg \tilde{G} \le n-1$ and $\|\tilde{G}\| \le \|G_0\| + \|E\| \cdot \|G_1\| \le \|G_0\| + \|G_1\| = \|G\|$. Let $\tilde{G} = \tilde{Q} \cdot F + \tilde{R}$ with (unique) $\tilde{Q} \in \Pi_{n-k-1}$ and $\tilde{R} \in \Pi_{k-1}$. Then $G = \tilde{G} + F \cdot G_1 = (\tilde{Q} + G_1) \cdot F + \tilde{R}$, hence $R = \tilde{R}$ and $Q = \tilde{Q} + G_1$. With the induction hypothesis, applied to $\tilde{G}$, this yields $\|R\| = \|\tilde{R}\| \le \|\tilde{G}\| \le \|G\|$ and $\|Q\| \le \|\tilde{Q}\| + \|G_1\| \le (n-k) \cdot \|G\| + \|G_1\| \le (n-k+1) \cdot \|G\|$.

The example $z^n = \left(z^{n-k} + z^{n-k-1} + \cdots + z^2 + z + 1 \right) \cdot \left(z^k - z^{k-1} \right) + z^{k-1}$ shows that both estimates are sharp. $\square$

5   A Newton Step

This section provides a detailed analysis of the numerical performance of the Newton step as outlined in Sect. 3.3. To make the idea clear, the Newton step is first discussed in its theoretical form, ignoring the problem how to compute the Newton correction and in particular the question how the error of $\mathcal{H}(F)$ is involved.

Lemma 5.1 *Let $0 < \epsilon < 1$, $M \ge 1$, $F \in \Pi_k^0(1 - M\epsilon)$, $Q = \mathcal{Q}(F)$ and $R = \mathcal{R}(F)$. Assume that F and Q are relatively prime and let $H = \mathcal{H}(F)$. Finally, assume that $\|R\| \le \epsilon$ and $\|H\| \le M$.*

Define $f \in \Pi_{k-1}$ by the congruence $f \cdot Q \equiv R \bmod F$ and let $\hat{F} := F + f$. Then $\|\mathcal{R}(\hat{F})\| \le M^2 \cdot n^2 \cdot \epsilon^2$.

Proof.. The congruence $f \equiv R \cdot H \bmod F$ and Lemma 4.1 imply

$$\|f\| \ \le \ \|R\| \cdot \|H\| \ \le \ M \cdot \epsilon \ ,$$

which in turn implies $\|\hat{F} - z^k\| \le M \cdot \epsilon + \|f\| \le 1$ and hence $\hat{F} \in \Pi_k^0(1)$. Therefore Lemma 4.1 can be applied with $\hat{F}$ in the place of F.

Now let $g \in \Pi_{m-1}$ be such that $R = f \cdot Q + g \cdot F$ and define $\hat{G} = Q + g$. Then

$$P - \hat{F} \cdot \hat{G} \ = \ P - F \cdot Q - (f \cdot Q + g \cdot F) - f \cdot g \ = \ -f \cdot g \ .$$

Let $-f \cdot g = \phi \cdot \hat{F} + \tilde{R}$ with unique $\phi \in \Pi_{m-2}$ and $\tilde{R} \in \Pi_{k-1}$. Then $P - \hat{F} \cdot (\hat{G} + \phi) = \tilde{R}$ and hence $\mathcal{Q}(\hat{F}) = \hat{G} + \phi$ and $\mathcal{R}(\hat{F}) = \tilde{R}$. Lemma 4.1 implies

$$\|\mathcal{R}(\hat{F})\| \ = \ \|\tilde{R}\| \ \le \ \|f \cdot g\| \ \le \ \|f\| \cdot \|g\| \ .$$

The equation $P = F \cdot Q + R$ and Lemma 4.1 imply $\|Q\| \leq (n-k+1) \cdot \|P\| < n$, the latter bound being used for the sake of simplicity. The equation $-f \cdot Q = g \cdot F - R$ and Lemma 4.1 imply $\|g\| \leq (n-k) \cdot \|f\| \cdot \|Q\| < n^2 \cdot M \cdot \epsilon$. Altogether the asserted estimate $\|\mathcal{R}(\hat{F})\| \leq n^2 \cdot M^2 \cdot \epsilon^2$ follows. $\qquad\square$

The analysis of the algorithm is based on an extension of Lemma 5.1 which takes into account that H is only an approximation for $\mathcal{H}(F)$. In addition, it must be investigated how well H approximates $\mathcal{H}(\hat{F})$:

Lemma 5.2 *Let M, ϵ, F, Q, and R as in Lemma 5.1. Let $H \in \Pi_{k-1}$ and $D \in \Pi_{k-1}$ be such that $H \cdot Q \equiv 1 - D \bmod F$. Furthermore, let $0 < \eta < 1$ and assume that $\|R\| \leq \epsilon$, $\|H\| \leq M$, and $\|D\| \leq \eta$.*
Define $f \in \Pi_{k-1}$ by $f \equiv R \cdot H \bmod F$ and let $\hat{F} := F + f$. Then the error of the new splitting is at most

$$\|\mathcal{R}(\hat{F})\| \leq \epsilon \cdot \eta + n^2 \cdot M^2 \cdot \epsilon^2 . \tag{5.1}$$

Let $\hat{D} \in \Pi_{k-1}$ be such that $H \cdot \mathcal{Q}(\hat{F}) \equiv 1 - \hat{D} \bmod F$, and assume in addition that $\epsilon \leq 1/n$. Then

$$\|\hat{D}\| \leq \eta + 3 \cdot n^2 \cdot M^2 \cdot \epsilon . \tag{5.2}$$

Proof.. For abbreviation, let $\hat{Q} := \mathcal{Q}(\hat{F})$ and $\hat{R} := \mathcal{R}(\hat{F})$. Like in the proof of Lemma 5.1, it follows that $\|f\| \leq M \cdot \epsilon$ and therefore $\hat{F} \in \Pi_k^0(1)$. Now

$$f \cdot Q \equiv R \cdot H \cdot Q \equiv R \cdot (1 - D) \bmod F .$$

Let $\nu := \max\{n - k, k - 1\}$. Then there is $g \in \Pi_{\nu-1}$ such that

$$R - R \cdot D = f \cdot Q + g \cdot F .$$

Let $\hat{G} = Q + g$. Then

$$P - \hat{F} \cdot \hat{G} = P - F \cdot Q - (f \cdot Q + g \cdot F) - f \cdot g = R \cdot D - f \cdot g .$$

Let $R \cdot D - f \cdot g = \phi \cdot \hat{F} + \tilde{R}$ with unique $\phi \in \Pi_{\nu-1}$ and $\tilde{R} \in \Pi_{k-1}$. Then $P - \hat{F} \cdot (\hat{G} + \phi) = \tilde{R}$ and hence $\hat{Q} = \hat{G} + \phi$ and $\hat{R} = \tilde{R}$. Lemma 4.1 implies

$$\|\hat{R}\| = \|\tilde{R}\| \leq \|R \cdot D - f \cdot g\| \leq \epsilon \cdot \eta + \|f\| \cdot \|g\| .$$

The equation $P = F \cdot Q + R$ and Lemma 4.1 imply the estimate

$$\|Q\| \leq (n - k + 1) \cdot \|P\| \leq n - 1 .$$

From $f \cdot Q + R \cdot D = g \cdot F + R$ and Lemma 4.1 it follows that

$$\|g\| \leq \nu \cdot (\epsilon \cdot \eta + \|f\| \cdot \|Q\|) < (n-1)^2 \cdot M \cdot \epsilon .$$

Altogether this yields the asserted estimate (5.1) for $\|\hat{R}\|$.
For the estimate of $\hat{D}$, let $K \in \Pi_{m-1}$ with $H \cdot Q + K \cdot F = 1 - D$. Then

$$H \cdot \hat{Q} + K \cdot \hat{F} = H \cdot Q + K \cdot F + H \cdot (g + \phi) + K \cdot f = 1 - D + H \cdot (g + \phi) + K \cdot f ,$$

hence
$$\hat{D} \equiv D - H \cdot (g + \phi) - K \cdot f \mod \hat{F} ,$$

which with Lemma 4.1 implies:
$$\|\hat{D}\| \le \|D\| + \|H\| \cdot (\|g\| + \|\phi\|) + \|K\| \cdot \|f\| .$$

Again with Lemma 4.1, the equation $H \cdot Q = K \cdot F + (1 - D)$ implies the estimate
$$\|K\| \le (n - k) \cdot \|H\| \cdot \|Q\| < n^2 \cdot M .$$

The equation $R \cdot D - f \cdot g = \phi \cdot \hat{F} + \tilde{R}$ implies
$$\|\phi\| \le \nu \cdot (\|R\| \cdot \|D\| + \|f\| \cdot \|g\|) \le \nu \cdot (\epsilon \cdot \eta + (n-1)^2 \cdot M^2 \cdot \epsilon^2) < n^2 \cdot M \cdot \epsilon$$

(use $\epsilon \le 1/n$ for the last estimate). Together this yields (5.2).   □

Now one step of the quadratic iteration for H is analyzed quantitatively:

Lemma 5.3 *Let $F \in \Pi_k^0(1)$ and $Q = \mathcal{Q}(F)$. Let $H \in \Pi_{k-1}$ and $D \in \Pi_{k-1}$ be such that $H \cdot Q \equiv 1 - D \mod F$.*
   Define $\hat{H} \in \Pi_{k-1}$ by $\hat{H} \equiv H \cdot (1 + D) \mod F$ and let $\hat{D} \in \Pi_{k-1}$ be such that $\hat{H} \cdot Q \equiv 1 - \hat{D} \mod F$. Then $\|\hat{H}\| \le \|H\| \cdot (1 + \|D\|)$ and $\|\hat{D}\| \le \|D\|^2$.

Proof.. The estimate of $\|\hat{H}\|$ follows from the definition of $\hat{H}$ and Lemma 4.1. The congruence $\hat{H} \cdot Q \equiv 1 - D^2 \equiv 1 - \hat{D} \mod F$ implies $\hat{D} \equiv D^2 \mod F$, which with Lemma 4.1 yields $\|\hat{D}\| \le \|D\|^2$.   □

6   Newton Iteration

6.1   Organization of the Newton Step

This section describes the organization of a Newton step and discusses some quantitative details. As some peculiarities will only become clear in the description of the overall algorithm, a detailed description of the *implementation* of the Newton step is postponed to Sect. 6.3. There it is also discussed which of the data mentioned below are actually used in the computations.

Lemma 5.2 suggests to organize the Newton step as follows: Let M, ϵ, F, Q, R, H, D as in Lemma 5.2, where all polynomials but D are given by their coefficients, and D is known to fulfil the estimate

$$\|D\| \le n^2 \cdot M^2 \cdot \epsilon . \tag{6.1}$$

This can be achieved before starting the Newton iteration with at most $\lceil \log(1/(n^2 \cdot M^2 \cdot \epsilon)) \rceil$ steps of the quadratic iteration according to Lemma 5.3.
   According to Lemma 5.2, the Newton step applied to F, Q, R, and H yields a new decomposition $P = \hat{F} \cdot \hat{Q} + \hat{R}$ with

$$\|\hat{R}\| \le \hat{\epsilon} := 2 \cdot n^2 \cdot M^2 \cdot \epsilon^2 . \tag{6.2}$$

If $\epsilon \leq 1/(4 \cdot n^2 \cdot M^2)$, then $\hat{\epsilon} < \epsilon/2$ i.e., one bit is gained for the accuracy of the factorization. With smaller "starting" ϵ, more substantial progress, i.e., better than linear convergence, can be achieved. E.g., $\epsilon \leq 1/(4 \cdot n^4 \cdot M^4)$ would guarantee $\hat{\epsilon} \leq \epsilon^{3/2}$.

Now the "defect squaring" method from Lemma 5.3 is used to replace H with a better approximation $\hat{H}$ for $\mathcal{H}(\hat{F})$. According to Lemma 5.2, the defect $\tilde{D} \in \Pi_{k-1}$ in the congruence $H \cdot \hat{Q} \equiv 1 - \tilde{D} \bmod \hat{F}$ is bounded by

$$\|\tilde{D}\| \leq 4 \cdot n^2 \cdot M^2 \cdot \epsilon . \tag{6.3}$$

Replacing H with $\hat{H} := H \cdot (1 + \tilde{D}) \bmod F$ yields $\hat{H} \cdot \hat{Q} \equiv 1 - \hat{D} \bmod \hat{F}$ with

$$\|\hat{D}\| \leq \|\tilde{D}\|^2 \leq 16 \cdot n^4 \cdot M^4 \cdot \epsilon^2 = 8 \cdot n^2 \cdot M^2 \cdot \hat{\epsilon} .$$

Up to the factor 8, this estimate is sufficient to start the next step, compare (6.1). One way to get rid of this factor would be a second defect squaring step. This would involve, however, some more polynomial multiplications and another division by $\hat{F}$ (see below for details). Another way simplifies the analysis and will eventually be more efficient: "Forgetting" three bits of accuracy gained by the Newton step, i.e., setting $\hat{\epsilon} := 8 \cdot n^2 \cdot M^2 \cdot \epsilon^2$ will also make $\hat{H}$ "suitable".

This discussion has ignored the fact that $\|\hat{H}\|$ may be slightly larger than M. The growth of the size of $\|H\|$ and of $\|F - z^k\|$ during the iteration must be controlled.

6.2 A Recursive Newton Algorithm

This section discusses quantitative details of the Newton algorithm. Although Newton's method is often referred to as "Newton *iteration*", it is favourable to organize the algorithm *recursively*.

The *starting situation* is the following: Let $F_0 \in \Pi_k^0$, $Q_0 = \mathcal{Q}(F_0)$ and $R_0 := \mathcal{R}(F_0)$. Moreover, let $H_0, D_0 \in \Pi_{k-1}$ with $H_0 \cdot Q_0 \equiv 1 - D_0 \bmod F_0$. Finally, let $0 < \epsilon_0 < 1$ and $M \geq 1$ be such that

$$\begin{array}{lll} (\mathbf{A_0}) & F_0 \in \Pi_k^0(\tau_0) & \text{with} \quad \tau_0 = 1/2 , \\ (\mathbf{B_0}) & \|R_0\| \leq \epsilon_0 , \\ (\mathbf{C_0}) & \|H_0\| \leq M , \\ (\mathbf{D_0}) & \|D_0\| \leq n^2 \cdot M^2 \cdot \epsilon_0 . \end{array}$$

It is not necessary to have all these data available for the computations. What is needed are the coefficients of F_0, R_0, and H_0 and the bounds ϵ_0 and M. The coefficients of Q_0 and D_0 are not used, but knowledge of the above estimates is of course crucial. See Sect. 6.3 for further implementation details.

The initial data are subject to quantitative restrictions which guarantee the algorithm to work. A first condition is that the bound for $\|D_0\|$ should be less than 1, hence $\epsilon_0 < 1/(n^2 \cdot M^2)$ is required. Further restrictions are derived below.

Now some quantities involved in the algorithm are defined, and some more restrictions are imposed: Let

$$C := 64 \cdot n^2 \cdot M^2 \qquad (6.4)$$

(this is just an abbreviation) and

$$M_0 := M \cdot \frac{1 + C \cdot \epsilon_0/8}{1 - C \cdot \epsilon_0} \ .$$

For $0 < \epsilon \le \epsilon_0$ define a bound M_ϵ for H depending on the bound ϵ for R,

$$M_\epsilon := M_0 \cdot (1 - C \cdot \epsilon) \ , \qquad (6.5)$$

and a bound τ_ϵ for the coefficients of $\|F - z^k\|$,

$$\tau_\epsilon := 1 - \frac{\sqrt{2} \cdot M}{C} \cdot \sum_{j=0}^{\infty} (C \cdot \epsilon)^{2^j} \ . \qquad (6.6)$$

For all $\epsilon \le \epsilon_0$, these parameters shall satisfy the estimates

$$M_\epsilon < M_0 \le \sqrt{2} \cdot M \qquad (6.7)$$

and

$$0 \le \tau_\epsilon \le 1 \ . \qquad (6.8)$$

In the following it is required that

$$\epsilon_0 \le \frac{\sqrt{2} - 1}{\sqrt{2} + 1/8} \cdot \frac{1}{C} \ \doteq \ \frac{0.2691}{C} \ . \qquad (6.9)$$

This is a sufficient condition for (6.7). The inequality $\sqrt{2} \cdot M \cdot \epsilon/(1 - C \cdot \epsilon) \le 1$ is sufficient for (6.8). (Estimate the sum in (6.6) with a geometric series.) This condition is equivalent to $\epsilon \le 1/(C + \sqrt{2} \cdot M)$, which is already guaranteed by (6.9) and (6.4).

The *recursive version of the Newton algorithm computes* a better approximate factor $\hat{F} \in \Pi_k^0$ of P, the corresponding quotient $\hat{Q} = \mathcal{Q}(\hat{F})$ and remainder $\hat{R} = \mathcal{R}(\hat{F})$, and an approximation $\hat{H} \in \Pi_{k-1}$ for for $\mathcal{H}(\hat{F})$ such that for the "target" error bound $\hat{\epsilon} \le \epsilon_0$ and the defect polynomial $\hat{D} \in \Pi_{k-1}$ defined by $\hat{H} \cdot \hat{Q} \equiv 1 - \hat{D} \bmod \hat{F}$, the following conditions hold:

$$(\hat{\mathbf{A}}) \quad \hat{F} \in \Pi_k^0(\tau_{\hat{\epsilon}}) \ ,$$
$$(\hat{\mathbf{B}}) \quad \|\hat{R}\| \le \hat{\epsilon} \ ,$$
$$(\hat{\mathbf{C}}) \quad \|\hat{H}\| \le M_{\hat{\epsilon}} \ ,$$
$$(\hat{\mathbf{D}}) \quad \|\hat{D}\| \le n^2 \cdot M^2 \cdot \hat{\epsilon} \ .$$

The algorithm works as follows: If $\hat{\epsilon} = \epsilon_0$, then $\tau_{\hat{\epsilon}} = \tau_{\epsilon_0} = \tau_0$ and $\|H_0\| \le M < M_{\epsilon_0} = M_{\hat{\epsilon}}$ by definition, so there is nothing to do. Otherwise let

$$\epsilon := \min\left\{\epsilon_0, \sqrt{\hat{\epsilon}/C}\right\} \qquad (6.10)$$

and call the algorithm recursively with the new (larger) target error bound ϵ. This produces $F \in \Pi_k^0$, $R = \mathcal{R}(F)$, $Q = \mathcal{Q}(F)$, and $H, D \in \Pi_{k-1}$ with $H \cdot Q \equiv 1 - D \bmod F$ such that the following conditions hold:

$$
\begin{aligned}
&\textbf{(A)} \quad && F \in \Pi_k^0(\tau_\epsilon) \ , \\
&\textbf{(B)} \quad && \|R\| \ \leq \ \epsilon \ , \\
&\textbf{(C)} \quad && \|H\| \ \leq \ M_\epsilon \ , \\
&\textbf{(D)} \quad && \|D\| \ \leq \ n^2 \cdot M^2 \cdot \epsilon \ .
\end{aligned}
$$

Now the Newton step as described in Sect. 6.1 is applied to F, R, and H: Let $f := (R \cdot H) \bmod F$, $\hat{F} := F + f$ and $\hat{R} := \mathcal{R}(\hat{F})$. These polynomials are now shown to satisfy $(\hat{\textbf{A}})$ and $(\hat{\textbf{B}})$.

Conditions $(\textbf{A})$, $(\textbf{C})$, (6.7), and Lemma 4.1 imply that the norm of the Newton correction f is at most

$$
\|f\| \ \leq \ M_\epsilon \cdot \epsilon \ \leq \ \sqrt{2} \cdot M \cdot \epsilon \ .
$$

Hence $\hat{F} \in \Pi_k^0(\tau)$, where $\tau = \tau_\epsilon + \sqrt{2} \cdot M \cdot \epsilon$. If $\epsilon = \sqrt{\hat{\epsilon}/C}$, then a straightforward computation shows that $\tau_\epsilon + \sqrt{2} \cdot M \cdot \epsilon = \tau_{\hat{\epsilon}}$. If $\epsilon = \epsilon_0$, then even $\tau \leq 1/2 + \|f\| \leq 1/2 + M \cdot \epsilon_0$. On the other hand it is easy to see that $\tau_{\epsilon_0} \geq 1 - 2 \cdot M \cdot \epsilon_0$. Now (6.9) and (6.4) imply $\epsilon_0 \leq 1/(6M)$ which yields $1/2 + M \cdot \epsilon_0 \leq \tau_{\epsilon_0} \leq \tau_{\hat{\epsilon}}$. This proves $(\hat{\textbf{A}})$. Conditions (6.2), (6.4), (6.7), and (6.10) yield the estimate

$$
\|\hat{R}\| \ \leq \ 2 \cdot n^2 \cdot M_\epsilon^2 \cdot \epsilon^2 \ \leq \ 4 \cdot n^2 \cdot M^2 \cdot \hat{\epsilon}/C \ \leq \ \hat{\epsilon}/16 \ .
$$

Obviously this implies $(\hat{\textbf{B}})$.

When four bits of the gained accuracy are ignored, H can be adapted to fit into the new situation with only one application of Lemma 5.3: H approximates $\mathcal{H}(\hat{F})$ as follows: Define $\hat{Q} := \mathcal{Q}(\hat{F})$ and let $\tilde{D} \in \Pi_{k-1}$ with $H \cdot \hat{Q} \equiv 1 - \tilde{D}$. Then

$$
\|\tilde{D}\| \ \leq \ 4 \cdot n^2 \cdot M_\epsilon^2 \cdot \epsilon \ \leq \ 8 \cdot n^2 \cdot M^2 \cdot \epsilon \ \leq \ n \cdot M \cdot \sqrt{\hat{\epsilon}}. \tag{6.11}
$$

Now let $\hat{H} := H \cdot (1 + \tilde{D}) \bmod \hat{F}$. If $\epsilon = \sqrt{\hat{\epsilon}/C}$, then it follows with Lemma 5.3 $(\textbf{C})$, (6.11), (6.5), and (6.10) that

$$
\begin{aligned}
\|\hat{H}\| \ &\leq \ \|H\| \cdot (1 + \|\tilde{D}\|) \ \leq \ M_\epsilon \cdot (1 + 8n^2 \cdot M^2 \cdot \epsilon) \\
&< \ M_0 \cdot (1 - C \cdot \epsilon) \cdot (1 + C \cdot \epsilon) \ = \ M_0 \cdot (1 - C^2 \cdot \epsilon^2) \\
&\leq \ M_0 \cdot (1 - C \cdot \hat{\epsilon}) \ = \ M_{\hat{\epsilon}} \ .
\end{aligned}
$$

If $\epsilon = \epsilon_0$, then one estimates

$$
\|\hat{H}\| \ \leq \ \|H_0\| \cdot (1 + \|\tilde{D}\|) \ \leq \ M \cdot (1 + 8n^2 \cdot M^2 \cdot \epsilon_0) \ \leq \ M_{\hat{\epsilon}}
$$

instead. In any case this establishes condition $(\hat{\textbf{C}})$.

Finally, condition $(\hat{\textbf{D}})$ follows from (6.3) which states that that norm of the defect $\hat{D} \in \Pi_{k-1}$ in the congruence $\hat{H} \cdot \hat{Q} \equiv 1 - \hat{D} \bmod \hat{F}$ is at most

$$
\|\hat{D}\| \ \leq \ \|\tilde{D}\|^2 \ = \ n^2 \cdot M^2 \cdot \epsilon.
$$

Altogether this establishes partial correctness of the algorithm. The restriction (6.9) on ϵ_0 implies $\epsilon_0 < 1/(3C)$. Thus if $\epsilon = \sqrt{\hat{\epsilon}/C} < \epsilon_0$, then $\epsilon > 3 \cdot \hat{\epsilon}$. This estimate shows that the recursion terminates.

The convergence result is summarized in the following theorem:

Theorem 6.1 *Let $M \leq 1$, $\epsilon_0 > 0$, $F_0 \in \Pi_k^0(1/2)$ (for example $F_0(z) = z^k$) and $\mathcal{R}(F_0)$ be given such that*

$$\|\mathcal{R}(F_0)\| \leq \epsilon_0 \leq \frac{1}{256 \cdot n^2 \cdot M^2}. \tag{6.12}$$

Moreover, let an approximation $H_0 \in \Pi_{k-1}$ for $\mathcal{H}(F_0)$ be given such that $D_0 \in \Pi_{k-1}$ defined by the congruence $H_0 \cdot \mathcal{Q}(F_0) \equiv 1 - D_0 \bmod F_0$ is bounded by $\|D_0\| \leq n^2 \cdot M^2 \cdot \epsilon_0^2$. Finally, let $\hat{\epsilon} = 2^{-s} < \epsilon_0$.

Then $\hat{F} \in \Pi_k^0(1)$ with $\|\mathcal{R}(\hat{F})\| \leq \hat{\epsilon}$ can be computed with at most $\lceil \log_{4/3}(s) \rceil + \lceil \log_3(8 \cdot M \cdot n) \rceil$ steps of the above iteration.

Corollary 6.2 *Fix $0 < \alpha < 1$ and let $\epsilon_0 \leq C^{-1/(1-\alpha)}$. Then $\lceil \log_{1+\alpha}(s) \rceil$ steps are sufficient to achieve $\|\mathcal{R}(\hat{F})\| \leq 2^{-s}$. E.g., $\epsilon_0 \leq 1/(512 \cdot n^3 \cdot M^3)$ is sufficient for $\alpha = 1/3$, i.e., $\lceil \log_{4/3}(s) \rceil$ steps.*

Proof.. Each step of the iteration improves the accuracy by at least a factor $1/3$. Hence an error bound $\epsilon_1 \leq 1/(8 \cdot M \cdot n)^3$ is achieved after at most $\lceil \log_3(8 \cdot M \cdot n) \rceil$ steps. If $\epsilon \leq \epsilon_1$, then geometric convergence is guaranteed: $C \cdot \epsilon^2 \leq \epsilon^{4/3}$. Therefore $\lceil \log_{4/3}(s) \rceil$ steps are sufficient for the final error bound.

The exponent $4/3$ has been chosen for the sake of clarity. The same argument works with any exponent $1 + \alpha$, where $0 < \alpha < 1$. This proves the corollary. □

6.3 Some Implementation Details

The Newton step with the correction step for H (short: the *"extended Newton step"*) as described in Sect. 6.1 can be implemented in various ways. Table 1 lists the operations involved with their parameters. Note that squaring is not treated

Abbreviation:	compute ...	with ...	
$PADD(\mu, \nu)$	$U + V$	$U \in \Pi_\mu,$	$V \in \Pi_\nu$
$PMUL(\mu, \nu)$	$U \cdot V$	...	...
$PSQU(\mu)$	U^2	...	
$PREM(\mu, \nu, \tau)$	$U \bmod V$	...	$V \in \Pi_\nu^0(\tau)$
$PDIV(\mu, \nu, \tau)$	$U \bmod V$ and $U \operatorname{div} V$	...	...

Table 1. Abbreviations and parameters for polynomial operations

as a special case of multiplication, because it usually is faster. (E.g., a Fast Fourier Transform based squaring routine implementation would involve only two FFTs, while multiplication uses three). Likewise, remainder computation may be cheaper when the quotient is not needed.

Besides the degree and the bound τ for the divisor in *PREM* and *PDIV*, a complete analysis of the cost of the algorithm would also involve bounds for the coefficient size of the polynomials U and V. Like the discussion of rounding errors for the basic polynomial operations, this is beyond the scope of this paper.

Table 2 shows how $\hat{F}$, $\hat{R}$, and $\hat{H}$ with $(\hat{\mathbf{A}})$, $(\hat{\mathbf{B}})$, $(\hat{\mathbf{C}})$, and $(\hat{\mathbf{D}})$ can be computed from F, R, and H with $(\mathbf{A})$–$(\mathbf{D})$ in a straightforward way (for clarity, $k-1$ is replaced by k etc.). The polynomials computed are listed with the operations

(1)	$R \cdot H$	$PMUL(k,k)$	
	$f = (R \cdot H) \bmod F$	$PREM(2k,k,1)$	
	$\hat{F} = F + f$	$PADD(k,k)$	
(2)	$\hat{R} = P \bmod \hat{F}$ and $\hat{Q} = P \operatorname{div} \hat{F}$	$PDIV(n,k,1)$	$(*)$
(3)	$H \cdot \hat{Q}$	$PMUL(k,m)$	$(*)$
	$\tilde{D} = 1 - (H \cdot \hat{Q}) \bmod \hat{F}$	$PREM(n,k,1)$	$(*)$
(4)	$H \cdot \tilde{D}$	$PMUL(k,k)$	
	$h = (H \cdot \tilde{D}) \bmod \hat{F}$	$PREM(2k,k,1)$	
	$\hat{H} = H + h$	$PADD(k,k)$	

Table 2. Extended Newton step

computing them. Three of these steps $(*)$ involve n coefficients. If k is small compared with n, involving more than $O(k)$ coefficients should be avoided in as many steps as possible. In fact, it is sufficient to involve n coefficients only once in each Newton step. Table 3 shows a corresponding patch.

(2')	$\hat{F}^2$	$PSQU(k)$
	$S = P \bmod \hat{F}^2$	$PREM(n,2k,1)$
	$\hat{R} = S \bmod \hat{F}$ and $\tilde{Q} = S \operatorname{div} \hat{F}$	$PDIV(2k,k,1)$
(3')	$H \cdot \tilde{Q}$	$PMUL(k,k)$
	$\tilde{D} = 1 - (H \cdot \tilde{Q}) \bmod \hat{F}$	$PREM(2k,k,1)$

Table 3. Patch for the extended Newton step

Another variant is to compute h from the formula $h = ((1-H \cdot \tilde{Q}) \cdot H) \bmod \hat{F}$ instead of $h = ((1-(H \cdot \tilde{Q}) \bmod \hat{F}) \cdot H) \bmod \hat{F}$, thereby trading two $PDIV(2k,k)$ and one $PMUL(k,k)$ for one $PDIV(3k,k)$ and one $PMUL(2k,k)$. It is not clear whether this is cheaper in practice.

7   Starting values

This section discusses how the the starting condition (6.12) from Theorem 6.1 can be checked from the coefficients of P. As root clusters will not occur frequently in most applications, starting value criteria must be simple and fast to compute.

Let $P(z) = a_0 + a_1 z + \cdots + a_n z^n$. An index k with $2 \le k < n$ is called an *admissible cluster size* for P, if (6.12) is fulfilled with $F_0(z) = z^k$. For this particular F_0, $R_0 := \mathcal{R}(F_0) = a_0 + a_1 z + \cdots + a_{k-1} z^{k-1}$, and $H_0 := \mathcal{H}(F_0)$ is defined by the lower k coefficients of $Q_0 := \mathcal{Q}(F_0)$, $H_0 = (\tilde{Q}_0)^{-1} \bmod z^k$ with $\tilde{Q}_0(z) = a_k + a_{k+1} z + \cdots + a_{2k-1} z^{k-1}$. Now (6.12) reads:

$$256 \cdot n^2 \cdot \|H_0\|^2 \cdot \|R_0\| \le 1 \ . \tag{7.1}$$

The coefficients of $H_0(z) = b_0 + b_1 z + \cdots + b_{k-1} z^{k-1}$ are given recursively by

$$b_0 = \frac{1}{a_k} \quad \text{and} \quad b_{l+1} = -\frac{1}{a_k} \cdot \sum_{j=1}^{l} a_{k+j} \cdot b_{l-j} \quad \text{for} \quad 0 < l < k \ .$$

With $\|P\| = 1$ this yields lower bounds for M: Let $M_0 = 1$ and $M_l = |b_0| + \cdots + |b_{l-1}|$ for $2 \le l \le k$. Then

$$1 \le \frac{1}{|a_k|} \le \frac{|a_k| + |a_{k+1}|}{|a_k|^2} \le M_l \le M \quad (2 \le l \le k) \ , \tag{7.2}$$

Now let $\sigma_k := |a_0| + \cdots + |a_{k-1}|$. Using only $1 \le M_1 \le M_2 \le M$, (7.2) implies the following necessary conditions for k to be an admissible cluster size:

$$256 \cdot n^2 \cdot \sigma_k \le |a_k|^2 / (|a_k| + |a_{k+1}|) \le |a_k| \le 1 \ .$$

This suggests the following simple test procedure: Starting with $k = 2$, test whether $256 \cdot n^2 \cdot \sigma_k \le 1$, and if so, compare with $|a_k|$; if this does not rule out k, compare with $|a_k|^2 / (|a_k| + |a_{k+1}|)$. One could continue this, computing b_l recursively and checking whether $256 \cdot n^2 \cdot \sigma_k \le M_l$ for $l = 2, \ldots, k - 1$. This procedure would simultaneously compute the coefficients of H_0.

There are two points where this test procedure should be modified for practical implementations: First, using the recursion may not be the most efficient way to compute H. Divide and conquer algorithms like the one described in [1, p. 22] or (at least asymptotically) FFT based polynomial division algorithms as in [20, § 4] will be faster.

Second, for a first fast check one will only use crude approximations (or even only the bit length) of the numbers involved, using, e.g., a *size routine* which computes $\lceil \log(|\alpha + i\beta|) \rceil$ or even $\lceil \log(\max\{|\alpha|, |\beta|\}) \rceil$. An example for such a size routine is *CLGSIZE*, see [24, section 8.1.6].

This suggests the following approach: Let $\nu := \lceil \log n \rceil$, $\gamma_k = \lceil \log \sigma_k \rceil$, and $\mu_l := \lfloor \log M_l \rfloor$. Then for k to be an admissible cluster size it is necessary that

$$\gamma_k + \nu + 8 \le \mu_l \quad \text{for} \quad 0 \le l \le k \ .$$

This necessary condition is now checked for $l = 0, 1, \ldots, l_0$ with small l_0, say, $l_0 = 2$. The choice of l_0 depends on the peculiarities of the implementation. If k passes this test, then one computes H_0, e.g. with one of the algorithms mentioned above, and $M = \|H_0\|$ and checks whether (7.1) holds.

8 Root Radius Conditions

For some applications it is useful to have starting value criteria for the Newton algorithm in terms of the roots of P rather than the coefficients.

Theorem 8.1 *Let $P(z) = \alpha \cdot (z - u_1) \cdots (z - u_k) \cdot (z - v_1) \cdots (z - v_m)$ with k small roots, $|u_1|, \ldots, |u_k| \leq r < 1$, and m large roots, $|v_1|, \ldots, |v_m| \geq R > 1$, and let $\alpha > 0$ be a normalizing constant such that $\|P\| = 1$. Then the starting value condition (6.12) of Thm. 6.1 is fulfilled if*

$$\frac{r}{R} \leq \frac{1}{11431 \cdot k \cdot m \cdot n^2} \cdot \qquad (8.1)$$

The corresponding condition in Malajovich and Zubelli's Main Theorem [13] states that their Newton algorithm converges quadratically if $R \geq 64n^3$ and $r \leq 1/(64n^3)$, i.e., $r/R \leq 1/(4096n^6)$. This should be compared with a condition like in Cor. 6.2, which guarantees a n error bound $\hat{\epsilon}$ in $O(\log \log(1/\epsilon))$ steps. E.g., choosing $\alpha = 1/3$ in Cor. 6.2 yields the condition $r/R \leq 1/(22462 \cdot k \cdot m \cdot n^3)$ instead of (8.1). Because of $k \cdot m = k \cdot (n - k) \leq n^2/4$, this implies the sufficient condition $r/R \leq 1/(5617 \cdot n^5)$, which is slightly weaker than the one in [13].

The remainder of this chapter is devoted to the proof of Thm. 8.1. With proper scaling it can be achieved that

$$R \geq 3m \quad \text{and} \quad r \leq \frac{1}{3810.24 \cdot k \cdot n^2} \cdot \qquad (8.2)$$

Namely, if one of these conditions does not hold, then $P(z)$ is replaced with $P_1(z) = \max\{1, \rho^n\} \cdot P(z/\rho)$, where $\rho = 3m/R$. Then the large roots of P_1 have modulus at least $\rho \cdot R = 3m$, and the small roots of P_1 are bounded by $\rho \cdot r \leq 3m \cdot r/R \leq 1/(3810.24 \cdot k \cdot n^2)$. ¿From now on, (8.2) is assumed w.l.o.g. Some intermediate estimates use the bound $r \leq 3/(64k)$.

Lemma 8.2 $(1 + r)^k \leq (1 + 3/(64k))^k < e^{3/64}$, $(1 + 1/R)^m \leq (1 + 1/(3m))^m < e^{1/3}$, *and* $(1 + r)^k \cdot (1 + 1/R)^m < e^{3/64 + 1/3} < \frac{3}{2}$.

Proof.. This follows immediately from $r \leq 3/(64k)$ and $R \geq 3m$. □

Definition 8.3 *Let $F(z) = (z - u_1) \cdots (z - u_k) = f_0 + f_1 z + \cdots + f_k z^k$ and $G(z) = (z - v_1) \cdots (z - v_m) = g_0 + g_1 z + \cdots + g_m z^m$ denote the (monic) factors of P corresponding to the small resp. large roots, and let $P_0(z) = F(z) \cdot G(z) = P(z)/\alpha = c_0 + c_1 z + \cdots + c_n z^n$.*

Definition 8.4 *Let $V := v_1 \cdots v_m$ and $\beta := |V|$.*

Lemma 8.5 $\|P_0\| < \frac{3}{2} \cdot \beta$.

Proof.. Vieta's theorem implies the estimates $f_j \leq \binom{k}{j} \cdot r^j$ and $g_j \leq \binom{m}{j} \cdot \beta/R^j$. Summing up yields $\|F\| \leq (1+r)^k$ and $\|G\| \leq \beta \cdot (1+1/R)^m$. The asserted estimate follows with $\|P_0\| = \|F \cdot G\| \leq \|F\| \cdot \|G\|$ and Lemma 8.2. $\square$

Lemma 8.6 $\epsilon_0 := \|P \bmod z^k\| \leq \alpha \cdot \beta \cdot k \cdot r \cdot (1+1/R)^m \cdot (1+r)^{k-1}$.

Proof.. This is also proved with Vieta's theorem, but is technically more complicated. The lower coefficients $c_0, \ldots, c_{k-1}$ of P_0 are bounded as follows:

$$|c_0| \leq r^k \cdot \beta,$$

$$|c_1| \leq r^k \cdot m \cdot \frac{\beta}{R} + k \cdot r^{k-1} \cdot \beta,$$

$$|c_2| \leq r^k \cdot \binom{m}{2} \cdot \frac{\beta}{R^2} + k \cdot r^{k-1} \cdot m \cdot \frac{\beta}{R} + \binom{k}{2} \cdot r^2 \cdot \beta,$$

$$|c_3| \leq r^k \cdot \binom{m}{3} \cdot \frac{\beta}{R^3} + k \cdot r^{k-1} \cdot \binom{m}{2} \cdot \frac{\beta}{R^2} + \binom{k}{2} \cdot r^2 \cdot m \cdot \frac{\beta}{R} + \binom{k}{3} \cdot r^3 \cdot \beta,$$

etc., which implies (summing up the columns first)

$$\|P_0 \bmod z^k\| \leq \beta \cdot \left(1+\frac{1}{R}\right)^m \cdot \left(r^k + k \cdot r^{k-1} + \binom{k}{2} \cdot r^{k-2} + \cdots + k \cdot r\right)$$

$$\leq \beta \cdot \left(1+\frac{1}{R}\right)^m \cdot \left((1+r)^k - 1\right).$$

The assertion follows with $P = \alpha \cdot P_0$ and $(1+r)^k - 1 \leq k \cdot r \cdot (1+r)^{k-1}$. $\square$

Lemma 8.7 $|c_k| > \frac{62}{63} \cdot \beta$.

Proof.. Vieta's formula

$$c_k = V + \sum_{j=1}^{k} u_j \sum_{l=1}^{m} \frac{V}{v_l} + \sum_{j_1 \neq j_2} u_{j_1} u_{j_2} \sum_{l_1 \neq l_2} \frac{V}{v_{l_1} v_{l_2}} + \cdots$$

implies

$$|c_k| \geq \beta - k \cdot r \cdot m \cdot \frac{\beta}{R} - \binom{k}{2} \cdot r^2 \cdot \binom{m}{2} \cdot \frac{\beta}{R^2} - \binom{k}{3} \cdot r^3 \cdot \binom{m}{3} \cdot \frac{\beta}{R^3} - \cdots.$$

For convenience, the binomial coefficients are estimated with $\binom{k}{j} \leq k^j$ and $\binom{m}{l} \leq m^l$. Then with $\gamma := k \cdot m \cdot r/R \ (\leq 1/64)$:

$$|c_k| \geq \beta \cdot (1 - \gamma - \gamma^2 - \gamma^3 - \cdots) < \beta \cdot \left(1 - \frac{\gamma}{1-\gamma}\right) \geq \frac{62}{63} \cdot \beta. \quad \square$$

Corollary 8.8 $\frac{2}{3} \cdot 1/\beta < \alpha < \frac{63}{62} \cdot 1/\beta$.

Proof.. Lemmas 8.5 and 8.7 imply $\frac{2}{3} \cdot 1/\beta < \alpha = 1/\|P_0\| \leq 1/|c_k| < \frac{63}{62} \cdot 1/\beta$.   □

Corollary 8.9 $\epsilon_0 < \frac{3}{2} \cdot k \cdot r$.

Proof.. Lemmas 8.6 and 8.2, and Cor. 8.8 imply $\epsilon_0 \leq k \cdot r \cdot \frac{63}{62} \cdot e^{1/3 + 3/64} < \frac{3}{2} \cdot k \cdot r$.

□

Lemma 8.10 $|c_{k+j}| < \dfrac{64}{63} \cdot \dbinom{m}{j} \cdot \dfrac{\beta}{R^j}$ *for* $j = 1, \ldots, m$.

Proof.. The Vieta formulae

$$c_{k+1} = \sum_{l=1}^{m} \frac{V}{v_l} + \sum_{j=1}^{k} u_j \sum_{l_1 \neq l_2} \frac{V}{v_{l_1} v_{l_2}} + \sum_{j_1 \neq j_2} u_{j_1} u_{j_2} \sum_{l_1, l_2, l_3} \frac{V}{v_{l_1} v_{l_2} v_{l_3}} + \cdots ,$$

$$c_{k+2} = \sum_{l_1 \neq l_2} \frac{V}{v_{l_1} v_{l_2}} + \sum_{j=1}^{k} u_j \sum_{l_1, l_2, l_3} \frac{V}{v_{l_1} v_{l_2} v_{l_3}}$$

$$+ \sum_{j_1 \neq j_2} u_{j_1} u_{j_2} \sum_{l_1, \ldots, l_4} \frac{V}{v_{l_1} v_{l_2} v_{l_3} v_{l_4}} + \cdots ,$$

etc., yield the estimates

$$|c_{k+j}| \leq \sum_{t=0}^{m-j} \binom{m}{j+t} \cdot \frac{\beta}{R^{j+t}} \cdot \binom{k}{t} \cdot r^t$$

$$= \binom{m}{j} \cdot \frac{\beta}{R^j} \cdot \sum_{t=0}^{m-j} \left[\binom{m}{j+t} / \binom{m}{j} \right] \cdot \binom{k}{t} \cdot \left(\frac{r}{R} \right)^t .$$

With $\binom{m}{j+t} / \binom{m}{j} \leq m^t$, $\binom{k}{t} \leq k^t$, and $\gamma = k \cdot m \cdot r/R$, this implies

$$|c_{k+j}| \leq \binom{m}{j} \cdot \frac{\beta}{R^j} \cdot \sum_{t=0}^{m-j} \gamma^j < \frac{1}{1-\gamma} \cdot \binom{m}{j} \cdot \frac{\beta}{R^j} \leq \frac{64}{63} \cdot \binom{m}{j} \cdot \frac{\beta}{R^j} . \quad □$$

Lemma 8.11 *Let* $Q(z) = c_k + c_{k+1}z + \cdots + c_n z^m$, *and let* $H(z) = h_0 + h_1 z + \cdots + h_{k-1} z^{k-1}$ *denote the inverse of* Q *modulo* z^k. *Then* $\|H\| < \frac{21}{10} \cdot 1/\beta$.

Proof.. The coefficients of H are given recursively by $h_0 = 1/c_k$ and $h_j = -\left(\sum_{t=0}^{j-1} h_t \cdot c_{k+j-t} \right) / c_k$ for $j \leq 1$. With the bounds from Lemmas 8.7 and 8.10 it can be shown by induction that

$$|h_0| < \frac{63}{62} \cdot \frac{1}{\beta} \quad \text{and} \quad |h_j| \leq \frac{16}{31} \cdot \frac{1}{\beta} \cdot \left(\frac{63}{31} \right)^j \cdot \left(\frac{m}{R} \right)^j \quad \text{for} \quad j \leq 1 .$$

The induction is straightforward (but tedious) and is therefore omitted. Summing up the bounds for the h_j and using $R \geq 3m$ implies (after a straightforward computation, which is omitted, too) the asserted bound for $\|H\|$.   □

Corollary 8.12 $M := \|\mathcal{H}(z^k)\| < 63/20.$

Proof.. Note that $\mathcal{Q}(z^k) = \alpha \cdot Q$ and hence $\mathcal{H}(z^k) = H/\alpha$. Therefore Lemma 8.11 and Cor. 8.8 imply $M \leq \|H\|/\alpha < \frac{3}{2} \cdot \beta \cdot \frac{21}{10} \cdot (1/\beta) = \frac{63}{20}.$ □

Proof. of Thm. 8.1. Corollaries 8.9 and 8.12 show that the convergence criterion $\epsilon_0 \leq 1/(256n^2 M^2)$ from Thm. 6.1 is satisfied if $\frac{3}{2} \cdot k \cdot r \leq \left(256 \cdot n^2 \cdot \left(\frac{63}{20}\right)^2\right)^{-1}$, equivalently $r \leq \frac{25}{95256} \cdot 1/\left(k \cdot n^2\right)$. This is exactly the bound for r which was assumed in (8.2). □

9   Examples

This chapter complements the general result from Chap. 8 with some explicit quantitative examples. Let $P(z) = \alpha \cdot F(z) \cdot G(z)$, where as before $F \in \Pi_k^0$ has roots of modulus at most $r < 1$, $G \in \Pi_m^0$ has all its roots far away from the origin, and $\alpha > 0$ is such that $\|P\| = 1$.

As a standard example let us choose $k = 4$, $m = 4, 12$, resp. 28, and $R = 1$. Then Thm. 8.1 implies that the Newton algorithm converges if $r \leq 2^{-24}$, $r \leq 2^{-27}$, and $r \leq 2^{-31}$, respectively.

These worst case bounds are independent of the distribution of the "large" roots of P. The following simple analysis shows that much better bounds are achieved when the roots of G are "well distributed".

For the sake of simplicity higher order terms in r are neglected. This does not affect the results very much, as r will be small anyway. It follows from $|P^{(k-1)}(0)| = \alpha|F^{(k-1)}(0)| \cdot |G(0)| + O(r^2) = \alpha \cdot k! \cdot r + O(r^2)$ and $|P^{(j)}(0)| = O(r^2)$ for $0 \leq j \leq k-2$ that the small coefficients of $R_0 = P \bmod z^k$ are bounded by $\epsilon_0 = \alpha \cdot k \cdot r + O(r^2).$

The coefficients of $z^k, \ldots, z^{2k-1}$ in P and the coefficients of $z^0, \ldots, z^{k-1}$ in $\alpha \cdot G$ differ by an error of order $O(r)$. Therefore M is approximated up to an error of $O(r)$ by M_0/α where $M_0 = \|G^{-1} \bmod z^k\|$. Finally, $\alpha = 1/\|G\| + O(r)$. Altogether this yields the condition

$$r \leq \frac{1}{256 \cdot n^2 \cdot N^2 \cdot \|G\|} \tag{9.1}$$

A favourable situation is the case when the roots of G are "well distributed". A prominent example are the polynomials $G(z) = z^m - 1$. Then $\|G\| = 2$. If $m \geq k$, then $N = 1$, because $G \equiv 1 \bmod z^k$. This shows that Newton iteration for F will work if $r \leq 2^{-9}/n^2$ independent of k. For the above examples this is $r \leq 2^{-15}$, $r \leq 2^{-17}$, and $r \leq 2^{-19}$, respectively.

For polynomials with clustered zeros, (9.1) is much worse than Thm. 8.1. As an extreme example take $G(z) = (z-1)^m$. Then $\|G\| = 2^m$ and $N = 35, 455$, and 4495, respectively for the standard examples. Condition (9.1) guarantees convergence if $r \leq 2^{-30}$, $r \leq 2^{-46}$ resp. $r \leq 2^{-72}$. These bad bounds are due to the artificial factor $\|G\|$ in the denominator of (9.1), which is caused by the simplified analysis.

10 Conclusion

Some things remain to do for an implementation. First, the effect of rounding errors must be analyzed. Second, some of the error estimates provided here can be improved for the price of a more complicated presentation. Third, a priori error bounds have been used throughout. An efficient implementation should work adaptively instead.

The starting value condition is still very restrictive. For a generalization, one could think of constructing a homotopy algorithm as follows: Let $P_\alpha(z) = z^k \cdot U(z) + \alpha \cdot V(z)$ with $U \in \Pi_m^0$, $V \in \Pi_k$, and $0 \le \alpha \le 1$. Then z^k is a factor of P_0. Can one split $P_1 = F \cdot G$ into factors of degrees k and m, respectively, by using an extension of the Newton algorithm presented here, to successively compute factors of P_{α_j} for $\alpha_j \nearrow 1$?

Acknowledgements

I would like to thank A. Schönhage, T. Ahrendt, T. Lickteig, and E. Vetter for helpful suggestions and comments.

References

1. BINI, D. AND PAN, V. *Polynomial and Matrix Computations, Volume 1: Fundamental Algorithms.* Birkhäuser, Boston, 1994.
2. FRIEDMAN, J. On the convergence of Newton's method. *J. of Complexity* **5** (1989), 12–33.
3. GOURDON, X. Algorithmique du théorème fondamental de l'algèbre. Rapport de Recherche 1852, INRIA, February 1993.
4. GOURDON, X., November 1995. Personal communication.
5. GRAU, A. A. The simultaneous improvement of a complete set of approximate factors of a polynomial. *SIAM J. Numer. Analysis* **8** (1971), 425–438.
6. HENRICI, P. *Applied and Computational Complex Analysis, vol. 1.* Wiley & Sons, New York, 1974.
7. JENKINS, M. A. AND TRAUB, J. F. A three-stage variable-shift iteration for polynomial zeros and its relation to generalized rayleigh iteration. *Numer. Math.* **14** (1970), 252–263.
8. KANTOROVICH, L. AND AKILOV, G. *Functional Analysis in Normed Spaces.* MacMillan, New York, 1964.
9. KIM, M.-H. On approximate zeros and rootfinding algorithms for a complex polynomial. *Math. Comp.* **51** (1988), 707–719.
10. KIRRINNIS, P. *Zur Berechnung von Partialbruchzerlegungen und Kurvenintegralen rationaler Funktionen.* Dissertation, Univ. Bonn, Institut für Informatik, January 1993.
11. KIRRINNIS, P. Partial fraction decomposition in $\mathbf{C}(z)$ and simultaneous Newton iteration for factorization in $\mathbf{C}[z]$. Preprint, Univ. Bonn. Submitted for publication, September 1995.
12. LEHMER, D. H. A machine method for solving polynomial equations. *J. Assoc. Comput. Mach.* **8** (1961), 151–162.

13. MALAJOVICH, G. AND ZUBELLI, J. P. A fast and stable algorithm for splitting polynomials. Preprint, Dept. de Matematica Aplicada, Univ. Federal do Rio de Janeiro, January 1996.

14. NEFF, C. A. AND REIF, J. H. An $O(n^{1+\epsilon} \log b)$ algorithm for the complex root problem. In *Proc. 35th Ann. IEEE Symp. on Foundations of Computer Science* (1994), pp. 540–547.

15. PAN, V. Y. Sequential and parallel complexity of approximate evaluation of polynomial zeros. *Comput. Math. Appl.* **14** (1987), 591–622.

16. PAN, V. Y. New techniques for approximating complex polynomial zeros. In *Proc. 5th Ann. ACM-SIAM Symp. on Discrete Algorithms* (1994), pp. 260–270.

17. PAN, V. Y. Optimal (up to polylog factors) sequential and parallel algorithms for approximating complex polynomial zeros. In *Proc. 27th Ann. ACM Symp. on the Theory of Computing (STOC '95)* (1995), pp. 741–750.

18. RENEGAR, J. On the cost of approximating all roots of a complex polynomial. *Math. Programming* **32** (1985), 319–336.

19. RENEGAR, J. On the worst-case complexity of approximating zeros of polynomials. *J. of Complexity* **3** (1987), 90–113.

20. SCHÖNHAGE, A. Asymptotically fast algorithms for the numerical multiplication and division of polynomials with complex coefficients. In *Computer Algebra EUROCAM '82 (Marseille 1982)* (1982), J. Calmet, Ed., vol. 144 of *Springer Lect. Notes in Computer Science*, Springer, pp. 3–15.

21. SCHÖNHAGE, A. The fundamental theorem of algebra in terms of computational complexity. Tech. rep., Univ. Tübingen, 1982.

22. SCHÖNHAGE, A. Equation solving in terms of computational complexity. In *Proc. Int. Congress of Mathematicians, Berkeley, CA* (1986), pp. 131–153.

23. SCHÖNHAGE, A. Ordinary root finding. Preprint, Univ. Bonn, January 1996.

24. SCHÖNHAGE, A., GROTEFELD, A. F. W., AND VETTER, E. *Fast Algorithms: A Multitape Turing Machine Implementation.* B.I. Wissenschaftsverlag, Mannheim, 1994.

25. SHUB, M. AND SMALE, S. Computational complexity: On the geometry of polynomials and a theory of cost: Part I. *Ann. scient. Ec. Norm. Sup., 4e série* **18** (1985), 107–142.

26. SHUB, M. AND SMALE, S. Computational complexity: On the geometry of polynomials and a theory of cost: II. *SIAM J. Comput.* **15** (1986), 145–161.

27. SMALE, S. The fundamental theorem of algebra and complexity theory. *Bull. (New Series) Amer. Math. Soc.* **4** (1981), 1–36.

28. SMALE, S. Newton's method estimates from data at one point. In *The Merging of Disciplines: New Directions in Pure, Applied, and Computational Mathematics* (1985), R. E. Ewing, K. I. Gross, and C. F. Martin, Eds., Springer, pp. 185–196.

29. SMALE, S. On the efficiency of algorithms of analysis. *Bull. (New Series) Amer. Math. Soc.* **13** (1985), 87–121.

30. SMALE, S. Algorithms for solving equations. In *Proc. Int. Congress of Mathematicians, Berkeley, CA* (1986), pp. 172–195.

31. WEYL, H. Randbemerkungen zu Hauptproblemen der Mathematik. II. Fundamentalsatz der Algebra und Grundlagen der Mathematik. *Math. Zeitschr.* **20** (1924), 142–150.

Szemerédi's Regularity Lemma
for Sparse Graphs

Y. Kohayakawa*

Abstract. A remarkable lemma of Szemerédi asserts that, very roughly speaking, *any* dense graph can be decomposed into a bounded number of pseudorandom bipartite graphs. This far-reaching result has proved to play a central rôle in many areas of combinatorics, both 'pure' and 'algorithmic.' The quest for an equally powerful variant of this lemma for sparse graphs has not yet been successful, but some progress has been achieved recently. The aim of this note is to report on the successes so far.

1   Introduction

Szemerédi's celebrated proof [39] of the conjecture of Erdős and Turán [10] on arithmetic progressions in dense subsets of integers is certainly a masterpiece of modern combinatorics. An auxiliary lemma in that work, which has become known in its full generality [40] as *Szemerédi's regularity lemma*, has turned out to be a powerful and widely applicable combinatorial tool. For an authoritative survey on this subject, the reader is referred to the recent paper of Komlós and Simonovits [29]. For the algorithmic aspects of this lemma, the reader is referred to the papers of Alon, Duke, Lefmann, Rödl, and Yuster [1] and Duke, Lefmann, and Rödl [8].

Very roughly speaking, the lemma of Szemerédi says that *any* graph can be decomposed into a bounded number of pseudorandom bipartite graphs. Since pseudorandom graphs have a predictable structure, the regularity lemma is a powerful tool for introducing 'order' where none is visible at first. The notion of pseudorandomness that appears in the lemma has to do with distribution of edges. In fact, the bipartite graphs that are used to decompose a given graph are guaranteed to have its edges uniformly distributed with an error term that is quadratic in the number of vertices of the graph, but with an arbitrarily small multiplicative constant. Therefore, we have a handle on the edge distribution of the original graph as long as it has a quadratic number of edges. If the original graph has a subquadratic number of edges, however, Szemerédi's lemma will tell us nothing. In this note, we focus our attention on certain closely related variants of the regularity lemma that can handle this 'sparse' case with some success.

* Partially supported by FAPESP (Proc. 96/4505–2) and CNPq (Proc. 300334/93–1 and ProTeM-CC-II Project ProComb).

An unpublished manuscript of the present author [23] dealt with one such variant. As mentioned there, these variants were also independently discovered by Professor V. Rödl.

These sparse versions of the regularity lemma have already been used for studying extremal properties of graphs [19, 20], of random graphs [21, 22, 24, 26], and of random sets of integers [27], and also for developing an optimal algorithm for checking pseudorandomness of graphs [28]. In the sequel, we shall discuss these applications and we shall also state a few related problems.

Before we close the introduction, let us mention that, since the paper of Komlós and Simonovits [29] went to press, another nice application of the regularity lemma, together with a new variant, has appeared. Frieze and Kannan [14] have approached some **MAXSNP**-hard problems by means of the regularity lemma. Their method gives polynomial time approximation schemes for many graph partitioning problems if the instances are restricted to dense ones. Their paper thus provides yet another motivation for investigating sparse versions of the fascinating lemma of Szemerédi.

This note is organised as follows. The statement of a version of the regularity lemma for sparse graphs, Theorem 1, is given in Section 2 below. A variant of Theorem 1 is also discussed in that section. We then present in Section 3 an application of Theorem 1 to the study of pseudorandom graphs, highlighting an algorithmic consequence. In Section 4 we present applications of Theorem 1 to graph theory and to combinatorial number theory. A proof of Theorem 1 is outlined in Section 5. We close this note with some remarks and open problems.

Caveat. The regularity lemma is a powerful tool, but, naturally, several of its applications involve many further additional ideas and techniques. Although crucial in the applications we shall discuss below, the sparse version of this lemma is still far weaker than one would wish, and one has to fight quite hard to prove the results in question. Therefore, in this note, we shall not be able to state precisely how this lemma comes in in the proofs of the theorems under discussion. However, we shall try to hint what the rôle of the regularity lemma is in each case.

2   Sparse Variants of the Regularity Lemma

2.1   Preliminary Definitions

Let a graph $G = G^n$ of order $|G| = n$ be fixed. For U, $W \subset V = V(G)$, we write $E(U, W) = E_G(U, W)$ for the set of edges of G that have one endvertex in U and the other in W. We set $e(U, W) = e_G(U, W) = |E(U, W)|$. Now, let a partition $P_0 = (V_i)_1^\ell$ ($\ell \geq 1$) of V be fixed. For convenience, let us write $(U, W) \prec P_0$ if $U \cap W = \emptyset$ and either $\ell = 1$ or else $\ell \geq 2$ and for some $i \neq j$ ($1 \leq i, j \leq \ell$) we have $U \subset V_i$, $W \subset V_j$.

Suppose $0 \leq \eta \leq 1$. We say that G is (P_0, η)-*uniform* if, for some $0 \leq p \leq 1$, we have that for all U, $W \subset V$ with $(U, W) \prec P_0$ and $|U|$, $|W| \geq \eta n$, we have

$$\left| e_G(U, W) - p|U||W| \right| \leq \eta p |U||W|. \tag{1}$$

We remark that the partition P_0 is introduced to handle the case of ℓ-partite graphs ($\ell \geq 2$). If $\ell = 1$, that is, if the partition P_0 is trivial, then we are thinking of the case of ordinary graphs. In this case, we shorten the term (P_0, η)-uniform to η-*uniform*.

The prime example of an η-uniform graph is of course a *random graph* $G_p = G_{n,p}$. Note that for $\eta > 0$ a random graph G_p with $p = p(n) = C/n$ is almost surely η-uniform provided $C \geq C_0 = C_0(\eta)$, where $C_0(\eta)$ depends only on η. Here and in the sequel, we use standard definitions and notation concerning random graphs. Let $0 < p = p(n) \leq 1$ be given. The standard binomial random graph $G_p = G_{n,p}$ has as vertex set a fixed set $V(G_p)$ of cardinality n and two such vertices are adjacent in G_p with probability p, with all such adjacencies independent. For concepts and results concerning random graphs, see, *e.g.*, Bollobás [3].

2.2   A Regularity Lemma for Sparse Graphs

We first introduce a few further definitions that will allow us to state a version of Szemerédi's lemma for sparse graphs. Let a graph $G = G^n$ be fixed as before. Let $H \subset G$ be a spanning subgraph of G. For $U, W \subset V$, let

$$d_{H,G}(U,W) = \begin{cases} e_H(U,W)/e_G(U,W) & \text{if } e_G(U,W) > 0 \\ 0 & \text{if } e_G(U,W) = 0. \end{cases}$$

Suppose $\varepsilon > 0$, $U, W \subset V$, and $U \cap W = \emptyset$. We say that the pair (U, W) is (ε, H, G)-*regular*, or simply ε-*regular*, if for all $U' \subset U, W' \subset W$ with $|U'| \geq \varepsilon|U|$ and $|W'| \geq \varepsilon|W|$, we have

$$|d_{H,G}(U',W') - d_{H,G}(U,W)| \leq \varepsilon.$$

We say that a partition $Q = (C_i)_0^k$ of $V = V(G)$ is (ε, k)-*equitable* if $|C_0| \leq \varepsilon n$, and $|C_1| = \ldots = |C_k|$. Also, we say that C_0 is the *exceptional* class of Q. When the value of ε is not relevant, we refer to an (ε, k)-equitable partition as a k-*equitable* partition. Similarly, Q is an *equitable* partition of V if it is a k-equitable partition for some k. If P and Q are two equitable partitions of V, we say that Q *refines* P if every non-exceptional class of Q is contained in some non-exceptional class of P. If P' is an arbitrary partition of V, then Q *refines* P' if every non-exceptional class of Q is contained in some block of P'. Finally, we say that an (ε, k)-equitable partition $Q = (C_i)_0^k$ of V is (ε, H, G)-*regular*, or simply ε-*regular*, if at most $\varepsilon\binom{k}{2}$ pairs (C_i, C_j) with $1 \leq i < j \leq k$ are not ε-regular. We may now state an extension of Szemerédi's lemma to subgraphs of (P_0, η)-uniform graphs.

Theorem 1 *Let $\varepsilon > 0$ and $k_0, \ell \geq 1$ be fixed. Then there are constants $\eta = \eta(\varepsilon, k_0, \ell) > 0$ and $K_0 = K_0(\varepsilon, k_0, \ell) \geq k_0$ satisfying the following. For any (P_0, η)-uniform graph $G = G^n$, where $P_0 = (V_i)_1^\ell$ is a partition of $V = V(G)$, if $H \subset G$ is a spanning subgraph of G, then there exists an (ε, H, G)-regular (ε, k)-equitable partition of V refining P_0 with $k_0 \leq k \leq K_0$.*   □

Remark. To recover the original regularity lemma of Szemerédi from Theorem 1, simply take $G = K^n$, the complete graph on n vertices.

2.3   A Second Regularity Lemma for Sparse Graphs

In some situations, the sparse graph H to which one would like to apply the regularity lemma is not a subgraph of some fixed η-uniform graph G. A simple variant of Theorem 1 may be useful in this case. For simplicity, we shall not state this variant for 'P_0-partite' graphs as we did in Section 2.2.

Let a graph $H = H^n$ of order $|H| = n$ be fixed. Suppose $0 < \eta \leq 1$ and $0 < p \leq 1$. We say that H is η-*upper-uniform* with *density* p if, for all U, $W \subset V$ with $U \cap W = \emptyset$ and $|U|$, $|W| \geq \eta n$, we have $e_H(U, W) \leq (1 + \eta)p|U||W|$. In the sequel, for any two disjoint non-empty sets U, $W \subset V$, let $d_{H,p}(U, W) = e_H(U, W)/2p|U||W|$.

Now suppose $\varepsilon > 0$, U, $W \subset V$, and $U \cap W = \emptyset$. We say that the pair (U, W) is (ε, H, p)-*regular*, or simply (ε, p)-*regular*, if for all $U' \subset U$, $W' \subset W$ with $|U'| \geq \varepsilon|U|$ and $|W'| \geq \varepsilon|W|$ we have

$$|d_{H,p}(U', W') - d_{H,p}(U, W)| \leq \varepsilon.$$

We say that an (ε, k)-equitable partition $P = (C_i)_0^k$ of V is (ε, H, p)-*regular*, or simply (ε, p)-*regular*, if at most $\varepsilon\binom{k}{2}$ pairs (C_i, C_j) with $1 \leq i < j \leq k$ are not (ε, p)-regular. We may now state a version of Szemerédi's regularity lemma for η-upper-uniform graphs.

Theorem 2 *For any given $\varepsilon > 0$ and $k_0 \geq 1$, there are constants $\eta = \eta(\varepsilon, k_0) > 0$ and $K_0 = K_0(\varepsilon, k_0) \geq k_0$ such that any η-upper-uniform graph H with density $0 < p \leq 1$ admits an (ε, H, p)-regular (ε, k)-equitable partition of its vertex set with $k_0 \leq k \leq K_0$.* □

Remarks. (*i*) A further variant of Theorem 1 concerns the existence ε-regular partitions with respect to a *collection* of graphs on the same vertex set. Such a variant is used [20].

(*ii*) Variants of Theorems 1 and 2 for sparse hypergraphs may be proved easily. However, we know of no applications of such results. For hypergraph versions of the regularity lemma, see [4, 12].

3   Checking Pseudorandomness of Graphs

The investigation of explicitly constructible graphs that are 'random-like' has proved to be very fruitful in providing examples for many extremal problems in graph theory. These graphs have also played a crucial rôle in algorithmic problems: the use of expanders for amplifying the power of random sources is but one example.

The study of pseudorandom graph properties, *i.e.*, properties that random graphs have and somehow capture their 'random nature,' can be traced back to Graham and Spencer [18], Rödl [33], and Frankl, Rödl, and Wilson [13]. Thomason [41, 42] and Chung, Graham, and Wilson [7] present further developments that have given this subject the status of a solid theory. (We do not go into the details, but we mention that similar theories for hypergraphs and subsets of $\mathbb{Z}_n$ are now available, see Chung and Graham [5, 6].)

In this section, we shall be concerned with a new pseudorandom graph property introduced by Rödl. This property characterises *quasi-random* graphs in the sense of Chung, Graham, and Wilson [7]. As a consequence of this characterisation, we shall have an optimal algorithm for checking quasi-randomness of graphs. For the proofs of the results in this section, the reader is referred to [28].

3.1   Preliminary Definitions

Let reals $0 < \varepsilon \leq 1$ and $0 < \delta \leq 1$ be given. We shall say that a graph G is $(1/2, \varepsilon, \delta)$-*quasi-random* if, for all U, $W \subset V(G)$ with $U \cap W = \emptyset$ and $|U|$, $|W| \geq \delta n$, we have

$$\left| e_G(U, W) - \frac{1}{2}|U||W| \right| \leq \frac{1}{2}\varepsilon|U||W|. \tag{2}$$

If $0 < \varrho \leq 1$ and A are reals, we say that an n-vertex graph $J = J^n$ is (ϱ, A)-*uniform* if, for all U, $W \subset V(J)$ with $U \cap W = \emptyset$, we have

$$\left| e_J(U, W) - \varrho|U||W| \right| \leq A\sqrt{r|U||W|}, \tag{3}$$

where $r = \varrho n$. In the sequel, we may take the graph J to be a Lubotzky–Phillips–Sarnak Ramanujan graph $X_{p,q}$ for some values of p and q, see [30]. If $J = X_{p,q}$, we have $r = p + 1$ and the graph J is r-regular. Moreover, in this case, inequality (3) holds with $A = 2$, and in fact r on the right hand side of (3) may be replaced by $r - 1$.

We shall now define a property for n-vertex graphs $G = G^n$, based on a fixed (ϱ, A)-uniform graph $J = J^n$ with the same vertex set as G. Let $0 < \varepsilon \leq 1$ be a real. We say that G satisfies property $P_{J,\triangle}(\varepsilon)$ if we have

$$\sum_{ij \in E(J)} \left| |\Gamma_G(i) \triangle \Gamma_G(j)| - \frac{1}{2}n \right| \leq \frac{1}{2}\varepsilon n e(J), \tag{4}$$

where, as usual, we write $\Gamma_G(x)$ for the G-neighbourhood of a vertex x of G, and we write $A \triangle B$ for the symmetric difference $(A \setminus B) \cup (B \setminus A)$. Moreover, in (4) and in the sequel, $e(J)$ denotes the number of edges in J. As we shall see in Section 3.2, inequality (4), which may be checked in time $O(n^2)$ if we take J to be a graph with $e(J) = O(n)$, turns out to be a quasi-random property in the sense of [7].

For technical reasons, we need to introduce a variant of property $P_{J,\triangle}(\varepsilon)$. Suppose $0 < \gamma \leq 1$ and $0 < \varepsilon \leq 1$ are two reals and $G = G^n$ is an n-vertex graph. We shall say that G satisfies property $P'_{J,\triangle}(\gamma, \varepsilon)$ if the inequality

$$\left| |\Gamma_G(i) \triangle \Gamma_G(j)| - \frac{1}{2}n \right| \leq \frac{1}{2}\varepsilon n \tag{5}$$

fails for at most $\gamma e(J)$ edges $ij \in E(J)$ of J. As a quick argument shows, properties $P_{J,\triangle}(\varepsilon)$ and $P_{J,\triangle}(\gamma, \varepsilon)$ are equivalent under suitable assumptions on the parameters, see Lemma 5.

3.2   The Equivalence Results

The following three results express the equivalence between quasi-randomness, in the technical sense of Chung, Graham, and Wilson [7], and property $P_{J,\triangle}$. Loosely speaking, quasi-randomness in the sense of [7] is equivalent to the property of being $(1/2, o(1), o(1))$-quasi-random, which, as the results below show, is equivalent to property $P_{J,\triangle}(o(1))$.

Theorem 3 *Let an r-regular (ϱ, A)-uniform graph $J = J^n$ be fixed, where $0 < \varrho = r/n \le 1$ and A is an absolute constant. Let constants $0 < \varepsilon \le 1$ and $0 < \delta \le 1$ be given. Then if $0 < \varepsilon' \le \varepsilon^2 \delta^3/8$ and $r \ge 2^{10} A^2 \varepsilon^{-2} \delta^{-2}$, we have that any graph $G = G^n$ on the same vertex set as J satisfying property $P_{J,\triangle}(\varepsilon')$ is $(1/2, \varepsilon, \delta)$-quasi-uniform.*   □

Theorem 4 *Let an r-regular (ϱ, A)-uniform graph $J = J^n$ be fixed, where $0 < \varrho = r/n \le 1$ and A is an absolute constant. Let constants $0 < \gamma \le 1$ and $0 < \varepsilon \le 1$ be given. Then there exist constants $0 < \varepsilon_0 = \varepsilon_0(\gamma, \varepsilon) \le 1$, $0 < \delta_0 = \delta_0(\gamma, \varepsilon) \le 1$, and $r_0 = r_0(\gamma, \varepsilon) \ge 1$, which depend only on γ and ε, such that any $(1/2, \varepsilon', \delta')$-quasi-uniform graph G on the same vertex set as J satisfies property $P'_{J,\triangle}(\gamma, \varepsilon)$ as long as $\varepsilon' \le \varepsilon_0(\gamma, \varepsilon)$, $\delta' \le \delta_0(\gamma, \varepsilon)$, and $r \ge r_0(\gamma, \varepsilon)$.*   □

Lemma 5 *Let a (ϱ, A)-uniform graph $J = J^n$ be given, where $0 < \varrho \le 1$ and A is an absolute constant. The following assertions hold.*

(i) Let $G = G^n$ be a graph on $V(J)$ satisfying property $P'_{J,\triangle}(\gamma, \varepsilon)$, where $0 < \gamma \le 1$ and $0 < \varepsilon \le 1$ satisfy $\gamma + \varepsilon \le 1$. Then G has property $P_{J,\triangle}(\varepsilon + \gamma)$.
(ii) Let $G = G^n$ be a graph on $V(J)$ satisfying property $P_{J,\triangle}(\varepsilon)$ and suppose $\varepsilon \le \varepsilon' \le 1$. Then G satisfies property $P'_{J,\triangle}(\varepsilon/\varepsilon', \varepsilon')$.   □

An immediate corollary to the above results is as follows.

Corollary 6 *Given an n-vertex graph G, we can decide in time $O(n^2)$ whether or not G is a quasi-random graph.*   □

Previously, the fastest known method for checking quasi-randomness was based on the quasi-random property $P_{J,\triangle}$ with J the complete graph K^n on n vertices. This gave an algorithm of time complexity $O(M(n))$, where $M(n) = O(n^{2.376})$ is the time needed for multiplying two n by n matrices with 0–1 entries over the integers.

The Role of the Regularity Lemma. Theorem 1 is used in the proof of Theorem 4. Roughly speaking, one takes a graph G as in the statement of Theorem 4, and assumes that it does not satisfy $P'_{J,\triangle}(\gamma, \varepsilon)$. One then takes the spanning subgraph H of J whose edges are the 'violating edges' $ij \in E(J)$, where we call an edge ij *violating* if (5) fails for ij. Note that a violating edge may be of two types, corresponding to the two inequalities expressed in (5). Thus, we may naturally split the edges of H into two groups; say $H = H_- \cup H_+$. Assume that, say, $e(H_+) \geq (1/2)e(H) \geq (\gamma/2)e(J)$. We may then apply Theorem 1 to the pair $H_+ \subset J$. Once an H_+-dense regular pair (V_i, V_j) is shown to exist, one may prove that there is a pair (U, W) of suitably large sets that violate (2). Details are given in [28].

4   Further Applications

4.1   Applications in Graph Theory

A classical area of extremal graph theory investigates numerical and structural problems concerning H-*free graphs*, namely graphs that do not contain a copy of a given fixed graph H as a subgraph. Let $\mathrm{ex}(n, H)$ be the maximal number of edges that an H-free graph on n vertices may have. A basic question is then to determine or estimate $\mathrm{ex}(n, H)$ for any given H and large n. A solution to this problem is given by the celebrated Erdős–Stone–Simonovits theorem, which states that, as $n \to \infty$, we have

$$\mathrm{ex}(n, H) = \left(1 - \frac{1}{\chi(H) - 1} + o(1)\right) \binom{n}{2}, \qquad (6)$$

where as usual $\chi(H)$ is the chromatic number of H. Furthermore, as proved independently by Erdős and Simonovits, every H-free graph $G = G^n$ that has as many edges as in (6) is in fact 'very close' (in a certain precise sense) to the densest n-vertex $(\chi(H) - 1)$-partite graph. For these and related results, see, for instance, Bollobás [2].

Here we are interested in a variant of the function $\mathrm{ex}(n, H)$. Let G and H be graphs, and write $\mathrm{ex}(G, H)$ for the maximal number of edges that an H-free subgraph of G may have. Formally, $\mathrm{ex}(G, H) = \max\{e(J): H \not\subset J \subset G\}$, where $e(J)$ stands for the size $|E(J)|$ of J as before. Clearly $\mathrm{ex}(n, H) = \mathrm{ex}(K^n, H)$.

One problem is, then, to study $\mathrm{ex}(G, H)$ when G is a 'typical' graph, by which we mean a *random graph*. In other words, we wish to investigate the random variable $\mathrm{ex}(G_{n,p}, H)$.

Let H be a graph of order $|H| = |V(H)| \geq 3$ and size $e(H) > 0$. Let us write $d_2(H)$ for the 2-*density* of H, that is,

$$d_2(H) = \max\left\{\frac{e(J) - 1}{|J| - 2}: J \subset H, |J| \geq 3\right\}.$$

Given a real $0 \leq \varepsilon \leq 1$ and an integer $r \geq 2$, let us say that a graph J is ε-*quasi r-partite* if J may be made r-partite by the deletion of at most $\varepsilon e(J)$ of its edges. A general conjecture concerning $\mathrm{ex}(G_{n,p}, H)$ is as follows (cf. [26]). As is usual in the theory of random graphs, we say that a property P holds *almost surely* or that *almost every* random graph $G_{n,p}$ satisfies P if P holds with probability tending to 1 as $n \to \infty$.

Conjecture 7 *Let H be a non-empty graph of order at least 3, and let $0 < p = p(n) \leq 1$ be such that $pn^{1/d_2(H)} \to \infty$ as $n \to \infty$. Then the following assertions hold.*

(i) *Almost every $G_{n,p}$ satisfies*

$$\text{ex}(G_{n,p}, H) = \left(1 - \frac{1}{\chi(H) - 1} + o(1)\right) e(G_{n,p}). \qquad (7)$$

(ii) *Suppose $\chi(H) \geq 3$. Then for any $\varepsilon > 0$ there is a constant $\delta = \delta(\varepsilon) > 0$ such that almost every $G_{n,p}$ has the property that any H-free subgraph $J \subset G_{n,p}$ of $G_{n,p}$ with $e(J) \geq (1 - \delta)\,\text{ex}(G_{n,p}, H)$ is ε-quasi $(\chi(H) - 1)$-partite.* $\qquad\square$

Recall that any graph G contains an r-partite subgraph $J \subset G$ with $e(J) \geq (1 - 1/r)e(G)$. Thus the content of Conjecture 7(i) is that $\text{ex}(G_{n,p}, H)$ is at most as large as the right-hand side of (7). There are a few results in support of Conjecture 7(i).

Any result concerning the tree-universality of expanding graphs, or else a simple application of Theorem 2, gives Conjecture 7(i) for forests. The cases in which $H = K^3$ and $H = C^4$ are essentially proved in Frankl and Rödl [11] and Füredi [15], respectively, in connection with problems concerning the existence of some graphs with certain extremal properties. The case in which H is a general cycle was settled by Haxell, Kohayakawa, and Łuczak [21, 22] (see also Kohayakawa, Kreuter, and Steger [25]), and the case in which $H = K^4$ was settled by Kohayakawa, Łuczak, and Rödl [26]. Conjecture 7(ii) in the case in which $0 < p \leq 1$ is a constant follows easily from Szemerédi's original regularity lemma. Theorem 2 and a lemma from Kohayakawa, Łuczak, and Rödl [27] concerning induced subgraphs of bipartite graphs may be used to verify Conjecture 7 for $H = K^3$ in full, and, more generally, Conjecture 7 may be proved for the case in which H is a cycle by making use of a lemma in Kohayakawa and Kreuter [24].

We must at this point mention that beautiful and very general results concerning *Ramsey* properties of random graphs in the spirit of Conjecture 7 were proved by Rödl and Ruciński [35, 36]. These authors used the original lemma of Szemerédi. A related result, of a much more restricted scope but apparently not accessible through the techniques in [35, 36], is proved in Kohayakawa and Kreuter [24]. Theorem 2 is crucial in [24].

The Role of the Regularity Lemma. A moment's thought reveals that Theorem 1 is immediately applicable to the situation given in Conjecture 7. Indeed, we have a subgraph J of a random and hence η-uniform graph $G_{n,p}$ at hand and therefore we may invoke Theorem 1 to obtain an $(\varepsilon, J, G_{n,p})$-regular partition P of $V = V(G_{n,p})$, for any constants $\varepsilon > 0$ and k_0. If we further know that

$$e(J) \geq \left(1 - \frac{1}{\chi(H) - 1} + \delta\right) e(G_{n,p}),$$

where $\delta > 0$ is some constant, it is a simple matter to find a set of $h = |H|$ classes V_i in the partition P that 'form a copy of H' (here we need to have ε suitably small and k_0 suitably large with respect to δ). If we are dealing with $0 < p \leq 1$ that is a constant, independent of n, then we are finished, since these h

classes may be shown to span a copy of H. If $p = p(n) \to 0$ as $n \to \infty$, however, we are stuck, since this last statement does not necessarily hold. In [21, 22, 24, 26], we are forced to take different approaches, all of them based on more involved applications of our sparse variants of the regularity lemma. We shall not go into the details. We shall, however, discuss a conjecture (cf. [26]) from which, if true, one may deduce Conjecture 7 through the standard approach described above.

Suppose H has vertices $v_1, \ldots, v_h$ ($h \geq 3$) and let $0 < p = p(m) \leq 1$ be given. Let also $\mathbf{V} = (V_i)_{i=1}^h$ be a family of h pairwise disjoint sets, each of cardinality m. Suppose reals $0 < \varepsilon \leq 1$ and $0 < \gamma \leq 1$ and an integer T are given. We say that an h-partite graph F with h-partition $V(F) = V_1 \cup \cdots \cup V_h$ and size $e(F) = |F| = T$ is an $(\varepsilon, \gamma, H; \mathbf{V}, T)$-graph if the pair (V_i, V_j) is (ε, F, p)-regular and has p-density $\gamma \leq d_{F,p}(V_i, V_j) \leq 1$ whenever $v_i v_j \in E(H)$.

Conjecture 8 *Let constants $0 < \alpha \leq 1$ and $0 < \gamma \leq 1$ be given. Then there exist constants $\varepsilon = \varepsilon(\alpha, \gamma) > 0$ and $C = C(\alpha, \gamma)$ such that, if $p = p(m) \geq Cm^{-1/d_2(H)}$, the number of H-free $(\varepsilon, \gamma, H; \mathbf{V}, T)$-graphs is at most*

$$\alpha^T \binom{\binom{h}{2}m^2}{T}$$

for all T and all sufficiently large m. □

If H above is a forest, Conjecture 8 holds trivially, since, in this case, all $(\varepsilon, \gamma, H; \mathbf{V}, T)$-graphs contain a copy of H. A lemma in Kohayakawa, Łuczak, and Rödl [27] may be used to show that Conjecture 8 holds for the case in which $H = K^3$. The general case of cycles is established in Kohayakawa and Kreuter [24].

Remark. Other graph theoretical applications of the regularity lemma for sparse graphs are given in Haxell and Kohayakawa [19] and Haxell, Kohayakawa, and Łuczak [20]. These applications concern Ramsey and anti-Ramsey properties of random and pseudorandom graphs.

4.2   An Application in Combinatorial Number Theory

We now turn to a problem concerning arithmetic progressions of integers. Here we are interested in the existence of a 'small' and 'sparse' set $R \subset [n] = \{1, \ldots, n\}$ with the property that every subset $A \subset R$ that contains a fixed positive fraction of the elements of R contains also a 3-term arithmetic progression. The measure of sparseness here should reflect the fact that R is locally poor in 3-term arithmetic progressions. Clearly, a natural candidate for such a set R is an M-element set R_M uniformly selected from all the M-element subsets of $[n]$, where $1 \leq M = M(n) \leq n$ is to be chosen suitably. The main result of Kohayakawa, Łuczak, and Rödl [27] confirms this appealing and intuitive idea.

For integers $1 \leq M \leq n$, let $\mathcal{R}(n, M)$ be the probability space of all the M-element subsets of $[n]$ equipped with the uniform measure. In the sequel, given $0 < \alpha \leq 1$ and a set $R \subset [n]$, write $R \to_\alpha 3$ if any $A \subset R$ with $|A| \geq \alpha |R|$

contains a 3-term arithmetic progression. The main result of [27] may then be stated as follows.

Theorem 9 *For every constant $0 < \alpha \leq 1$, there exists a constant $C = C(\alpha)$ such that if $C\sqrt{n} \leq M = M(n) \leq n$ then the probability that $R_M \in \mathcal{R}(n, M)$ satisfies $R_M \rightarrow_\alpha 3$ tends to 1 as $n \rightarrow \infty$.* □

Note that Theorem 9 is, in a way, close to being best possible: if $M = M(n) = \lfloor \varepsilon\sqrt{n} \rfloor$ for some fixed $\varepsilon > 0$ then the number of 3-term arithmetic progressions in $R_M \in \mathcal{R}(n, M)$ is, with large probability, smaller than $2\varepsilon^2 |R_M|$, and hence all of them may be destroyed by deleting at most $2\varepsilon^2 |R_M|$ elements from R_M; in other words, with large probability the relation $R_M \rightarrow_\alpha 3$ does *not* hold for $\alpha = 1 - 2\varepsilon^2$.

Theorem 9 immediately implies the existence of 'sparse' sets $S = S_\alpha$ such that $S \rightarrow_\alpha 3$ for any fixed $0 < \alpha \leq 1$. The following result makes this assertion precise.

Corollary 10 *Suppose that $s = s(n) = o(n^{1/8})$ and $g = g(n) = o(\log n)$ as $n \rightarrow \infty$. Then, for every fixed $\alpha > 0$, there exist constants C and N such that for every $n \geq N$ there exists $S \subset [n]$ satisfying $S \rightarrow_\alpha 3$ for which the following three conditions hold.*

(i) *For every $k \geq 0$ and $\ell \geq 1$ the set $\{k, k+\ell, \ldots, k+s\ell\}$ contains at most three elements of S.*

(ii) *Every set $\{k, k+\ell, \ldots, k+m\ell\}$ with $k \geq 0$, $\ell \geq 1$, and $m \geq \sqrt{n}\log n$ contains at most $Cm/\sqrt{n}$ elements of S.*

(iii) *If $\mathcal{F} = \mathcal{F}(S)$ is the 3-uniform hypergraph on the vertex set S whose hyperedges are the 3-term arithmetic progressions contained in S, then $\mathcal{F}$ has no cycle of length smaller than g.* □

In words, conditions (i) and (ii) above say that the set S intersects any arithmetic progression in a small number of elements. In particular, S contains no 4-term arithmetic progressions. Condition (iii) is more combinatorial in nature, and says that the 3-term arithmetic progressions contained in S form, locally, a tree-like structure, which makes the property $S \rightarrow_\alpha 3$ somewhat surprising.

Let us remark that the following extension of Szemerédi's theorem related to Corollary 10 was proved by Rödl [34], thereby settling a problem raised by Spencer [38]. Let k, $g \geq 3$ be fixed integers and $0 < \alpha \leq 1$ a fixed real. Theorem 4.3 in [34] asserts that then, for any large enough n, there exists a k-uniform hypergraph $\mathcal{F}$ on $[n]$, all of whose hyperedges are k-term arithmetic progressions, such that $\mathcal{F}$ contains no cycle of length smaller than g but each subset $A \subset [n]$ with $|A| \geq \alpha n$ contains a hyperedge of $\mathcal{F}$. For other problems and results in this direction, see Graham and Nešetřil [16], Nešetřil and Rödl [31], and Prömel and Voigt [32]. Note that Corollary 10 strengthens the above result of [34] in the case in which $k = 3$.

The Role of the Regularity Lemma. A weak version of Roth's theorem [37], namely, a version stating that any sequence of integers with *positive* upper density contains a 3-term arithmetic progression, may be proved by means of a more-or-less direct application of the regularity lemma, see, *e.g.*, Erdős, Frankl, and Rödl [9] and Graham and Rödl [17]. Since Theorem 9 above deals with sparse sets of integers, a similar approach makes results such as Theorem 2 come into play. The difficulties that arise are, however, quite substantial. We close by mentioning that an application of a variant of Theorem 2 and the proof of Conjecture 8 for $H = K^3$ are at the heart of the proof of Theorem 9.

5   Proof of Theorem 1

We now proceed to outline the proof Theorem 1, but, before we proceed, we stress that the argument below follows the one of Szemerédi [40] very closely. In particular, as in [40], the following 'defect' form of the Cauchy–Schwarz inequality will be important.

Lemma 11 *Let reals* $y_1, \ldots, y_v \geq 0$ *be given. Suppose* $0 \leq \varrho = u/v < 1$ *and* $\sum_{1 \leq i \leq u} y_i = \alpha \varrho \sum_{1 \leq i \leq v} y_i$. *Then*

$$\sum_{1 \leq i \leq v} y_i^2 \geq \frac{1}{v} \left(1 + (\alpha - 1)^2 \frac{\varrho}{1 - \varrho} \right) \left\{ \sum_{1 \leq i \leq v} y_i \right\}^2. \qquad \square$$

We now fix $G = G^n$ and put $V = V(G)$. Also, we assume that $P_0 = (V_i)_1^\ell$ is a fixed partition of V, and that G is (P_0, η)-uniform for some $0 \leq \eta \leq 1$. Moreover, we let $p = p(G)$ be as in (1). The following 'continuity' results for $d_{H,G}$ and $d_{H,G}^2$ may be proved in a straightforward manner.

Lemma 12 *Let* $0 < \delta \leq 10^{-2}$ *be fixed. Let* $U, W \subset V(G)$ *be such that* $(U, W) \prec P_0$, *and* $\delta|U|, \delta|W| \geq \eta n$. *If* $U^* \subset U$, $W^* \subset W$, $|U^*| \geq (1 - \delta)|U|$, *and* $|W^*| \geq (1 - \delta)|W|$, *then*

(i) $|d_{H,G}(U^*, W^*) - d_{H,G}(U, W)| \leq 5\delta$,
(ii) $|d_{H,G}(U^*, W^*)^2 - d_{H,G}(U, W)^2| \leq 9\delta$. $\qquad \square$

In the sequel, a constant $0 < \varepsilon \leq 1/2$ and a spanning subgraph $H \subset G$ of G is fixed. Also, we let $P = (C_i)_0^k$ be an (ε, k)-equitable partition of $V = V(G)$ refining P_0, where $4^k \geq \varepsilon^{-5}$. Moreover, we assume that $\eta \leq \eta_0 = \eta_0(k) = 1/k4^{k+1}$ and that $n = |G| \geq n_0 = n_0(k) = k4^{1+2k}$.

We now define an equitable partition $Q = Q(P)$ of $V = V(G)$ from P as follows. First, for each (ε, H, G)-irregular pair (C_s, C_t) of P with $1 \leq s < t \leq k$, we choose $X = X(s, t) \subset C_s$, $Y = Y(s, t) \subset C_t$ such that (i) $|X|, |Y| \geq \varepsilon|C_s| = \varepsilon|C_t|$, and (ii) $|d_{H,G}(X, Y) - d_{H,G}(C_s, C_t)| \geq \varepsilon$. For fixed $1 \leq s \leq k$, the sets $X(s, t)$ in

$$\{ X = X(s, t) \subset C_s : 1 \leq t \leq k \text{ and } (C_s, C_t) \text{ is not } (\varepsilon, H, G)\text{-regular} \}$$

define a natural partition of C_s into at most 2^{k-1} blocks. Let us call such blocks the *atoms* of C_s. Now let $q = 4^k$ and set $m = \lfloor |C_s|/q \rfloor$ $(1 \leq s \leq k)$. Note that $\lfloor |C_s|/m \rfloor = q$ as $|C_s| \geq n/2k \geq 2q^2$. Moreover, for later use, note that $m \geq \eta n$. We now let Q' be a partition of $V = V(G)$ refining P such that (i) C_0 is a block of Q', (ii) all other blocks of Q' have cardinality m, except for possibly one, which has cardinality at most $m - 1$, (iii) for all $1 \leq s \leq k$, every atom $A \subset C_s$ contains exactly $\lfloor |A|/m \rfloor$ blocks of Q', (iv) for all $1 \leq s \leq k$, the set C_s contains exactly $q = \lfloor |C_s|/m \rfloor$ blocks of Q'.

Let C_0' be the union of the blocks of Q' that are not contained in any class C_s $(1 \leq s \leq k)$, and let C_i' $(1 \leq i \leq k')$ be the remaining blocks of Q'. We are finally ready to define our equitable partition $Q = Q(P)$: we let $Q = (C_i')_0^{k'}$. The following lemma is easy to check.

Lemma 13 *The partition $Q = Q(P) = (C_i')_0^{k'}$ defined from P as above is a k'-equitable partition of $V = V(G)$ refining P, where $k' = kq = k4^k$, and $|C_0'| \leq |C_0| + n4^{-k}$.* □

In what follows, for $1 \leq s \leq k$, we let $C_s(i)$ $(1 \leq i \leq q)$ be the classes of Q' that are contained in the class C_s of P. Also, for all $1 \leq s \leq k$, we set $C_s^* = \bigcup_{1 \leq i \leq q} C_s(i)$. Now let $1 \leq s \leq k$ be fixed. Note that $|C_s^*| \geq |C_s| - (m-1) \geq |C_s| - q^{-1}|C_s| \geq |C_s|(1 - q^{-1})$. As $q^{-1} \leq 10^{-2}$ and $q^{-1}|C_s| \geq m \geq \eta n$, by Lemma 12 we have, for all $1 \leq s < t \leq k$,

$$|d_{H,G}(C_s^*, C_t^*) - d_{H,G}(C_s, C_t)| \leq 5q^{-1} \tag{8}$$

and

$$|d_{H,G}(C_s^*, C_t^*)^2 - d_{H,G}(C_s, C_t)^2| \leq 9q^{-1}. \tag{9}$$

As in [40], we define the *index* $\mathrm{ind}(R)$ of an equitable partition $R = (V_i)_0^r$ of $V = V(G)$ to be

$$\mathrm{ind}(R) = \frac{2}{r^2} \sum_{1 \leq i < j \leq \ell} d_{H,G}(V_i, V_j)^2.$$

Note that trivially $0 \leq \mathrm{ind}(R) < 1$. The next two lemmas show that, for $Q = Q(P)$ defined as above, we have $\mathrm{ind}(Q) \geq \mathrm{ind}(P) + \varepsilon^5/100$. The proof of the first lemma is based on the Cauchy–Schwarz inequality.

Lemma 14 *Suppose $1 \leq s < t \leq k$. Then*

$$\frac{1}{q^2} \sum_{i,j=1}^{q} d_{H,G}(C_s(i), C_t(j))^2 \geq d_{H,G}(C_s, C_t)^2 - \frac{\varepsilon^5}{100}. \quad □$$

The inequality in Lemma 14 may be improved if (C_s, C_t) is an (ε, H, G)-irregular pair. The following lemma, which is proved by invoking the defect form of the Cauchy–Schwarz inequality, Lemma 11, makes this precise.

Lemma 15 *Let $1 \leq s < t \leq k$ be such that (C_s, C_t) is not (ε, H, G)-regular. Then*

$$\frac{1}{q^2} \sum_{i,j=1}^{q} d_{H,G}(C_s(i), C_t(j))^2 \geq d_{H,G}(C_s, C_t)^2 + \frac{\varepsilon^4}{40} - \frac{\varepsilon^5}{100}. \quad □$$

The following result, which is the main lemma in the proof of Theorem 1, follows from Lemmas 14 and 15.

Lemma 16 *Suppose $k \geq 1$ and $0 < \varepsilon \leq 1/2$ are such that $4^k \geq 1800\varepsilon^{-5}$. Let $G = G^n$ be a (P_0, η)-uniform graph of order $n \geq n_0 = n_0(k) = k4^{2k+1}$, where $P_0 = (V_i)_1^\ell$ is a partition of $V = V(G)$, and assume that $\eta \leq \eta_0 = \eta_0(k) = 1/k4^{k+1}$. Let $H \subset G$ be a spanning subgraph of G. If $P = (C_i)_0^k$ is an (ε, H, G)-irregular (ε, k)-equitable partition of $V = V(G)$ refining P_0, then there is a k'-equitable partition $Q = (C_i')_0^{k'}$ of V such that (i) Q refines P, (ii) $k' = k4^k$, (iii) $|C_0'| \leq |C_0| + n4^{-k}$, and (iv) $\mathrm{ind}(Q) \geq \mathrm{ind}(P) + \varepsilon^5/100$.* □

Proof. of Theorem 1. (Outline) Let $\varepsilon > 0$, $k_0 \geq 1$, and $\ell \geq 1$ be given. We may assume that $\varepsilon \leq 1/2$. Pick $s \geq 1$ such that $4^{s/4\ell} \geq 1800\varepsilon^{-5}$, $s \geq \max\{2k_0, 3\ell/\varepsilon\}$, and $\varepsilon 4^{s-1} \geq 1$. Let $f(0) = s$, and put inductively $f(t) = f(t-1)4^{f(t-1)}$ $(t \geq 1)$. Let $t_0 = \lfloor 100\varepsilon^{-5} \rfloor$ and set $N = \max\{n_0(f(t)): 0 \leq t \leq t_0\} = f(t_0)4^{2f(t_0)+1}$, $K_0 = \max\{6\ell/\varepsilon, N\}$, and $\eta = \eta(\varepsilon, k_0, \ell) = \min\{\eta_0(f(t)): 0 \leq t \leq t_0\} = 1/4f(t_0+1) > 0$. It is now straightforward to check that η and K_0 as defined above will do. As in [40], the proof is simply based on the fact that the index of any partition is bounded, whereas, as stated in Lemma 16(iv), the index increases by a fixed amount every time we suitably refine an irregular partition. We omit the details. □

6   Final Remarks and Open Problems

It would be of interest to elucidate the algorithmic aspects of Theorem 1. For instance, can one find the partition guaranteed to exist in that result in time, say, $O(n^2)$, if we are concerned with subgraphs H of r-regular η-uniform graphs $G = G^n$, where r is a constant independent of n? If this turns out to be the case, many algorithms developed in [1, 8] may be improved to optimal algorithms.

We hope that the above applications of the regularity lemma for sparse graphs reveal the potential of such variants of Szemerédi's lemma. The applications also illustrate that the 'right' variant has not yet been found, since the successes are somewhat modest when compared with what the original lemma can achieve in the dense case. The problem of finding the 'right' variant of Szemerédi's lemma for the sparse case is certainly of great interest.

References

1. N. Alon, R.A. Duke, H. Lefmann, V. Rödl, and R. Yuster. The algorithmic aspects of the regularity lemma. *Journal of Algorithms*, 16(1):80–109, 1994.
2. B. Bollobás. *Extremal Graph Theory*. Academic Press, London, 1978.
3. B. Bollobás. *Random Graphs*. Academic Press, London, 1985.
4. F.R.K. Chung. Regularity lemmas for hypergraphs and quasi-randomness. *Random Structures and Algorithms*, 2(1):241–252, 1991.
5. F.R.K. Chung and R.L. Graham. Quasi-random set systems. *Journal of the American Mathematical Society*, 4(1):151–196, 1991.

6. F.R.K. Chung and R.L. Graham. Quasi-random subsets of $\mathbb{Z}_n$. *Journal of Combinatorial Theory, Series A*, 61(1):64–86, 1992.

7. F.R.K. Chung, R.L. Graham, and R.M. Wilson. Quasi-random graphs. *Combinatorica*, 9(4):345–362, 1989.

8. R.A. Duke, H. Lefmann, and V. Rödl. A fast approximation algorithm for computing the frequencies of subgraphs in a given graph. *SIAM Journal on Computing*, 24(3):598–620, 1995.

9. P. Erdős, P. Frankl, and V. Rödl. The asymptotic number of graphs not containing a fixed subgraph and a problem for hypergraphs having no exponent. *Graphs and Combinatorics*, 2(2):113–121, 1986.

10. P. Erdős and P. Turán. On some sequences of integers. *Journal of the London Mathematical Society*, 11(2):261–264, 1936.

11. P. Frankl and V. Rödl. Large triangle-free subgraphs in graphs without K_4. *Graphs and Combinatorics*, 2(2):135–144, 1986.

12. P. Frankl and V. Rödl. The uniformity lemma for hypergraphs. *Graphs and Combinatorics*, 8(4):309–312, 1992.

13. P. Frankl, V. Rödl, and R.M. Wilson. The number of submatrices of a given type in a Hadamard matrix and related results. *Journal of Combinatorial Theory, Series B*, 44(3):317–328, 1988.

14. A. Frieze and R. Kannan. The regularity lemma and approximation schemes for dense problems. In *Proc. 37th Ann. IEEE Symp. on Foundations of Computer Science*. IEEE, October 1996.

15. Z. Füredi. Random Ramsey graphs for the four-cycle. *Discrete Mathematics*, 126:407–410, 1994.

16. R.L. Graham and J. Nešetřil. Large minimal sets which force long arithmetic progressions. *Journal of Combinatorial Theory, Series A*, 42(2):270–276, 1986.

17. R.L. Graham and V. Rödl. Numbers in Ramsey theory. In C. Whitehead, editor, *Surveys in Combinatorics 1987*, volume 123 of *London Mathematical Society Lecture Note Series*, pages 112–153. Cambridge University Press, Cambridge–New York, 1987.

18. R.L. Graham and J. Spencer. A constructive solution to a tournament problem. *Canadian Mathematics Bulletin*, 14:45–48, 1971.

19. P.E. Haxell and Y. Kohayakawa. On an anti-Ramsey property of Ramanujan graphs. *Random Structures and Algorithms*, 6(4):417–431, 1995.

20. P.E. Haxell, Y. Kohayakawa, and T. Łuczak. The induced size-Ramsey number of cycles. *Combinatorics, Probability, and Computing*, 4(3):217–239, 1995.

21. P.E. Haxell, Y. Kohayakawa, and T. Łuczak. Turán's extremal problem in random graphs: forbidding even cycles. *Journal of Combinatorial Theory, Series B*, 64:273–287, 1995.

22. P.E. Haxell, Y. Kohayakawa, and T. Łuczak. Turán's extremal problem in random graphs: forbidding odd cycles. *Combinatorica*, 16(1):107–122, 1996.

23. Y. Kohayakawa. The regularity lemma of Szemerédi for sparse graphs, manuscript, August 1993, 10 pp.

24. Y. Kohayakawa and B. Kreuter. Threshold functions for asymmetric Ramsey properties involving cycles. *Random Structures and Algorithms*. Submitted, 1996, 34 pp.

25. Y. Kohayakawa, B. Kreuter, and A. Steger. An extremal problem for random graphs and the number of graphs with large even-girth. *Combinatorica*. Submitted, 1995, 18 pp.

26. Y. Kohayakawa, T. Łuczak, and V. Rödl. On K^4-free subgraphs of random graphs. *Combinatorica*. Submitted, 1995, 35 pp.
27. Y. Kohayakawa, T. Łuczak, and V. Rödl. Arithmetic progressions of length three in subsets of a random set. *Acta Arithmetica*, LXXV(2):133–163, 1996.
28. Y. Kohayakawa and V. Rödl. Checking pseudorandomness of graphs fast (tentative title). In preparation, 1996.
29. J. Komlós and M. Simonovits. Szemerédi's regularity lemma and its applications in graph theory. In D. Miklós, V.T. Sós, and T. Szőnyi, editors, *Combinatorics—Paul Erdős is eighty*, vol. 2, Bolyai Society Mathematical Studies, pages 295–352. János Bolyai Mathematical Society, Budapest, 1996.
30. A. Lubotzky, R. Phillips, and P. Sarnak. Ramanujan graphs. *Combinatorica*, 8:261–277, 1988.
31. J. Nešetřil and V. Rödl. Partite construction and Ramseyan theorems for sets, numbers and spaces. *Commentationes Mathematicae Universitatis Carolinae*, 28(3):569–580, 1987.
32. H.-J. Prömel and B. Voigt. A sparse Graham–Rothschild theorem. *Transactions of the American Mathematical Society*, 309(1):113–137, 1988.
33. V. Rödl. On universality of graphs with uniformly distributed edges. *Discrete Mathematics*, 59(1-2):125–134, 1986.
34. V. Rödl. On Ramsey families of sets. *Graphs and Combinatorics*, 16(2):187–195, 1990.
35. V. Rödl and A. Ruciński. Lower bounds on probability thresholds for Ramsey properties. In D. Miklós, V.T. Sós, and T. Szőnyi, editors, *Combinatorics—Paul Erdős is eighty*, vol. 1, Bolyai Society Mathematical Studies, pages 317–346. János Bolyai Mathematical Society, Budapest, 1993.
36. V. Rödl and A. Ruciński. Threshold functions for Ramsey properties. *Journal of the American Mathematical Society*, 8(4):917–942, 1995.
37. K.F. Roth. On certain sets of integers. *Journal of the London Mathematical Society*, 28:104–109, 1953.
38. J. Spencer. Restricted Ramsey configurations. *Journal of Combinatorial Theory, Series A*, 19(3):278–286, 1975.
39. E. Szemerédi. On sets of integer sets containing no k elements in arithmetic progression. *Acta Arithmetica*, 27:111–116, 1975.
40. E. Szemerédi. Regular partitions of graphs. In *Problèmes Combinatoires et Théorie des Graphes*, pages 399–401, Orsay, 1976. Colloques Internationaux CNRS n. 260.
41. A.G. Thomason. Pseudorandom graphs. In *Random graphs '85 (Poznán, 1985)*, volume 144 of *North-Holland Math. Stud.*, pages 307–331. North-Holland, Amsterdam-New York, 1987.
42. A.G. Thomason. Random graphs, strongly regular graphs pseudorandom graphs. In C. Whitehead, editor, *Surveys in Combinatorics 1987*, volume 123 of *London Mathematical Society Lecture Note Series*, pages 173–195. Cambridge University Press, Cambridge-New York, 1987.

Questions on Attractors of 3-Manifolds

Sóstenes Lins

Abstract. The *attractor* of a 3-manifold M^3 is the set of all *3-gems* which have a minimum number of vertices and induce M^3. A *gem* (graph-encoded manifold) is a special edge graph which encodes a ball complex whose underlying space is a manifold. Every 3-manifold is induced by a 3-gem. In this article I briefly recall the definitions and terminology of 3-gems, state some of the properties of attractors and list a number basic open questions concerning them. The characteristic of this approach to 3-manifolds is the massive use of computers and so, most of the open questions here stated simply await proper implementation of algorithms to be answered.

1   Introduction

Advances in computer technology is continuosly expanding our computational capability. With the widespread dissemination of computers in mathematics some of the classical methods of combinatorial topology are being revisited with some new emphasis. In particular the concepts *verifiability* and *recognition of structures* can be, with great profit, left to the machines.

Presently there is a good heuristics to decide, using a computer, whether two given *small* 3-manifolds are homeomorphic or not. By small I mean induced by a blackboard framed link [10] of not very many crossings. From this presentation we can effectively compute the WRT-invariants as done in [10], proving that they are non homeomorphic, or else apply the combinatorial dynamics studied in [11] to explicitly present a homeomorphism between two 3-gems induced by the blackboard framed links, Chapter 13 of [10], or Section 2.8 of [11]. Indeed to improve this heuristics and stay at the 3-gem level (which have a substantailly better simplification theory than the framed links) we need an implementable algorithm to obtain from a 3-gem a blakboard framed link inducing the same 3-manifold.

In spite of the success of 3-gems as a computational theory, the essence of the mathematical problem remains unconquerable: we know very few about minimal 3-gems. Indeed we would like to obtain general basic facts on them which are true, elegant, and useful in other contexts. And we look for mathematical proofs of these facts, not only verification in a finite yet enormous number of cases. I think the computer has also a role to play in this more theoretical realm: to suggest evidence for possible worthwhile paths to try in the seemingly hopeless maze of caleidoscopical shapes in which any M^3 presents itself to us. The last section of this paper contains some questions which have been suggested to me in this way along the years. The others sections prepare and motivate the last one.

2   Brief Review of 3-Gems

An *(n+1)-graph* is a finite graph G where at each vertex meet exactly $n + 1$ differently colored edges. The total number of colors is also $n + 1$. An *m-residue* is a connected component of a subgraph generated by m specified colors. Note that the 2-residues are even sided *bi*colored poly*gons* in G and are also called *bigons*. Let G^2 be the 2-complex obtained from G by attaching a 2-cell to each 2-residue. A *3-gem* is a (3+1)-graph satisfying the following condition:

$$v + t = b,$$

where v, t, b, stand for the number of 0-, 3- and 2-residues of G. If the above arithmetical condition holds then each 3-residue with its attached disks form a topological 2-sphere [6]. Let G^3 be the 3-complex obtained from G^2 by attaching a 3-cell to each such 2-sphere. It can be easily shown that $|G| = |G^3|$ is a closed 3-manifold and (from the triangulability of such 3-manifolds) that any of them arises in this from a 3-gem, [3].

The dual of a gem is a *colored triangulation*. In dimension 3 these are the triangulations of a 3-manifold with the property that each tetrahedron has four differently labeled vertices taken from the set of *colors* $\{0, 1, 2, 3\}$. Each 2-face of a colored triangulation has three diferently colored vertices. Color the face itself with the fourth color, the one which is missing among the three colors of its vertices. In this way, we get a coloration of the faces so that every tetrahedron has four differently colored faces. By taking the dual of the colored triangulation, the edges of the dual complex correspond to the colored faces. Let these colors be inherited. It can be easily checked that the edge-colored 1-skeleton of the dual of a colored triangulation is a 3-gem.

A *1-dipole* in a 3-gem is a pair of vertices linked by an h-colored edge and which are in distinct $\{i, j, k\}$-residues for (h, i, j, k) a permutation of $(0, 1, 2, 3)$. The *cancellation of a 1-dipole* is *fusion* of the constituent vertices, where, in general, the *fusion of a pair of vertices* in an $(n + 1)$-graph is the deletion of the pair together with all the possible edges incident to both vertices followed by the identification of the pairs of the free ends (of the edges incident to one vertex of the pair) along edges of the same color. The fusion on an (n+1)-graph G at vertices x, y is denoted by $G_{x,y}^{fus}$. It is easy to check that the cancellation of a 1-dipole or its inverse, the *creation of a 1-dipole* do not change the homeomorphism type of the induced 3-manifolds. By keeping cancelling 1-dipoles, we get a 3-gem without them. These objects are called *crystallizations*, [2], [5].

A *2-dipole* in a 3-gem is a pair of vertices linked by two edges in color h and i which are in distinct $\{j, k\}$-residues (jk-gons). The *cancellation of a 2-dipole* if the fusion of the constituent vertices. The cancellation of a 2-dipole or its inverse, the *creation of a 2-dipole* also do not change the homeomorphism type of the induced 3-manifold. The creation or cancellation of an i−dipole (i=1,2) is called an *i−dipole move*. These moves permit a combinatorial counterpart of homeomorphisms. Two 3-gems are *equivalent* if they induce the same 3-manifold.

Theorem 1 *Two 3-gems are equivalent if and only if they are linked by a finite set of i-dipole moves (i=1,2).*

Proof.. See [2]

The main advantage of the 3-gems over the dual objects, the colored triangulations, is that there are no higher dimensional cells. Just the colored 1-skeleton, a graph. The additional structure, the colors on the edges, induce a rich and useful combinatorics. which has enabled us to get a *classification up to homeomorphism* of **all** 3-manifolds representable by 3-gems up to 28 vertices, [7], [8], [9], [11]. The classification of the 30-vertex 3-gems is complete but not yet published.

An *attractive* 3-gem is a 3-gem inducing a prime 3-manifold and belonging to an attractor.

3   Properties of Attractive Gems

In this section we study various subconfigurations on 3-gems which permit their simplification. Thus they are forbidden subconfigurations on attractive gems.

3.1 ρ_i-pairs

Henceforth we restrict to orientable 3-manifolds. The orientibility of the manifolds correspond to the bipartiteness of the gems. A ρ_i-pair in a (bipartite) 3-gem G is a pair of equaly colored edges that appear together in $i(i = 2$ or $i = 3)$ bigons. The switching of a ρ_i-pair $\{a, b\}$ is the passage from G to $G_{a,b}^{swt}$, obtaining by replacing $\{a, b\}$ by new edges $\{a', b'\}$ having the same ends and preserving the bipartition.

Proposition 2 *If G is a connected bipartite gem so is $G_{a,b}^{swt}$. Moreover, it induces the same manifold as G and has one more 3-residue. Thus $G_{a,b}^{swt}$ has 1-dipoles.*

Proof.. See Lemma 4 of [11].

Corollary 3 *An attractive gem G has no ρ_2-pairs.*

Proof.. If G has a ρ_2-pair $\{a, b\}$, then $G_{a,b}^{swt}$ is a gem equivalent to G and with 1-dipoles. The cancellation of one of such dipole produces a gem H equivalent to G and having less vertices. Therefore G is not attractive.

When a gem is the only member of an attractor it is called the *superattractor* for the 3-manifold that it induces. Here is the superattractor of $S^1 \times S^2$, denoted $s^1 \times s^2$. The color of each edge is indicated by the number of marks attached to it.

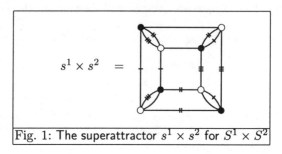

$$s^1 \times s^2 \quad =$$

Fig. 1: The superattractor $s^1 \times s^2$ for $S^1 \times S^2$

Proposition 4 *Let $\{a, b\}$ be a ρ_3-pair in a gem G. Then if $G_{a,b}^{swt}$ is connected,*

$$|G| \cong |G_{a,b}^{swt}| \# |s^1 \times s^2|.$$

If $G_{a,b}^{swt}$ is disconnected then it is the union of two connected gems G_1 and G_2. Moreover,

$$|G| \cong |G_1| \# |G_2|.$$

Proof.. See Proposition 20 of [11].

Thus, a ρ_3-pair induces either a connected sums with $S^1 \times S^2$ which can be detected at the combinatorial level or a factorizations of the manifold as a connected sum of manifolds induced by smaller gems.

If G_1 and G_2 are disjoint gems, x a vertex of G_1, y is G_2, then we denote $(G_1 \cup G_2)_{x,y}^{fus}$ by $G_1^x \# {}^y G_2$.

Corollary 5 *If G is attractive and it has a ρ_3-pair then $G = s^1 \times s^2$.*

Proof.. Let $\{a, b\}$ be a ρ_3-pair in a connected gem G. Then if $G_{a,b}^{swt}$ is connected,

$$|G| \cong |G_{a,b}^{swt}| \# |s^1 \times s^2|.$$

Since G is prime and has a minimum number of vertices, it follows that $G_{a,b}^{swt}$ is equivalent to s^3, the 3-gem with two vertices inducing S^3, and by the minimality of G it equals $s^1 \times s^2$.

If $G_{a,b}^{swt}$ is disconnected then it equals $G_1 \cup G_2$ for some connected gems G_1 and G_2 and $|G| \cong |G_1| \# |G_2|$. This union has the same number of vertices as G. But $|G_1| \# |G_2| \cong |G_1^x \# {}^y G_2|$ for some vertices x of G_1 and y of G_2. Since vertices x and y are cancelled by the fusion to produce $G_1^x \# {}^y G_2$, this gem has fewer vertices than G, a contradiction.

A 3-gem without ρ_2-pairs and without ρ_3-pairs is called a *rigid gem*. We have files (available upon request) with of all rigid bipartite 3-gems up to 30 vertices.

3.2 9-Clusters

There are two types of local 9-vertex configurations on 3-gems which permit simplification. Let (h, i, j, k) be a permutation of the colors $(0, 1, 2, 3)$. Suppose that the ij-gon, the jk-gon and the ki-gon which incide to v have size 4. Suppose that at least one of the hi-gon, the hj-gon, the hk-gon incident to v also has size 4, say the hi-gon. Let v_x for x in $\{0, 1, 2, 3\}$ be the neighbor of v by color x and let v_{xy} be the neighbor of v_x by color y. Then if the nine vertices $v, v_h, v_i, v_j, v_k,$ $v_{ij} = v_{ji}, v_{jk} = v_{kj}, v_{ki} = v_{ik}, v_{ih} = v_{hi}$ are all distinct, the subgem induced by them is called a *9-cluster of type I*.

Assume now that the hi-gon, the ij-gon, the jk-gon and the kh-gon incident to v are of size 4. Then if the nine vertices:

$v, v_h, v_i, v_j, v_k, v_{hi} = v_{ih}, v_{ij} = v_{ji}, v_{jk} = v_{kj}, v_{kh} = v_{hk}$ are distinct, the subgem induced by them is called a 9-cluster of type II.

Proposition 6 *Let G be a gem having a 9-cluster of type I or II. Then G is equivalent to a gem G' having two fewer vertices.*

Proof.. The proof follows from the fact that the two moves below ($(9 \mapsto 7)$-moves) are factorable as dipole moves.

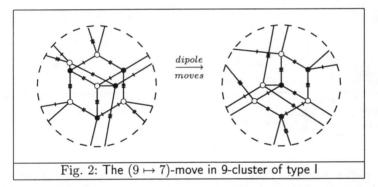

Fig. 2: The $(9 \mapsto 7)$-move in 9-cluster of type I

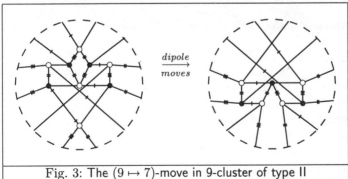

Fig. 3: The $(9 \mapsto 7)$-move in 9-cluster of type II

The dipole moves that realize the $(9 \mapsto 7)$-moves are displayed in Proposition 24 of [11].

Observation. In the proof of this proposition in [11] we have assumed implicity that the nine vertices involved are distinct. Without this hypothesis, the statement of Proposition 24 of [11] is just not true.

Corollary 7 *An attractive gem has no 9-clusters.*

Proof.. Straightforward.

Most rigid 3-gems have 9-clusters and so are not attractives. See Table 1.

3.3   Quartets

A *quartet in a 3-gem* is a set F of 4 differently colored edges. A *5-quartet* is a quartet so that except for one pair of its constituent edges all the other five pairs are in a some bigon. A *6-quartet* is a quartet in which in which all pairs of edges are in a same bigon. The set of edges incident to any vertex is a 6-quartet named *trivial 6-quartet*.

The operation of *breaking a quartet* F in a 3-gem G is the one that deletes the four edges introduces two new vertices x, y and eight new edges two of each color so as to form a bipartite $(3 + 1)$-graph $H = G_F^{brk}$ so that $H_{x,y}^{fus} = G$.

Proposition 8 *If an attractive gem G contains a non-trivial 6-quartet then $G = s^1 \times s^2$.*

Proof.. If F is a non-trivial 6-quartet in G then $G_F^{brk} = G_1 \# G_2$ in the case that G_F^{brk} is disconnected or else $|G| = |G_F^{brk} \# s^1 \times s^2|$ in the case that G_F^{brk} is connected. The conclusion is straightforward in the second case. In the first case, since $|G|$ is prime one of the summands, say, G_1 induces S^3 and it has at least 4 vertices. It would follow that G and G_2 are equivalent and G_2 has fewer vertices than G, a contradiction.

As we might expect there are many 3-gems free of 9-clusters and having non-trivial 6-quartets. These are also not attractive by definition. See table 1.

Proposition 9 *An attractive gem has no 5-quartets.*

Proof.. Let F be a 5-quartet in a gem G. We claim that G_F^{brk} is a 3-gem equivalent to G and having two more 3-residues than G. To prove the claim, assume that the edges of F are e_0, e_1, e_2, e_3 and that only e_0 and e_2 are not in the same bigon. Let J be the graph obtained from G by breaking the trio od edges e_1, e_2, e_3 and introducing two new vertices x, y linked by a new edge of color 0, e_0' forming a 1-dipole. The cancellation of this dipole yields back G. Note that $\{e_0, e_0'\}$ is a ρ_2-pair and $J_{e_0,e_0'}^{swt} = G_F^{brk}$

From G to J and from J to G_F^{brk} the induced manifold is invariant and each passage increases by 1 the number of 3-residues. Therefore, in G_F^{brk} we can cancell two 1-dipoles arriving to a 3-gem H, equivalent to G and with two fewer vertices. Therefore, if it has 5-quartet, G is not attractive.

There are rigid gems without 9-clusters without non-trivial 6-quartets and yet having 5-quartets. The first level in which this appear is at 28 vertices. There are two of such gems in this level: r_{35}^{28} and r_{1235}^{28}. (We denote by r_p^n the p-th rigid gem in the catalogue of the rigid 3-gems with n vertices.)

A gem which is *pseudo attractive* if it has no ρ-pairs, no 9-clusters, no non-trivial 6-quartets and no 5-quartets. There are gems which are pseudo-attractive and not attractive, for instance, r_3^{26} which is equivalent to r_5^{24}, see Subsection 4.1.7 of [11]. The concept of pseudo-attractive gem is the current limit attained by the local simplifying structures. The following table performs successive filtering from rigid to pseudo-attractive gems.

No. vertices	Rigid bipartite gems	Free of 9-clusters	Free of non-trivial 6-quartets	Pseudo-attractive gems
2	1	1	1	1
8	2	2	2	2
12	1	1	1	1
16	3	3	3	3
18	4	2	1	1
20	23	15	13	13
22	44	18	5	5
24	262	106	94	94
26	1252	335	199	199
28	7760	1761	1471	1469
30	56912	9737	7467	7447

Table 1: Filtering from rigid to pseudo-attractive

A ρ-*move* in a 3-gem G is a move which does not change the induced 3-manifold and drops by two the number of vertices of G. The available *local* ρ-moves are:

- Cancellation of a 1- or of a 2-dipole.
- Switching of a ρ_2- or ρ_3-pair followed by the cancellation of a 1-dipole. In the case of ρ_3-pair we have removed a handle and so we add 1 to the handle-number of the current 3-manifold induced by the final 3-gem.
- A $(9 \mapsto 7)$-move in 9-clusters.
- The breaking of a 5-quartet followed by the cancellation of two 1-dipoles.

4 Combinatorial Simplifying Dynamics

To look for further simplifications we introduce two kinds of moves.

4.1 *TS*-Moves

If a 3-gem is pseudo-attractive we start looking for the availability of the configurations:

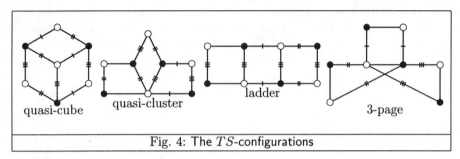

quasi-cube   quasi-cluster   ladder   3-page

Fig. 4: The *TS*-configurations

By the Euler formula the above configurations are very frequent in rigid 3-gems. They induce six involutory types of moves called *TS*-moves. The first three of these moves, TS_1, TS_2 and TS_3 , are available when the first configuration of three squares occurs. The configuration is named a *quasi-cube*.

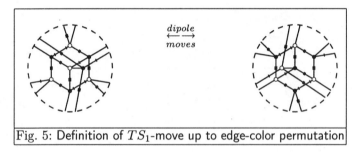

Fig. 5: Definition of TS_1-move up to edge-color permutation

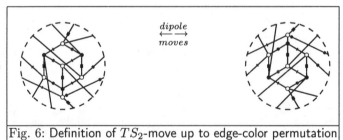

Fig. 6: Definition of TS_2-move up to edge-color permutation

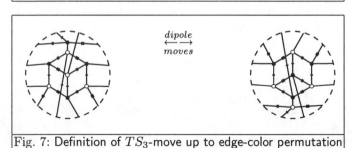

Fig. 7: Definition of TS_3-move up to edge-color permutation

The fourth TS-move, TS_4-move, is available whenever three squares meet as in the *quasi-cluster* of Fig. 4. The corresponding move is defined below:

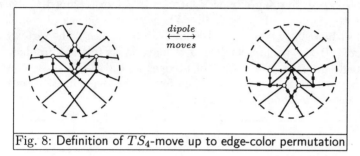

Fig. 8: Definition of TS_4-move up to edge-color permutation

The fifth TS-move, TS_5-move, is available whenever three squares meet as in the *ladder* of Fig. 4. The corresponding move is defined below:

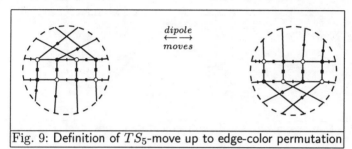

Fig. 9: Definition of TS_5-move up to edge-color permutation

The sixth TS-move, TS_6-move, is available whenever three squares meet as in the *3-page* of Fig. 4. The corresponding move is defined below:

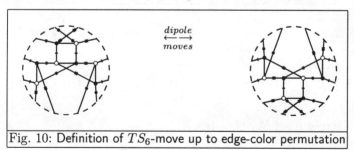

Fig. 10: Definition of TS_6-move up to edge-color permutation

For the above factorizations as dipole moves see Section 4.1 of [11]. Starting with a pseudo-attractive gem G we form a finite graph denoted Γ_G^{TS}. The vertices of this graph are in 1-1 correspondence with the 3-gems that are obtained from G by a finite number of TS-moves. An edge in this graph corresponds to a single TS-move. All the vertices of Γ_G^{TS} are 3-gems with the same number of vertices inducing the same 3-manifold. If there is a vertex H in Γ_G^{TS} which is not pseudo-attractive a smaller 3-gem inducing the same manifold (up to handles) is easily produced from H by a ρ-move. If such a gem is found the whole process starts over again with a smaller gem. This is the central part of the combinatorial simplifying dynamics.

4.2   The U-Move

The last piece in our simplifying dynamics is a move, named U-*move*, which increases the number of vertices (!). Whenever two bigons of complementary colors meet in a single vertex v, a U-move can be applied. The vertex v is called a *monopole*. Below we present how a U-move looks in the $\hat{0}$-residue. Vertex v is the single meeting of a 01-gon of 6 edges and a 23-gon of 6 edges. The 0-colored edges are presented in a dashed form. The generalization to other sizes of complementary bigons is straightforward.

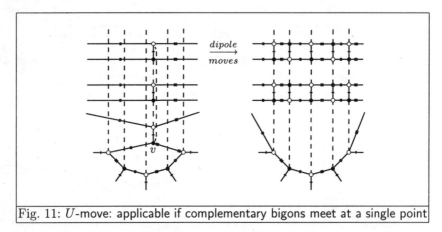

Fig. 11: U-move: applicable if complementary bigons meet at a single point

The inverse of a U-move is the counterpart in the theory of 3-gems of the Reidemeister-Singer stabilization move in the theory of Heegaard decompositions. The U-move corresponds to the cancellation of a pair of complementary handles.

The force of the U-move is that it induces many TS-configurations which in turn may provide various simplifications. In conjunction with the local ρ-moves and the TS-moves, the U-moves achieve the complete topological classification of 3-gems up to 30 vertices.

5   Twistors and Antipoles

Let $(0, i, j, k)$ be a permutation of the color set $(0, 1, 2, 3)$. A ^{i}*twistor* in a bipartite $(3+1)$-graph is a pair of vertices in the same class of the bipartition which are in the same $0i$-gon, the same jk-gon and in distinct $0j$-, $0k$-, ij- and ik-gons. In dual terms an ^{i}twistor in a 3-gem corresponds to a pair of tetrahedra having precisely one pair of opposite edges in common. The regular neighborhood of such a pair of tetrahedra is a solid torus. The ^{0j}recoupling of an ^{i}twistor $\{x, y\}$ is the operation of switching the 0-neighbors and the j-neighbors of the vertices x, y of G. Note that after the operation $\{x, y\}$ becomes a ^{k}twistor. The resulting gem is denoted by $G_{x,y}^{rec_{ij}}$. By the dual interpretation we see that $|G_{x,y}^{rec_{ij}}|$ is obtained from $|G|$ by a Dehn-Lickorish surgery replacing a solid torus by another attached differently.

A move M on 3-gems is called *universal* if it is internal to the class of 3-gems and given any two 3-gems they are linked by a finite sequence of moves each of which is a 1-dipole move, a 2-dipole move of an M-move.

Theorem 10 *The ^{0j} recoupling on ^{i} twistors is a universal operation on 3-gems.*

Proof.. See [12].

A *4-quartet* of type i (i=1,2,3) or an ^{i}quartet is a quartet in which the pair of edges of colors 0 and i and the pair of edges of color j and k are not in the same bigon. All of the other pairs are in the same bigon.

An ^{i}*antipole* $\{x, y\}$ is like an ^{i}twistor differing on the fact that the two vertices x and y are in distinct classes of the bipartition. The cancelling of an antipole $\{x, y\}$ is the fusion of this pair.

Proposition 11 *The cancelling of an ^{i}antipoles and its inverse operation breaking of ^{i}quartets are universal operations on 3-gems.*

Proof.. The $0j$-recoupling on an ^{i}twistor is factorable as the creation of a 2-dipole followed by the cancellation of an antipole. See Proposition 4 of [12]. The cancelling of an ^{i}antipole is factorable as the breaking of an ^{i}quartet up to dipole moves. See Proposition 5 of [12].

In dual terms an ^{i}antipole is a pair of tetrahedra sharing a pair of opposite edges and its regular neighborhood is a solid torus. The proof of Theorem 10 given in [12] is entirely combinatorial but it is rather involved because the technique to get from an arbitrary gem G to the gem s^3 with 2 vertices makes use of U-moves (previous section) which increases the number of vertices in the intermediate stages.

6   A Class of 3-Manifolds from Plane Graphs

Given a graph G embedded in the plane, we can proceed to get a crystallization $\Psi(G)$ as follows. Take the dual of the barycentric subdivision of the 2-cell complex induced by the plane embedding of G. This dual is a cubic bipartite plane graph which is naturally 3-edge colored. Indeed, the original edges correspond to square bigons which we paint with colors 1 and 2. We use color 3 to paint the other edges which correspond to adjacent pairs of vetices in the original edges. To get the associated bipartite crystallization introduce the 0-colored edges, as follows: the other end of a black vertex by a 0-colored edge, is the terminus of a 3-edge path which starting at the vertex uses a 1-colored edge, followed by a 2-colored edge, followed by a 3-colored edge. The (3+1)-graph so obtained is a crystallization ([6]) which we denote by $\Psi(G)$.

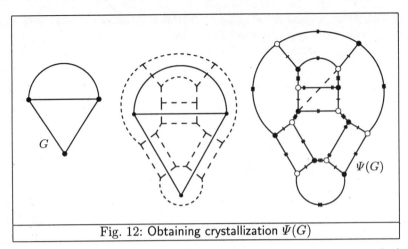

Fig. 12: Obtaining crystallization $\Psi(G)$

This construction has the property that the first homology group of the $|\Psi(G)|$ is finite and its order equals the number of spanning trees of the originating plane graph G, 5 in the above case. This property is proved in [6].

The associated class of 3-manifolds has been recognized by M. Ferri [4] by using the central idea in [1] as composed of 2-fold branched coverings of S^3 branched along the alternating links, $L(G)$, defined by the *medial* of the plane graphs G.

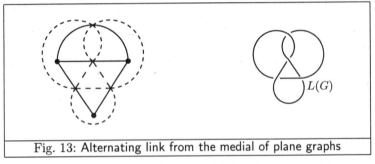

Fig. 13: Alternating link from the medial of plane graphs

Thus, the 3-manifold induced by the above $\Psi(G)$ is homeomorphic to the 2-fold branched covering over the figure 8 knot, namely the lens space $L_{5,2}$.

If G is a tree, then $|\Psi(G)|$ is S^3. It seems that a 2-connected G yields an attractive gem $\Psi(G)$ and that if G is 3-connected then $\Psi(G)$ is a superattractor. This issue is reinforced in Problems 4 and 5 of next section.

7   Open Questions

1. A u^0-*move* on a 3-gem is either a ρ-move or a TS-move, whereas a u^0_*-*move* is the identity or a finite sequence of u^0-moves. A u^1-*move* is a move of type Uu^0_* *which may decrease but does not increase the number of vertices.* A pseudo-attractive gem is u^1-*irreducible* if all the gems obtained from it by u^1-moves are pseudo-attractives.

Strong Conjecture on 3-gems: *(a) A 3-gem is attractive if and only if it is u^1-irreducible. (b) Two attractive equivalent 3-gems are linked by a finite number of u^1-moves.*

A positive solution of this Conjecture solves the homeomorphism problem for closed orientable 3-manifolds.

2. A ρ-twistor is an ^{i}twistor x, y so that the x and y are linked by two edges of colors 0 and i or by two edges of colors j and k. After the ^{0j}recoupling on x, y we get a 3-gem with at least one 2-dipole, which can be simplified.

   Is it true that a pseudo-attractive gem has either a ρ-twistor or an antipole? An affirmative answer to this question would make straightforward the proof of Theorem 10 and would be important as an inductive tool.

3. A *quadricolor* in a gem is a polygon with four differently colored edges. The tripling of a quadricolor is the operation deficted below. The result is another gem.

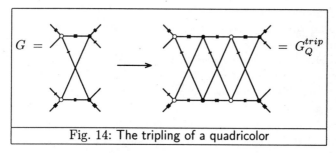

Fig. 14: The tripling of a quadricolor

Let Q be a quadricolor in G. Decide the equivalence: *G is attractive iff G_Q^{trip} is attractive.*

If this question has a positive answer it would be easy to get attractive gems of arbitrary size. Note that to decide whether a 3-gem is attractive by the definition is a huge (and unsolved) general problem which nevertheless we have solved for 3-gems up to 30 vertices.

4. **Conjecture on "plane" 3-manifolds:** *If G is 2-connected, then $\Psi(G)$ is an attractive gem. If G is 3-connected, then $\Psi(G)$ is the superattractor for $|\Psi(G)|$.*

5. The graphic matroid of G seems to be a complete invariant for the oriented form of $|\Psi(G)|$. Changing in orientation corresponds to take the dual matroid.

   How to associate an invariant matroid to a general 3-manifold in a way to extend this association in the case of plane manifolds?

   In problems 6 to 15 G_1 and G_2 are two arbitrary but fixed attractive equivalent gems.

6. Define $\mathcal{M}_{fa}(G)$ to be the set of all manifolds obtained as $|G_{x,y}^{fus}|$ for some antipole $\{x, y\}$ in G.
   Is it true that $\mathcal{M}_{fa}(G_1) = \mathcal{M}_{fa}(G_2)$?

7. Define $\mathcal{M}_{rt}(G)$ to be the set of all manifolds obtained as $|G_{x,y}^{rec_{ij}}|$ for some ^{i}twistor $\{x, y\}$ of G.
   Is it true that $\mathcal{M}_{rt}(G_1) = \mathcal{M}_{rt}(G_2)$?

8. Define $\mathcal{M}_{b4q}(G)$ to be the set of manifolds obtained as $|G_F^{brk}|$ for some 4-quartet F on G.
   Is it true that $\mathcal{M}_{b4q}(G_1) = \mathcal{M}_{b4q}(G_2)$?

9. A 12-*octet* is a pair of 4-quartets of the same type. Moreover an edge in one of the 4-quartets has precisely a *mate* in the other 4-quartet. Two edges in distinct 4-quartets are in the same bigon iff they are mate. In dual terms the eight edges of a 12-octet correspond to 8 triangles forming an embedded torus or Klein bottle. This surface may be separable or not. The breaking of a 12-octet O in G is the breaking of the pair of 4-quartets and is internal to the class of gems, providing a gem G_O^{brk}. Define $\mathcal{M}_{b12o}(G)$ to be the set of manifolds obtained as $|G_O^{brk}|$ for some 12-octet O in G.
   Is it true that $\mathcal{M}_{b12o}(G_1) = \mathcal{M}_{b12o}(G_2)$?

10. Define $\mathcal{M}_{trqd}(G)$ to be the set of manifolds obtained as $|G_Q^{trip}|$ for some quadricolor Q of G.
    Is it true that $\mathcal{M}_{trqd}(G_1) = \mathcal{M}_{trqd}(G_2)$?

11. The *smoothing on a quadricolor Q* is the operation which deletes the four edges and the four vertices of Q and identifies the 8 free ends of edges along edges of the same color yielding a gem G_Q^{sm}. It can be shown that this is a universal operation for 3-gems. Define $\mathcal{M}_{sqd}(G)$ to be the set of manifolds induced as $|G_Q^{sm}|$ for some quadricolor Q of G.
    Is it true that $\mathcal{M}_{sqd}(G_1) = \mathcal{M}_{sqd}(G_2)$?

12. The *antipodal identification on a quadricolor Q* is the operation which deletes the four edges of Q and identifies the antipodal vertices of Q yielding a gem G_Q^{ai}. It can be shown that this is also a universal operation for 3-gems. Define $\mathcal{M}_{aiqd}(G)$ to be the set of manifolds induced as $|G_Q^{ai}|$ for some quadricolor Q of G.
    Is it true that $\mathcal{M}_{aiqd}(G_1) = \mathcal{M}_{aiqd}(G_2)$?

13. The *torsion on a quadricolor Q* is the operation which interchanges the neighbors of the pairs of edges of the same color in the coboundary of Q. The resulting graph is a gem and is denoted denoted by G_Q^{trs}. The homology mod 2 is invariant. It can be shown that this is a universal operation for Z_2-homology spheres: that is, starting with s^3 and making only i-dipole moves ($i = 1, 2$) and torsions on quadricolors we obtain all Z_2-homology spheres. A proof will appear elsewhere. Define $\mathcal{M}_{tsqd}(G)$ to be the set of manifolds induced as $|G_Q^{trs}|$ for some quadricolor Q of G.
    Is it true that $\mathcal{M}_{tsqd}(G_1) = \mathcal{M}_{tsqd}(G_2)$?

14. If G and H are disjoint gems, a, b is an ^{i}twistor or an ^{i}antipole in G, b, d is also an ^{i}twistor or an ^{i}antipole in H (same i), then we denote $((G \cup H)_{a,c}^{fus})_{b,d}^{fus}$ by $G_c^a \#_d^b H$. This $(3+1)$-graph is a 3-gem and we call it the *biconnected sum of G and H along twistors (or antipoles) (a, c) and (b, d)*. Define $\mathcal{M}_{bst}(G, H)$ to be the set of manifolds of the form $|G_b^a \#_d^c H|$ along twistors.
    Is it true that $\mathcal{M}_{bst}(G_1, G_2) = \mathcal{M}_{bst}(G_1', G_2')$?

15. Define $\mathcal{M}_{bsa}(G, H)$ to be the set of manifolds of the form $|G_b^a \#_d^c H|$ along antipoles.
    Is it true that $\mathcal{M}_{bsa}(G_1, G_2) = \mathcal{M}_{bsa}(G_1', G_2')$?

16. *Find an implementable algorithm whose input is a 3-gem G and whose outup is a framed link inducing the same 3-manifold as G.*

17. *Find an implementable algorithm whose input is a 3-crystallization and a proof that it induces S^3, given by a monotone sequence of reducing moves down to s^3 and whose outup is a piecewise linear embedding in S^3 of the colored triangulation dual to G (in the spherical geometry of S^3, i.e., the lines, faces and 3-cells are made up of a finite number of geodesic pieces).*
    This seems to be a necessary step in providing the previous algorithm.

18. The number of vertices of a graph G is denoted by v_G.
    Narrow Band Conjecture: *There exists a fixed integer $k > 1$ (and hopefully small) so that given any attractive gem G with either a ρ-twistor x, y or an antipole x, y and there exists another attractive gem H so that $2 \leq v_G - v_H \leq 2k$, and either $H \equiv G_{x,y}^{rec_{ij}}$ for the ^{i}twistor x, y or else $H \equiv G_{x,y}^{fus}$ for the antipole x, y.*
    The existence of such a bound would imply a positive answer for question 2 and would be relevant to a better understanding of attractors.

References

1. Ferri, M.: Crystallizations of 2-fold branched covering of S^3. Proc. Amer. Math. Soc. **73** (1979) 271-276

2. Ferri, M., Gagliardi, C.: Crystallization moves. Pacific J. Math **100** (1982) 85–103

3. Lins, S., Mandel, A.: Graph-Encoded 3-Manifolds. Discrete Mathematics **57** (1985) 261–284

4. Ferri, M.: Personal Communication (1985)

5. Ferri, M., Gagliardi, C., Graselli, L.: A graph-theoretical representation of PL-manifolds: a survey on crystallizations. Aequationes Math. **31** (1986) 121–141

6. Lins, S.: On the fundamental group of 3-gems and a "planar" class of 3–manifolds. Europ. J. Combinatorics, **9** (1988) 291–305

7. Lins, S., Durand, C.: A complete catalogue of rigid graph-encoded orientable 3-manifolds up to 28 vertices. Notas Com. Mat. UFPE **168** (1989)

8. Lins, S., Durand, C.: Topological classification of small graph-encoded orientable 3-manifolds. Notas Com. Mat. UFPE **177** (1991)

9. Durand, C.: Geração e classificação de 3-variedades. Master Thesis presented at UFPE, (1992)

10. Kauffman, L.H.,Lins, S.: Temperley-Lieb recoupling theory and invariants of 3-manifolds. Annals of Mathematical Studies, **134**, Princeton Univ. Press (1994)

11. Lins, S.: Gems, Computers and Attractors for 3-Manifolds. K & E Series on Knots and Everything **5** (1995)

12. Lins, S.: Twistors: Bridges Among 3-Manifolds. To appear on Discrete Mathematics (1996)

A Trust-Region SLCP Model Algorithm for Nonlinear Programming *

José Mario Martínez

Abstract. We introduce a new model algorithm for solving nonlinear programming problems. At each iteration, the method solves (approximately) linearly constrained optimization problems. For this reason, it belongs to the class of SLCP (Sequential Linearly Constrained Programming) methods. Each iteration begins with a Restoration Phase, where feasibility of the current iterate is improved and follows with a Minimization Phase of Trust-Region type. In the Minimization Phase the objective function is reduced within an approximate (linearized) feasible set. The current point and the trial point obtained in the Minimization Phase are compared on the basis of a nonsmooth merit function that combines feasibility and optimality. We prove global convergence results. **Keywords:** Nonlinear programming, trust regions, GRG methods, SGRA methods, projected gradients, SQP methods, global convergence.

1 Introduction

We wish to introduce a model algorithm for potentially large-scale nonlinearly constrained optimization that makes full use of available methods for linearly constrained problems. Let us state the nonlinear programming problem in the form

$$\text{Minimize } f(x)$$

$$\text{subject to } C(x) \leq 0, \quad x \in \Omega, \tag{1}$$

where $f : \mathbb{R}^n \rightarrow \mathbb{R}$ and $C : \mathbb{R}^n \rightarrow \mathbb{R}^m$ are continuously differentiable and $\Omega \subset \mathbb{R}^n$ is closed and convex. In practice, we are mostly interested in the case in which Ω is a polytope.

The new model algorithm generates feasible iterates with respect to Ω ($x^k \in \Omega$ for all $k = 0, 1, 2, \ldots$). Each iteration involves two different procedures: Restoration and Minimization. In the Restoration Step (which is executed once per iteration) an intermediate point $x_{af}^k \in \Omega$ is found such that the infeasibility at x_{af}^k is a fraction of the infeasibility at x^k ("af" means "approximate feasible"). Immediately after Restoration we construct an approximation π_k of the feasible region using available information at x_{af}^k. In the Minimization Step we compute a trial point $x_{trial}^{k,i} \in \pi_k$ such that $f(x_{trial}^{k,i}) << f(x_{af}^k)$ and $\|x_{trial}^{k,i} - x_{af}^k\| \leq \delta_{k,i}$,

* Work sponsored by FAPESP (Grant 90-3724-6), CNPq and FAEP-UNICAMP.

where $\delta_{k,i}$ is a trust-region radius. The trial point $x_{trial}^{k,i}$ is accepted as new iterate if the value of a nonsmooth (exact penalty) merit function at $x_{trial}^{k,i}$ is sufficiently smaller than its value at x^k. If $x_{trial}^{k,i}$ is not acceptable, the trust-region radius is reduced.

When Ω is a polytope, the approximate feasible region π_k is a polytope as well. So, if $\| \cdot \|$ is the sup-norm, the Minimization Step consists of an inexact (approximate) minimization of f with linear constraints. In that case, the Restoration Step also represents an inexact minimization of infeasibility with linear constraints. Therefore, available algorithms for (large-scale) linearly constrained minimization can be fully exploited.

The new algorithm is related to classical *feasible* methods for nonlinear programming, such as the Generalized Reduced Gradient (GRG) method and the family of Sequential Gradient Restoration algorithms (SGRA). See [3], [4] and references therein. However, in our approach the successive approximations to the solution of (1) are not necessarily feasible (or nearly feasible) with respect to $C(x) \leq 0$. In spite of that, the necessity of considering and probably improving feasibility is taken actively into account at all the iterations. This feature constrasts with the strategy adopted in Sequential Quadratic Programming (SQP) algorithms, where the trial point at each iteration is obtained after considering only a linear model of the constraints.

The convergence theory developed in this paper has several points in common with global convergence theories for different SQP-like algorithms with trust-regions, in particular the one developed in [1]. The new model algorithm is also related to the method introduced in [2] for problems where the constraints are given in the form $C(x) = 0, \quad x \in \Omega$. Of course, an inequality constraint can be transformed into an equality constraint plus a simple bound by means of the introduction of a slack variable but, in critical cases, this device can increase the complexity of the problem in an undesirable way. In [2] the merit function is an augmented Lagrangian, while here we use an exact penalty-like merit function. Moreover, the algorithm introduced in this paper use trust-regions centered on the intermediate point x_{af}^k instead of the more usual trust-regions centered on the current point x^k. Consequently, only the Minimization Step is repeated after a reduction of the trust-region radius.

Notation.

In this work we use two (perhaps different) norms. We denote $|\cdot|$ a monotone norm on $I\!\!R^m$ ($|v| \leq |w|$ whenever $0 \leq v \leq w$) and $\|\cdot\|$ an arbitrary norm on $I\!\!R^n$.

We denote $C'(x) \in I\!\!R^{m \times n}$ the Jacobian matrix of $C(x)$ and $C_j'(x) = \nabla C_j(x)^T$ for all $j = 1, \ldots, m$

We also denote $C_j^+(x) = \max\{C_j(x), 0\}$ and $C^+(x) = (C_1^+(x), \ldots, C_m^+(x))^T$.

2   The Model Algorithm

2.1   Restoration Step

As we mentioned in the Introduction, given the current iterate $x^k \in \Omega$, the model algorithm begins computing and intermediate "more feasible" point $x_{af}^k \in \Omega$. The conditions that must be satisfied by x_{af}^k are

$$|C^+(x_{af}^k)| \le r|C^+(x^k)| \tag{2}$$

$$\|x_{af}^k - x^k\| \le \beta|C^+(x^k)|. \tag{3}$$

where $r \in [0,1)$ and $\beta > 0$ are parameters given independently of k. Condition (2) states the necessity of having an intermediate point at least as feasible as x^k. Condition (3) imposes that x_{af}^k must be equal to x^k if the current point is feasible. For finding x_{af}^k we can use any available algorithm that aims to solve the problem

$$\text{Minimize } \|C^+(x)\|_2^2 \text{ subject to } \|x - x^k\| \le \beta|C^+(x^k)| \text{ and } x \in \Omega. \tag{4}$$

If Ω is a polytope and $\|\cdot\|$ is the sup-norm, (4) turns out to be a smooth linearly constrained minimization problem, for which we wish to obtain the approximate solution x_{af}^k, using x^k as initial guess. We can also obtain x_{af}^k by means of an algorithm especially designed for solving feasibility problems. With both approaches we will be able to obtain (2) and (3) if $C(x)$ satisfies some regularity conditions. In general, only the presence of local-nonglobal minimizers of $|C^+(x)|$ will inhibit the success of usual methods when (1) is a feasible problem. Of course, if the feasible region of (1) is empty, we will fatally find a point $x^k \in \Omega$ starting of which x_{af}^k cannot be found.

2.2   The Approximate (Linearized) Feasible Region

After the computation of x_{af}^k with the conditions (2) and (3) we define a linear approximation of the feasible region of (1), containing the intermediate point x_{af}^k. This auxiliary region is given by

$$\pi_k = \{x \in \Omega \mid C_j(x_{af}^k) + C_j'(x_{af}^k)(x - x_{af}^k) \le C_j^+(x_{af}^k) \text{ whenever } C_j(x_{af}^k) \ge -p\}, \tag{5}$$

where $p > 0$ is a parameter given independently of the iteration index k. So, π_k is the intersection of Ω with the linear approximations of the sets $C_j(x) \le C_j^+(x_{af}^k)$, excluding the indices j that correspond to constraints that are strongly satisfied at x_{af}^k, according to the tolerance p. If p is large the approximate feasible region takes into account all the constraints $C_j(x) \le 0$, independently of $C_j(x_{af}^k)$. On the other hand, if p is small, only the constraints violated at x_{af}^k tend to be considered in the definition of π_k. In other words, if $C_j(x_{af}^k) < -p$, it is considered that the approximation of the set $C_j(x) \le 0$ that uses information at x_{af}^k is the whole space $I\!\!R^n$. In principle, it should be better to use a large p, for this gives a more faithful representation of the true feasible region. However, the subproblem involved in the Minimization Step is simpler when p is small.

2.3 Minimization Step

The objective of the Minimization Step is to obtain $x_{trial}^{k,i} \in \pi_k \cap I\!\!B_{k,i}$ such that $f(x_{trial}^{k,i}) << f(x_{af}^k)$, where

$$I\!\!B_{k,i} = \{x \in I\!\!R^n \mid \|x - x_{af}^k\| \le \delta_{k,i}\}, \qquad (6)$$

and $\delta_{k,i} > 0$ is a trust-region radius. The first trial point at each iteration is obtained using with a trust-region radius $\delta_{k,0}$. Successive trust-region radius are tried until a point $x_{trial}^{k,i}$ is found such that the merit function at this point is sufficiently smaller than the merit function at x^k.

The minimization step is preceded by the computation of the Cauchy-like direction (independent of i)

$$d_{tan}^k = P_k(x_{af}^k - \eta \nabla f(x_{af}^k)) - x_{af}^k, \qquad (7)$$

where $P_k(z)$ denotes the orthogonal projection of z on π_k and and $\eta > 0$ is an arbitrary scaling parameter independent of k. It turns out that d_{tan}^k is a feasible descent direction of f on π_k. Its norm will be used to define a convergence criterion for the algorithm. The trial point $x_{af}^k + d_{tan}^k$ belongs to π_k but it does not necessarily belong to $I\!\!B_{k,i}$. So, we define the breakpoint $x_{af}^k + t_{break}^{k,i} d_{tan}^k$ by

$$t_{break}^{k,i} = \sup \{t \in [0,1] \mid [x_{af}^k, x_{af}^k + td_{nor}^k] \subset I\!\!B_{k,i}\}. \qquad (8)$$

Moreover, the value of the objective function f at $x_{af}^k + t_{break}^{k,i} d_{tan}^k$ is not necessarily smaller than $f(x_{af}^k)$, therefore a sufficiently smaller functional value $f(x_{af}^k + t_{dec}^{k,i} d_{tan}^k)$ must be obtained using a classical backtracking procedure. Finally, $x_{trial}^{k,i} \in \pi_k \cap I\!\!B_{k,i}$ will be any point such that $f(x_{trial}^{k,i}) \le f(x_{af}^k + t_{dec}^{k,i} d_{tan}^k)$. Alternatively, $x_{trial}^{k,i}$ can be any point of $\pi_k \cap I\!\!B_{k,i}$ such that $f(x_{trial}^{k,i}) \le f(x_{af}^k) - \tau_1 \delta_{k,i}$ or $f(x_{trial}^{k,i}) \le f(x^k) - \tau_2$, where τ_1 and τ_2 are nonnegative parameters of the algorithm. This means that, for computing the trial point $x_{trial}^{k,i}$ in an efficient way, we can apply any reasonable algorithm (with a mild convergence criterion) to the resolution of the minimization problem

$$\text{Minimize } f(x) \quad \text{subject to} \quad x \in \pi_k \cap I\!\!B_{k,i}. \qquad (9)$$

2.4   Merit Function and Penalty Parameter

The comparison of $x_{trial}^{k,i}$ and x^k involves the evaluation of a merit function at both points. We decided to use the exact penalty-like nonsmooth merit function, given by

$$\psi(x, \theta) = \theta f(x) + (1 - \theta)|C^+(x)| \qquad (10)$$

where $\theta \in (0, 1]$ is a penalty parameter used to give different weights to the objective function and to the feasibility objective. The choice of the parameter θ at each iteration depends of practical and theoretical considerations. For example, if $|C^+(x^k)|$ is large, the weight assigned to $f(x)$ must be small, for it does not make sense to worry about the functional values if the current point is far from the feasible region. We will see that our choice of the penalty parameter automatically takes into account this practical necessity.

Roughly speaking, we wish that the merit function at the new point should be less than the merit function at the current point x^k. That is, we desire that $\mathbf{Ared}_{k,i} > 0$, where $\mathbf{Ared}_{k,i}$, the "actual reduction of the merit function", is defined by

$$\mathbf{Ared}_{k,i} = \psi(x^k, \theta_{k,i}) - \psi(x_{trial}^{k,i}, \theta_{k,i}). \qquad (11)$$

However, as so happens to occur in unconstrained optimization, a reduction of the merit function is not sufficient to guarantee convergence. In fact, we need a "sufficient reduction" of the merit function, that will be defined by the satisfaction of the following test

$$\mathbf{Ared}_{k,i} \geq 0.1\mathbf{Pred}_{k,i}, \qquad (12)$$

where $\mathbf{Pred}_{k,i}$ is a positive "predicted reduction" of the merit function between x^k and $x_{trial}^{k,i}$. In our case, we define

$$\mathbf{Pred}_{k,i} = \theta_{k,i}[f(x^k) - f(x_{trial}^{k,i})] + (1 - \theta_{k,i})[|C^+(x^k)| - |C^+(x_{af}^k)|]. \qquad (13)$$

The quantity $\mathbf{Pred}_{k,i}$ defined above can be nonpositive depending on the value of the penalty parameter. Fortunately, if $\theta_{k,i}$ is small enough, $\mathbf{Pred}_{k,i}$ is arbitrarily close to $|C(x^k)| - |C(x_{af}^k)|$ which is necessarily nonnegative. Therefore, we will always be able to choose $\theta_{k,i} \in (0, 1]$ such that

$$\mathbf{Pred}_{k,i} \geq \frac{1}{2}[|C^+(x^k)| - |C^+(x_{af}^k)|]. \qquad (14)$$

When the criterion (12) is satisfied, we accept $x^{k+1} = x_{trial}^{k,i}$. Otherwise, we reduce the trust-region radius.

2.5   Description of the Model Algorithm

Assume that $p > 0$, $\eta > 0$, $\beta > 0$, $r \in [0,1)$, $\delta_{min} > 0$, $\tau_1 > 0, \tau_2 > 0$ are algorithmic parameters given independently of k and $\sum_{k=0}^{\infty} \omega_k$ is a convergent series of nonnegative terms. Suppose that $x^0 \in \Omega$ is an initial approximation to the solution and that $\theta_{-1} \in (0,1)$ is an initialization of the penalty parameter. Given $x^k \in \Omega$, $\theta_{k-1} \in (0,1]$, $\delta_{k,0} \geq \delta_{min}$, the steps for computing x^{k+1} or for stopping the process are given by the following algorithm.

Algorithm 2.1
Step 1. *Compute* x_{af}^k, d_{tan}^k *and decide termination*

Compute $x_{af}^k \in \Omega$ such that (2) and (3) hold. If this is not possible, stop the execution of the algorithm declaring "failure in improving feasibility". Otherwise, set $i \leftarrow 0$,

$$\theta_{k,-1} = \min \{1, \min \{\theta_{-1}, \ldots, \theta_{k-1}\} + \omega_k\}.$$

and compute d_{tan}^k using (7). If $C^+(x^k) = 0$ and $d_{tan}^k = 0$ terminate the execution of the algorithm declaring "finite convergence".

Step 2. *Minimization Step*

Compute $t_{break}^{k,i}$ using (8). Define $t_{dec}^{k,i}$ as the first term t of the sequence $\{t_{k,1}, t_{k,2}, \ldots\}$ such that

$$f(x_{af}^k + td_{tan}^k) \leq f(x^k) + 0.1t\langle \nabla f(x^k), d_{tan}^k \rangle, \tag{15}$$

where $\{t_{k,j}\}$ is defined by $t_{k,1} = t_{break}^{k,i}$ and $t_{k,j+1} \in [0.1t_{k,j}, 0.9t_{k,j}]$ for all $j = 1, 2, \ldots$ Compute $x_{trial}^{k,i} \in \pi_k \cap B_{k,i}$ such that

$$f(x_{trial}^{k,i}) \leq \max \{f(x_{af}^k + t_{dec}^{k,i}d_{tan}^k), f(x_{af}^k) - \tau_1\delta_{k,i}, f(x_{af}^k) - \tau_2\}. \tag{16}$$

Step 3. *Choice of the penalty parameter*

Define, for all $\theta \in [0,1]$,

$$Pred_{k,i}(\theta) = \theta[f(x^k) - f(x_{trial}^{k,i})] + (1-\theta)[|C^+(x^k)| - |C^+(x_{af}^k)|].$$

Choose $\theta_{k,i}$ the supremum of the values of θ in the interval $[0, \theta_{k,i-1}]$ such that

$$Pred_{k,i}(\theta) \geq \frac{1}{2}[|C^+(x^k)| - |C^+(x_{af}^k)|]. \tag{17}$$

Step 4. *Acceptance or rejection of the trial point*

Define $\mathbf{Ared}_{k,i}$ and $\mathbf{Pred}_{k,i}$ as in (11) and (13) respectively. If the test (12) is satisfied, define $x^{k+1} = x_{trial}^{k,i}$, $\theta_k = \theta_{k,i}$, $iacc(k) = i$ ("*iacc*" means "accepted i") and finish the iteration. If (12) does not hold, choose $\delta_{k,i+1} \in [0.1\delta_{k,i}, 0.9\delta_{k,i}]$, set $i \leftarrow i+1$ and go to Step 2.

2.6   Some Remarks and Elementary Properties

By means of the introduction of the nonnegative parameters ω_k a "moderate" increase of the penalty parameter between different iterations is permitted. This prevents the possibility of inheriting very small penalty parameters from the very beginning of the algorithm, when they are not necessary for guaranteeing feasibility. It is easy to see that the sequence of penalty parameters finally used at each iteration $\{\theta_k\}$ is convergent. In fact, defining $\theta_k^{small} = \min\{\theta_{-1}, \ldots, \theta_k\}$ and $\theta_k^{large} = \theta_k^{small} + \omega_k$, we see that $\theta_{k+1} \leq \theta_k^{large}$ and $\theta_k \geq \theta_k^{small}$ for all k. Clearly, $\{\theta_k^{large}\}$ and $\{\theta_k^{small}\}$ are convergent to the same limit, so $\{\theta_k\}$ is also convergent. We can also prove, by induction, that $\theta_{k,i} > 0$ for all k, i.

It is easy to verify that d_{tan}^k is a descent direction. In fact, since $x_{af}^k \in \pi_k$, we have that

$$\|(x_{af}^k - \eta\nabla f(x_{af}^k)) - P_k(x_{af}^k - \eta\nabla f(x_{af}^k))\|_2 \leq \|(x_{af}^k - \eta\nabla f(x_{af}^k)) - x_{af}^k\|_2.$$

Therefore,

$$\|x_{af}^k - P_k(x_{af}^k - \eta\nabla f(x_{af}^k))\|_2^2 + \|\eta\nabla f(x_{af}^k)\|_2^2 + 2\eta\langle P_k(x_{af}^k - \eta\nabla f(x_{af}^k)) - x_{af}^k, \nabla f(x_{af}^k)\rangle$$

$$\leq \|\eta\nabla f(x_{af}^k)\|_2^2,$$

so,

$$\langle d_{tan}^k, \nabla f(x_{af}^k)\rangle \leq -\frac{1}{2\eta}\|d_{tan}^k\|_2^2 \leq -\frac{c}{2\eta}\|d_{tan}^k\|^2, \tag{18}$$

where $c > 0$ is a norm-dependent constant. Using classical arguments for justifying backtracking with Armijo-like conditions (see [?], Chapter 6), (18) implies that $t_{dec}^{k,i}$ is well defined at Step 2 of Algorithm 2.1. In other words, given the current point x^k and the trust-region radius $\delta_{i,k}$ it is possible to compute, in finite time, the trial point $x_{trial}^{k,i}$.

3   General Assumptions and Consequences

From now on, we will suppose that the nonlinear programming problem (1) satisfies the assumptions A1, A2 and A3 stated below.

A1. Ω is convex and compact.

A2. The Jacobian matrix of $C(x)$ exists and satisfies the Lipschitz condition

$$\|C'(y) - C'(x)\| \leq L_1\|y - x\| \text{ for all } x, y \in \Omega. \tag{19}$$

A3. The gradient of f exists and satisfies the Lipschitz condition

$$\|\nabla f(y) - \nabla f(x)\| \leq L_2\|y - x\| \text{ for all } x, y \in \Omega. \tag{20}$$

The assumption on the boundedness of Ω can be replaced by hypotheses that state boundedness of a set of quantities depending on the iterates. This is frequently done in global convergence theories for SQP algorithms. We prefer to state directly Assumption A1 seems it seems to be the only verifiable assumption *on the problem* that guarantees boundedness of the required magnitudes.

The following theorem is directly deduced from the general assumptions. It states a bounded deterioration result for the feasibility of $x_{trial}^{k,i}$ in relation to the feasibility of x_{af}^k. Briefly speaking, we prove that only a second order deterioration of feasibility can be expected for a trial point $x \in \pi_k$.

Theorem 3.1 *There exists $c_1 > 0$ (independent of k) such that, whenever $x_{af}^k \in \Omega$ is defined and $x \in \pi_k$, we have*

$$|C^+(x)| \le |C^+(x_{af}^k)| + c_1\|x - x_{af}^k\|^2 \tag{21}$$

In the next theorem we compute the decrease of the objective function that can be expected when we move from x_{af}^k to $x_{trial}^{k,i}$.

Theorem 3.2 *There exist $c_2 > 0$, $c_3 > 0$ (independent of k) such that, whenever $x_{af}^k \in \Omega$ is defined and $x_{trial}^{k,i}$ is computed at Step 2 of Algorithm 2.1, we have that*

$$f(x_{trial}^{k,i}) \le f(x_{af}^k) - \ min\ \{\tau_2, c_2\|d_{tan}^k\|^2, \tau_1\delta_{k,i}, c_3\|d_{tan}^k\|\|\delta_{k,i}\}.$$

The next two theorems of this section state that the algorithm is well defined (12) holds after a finite number of reductions of the trust-region radius) and that the actual reduction tends to zero.

Theorem 3.3 *Algorithm 2.1 is well defined.*

Theorem 3.4 *Suppose that Algorithm 2.1 generates an infinite sequence. Then*

$$\lim_{k\to\infty} \psi(x^k, \theta_k) - \psi(x^{k+1}, \theta_k) = 0$$

An easy consequence of Theorem 3.4 is that, when Algorithm 2.1 generates an infinite sequence (that is, it is not stopped at Step 1), we have that $\lim_{k\to\infty} |C^+(x^k)| = 0$. This means that points arbitrarily close to feasibility are eventually generated.

Theorem 3.5 *If Algorithm 2.1 does not stop at Step 1 for all $k = 0, 1, 2, \ldots$, then*

$$\lim_{k\to\infty} |C^+(x^k)| = 0.$$

(In particular, every limit point of $\{x^k\}$ is feasible.)

4   Convergence to Optimality

In the former section we saw that, if the algorithm does not stop at Step 1, it achieves approximate feasibility up to any desired precision. In this section we are going to see that, in that case, the optimality indicator $\|d_{tan}^k\|$ cannot be bounded away from zero. In practice, this implies that given arbitrarily small convergence tolerances $\varepsilon_{feas}, \varepsilon_{opt} > 0$, Algorithm 2.1 eventually finds an iterate x^k such that $\|C^+(x^k)\| \leq \varepsilon_{feas}$ and $\|d_{tan}^k\| \leq \varepsilon_{opt}$. For proving this result, we will proceed by contradiction, assuming that $\|d_{tan}^k\|$ is bounded away from zero for k large enough. From this assumption (stated as Assumption C below) we will deduce a number of intermediate results that, finally, lead us to a contradiction.

Assumption C. *Algorithm 2.1 generates an infinite sequence* $\{x^k\}$ *and there exists* $\varepsilon > 0$, $k_0 \in \{0, 1, 2, \ldots\}$ *such that*

$$\|d_{tan}^k\| \geq \varepsilon \quad \text{for all} \quad k \geq k_0.$$

Lemma 4.1 *Suppose that Assumption C holds. Then, there exist* $c_4, c_5 > 0$ *(independent of k) such that*

$$f(x_{af}^k) - f(x_{trial}^{k,i}) \geq \quad \min \{c_4, c_5 \delta_{k,i}\}$$

for all $i = 0, 1, \ldots, iacc(k)$

Lemma 4.2 *Suppose that Assumption C holds. Then, there exist* $\alpha, \varepsilon_1 > 0$, *independent of k and i, such that* $|C^+(x^k)| \leq \quad \min \{\varepsilon_1, \alpha \delta_{k,i}\}$ *implies that* $\theta_{k,i} = \theta_{k,i-1}$.

In the next Lemma, we state that, under Assumption C, the penalty parameters $\{\theta_k\}$ are bounded away from zero. It must be warned that this is a property of sequences that satisfy Assumption C (which, in turn, will be proved to be non-existent!) and not of *all* the sequences effectively generated by the model algorithm.

Lemma 4.3 *Suppose that Assumption C holds. Then, there exists* $\bar{\theta} > 0$ *such that* $\theta_k \geq \bar{\theta}$ *for all* $k \in \{0, 1, 2, \ldots\}$.

Finally, in Theorem 4.4, it is proved that Assumption C cannot be true.

Theorem 4.4 *Let* $\{x^k\}$ *be an infinite sequence generated by Algorithm 2.1. Then, there exists* K_2 *an infinite set of indices of* $\{0, 1, 2, \ldots\}$ *such that*

$$\lim_{k \to \infty} \|d_{tan}^k\| = 0. \tag{22}$$

References

1. F. M. Gomes, M. C. Maciel and J. M. Martínez, *Nonlinear programming algorithms using trust regions and augmented Lagrangians with nonmonotone penalty parameters*, Relatório Técnico, IMECC-UNICAMP, Campinas, 1995.
2. J. M. Martínez, *A two-phase trust-region model algorithm with global convergence for nonlinear programming*, Relatório Técnico, IMECC-UNICAMP, Campinas, 1996.
3. M. Rom and M. Avriel, *Properties of the sequential gradient-restoration algorithm (SGRA), Part 1: Introduction and comparison with related methods*, Journal of Optimization Theory and Applications, Vol. 62, pp. 77-98, 1989.
4. M. Rom and M. Avriel, *Properties of the sequential gradient-restoration algorithm (SGRA), Part 2: Convergence Analysis*, Journal of Optimization Theory and Applications, Vol. 62, pp. 99-126, 1989.

On the height used by additives BSS machines[*]

Martín Matamala

Abstract. We define a new complexity measure: the height of an additive machine. This quantity is the amount of space necessary to code in binary any number computed by the machine as a linear integral combination of its input and constants. We give some arguments in favor of the following thesis: height in additive machines is equivalent to space in Turing machines. Among our results we emphasize:

- The existence of a height hierarchy.
- The class of sets decidable in polynomial height equals the class of sets decidable in parallel polynomial time.
- A version of Savitch's theorem for the height.
- The existence of a hierarchy for nondeterministic height.

1 Introduction

Traditionally, space captures the notion of memory used by Turing machines. Restrictions on the space usable by these machines give rise to well known complexity classes. Another complexity measure that appears in the classical setting is running time. Among the most basic relationships between time and space polynomial time machines use polynomial space, and polynomial space machines can be supposed to run in exponential time [1].

Time and space restrictions can also be considered for other machine models such as nondeterministic and parallel machines. Each of these machine models gives rise to well known complexity classes. Nondeterministic machines yield NP and PH. Parallel machine models give rise to parallel polynomial time. Well known complexity classes are the corresponding to nondeterministic polynomial time —most generally, the classes on the polynomial hierarchy—, parallel polynomial time and nondeterministic polynomial space. Some well known relationships between the latter classes are

- Parallel Thesis: parallel time is equivalent to space.
- Savitch's theorem: any subset decidable by a nondeterministic machine using space $h(n)$ can be decided by a deterministic machine using space at most $O(h^2(n))$.

[*] This work was partially supported by the projects ECOS and Programa de Financiamiento de Investigadores en Formación, Fac. Cs. Fís. y Mat., Universidad de Chile.

- The polynomial hierarchy is contained in the class of sets decidable in polynomial space.
- Deciding the truth of first-order Boolean sentences is a complete problem for polynomial space.

In 1989 Blum, Shub and Smale [2] proposed a model of computation to deal with real numbers. Classes corresponding to deterministic, nondeterministic or parallel time are well defined here and most of the basic relationships known for these classes in the classical model hold here as well. A difference which stands out between the real and the classical model is the power of space. While on the latter setting the class of sets decidable in polynomial space is contained in the class of sets decidable in exponential time, in the former it has been proved by Michaux [8] that every recursive subset of $\mathbb{R}^\infty$, the disjoint union of $\mathbb{R}^n$ for $n \geq 1$, can be decided using a polynomial number of registers.

Our goal in this paper is to define a complexity measure for the BSS model whose properties are similar to those of space in the classical setting.

We focus our study on additive machines. These machines are a particular kind of real Turing machines where multiplications and divisions are forbidden. The input space of these machines is $\mathbb{R}^\infty$. Since we will deal with decision problems we only consider machines which return 0 or 1. Additive machines were studied in [5] and a detailed picture of their capabilities and the complexity classes they give rise to can be found in [6, 4].

A version of Michaux's result about the power of the space described above for additive machines was obtained by Koiran [6]. The latter work shows that any set decidable by an additive machine can be decided by another additive machine using at most linear space and with at most a polynomial slowdown [6].

The notion we will define here is called *height*. We establish that this resource is meaningful by proving in Section 3 the existence of a height hierarchy. From now on, we denote by P_{add}, $PHEIGHT_{add}$ and EXP_{add} the class of sets decidable by additive machines in polynomial time, in polynomial height, and in exponential time, respectively. We prove in Section 2 that previous complexity classes satisfy

$$P_{add} \subseteq PHEIGHT_{add} \subseteq EXP_{add}.$$

In [4], it was proposed a complexity class denoted by $\mathcal{E}$ which is defined as follows. A set A belongs to $\mathcal{E}$ if there exists an additive machine M accepting A which satisfies:

1. M works within exponential time.
2. M works within polynomial space.
3. At any moment of the computation of M over an input x, all the coordinates of the state space of M contain a linear combination of $x_1, ..., x_n, \alpha_1, ..., \alpha_p$ whose coefficients are integers of polynomial height. Here, $\alpha_1, ..., \alpha_p$ are the constants of M.

Denote by PAR_{add} the class of sets decidable in parallel polynomial time. Corollary 3 in [4] establishes that the class $\mathcal{E}$ and the class PAR_{add}, are the same. In [4] it is also shown that $\mathcal{E}$ possesses complete problems. Here, we will prove that $\mathcal{E} = \text{PHEIGHT}_{\text{add}}$ and that it is completely defined by conditions 2 and 3 above. From that, we will obtain an analogous to the parallel thesis

$$\text{PHEIGHT}_{\text{add}} = \text{PAR}_{\text{add}}.$$

In the last section, we will study the classes defined by nondeterministic machines with a height restriction. Our central result is a version of Savitch's theorem for nondeterministic height (see [7] for a proof of this theorem in the classical model)

$$\text{NHEIGHT}_{\text{add}}(h(n)) \subseteq \text{HEIGHT}_{\text{add}}(O(h^4(n)))$$

where $\text{NHEIGHT}_{\text{add}}(g(n))$ is the class of subsets of $\mathbb{R}^\infty$ decidable by nondeterministic additive machines using height at most $g(n)$ and $\text{HEIGHT}_{\text{add}}(g(n))$ is defined analogously but in terms of deterministic additive machines. Using this result, we will construct a nondeterministic hierarchy as it is done in the classical model.

2   Basic relationships between Time and Height

Recall that $\mathbb{R}^\infty$ denotes the disjoint union of $\mathbb{R}^n$ for $n \geq 1$. If $x \in \mathbb{R}^n$ we will say that n is the *size* of x.

Real Turing machines take as inputs elements of $\mathbb{R}^\infty$ and return elements of $\mathbb{R}^\infty$ too. They perform, with unit cost, the four arithmetic operations as well as order comparisons. For a formal definition see [2].

An *additive machine* is a real Turing machine which only performs additions and subtractions. In the sequel we refer to them simply as machines. For basic properties of these machines and the complexity classes they give rise to see [6, 4].

Let M be a machine and let $\alpha \in \mathbb{R}^p$ be the constants which appear in the program of M. Let $x \in \mathbb{R}^n$ its input. Then, at each step in the computation of M on input $x \in \mathbb{R}^n$, any register contains a linear integral combination z of x and α, i.e.

$$z = \sum_{j=1}^{n} u_j x_j + \sum_{j=1}^{p} v_j \alpha_j,$$

where $u_j, v_j \in \mathbb{Z}$. The state of the machine at any moment in the computation is completely determined by the current node (instruction) q, and a vector Z in the state space $\mathbb{R}^\infty$. We call the pair (q, Z) a *configuration* and the integers u_j, v_j the *coefficients* of the configuration.

Definition 1 Let the *height* of integer x be $\log(1 + |x|)$. The height of a rational number $z \in \mathbb{Q}$ is the addition of the height of p and q where $z = \frac{p}{q}$ and p and q are relatively prime.

The height of an integer, or a rational number, is roughly the number of bits necessary to write it in binary.

Definition 2 We define the *height* of a configuration (q, Z) of M over $x \in \mathbb{R}^n$ where $Z = (z_1, ..., z_s, 0, ...)$ to be the product of the number s of active variables (those different from zero), the number $n + p$ of real numbers involved, and the maximum height of its coefficients. That is, if h is this height then

$$h = s \times (n + p) \times \max\{m(x), m(\alpha)\}$$

where

$$m(x) = \max_{\substack{i=1,...,s \\ j=1,...,n}} \{\log(1 + |u_{ij}|)\}$$

$$m(\alpha) = \max_{\substack{i=1,...,s \\ j=1,...,p}} \{\log(1 + |v_{ij}|)\}$$

and

$$z_i = \sum_{j=1}^{n} u_{ij} x_j + \sum_{j=1}^{p} v_{ij} \alpha_j \quad i = 1, \ldots s.$$

The *height used by M on input x* will be the maximum height over all the configurations reached during the computation of M on input x.

We denote by $\mathrm{HEIGHT}_{\mathrm{add}}(h(n))$ the class of subsets of $\mathbb{R}^\infty$ accepted by a machine which uses height at most $h(n)$ for every input of size n. We denote by $\mathrm{PHEIGHT}_{\mathrm{add}}$ the union of $\mathrm{HEIGHT}_{\mathrm{add}}(n^k)$ for $k \geq 1$.

Our first result bounds the height of a configuration as a function of the number of steps used to reach this configuration.

Theorem 3 *The height of a configuration at time t is bounded by $(n + p)t^2$.*

Proof. It is easy to see that the space is bounded by t. On the other hand, at time t the absolute value of any coefficient is bounded by 2^t. The result follows.
□

The statement of the next result requires the following.

Definition 4 A function $f : \mathbb{N} \to \mathbb{N}$ is said to be *height constructible* if there exists a machine which takes $n \in \mathbb{N}$ as its input and returns $f(n)$ using height $O(f(n))$.

We now give a sort of converse of Theorem 3.

Theorem 5 *Let h be a height constructible function. Then, any subset decided in height $h(n)$ can be decided in time $2^{O(h(n))}$ and height $O(h(n))$.*

Proof. Let A be a set decided by a machine M using height $h(n)$. Let M' be a machine that computes $h(n)$. Then, we can design another machine M'' which computes $2^{h(n)}$. We observe that within height $h(n)$ the machine M can reach only $N2^{h(n)}$ different configurations for a given input of size n. Here N denotes the number of instructions in the program of M. Thus, if M runs for more than $N2^{h(n)}$ steps, it enters a loop. So, we can simulate $N2^{h(n)}$ steps of M's computation and then accept if M has accepted and reject if M has not.   □

Corollary 6 $P_{add} \subseteq PHEIGHT_{add} \subseteq EXP_{add}$.   □

We close this section with a lemma which will be useful in what remains of the paper. We only sketch its proof since the main idea is already written in [6].

Lemma 7 *Let S be a set decidable by a machine M using height $h(n)$. For every $n \in \mathbb{N}$ there exist a finite family of linear systems of strict and non-strict inequalities such that $x \in S \cap \mathbb{R}^n$ if and only if x satisfies one of these systems. Each system has the form*

$$Ax \le B\alpha$$
$$A'x < B'\alpha$$

where A and A' are matrices with n columns, B and B' are matrices with p columns, and $\alpha^t = (\alpha_1, ..., \alpha_p)^t \in \mathbb{R}^p$ are the constants of M. Moreover, the entries of A, A', B and B' are integers of height bounded by $h(n)$.

Proof. On inputs of size n the M's computation can be described by an additive decision tree. Each path in this tree is completely defined by the answers to tests of the form $z \ge 0$ where z is a register of M. At any step in the computation the contents of these registers are linear integer combinations of the input x and the constants α. For a path γ in this tree the set S_γ of points in $\mathbb{R}^n$ which follow the path γ is characterized by a system of inequalities of the form

$$Ax \le B\alpha$$
$$A'x < B'\alpha$$

where the coefficients of the matrices A, A', B and B' are integers of height bounded by $h(n)$. Since S is the union of the S_γ for those γ which lead to an accepting leaf, the result follows.   □

3   The Height Hierarchy

In this section we prove that the more height is allowed to be used, the more sets can be decided.

The proof of the next theorem is inspired in a similar result obtained by Cucker in [3].

Theorem 8 *Let h_1 and h_2 be height constructible functions such that $\forall n \in \mathbb{N} \ \ h_1(n) \leq h_2(n)$ and $\lim \frac{h_1(n)}{h_2(n)} = 0$. Then, the following strict inclusion holds*

$$\text{HEIGHT}_{\text{add}}(O(h_1(n))) \subset \text{HEIGHT}_{\text{add}}(O(h_2(n)))$$

Proof. The inclusion is immediate. In what follows we will prove that it is strict. Let S be the set defined by

$$S = \{x \in \mathbb{R}^\infty : \ \ 2^{h_2(n)}x_1 + x_2 = 0, \text{ where } n \text{ is the size of } x\}$$

Since h_2 is a height constructible function it is easy to see that S is decided using height $O(h_2(n))$. A machine M deciding S first computes $h_2(n)$ and then $m = 2^{h_2(n)}x_1$. Finally, it performs the addition $m + x_2$ and accepts if it is 0. Clearly, M uses height $O(h_2(n))$.

Assume that S is decided using height $O(h_1(n))$. From Lemma 7 the set S is a finite union of sets defined by systems of inequalities. Each system has the form

$$Ax \leq b$$
$$A'x < b'$$

where b and b' are real vectors and the entries of A and A' are integers of height $O(h_1(n))$. By separating non-strict inequalities into equalities and strict inequalities we can assume that S is a finite union of sets satisfying systems of the form

$$Bx = c$$
$$B'x < c'$$

where the entries of B and B' are still integers of height $O(h_1(n))$.

Since $S \cap \mathbb{R}^n$ has dimension $n - 1$ we can assume that one of these systems contains exactly one equation $ax = d$. The height of each coordinate of a is $O(h_1(n))$.

The restriction $S \cap \mathbb{R}^n$ is the hyperplane defined by the equation $2^{h_2(n)}x_1 + x_2 = 0$. Hence, from basic linear algebra we deduce that $d = 0$ and that $a = \lambda \cdot (2^{h_2(n)}, 1, 0, ..., 0)$ for some $\lambda \in \mathbb{R}$. The vector a has integer coordinates, thus λ must be an integer. Therefore the height of the coordinates of a is at least $h_2(n)$ which contradicts the fact that $\lim \frac{h_1(n)}{h_2(n)} = 0$. □

4   The Real Parallel Thesis

In this section we establish that PHEIGHT$_{\text{add}}$, the class of sets decidable within polynomial height, is equal to PAR$_{\text{add}}$, the class of sets decidable in parallel polynomial time. This equality is a real analogous of the parallel thesis in which height is considered as the real version of space.

The result is a consequence of Corollary 3 in [4] which states that the class PAR$_{\text{add}}$ is equal to the class $\mathcal{E}$. A set A belongs to $\mathcal{E}$ if there exists a machine M which on inputs of size n satisfies

- Any configuration of M has height $n^{O(1)}$.
- M uses space $n^{O(1)}$.
- M runs for time $2^{O(n)}$.

Theorem 9 PHEIGHT$_{add}$ = PAR$_{add}$ = $\mathcal{E}$. *Moreover, a set A belongs to $\mathcal{E}$ if and only if there exists a machine M which on inputs of size n satisfies*

- *Any configuration of M has height $n^{O(1)}$.*
- *M uses space $n^{O(1)}$.*

Proof. It is enough to prove that PHEIGHT$_{add}$ = $\mathcal{E}$. ¿From the definition of PHEIGHT$_{add}$, we know that the space and the height of the coefficients of any machine in PHEIGHT$_{add}$ are polynomially bounded. ¿From Corollary 1 we know that PHEIGHT$_{add}$ is contained in EXP$_{add}$. So, PHEIGHT$_{add}$ $\subseteq$ $\mathcal{E}$. On the other hand, since the space and height of these coefficients are polynomially bounded for machines in $\mathcal{E}$, the height used by these machines will be polynomially bounded and the reverse inclusion holds too. □

We give several corollaries of the previous theorem and the results obtained in [4].

Corollary 10 [Proposition 1 in [4]] *Let* PH$_{add}$ *be the polynomial hierarchy defined in [4]. Then, the following inclusion holds:* PH$_{add}$ $\subseteq$ PHEIGHT$_{add}$ □

Corollary 11 [Corollary 2 in [4]] *The following strict inclusion holds*

$$\text{PHEIGHT}_{add} \subset \text{EXP}_{add}$$

□

Consider the set DTRAO of first-order formulas of the form

$$\varphi(u_1, ..., u_n) = Q_1 x_1 \in \{0, 1\} \cdots Q_l x_l \in \{0, 1\} \ \ F(x_1, ..., x_l, u_1, ..., u_r)$$

which are true, where $Q_i \in \{\exists, \forall\}$ and F is a quantifier-free formula of the theory of the reals with addition and order (cf. [4]). Then, from Theorem 4 we get the following result.

Corollary 12 [Theorem 7 in [4]] *The set* DTRAO *is* PHEIGHT$_{add}$-*complete* □

Corollary 12 corresponds to the classical result concerning the complexity of deciding the truth of quantified Boolean formulas (see Theorem 3.9 in [1]). From this corollary we deduce that if PHEIGHT$_{add}$ = PH$_{add}$ then the polynomial hierarchy collapses.

5   Nondeterminism and Height

The notion of nondeterminism for machines over the reals was introduced in the original paper of Blum, Shub and Smale. Nondeterministic machines are endowed with the ability of guessing a real number with unit cost. As usual, an input is accepted if and only if there is a sequence of guesses leading to acceptance.

Afterwards the notion of digital nondeterminism was introduced to deal with combinatorial problems over the reals [5]. Here, the guesses are restricted to be zeros and ones.

For additive machines P. Koiran showed in [6] that both forms of nondeterminism have the same power under polynomial time restrictions. Here, we prove a real version of Savitch's theorem for the height (see Theorem 2.9 in [1] for the classical version).

Definition 13 We say that a set $B \subseteq \mathbb{R}^\infty$ belongs to NHEIGHT$_{\text{add}}(h(n))$ if there exists a machine M such that for all $n \geq 0$ and all $x \in \mathbb{R}^n$

$x \in B$ iff $\exists y \in \mathbb{R}^\infty$ s. t. (x, y) is accepted by M using height at most $h(n)$.

Since the machine M can not use more than $h(n)$ registers we can assume that the element y above belongs to $\mathbb{R}^{h(n)}$.

In our next theorem we use the following result due to Cucker and Koiran.

Lemma 14 [Theorem 2 in [4]] *Let P be a non-empty polyhedron of $\mathbb{R}^s$ defined by a system*

$$A_1 y \leq b_1 \qquad A_2 y < b_2$$

of N_1 non-strict inequalities and N_2 strict inequalities, where the entries of A_1 and A_2 are integers of height L. There exists $I_1 \subset \{1, ..., N_1\}$, $I_2 \subset \{1, ..., N_2\}$, and $y \in P$ of the form

$$y_i = \sum_{j \in I_1} u_{ij} b_{1j} + \sum_{j \in I_2} u_{ij} b_{2j} + w_i, \quad i = 1, ..., s,$$

such that $|I_1| + |I_2| = s$, and the coefficients u_{ij}, v_{ij}, w_i are rational numbers of height $(Ls)^{O(1)} = Ls + s \log(s)$. □

Theorem 15 *Let $h : \mathbb{N} \to \mathbb{N}$ be height constructible. Then*

$$\text{NHEIGHT}_{\text{add}}(h(n)) \subseteq \text{HEIGHT}_{\text{add}}(O(h^4(n))).$$

Proof. Let S be a set in NHEIGHT$_{\text{add}}(h(n))$ and M a machine deciding S. We will divide the proof in two parts. First, we prove that the guesses of M can be assumed to be zeros and ones with only a polynomial increase in the height used. Then, we avoid the use of nondeterminism. This second part will not increase the height used.

Let $x \in \mathbb{R}^n$. By Theorem 5 we can assume that M works in time $2^{ch(n)}$ for some positive constant c. Consider $m = ch(n)$ and L be a bound for the height of the coefficients in the configurations reached by M with input x. Denote by s the space used by M.

The point x belongs to S if and only if there is a $y \in \mathbb{R}^s$ such that (x, y) is accepted by M using height $ch(n)$.

Given $y \in \mathbb{R}^s$ the computation of M with input (x, y) follows a computation path given by the sequence of answers to the tests done along the computation. The set of y's which follow the same path is then described by a linear system of inequalities

$$A_1 y \leq b_1(x)$$
$$A_2 y < b_2(x)$$

where the entries of A_1 and A_2 are integers of height bounded by L, and both $b_1(x)$ and $b_2(x)$ are real vectors (linear integral combinations of the inputs and the machine constants).

Therefore, the j's, where $y \in \mathbb{R}^s$, which satisfy one of these systems either all lead to acceptance or all lead to rejection.

We conclude that $x \in S$ if and only if there exists a computation path leading to acceptance and a $y \in \mathbb{R}^s$ satisfying the system of inequalities associated to this path.

¿From Lemma 14 we know that if such a y exists it can be chosen with rational coordinates of height at most $k = Ls + s \log(s)$. Thus, to decide the input $x \in \mathbb{R}^n$ it suffices to guess a solution $y \in \{0, 1\}^{ks}$.

We now proceed to the second part of the proof. We design a machine M' which on input $x \in \mathbb{R}^n$ outputs 1 if x belongs to S and 0 otherwise. The machine M' evolves by steps. In each step M' generates a sequence of rational numbers of height k, as pairs of integers $(p_1, q_1), ..., (p_s, q_s)$. Then it copies $(x_1, ..., x_n, y_1, ..., y_s)$ and simulates the evolution of M over $(x_1, ..., x_n, y_1, ..., y_s)$. The machine M' keeps the value of any register z as a pair $(z^1, z^2) \in \mathbb{R} \times \mathbb{N}$. The second component z^2 has height at most ks. The first one z^1 can be written as a linear integer combination of x and α. The integer coefficients of such linear combination have height at most ksL. The machine M' performs a test over register z by testing the sign of z^1. Therefore M' uses height at most $L^2 s^3 + L s^3 \log(s)$. From Definition 2 we know that this quantity is bounded by $c'' h^4$. If M accepts $(x_1, ..., x_n, y_1, ..., y_s)$ then M' accepts x and it halts. At each step M' uses the height used by M plus the height of the pairs of integers $(p_1, q_1), ..., (p_s, q_s)$ which represents $(y_1, ..., y_s)$. Therefore, we conclude that the whole height is $O(h^4)$ $\square$

Using Theorem 15 we show the existence of a hierarchy for nondeterministic height. We need the following lemma.

Lemma 16 *Let* $f : \mathbb{N} \to \mathbb{N}$ *be a height constructible function. If*

$$\text{NHEIGHT}_{\text{add}}(s(n)) \subseteq \text{NHEIGHT}_{\text{add}}(h(n))$$

then

$$\text{NHEIGHT}_{\text{add}}(s(f(n))) \subseteq \text{NHEIGHT}_{\text{add}}(h(f(n))).$$

Proof. Let S be a subset in $\text{NHEIGHT}_{\text{add}}(s(f(n)))$ and M a machine deciding S. Define $S' = \cup_{m \geq 1} S'_m$ where $S'_m \subseteq \mathbb{R}^m$ is given by

$$S'_m = \{(x_1, ..., x_m) | \; x_{2i} \in \{0,1\}, \exists! j, x_{2j} = 1, f(j) = m, \; x_1 x_3 \cdots x_{2j-1} \in S\}$$

where $\exists!$ means "there is only one".

We will see that S' belongs to $\text{NHEIGHT}_{\text{add}}(s(n))$ by designing a machine M' which decides S' using the required height. The machine M' with input $(x_1, ..., x_m)$ looks for an even coordinate j which is one and verify that all other even coordinates are zero. Since f is height constructible, using height $cf(j)$, for some $c > 0$, it is possible to test whether $f(j) = m$. If $f(j) \neq m$ M' rejects. Otherwise M' M' writes $y = (x_1, x_3, ..., x_{2j+1})$ and gives it as input to M. If M accepts y then M' accepts x. Otherwise M' rejects. The height used by M' is at most $c(s(m) + cf(j))$ which is $O(s(m))$.

By hypothesis there exists a machine T proving the membership of S' to $\text{NHEIGHT}_{\text{add}}(h(m))$. We finish the proof showing a machine T' which decides S as a set in $\text{NHEIGHT}_{\text{add}}(h(f(n)))$.

Given $x \in \mathbb{R}^n$, T' computes $f(n)$ and sets $y = (x_1, 0, ..., x_n, 1, \underbrace{0, 0, ..., 0}_{f(n)-2n}) \in$ $\mathbb{R}^m$. Next, T' inputs y to T. If T accepts y then T' accepts x. Otherwise it rejects. Since $f(n)$ is height constructible, $f(n)$ can be computed using height at most $f(n)$. Moreover, T decides S' using height $h(m)$. So, the whole process uses height at most $h(f(n))$. $\quad\square$

Theorem 17 *For any $r \geq 0$ and $\epsilon > 0$ the following strict inclusion holds*

$$\text{NHEIGHT}_{\text{add}}(n^r) \subset \text{NHEIGHT}_{\text{add}}(n^{r+\epsilon})$$

Proof. The proof follows the same idea used in the classical setting. We refer the reader to Theorem 12.12 in [7]. $\quad\square$

6   Conclusion

The results proved throughout this paper suggest that the height in the real setting plays the role of the space in the classical one, at least for additive machines.

It is not known whether results similar to those obtained here hold for arbitrary machines (i.e. machines which perform multiplications and divisions) as well.

7   Acknowledged

The author of this paper wants to thank F. Cucker for his useful comments and the center L.I.P of the École Normale Superieure de Lyon, France, where this work was finished.

References

1. J. Balcázar, J. Díaz and J. Gabarró. *Structural Complexity I.* EATCS Monographs on Theoretical Computer Science. Springer Verlag, 1988.
2. L. Blum, M. Shub and S. Smale. On a theory of computation and complexity over the real numbers: NP-completeness, recursive functions and universal machines. *Bulletin of the Amer. Math. Soc.*, vol.21, n.1, 1-46 (1989).
3. F. Cucker. $P_{\mathbb{R}} \neq NC_{\mathbb{R}}$. *Journal of Complexity* 8, 230-238 (1992).
4. F. Cucker and P. Koiran. Computing over the reals with addition and order: higher complexity classes. *Journal of Complexity* 11, 358-376 (1995).
5. F. Cucker and M. Matamala. On Digital Nondeterminism. *To appear in Math. Syst. Theory.*
6. P. Koiran. Computing over the reals with addition and order. *Theoretical Computer Science* 133, 35-47 (1994).
7. J. Hopcroft and J. Ullman. *Introduction to Automata Theory, Languages, and Computation.* Addison-Wesley, Reading, Mass, 1979.
8. C. Michaux. Une remarque à propos des machines sur **R** introduites par Blum, Shub et Smale. *C. R. Acad. Sci. Paris*, t.309, Série I, pp.435-437, 1989.

The Space Complexity of Elimination Theory: Upper Bounds

Guillermo Matera and Jose Maria Turull Torres

Abstract. We use a theorem by Borodin relating parallel time with sequential space in order to obtain algorithms that require small space resources. We first apply this idea to some linear algebra problems. Then we reduce several problems of Elimination Theory to linear algebra computations and establish PSPACE bounds for all of them. Finally, we show how this strategy can be improved by means of probabilistic arguments.

1 Introduction

We use standard notions and notations for complexity models and complexity classes as can be found in [1, 24]. We recall that the classes NC^i are defined as the set of $O(\log^i n)$-uniform families boolean circuits of polynomial size and depth $O(\log^i n)$ with bounded fan-in. A family of boolean circuit is said to be uniform iff its standard encoding can be built in deterministic space $O(\log^i n)$.

Intuitively, the classes NC^i represent the functions which can significantly benefit with the use of *parallel* computation. In fact, Borodin [4] showed that uniform circuit depth or parallel time turns into *sequential* space. This means that for many functions the property of well parallelizability can be re-interpreted, in the scope of sequential computing, as the use of a very small amount of working space in their computation. Moreover, its proof is constructive, so that given the family of boolean circuits in NC^i and the algorithm which witnesses its uniformity, we can use the proof to build a sequential algorithm which works in space $\log^i n$.

It is our intention to promote the use of this important tool for the solution of concrete problems in Formal Calculus. Hence we show here an alternative method to deal with problems coming from Algebraic and Geometric Elimination Theory that uses the strategy of Borodin's theorem in order to obtain algorithms requiring small space resources.

The central idea of the method is to reduce elimination problems to linear algebra computations. Unfortunately, the matrices involved in these reductions have exponential size, probably due to the sintactical aspect of the kind of problems that we consider, that is, the codification of multivariate polynomials which are the basic objects of our language. This inhibits the application of the algorithms most frequently used at present, as gaussian elimination, because they require polynomial space, becoming unfeasible for big matrices.

Then we developpe algorithms to perform linear algebra tasks which work with very low space (namely polilogarithmic). For this purpose, we study arithmetic circuits with controlled parallel time, as Berkowitz's algorithm [3] for the computation of the characteristic polynomial. Based on it, we design arithmetic circuits for the computation of the rank and the resolution of linear equations systems as it is explained in section 2. Then, we translate these arithmetic circuits into boolean circuits in order to produce a suitable input for the application of Borodin's theorem.

Although it seems natural trying to make this translation from the basis of the operations $\{+, *\}$ in $\mathbb{Z}$ to the basis $\{\wedge, \vee, \neg\}$ in $\{0, 1\}$, we show in Section 2 that the translation from the basis $\{+, *, \sum\}$ is more efficient, where $\sum$ is implemented in NC^1 by using a Carry Save Adder circuit (see [24]). In this way we get linear algebra algorithms which have the same perfomance in parallel time as the best uniform algorithms known at present, as those in [5], but with some additional advantages from a programming point of view.

So we have good candidates for the application of Borodin's theorem, and this leads us to the subject of uniformity. In [16], in collaboration with A. Grosso, N. Herrera and M.E. Stefanoni we exhibit the full program which witnesses the logspace uniformity of our linear algebra algorithms. Then we use the proof of Borodin's theorem to build a $log^2 n$ space sequential algorithm for the computation of the rank of integer matrices and the resolution of linear equations systems on $\mathbb{Z}$.

In section 3 we show how to reduce several problem of Elimination Theory to linear algebra computations. Then, on the basis of the algorithms developed in the previous section, we build PSPACE algorithms for some of the main problems in Elimination Theory: the Ideal Triviality problem, the Ideal Membership problem for Complete Intersections, the Radical Membership problem, the General Elimination problem, the Noether normalization, the Decomposition in equidimensional components, the computation of the degree of a variety and an algorithmic version of Quillen-Suslin Theorem. We do this following the suggestions in [17], where it is conjectured as well as in others, like [7] for instance, that some of this problems are in the class PSPACE.

Finally, in section 4 we show in an example, the Zero Elimination problem, how these techniques can be refined in order to get a class of probabilistic algorithms with a better time perfomance. This is done by applying our ideas and the techniques in [20]. The main ingredients for this improvement is the use of correct test sequences for polynomials encoding by means of straight–line programs, in the sense of Heintz-Schnorr [18].

Let us remark that this last approach allows us to get algorithms whose time perfomance is at least as well as that of the existent software on the subject (the Gröbner basis solving). The main advantage of our algorithms is that they work with reasonable space resources (unlike the case of the Gröbner basis solving where the space requirements are exponential).

Some PSPACE results on Elimination Theory have been announced in [2, 9, 23]. However, in those papers the authors do not prove the membership of their algorithms to PSPACE. In particular, the related aspects of uniformity of their algorithms are not considered.

2 Linear algebra algorithms in low space

2.1 The computation of the characteristic polynomial and the rank

Let us suppose that we are given a $n \times m$ matrix A with integer coefficients to compute its rank. Since $B := A \cdot A^t$ has the same rank as A, and B is diagonalizable, $rk(A)$ can be read from the coefficients of the characteristic polynomial of B. This polynomial can be computed by means of Berkowitz algorithm [3].

By analyzing the whole circuit produced in this way we see that we have a family of Boolean circuits of size $O(n^6 h^2 \log^3 n)$ and depth $O(\log(n) \log(hn))$ which computes the rank of any $n \times m$ integer matrix with h-bit coefficients. As it is proved in [16], the family is uniform, so we apply Borodin theorem and we get:

Theorem 1 *There exists a deterministic Turing machine that works in space* $\log(n) \log(hn)$ *and computes the rank of any $n \times m$ integer matrix $(n \geq m)$ with h-bit coefficients, for every positive integers n, m, h.*

Let us remark that if the matrix A belongs to $R^{n \times m}$, being R any domain ($\mathbb{Z}[X_1, \ldots, X_n]$ for instance), the computation of $rk(A)$ can be reduced to the computation of a characteristic polynomial applying the techniques developped by Mulmuley [21].

2.2 The resolution of linear equations systems

Let A be an integer $n \times m$ matrix, X a $m \times 1$ vector of unknowns and b an integer $n \times 1$ vector. Let h be the maximal bit size of all the occurring integers. We want to solve the linear system $A \cdot X = b$.

The idea is to reduce our problem to the resolution of a system with nonsingular square matrix. The latter will be performed by means of Cramer's rule.

This reduction consists in computing a nonsingular square submatrix of A, say $\tilde{A}$, of maximal rank, and re-stating the system in terms of the latter. This means that we will consider the vector $\tilde{b}$ which we get by deleting from b the elements which do not correspond to the rows from A that we use to build $\tilde{A}$. And we set to zero all the elements of X which do not correspond to the columns from A that we use to build $\tilde{A}$. This leads to a particular solution of the original system which fullfills our requirements.

Now, we just need to compute the square submatrix $\tilde{A}$ of maximal rank. Let us denote by A_i the i-th row of A. We compute (in parallel) $rk(A_1, \ldots, A_i)$ for $i = 1, \ldots, n$. Every time that $rk(A_1, \ldots, A_{i-1}) < rk(A_1, \ldots, A_i)$ we keep the index i. These indices will correspond to the rows which form $\tilde{A}$. The same scheme can be used to determine the columns which we choose to build $\tilde{A}$.

¿From the results of section 2.1, we can conclude that there exists a uniform family of boolean circuits of size $(nh)^{O(1)}$ and depth $O\big(\log n \log(nh)\big)$ such that it solves the given system $A \cdot X = b$. Now, by applying the proof of Borodin theorem we have:

Theorem 2 *There exists a deterministic Turing machine that works in space* $\log(n)\log(hn)$ *such that*

- *it checks whether there exists a solution of the system $A \cdot X = b$, and*
- *in case of an affirmative answer in i), it computes numerators and denominators of a particular solution of the system.*

Let us remark that a slight modification of the former scheme allows us to construct a maximal independent affine set of solutions of our linear system. For this sake, we only must note that, in order to compute a particular solution, we have set to zero a subset $\mathcal{A} := \{X_{r+1}, \dots, X_n\}$ of the variables. This can be done because there is no restriction on the values the variables in $\mathcal{A}$ can take. Hence, if we let the vector $(X_{r+1}, \dots, X_n)$ be assigned values from a maximal affine independent set of $\mathbb{Q}^{n-r-1}$, we will obtain a set of solutions which form a maximal independent affine set of solutions of our linear equations systems.

3   Some PSPACE results in Elimination Theory

In this section we will show upper bounds for the space complexity of some algorithmic problems of classical Algebraic and Geometric Elimination Theory. The uniformity of the algorithms which we use to solve these problems will be an easy consequence of the uniformity of the algorithms developed in the previous section.

We start fixing some notation: let $X_1, \dots, X_n$ be indeterminates over $\mathbb{Q}$. We are going to consider polynomials $F_1, \dots, F_s \in \mathbb{Z}[X_1, \dots, X_n]$ $(n \geq 2)$ where the total degrees $\deg F_j$ $(1 \leq j \leq s)$ are bounded by an integer $d \geq 2$. We assume that all the coefficients of the F_i's have their bit-size bounded by h. We will denote by $(F_1, \dots, F_s)$ the ideal generated by $F_1, \dots, F_s$ in $\mathbb{Q}[X_1, \dots, X_n]$ and by $V := \{F_1 = 0, \dots, F_s = 0\} := \{x \in \mathbb{C}^n; F_1(x) = \dots = F_s(x) = 0\}$ the algebraic variety of $\mathbb{C}^n$ defined by these polynomials.

3.1   The ideal triviality problem

As an example of the general method to be applied, we will first consider a classical problem: the ideal triviality problem.

Decide whether $V = \emptyset$ holds and if this is the case find $P_1, \dots, P_s \in \mathbb{Q}[X_1, .., X_n]$ such that the identity $1 = P_1 F_1 + \dots + P_s F_s$ is satisfied.

The essential tools which we use are a single exponential version of the affine Nullstellensatz in characteristic 0 by D. Brownawell [6] and its generalization for an arbitrary field in [7, 8, 19]:

Theorem 3 *(An effective Nullstellensatz for ideal triviality)*
The ideal $(F_1, \ldots, F_s)$ is trivial iff there exist $P_1, \ldots, P_s \in \mathbb{Q}[X_1, \ldots, X_n]$ satisfying the conditions

$$ 1 = P_1 F_1 + \cdots + F_s \qquad and \qquad \max_{1 \le j \le s} \{\deg P_j F_j\} \le d^n $$

This turns our problem into a linear algebra one: once we have degree bounds for $P_1, \ldots, P_s$, the condition above can be rewritten as a linear equations system of size $d^{n^2} \times sd^{n^2}$ whose coefficients have at most h bits with all the coefficients of the P_i's as unknowns.

Applying the results of section 2.2 to our case we get a uniform family of boolean circuits of size $(hsd^{n^2})^{O(1)}$ and depth $O(n^4 \log^2(hsd))$ which decides whether the system is compatible and, if this is the case, computes a particular solution of it.

At this stage we apply the proof of Borodin theorem in order to evaluate the circuit. In conclusion, we have proved:

Theorem 4 *The ideal triviality problem can be solved with deterministic space polynomially bounded by $O(n^4 \log^2(hsd))$.*

3.2   The ideal membership problem for complete intersections

The problem to be considered is the following: suppose that $F_1, \ldots, F_s$ form a regular sequence of $\mathbb{Q}[X_1, \ldots, X_n]$ and let F be an integer polynomial with height bounded by h. We want to decide whether F belongs to the ideal generated by $F_1, \ldots, F_s$ in $\mathbb{Q}[X_1, \ldots, X_n]$ and if this is the case, we want to find $P_1, \ldots, P_s \in \mathbb{Q}[X_1, \ldots, X_n]$ such that $F = P_1 F_1 + \cdots + P_s F_s$ holds.

Here again, the effective affine Nullstellensätzse are the key point.

Theorem 5 *([10])* *(An effective Nullstellensatz for complete intersections)*
Suppose that $F_1, \ldots, F_s$ form a regular sequence in $\mathbb{Q}[X_1, \ldots, X_n]$. Then F belongs to $(F_1, \ldots, F_s)$ iff there exist polynomials $P_1, \ldots, P_s \in \mathbb{Q}[X_1, \ldots, X_n]$ such that $F = \sum_j P_j F_j$ and $\max_j \{\deg P_j F_j\} \le \deg F + d^s \le \deg F + d^n$ holds.

By following the same analysis as above, we find a linear equations system of size $O(d^{n^2} \times sd^{n^2})$ and bit-size h, such that its solutions yields the coefficients of the required polynomials P_i. Then we can conclude:

Theorem 6 *The ideal membership problem for complete intersections can be solved in deterministic space $n^4 \log^2(hsd)$.*

3.3   The radical membership problem

The radical membership problem consists in deciding whether F vanishes on V and if this is the case, finding $N \in \mathbb{N}$ and $P_1, \ldots, P_s \in \mathbb{Q}[X_1, \ldots, X_n]$ such that $F^N = P_1 F_1 + \cdots + P_s F_s$ holds.

In Remark 1.6 of [10] it is shown that this problem can be reduced to the resolution of a linear equations system of size $O\big(d^{n^2}(\deg(F) + 1)^n\big)$ and bit-size $O\big(d^{3n} h \log(\deg(F))\big)$. Hence, by using the strategy and the results of section 2.2, we can prove:

Theorem 7 *There exists a deterministic Turing machine that solves the radical membership problem in space $n^4 \log^2 \big(dh \deg(F)\big)$*

3.4   The general elimination problem

Let $0 \le m < n$ and let $\pi : \mathbb{C}^n \to \mathbb{C}^m$ be the projection map $\pi(x_1, \ldots, x_n) = (x_1, \ldots, x_m)$. Our problem is to find polynomials $Q_1, \ldots, Q_t \in \mathbb{Q}[X_1, \ldots, X_m]$ and a quantifier free formula Φ in the first order language of fields with constants from $\mathbb{Q}$, involving only the polynomials $Q_1, \ldots, Q_t$ as basic terms, such that Φ defines the set $\pi(V)$.

The procedure starts checking coherence on all the possible $(F_1, \ldots, F_s)$-cells in parallel, which can be achieved by means of rank computations of $s^{3n} d^{3n^2}$ matrices whose size is $O(d^{n^2})$ and whose bit-size is $O(s^2 hn \log d)$ (see Theorem 2 in [13] or Théorème 3 in [12]).

Afterwards, the elimination of the existencial quantifiers is performed again by rank computations on matrices of similar size whose coefficients belong to $\mathbb{Z}[X_{m+1}, \ldots, X_n]$. Here, applying the techniques of [21] as it is indicated in the references above, a similar treatment as in section 2 allows us to get a uniform family of boolean circuits of suitable complexity to solve this problem. Then we get the following:

Theorem 8 *The general elimination problem can be solved in deterministic space $n^5 \log^2(sdh)$.*

3.5   Noether normalization and dimension

An essential preprocessing technique used to prepare polynomial data is the Noether normalization. Let $r := dim(V)$. We say that the variables $X_1, \ldots, X_n$ are in *Noether position* with respect to V if for each $r < i \le n$ there exists a polynomial of $\mathbb{Q}[X_1, \ldots, X_r, X_i]$ which is monic in X_i and vanishes on V. The Noether normalization is performed by a linear change of variables such that the new variables are in Noether position.

Here we follow the scheme proposed in [10], combined with the techniques of *questors* sets as in [20]. The first step is to compute a system of independent variables with respect to V (we observe that the cardinality of this set is the dimension of V). By means of Proposition 1.7 in [10], this is related to the resolubility of a linear equations system of size $O(d^{2n} \times sd^{2n})$ and bit-size h.

Proposition 1.8 of [10] allows us to reduce the decision of which of the variables X_i with $i \in \{1, \ldots, n\} \setminus I$ are integer with respect to the new independent variables to rank computations of some $O(d^{2n} \times sd^{2n})$ matrices of bit-size h.

Finally, a recursive procedure over the remaining variables computes the suitable change of variables. Proposition 1.2 of [10] and Corollary 19 of [20] are the basic tools that allows us to restate the problem in terms of linear algebra (see also Proposition 38 in [20]). In fact, one has to solve in each recursive step $O(s^3 d^{7n^2})$ linear equations systems of size $O(d^{3n^2})$ and maximal bit-size $O(s^{2n} d^{3n^3} h)$.

Applying the same strategy as in section 2 we can conclude:

Theorem 9 *The Noether normalization and the computation of the dimension of V can be performed in deterministic space bounded by $n^4 \log^2(hsd)$.*

3.6 Decomposition in equidimensional components

A fundamental problem for the description of the variety V consists in the computation of its equidimensional decomposition. That is, starting from equations for the variety V, for every intermediate dimension i $(0 \le i \le dim(V))$ we have to produce a finite set of polynomials defining the union of the irreducible components of V of dimension i.

Let us denote by V_i and W_i the union of the irreducible components of V of dimension i and less than i respectively, and set $r := dim(V)$. The algorithm we will describe consists of at most r recursive steps. In the i-th step, we compute a system of polynomial equations for V_{r-i+1} and for W_{r-i+2}.

The i-th steps starts computing equations to describe V_{r-i+1}. This task is reduced by Theorem 4.1.2 and Proposition 4.2.2. in [14] to linear algebra computations. In fact, we have to produce a basis of solutions of an $O(sd^{3n^2} \times d^{3n^2})$ homogeneous linear equations system $O(nd^{4n^2})$ times in parallel, whose bit-size is bounded by $(sd)^{O(n^2)} h$; and then we reduce the computed equations by means of more linear equations system solving of similar features as the former.

Afterwards, Proposition 4.2.6 in [14] allows us to compute equations for W_{r-i+2} with a similar scheme to the one explained above.

By adding these complexities, and by using the strategy of section 2, we can prove the following statement:

Theorem 10 *There exists a deterministic Turing machine that computes the equidimensional decomposition of V within space bounded by $n^9 \log^2(hsd)$.*

3.7 Computation of the degree of a variety

Let $V = C_1 \cup \ldots \cup C_t$ be the decomposition of V in irreducible components. For $1 \le j \le t$ we define the degree of C_j as usually and denote this quantity by $\deg C_j$. The degree of V is defined as $\deg V := \sum_j \deg C_j$.

In order to compute the degree of V, we take suitable modified ideas from [8, 14]. First of all, after computing the decomposition of V in equidimensional components as it was indicated in section 3.6 above, we can suppose that our variety is equidimensional. Let r be the dimension of V. We intersect the variety with r generic hyperplanes such that the intersection is 0-dimensional. By means of Remark 1.6 of [10], we compute univariate polynomials P_i that annihilate X_i on the intersection for every $i = 1, \dots, n$ as the solution of some $d^{O(n^2)} \times sd^{O(n^2)}$ linear equations system with bit-size h.

Further, we compute the minimal polynomial of every X_i on the intersection by means of more linear algebra manipulation.

Finally, we have to perform some rank computations (as in Proposition 5.3.1 in [14]) with matrices of size $d^{O(n^6)}$ in order to compute the degree of V. In conclusion, we can state:

Theorem 11 *The computation of the degree of V can be made in deterministic space $n^{10} \log^2(hsd)$.*

3.8   Quillen-Suslin Theorem

Let F be a unimodular $r \times s$ matrix with entries being n-variate integer polynomials. Denote by $deg(F)$ the maximum of the degrees of the entries of F and let $d = 1 + deg(F)$. We want to compute an unimodular $s \times s$ matrix M such that $FM = [I_r, 0]$, where $[I_r, 0]$ denotes the $r \times s$ matrix obtained by adding to the $r \times r$ unit matrix I_r $s - r$ zero columns.

Following the scheme of [11], we perform a procedure in which every step is divided in four stages. At the first stage, we construct a sequence of $O((1+rd)^{2n})$ polynomials of degree bounded by $(rd)^2$ with certain properties (see Theorem 3.1 in [11]). By means of Proposition 5.6 in [11], this is reduced to determinantal computations with $O((r^2sd^r)^r)$ polynomial $r \times r$ matrices.

In the second stage we test ideal triviality and we get a representation of 1 in the ideal generated by the polynomials computed in the first stage.

During the third stage, $O(1 + rd)^{2n}$ $s \times s$ matrices are built, which are multiplied in the fourth stage to give rise to the $s \times s$ matrix which is the final result of the corresponding recursive step. Here we make use of Lemma 4.5 in [11] to perform the third stage by means of linear algebra computations.

After n parallel steps like the one described above, we get a uniform family of boolean circuits which fulfills the conditions required by Borodin theorem. Then, we can apply the strategy of section 2 and conclude:

Theorem 12 *The Quillen-Suslin Theorem can be perfomed algorithmically in deterministic space $n^4 \log^2(shd)$.*

4   A probabilistic result for the Zero Elimination problem

In this section we show how the scheme of the previous section can be refined in order to improve the time perfomance of our algorithms. We take the Zero Dimensional Elimination problem to illustrate the idea:

Suppose that $\dim V = 0$ and let Y be a given linear form of $\mathbb{Z}[X_1, \ldots, X_n]$ with bit-size h. Then the problem consists in computing a nonzero one-variate polynomial $Q \in \mathbb{Z}[T]$ such that $Q(Y)$ annihilates ov V.

First of all, we observe that V can be described as the zero set of n equations, say $G_1, \ldots, G_n \in \mathbb{Z}[X_1, \ldots, X_n]$, which in turn are linear combinations of the input polynomials $F_1, \ldots, F_s$.

Under the assumption that we have already computed the polynomials $G_1, \ldots, G_n$, we can apply Lemme 3.3.3 of [15] (or Lemma 24 in [20]) in order to translate the problem to a similar one in the projective space with "good" conditions.

At this point, we apply Lemma 24 and Proposition 23 in [20] (see also Lemme 3.2.1 in [15]) in order to compute the output polynomial Q by means of linear algebra manipulation of controlled size.

A key point is that the coefficients appearing in the linear combinations of $F_1, \ldots, F_s$ that give the polynomials $G_1, \ldots, G_n$ can be chosen with the condition of not satisfying a certain polynomial equation which can be computed in low parallel time. As it is proved in [18] (in the version of [20], section 3.1), we can generate these coefficients at random with a high probability of getting a right sequence of coefficients. Then, we can perform the strategy of Borodin's Theorem on the circuit, generating every bit of every coefficient we need randomly in advance. In conclusion, we have the following result:

Theorem 13 *The 0-dimensional problem can be solved by means of a probabilistic BPP algorithm requiring working space $n^2 \log(nd) \log(nhsd)$ and hence, time $(nd)^{O\left(n^2 \log(nhsd)\right)}$.*

Acknowledgements

The main ideas leading to this work, as with respect to [16] are due to Joos Heintz. We also thank to Marc Giusti, Joel Marchand, José Luis Montaña and Luis Miguel Pardo for the very useful discussions we had with them.

References

1. Balcázar J., Díaz J. and Gabarró J.: *Structural complexity II*. EATCS Monographs on Theoretical Computer Science **22**, Springer-Verlag (1990).
2. Ben-Or M., Kozen M. and Reif J.: *The complexity of elementary algebra and geometry*. J. Comp. and Sys. Sciences **32** (1986) 251–264.
3. Berkowitz S.J.: *On computing the determinant in small parallel time using a small number of processors*. Inf. Proc. Letter **18** (1984) 147–150.
4. Borodin A.: *On relating time and space to size and depth*. SIAM J. Comput. **6** (1977) 733–744.
5. Borodin A., Cook S. and Pippenger N.: *Parallel computation for well-endowed rings and space-bounded probabilistic machines*. Inf. and Control **58** (1983) 113–136.
6. Brownawell W.D.: *Bounds for the degrees in the Nullstellensatz*. Ann. Math. (Second Series) **126** (3) (1987) 577–591.

7. Caniglia L., Galligo A. and Heintz J.: *Borne simplement exponentielle pour les degrés dans le théorème des zéros sur un corps de caractéristique quelconque.* C. R. Acad. Sci. Paris, Série I, **307** (1988) 255–258.
8. Caniglia L., Galligo A. and Heintz J.: *Some new effectivity bounds in computational geometry.* Springer LNCS **357** (1989) 131–151.
9. Canny J.: *Some algebraic and geometric computations in PSPACE.* Proc. 20-th STOC (1988) 460-467.
10. Dickenstein A., Giusti M., Fitchas N. and Sessa C.: *The membership problem for unmixed polynomial ideals is solvable in single exponential time.* Discrete Appl. Math. **33** (1991) 73–94.
11. N. Fitchas: *Algorithmic aspects of Suslin's solution of Serre's Conjecture.* Computational Complexity **3** (1993) 31–55.
12. Fitchas N., Galligo A. and Morgenstern J.: *Algorithmes rapides en séquentiel et parallèle pour l'élimination des quantificateurs en géométrie élémentaire.* Sélection d'exposes 1986-1987, Vol I, Publ. Math. Univ. Paris 7 **32** 103–145.
13. Fitchas N., Galligo A. and Morgenstern J.: *Precise sequential and parallel complexity bounds for the quantifier elimination over algebraically closed fields.* J. Pure Appl. Algebra **67** (1990) 1-14.
14. Giusti M. and Heintz J.: *Algorithmes –disons rapides– pour la décomposition d'une variété algébrique en composantes irréductibles et équidimensionnelles.* Progress in Mathematics Vol. **94**, Birkhäuser (1991) 169–193.
15. Giusti M. and Heintz J.: *La détermination des points isolés et de la dimension d'une variété algébrique peut se faire en temps polynomial.* Symposia Matematica vol. XXXIV, Istituto Nazionale di Alta Matematica, Cambridge University Press (1993) 216–256.
16. Grosso A., Herrera N., Matera G., Stefanoni M.E. and Turull Torres J.M.: *Un algoritmo para el cálculo del rango de matrices enteras en espacio polilogarítmico.* Proc. 25-th JAIIO (1996).
17. Heintz J. and Morgenstern J.: *On the intrinsic complexity of elimination theory.* Journal of Complexity **9** (1993) 471–498.
18. Heintz J. and Schnorr C.P.: *Testing polynomials which are easy to compute.* Proc. 12-th STOC (1980) 262–280.
19. Kollár J.: *Sharp effective Nullstellensatz.* J. AMS **1** (1988) 963–975.
20. Krick T. and Pardo L.M.: *A Computational method for diophantine approximation.* To appear in *Proc. MEGA'94, Birkhäusser Verlag,* (1995).
21. Mulmuley K.: *A fast parallel algorithm to compute the rank of a matrix over an arbitrary field.* Proc. 18-th STOC (1986) 338–339.
22. Reif J.: *Logarithmic depth circuits for algebraic functions.* SIAM J. on Comp. **15** (1986) 231–242.
23. Renegar J.: *On the Computational Complexity and Geometry of the first Order theory of the Reals.* J. of Symbolic Comput. **13**(3) (1992) 255–352.
24. Wegener I.: *The Complexity of Boolean Functions.* Wiley-Teubner Series in Computer Science (1987).

Global Stochastic Recursive Algorithms*

Sanjoy K. Mitter

Abstract

In a series of papers Saul Gelfand and myself ([1], [2]) have developed a theory of discrete-time stochastic recursive algorithms for obtaining the global minima of functions. These algorithms are generalizations of stochastic approximation algorithms which usually exhibit only local convergence. The key idea in our work is the elucidation of the role of a stochastic differential equation in the analysis of the recursive algorithm. This stochastic differential equation plays the same role as the ordinary differential equation plays in the analysis of ordinary stochastic approximation algorithms. In this talk ([4], [5]) I present a wide ranging generalization of these ideas to recursive algorithms which arise in adaptive parameter estimation, various signal processing algorithms, maximum likelihood estimation and quantization. These algorithms are global versions of algorithms presented in Beneveniste, Metivier and Priouret [3]. This work sheds new light on the behavior of ordinary stochastic approximation algorithms.

References

1. S.B. Gelfand and S.K. Mitter (1991b), "Recursive Stochastic Algorithms for Global Optimization in R^d," *SIAM Journal on Control and Optimization,* Vol. 29, pp. 999-1018
2. S.B. Gelfand and S.K. Mitter (1993), "Metropolis-type Annealing Algorithms for Global Optimization in R^d," *SIAM Journal on Control and Optimization,* Vol. 31, pp. 111-131.
3. A. Beneveniste, M. Metivier and P. Priouret (1990), *Adaptive Algorithms and Stochastic Approximations,* Springer Verlag, NY.
4. S.B. Gelfand and S.K. Mitter, forthcoming.
5. C.R. Hwang and S.J. Sheu (preprint), "Large Time Behaviors of Perturbed Diffusion Markov Processes with Applications I: Large Time Behaviors and the Convergence Rate to the Invariant Measures," Institute of Statistics, Academia Sinica.

---

* This research has been supported by ARO grant DAAL03-92-G-0115 (Center for Intelligent Control Systems).

Dynamical Recognizers:
Real-time Language Recognition
by Analog Computers
(Extended Abstract)

Cristopher Moore

1 Introduction

We consider a model of analog computation which performs language recognition in real time. We encode an input word as a point in $\mathbb{R}^d$ by composing iterated maps, and then apply inequalities to the resulting point to test for membership in the language.

Each class of maps and inequalities, such as quadratic functions with rational coefficients, is capable of recognizing a particular class of languages. We use methods equivalent to the Vapnik-Chervonenkis dimension to separate some of these classes from each other: linear maps are less powerful than quadratic or piecewise-linear ones, polynomials are less powerful than elementary (trigonometric and exponential) maps, and deterministic polynomials of each degree are less powerful than their non-deterministic counterparts.

We relate this model to other models of analog computation; in particular, it can be seen as a real-time, constant-space, off-line version of Blum, Shub and Smale's real-valued machines.

Many additional results, showing inclusions with classical language classes and complexity and decidability results, are given in [15].

2 Definitions

Define a *dynamical recognizer* as follows (using slightly different notation from [17]):

Let A^* be the set of finite words in an alphabet A, with ϵ the empty word. The length of a word w is $|w|$, and a^k is a repeated k times. The concatenation of two words is $u \cdot v$ or simply uv.

Suppose that, on some space M, we have a map f_a for each symbol $a \in A$. Then for any word w, let $f_w = f_{w_{|w|}} \circ \cdots \circ f_{w_2} \circ f_{w_1}$ be the composition of all the f_{w_i}. Then $x_w = f_w(x_0)$ is the encoding of w into M, where x_0 is a given initial point; in other words, apply f_{w_1} to x_0, then f_{w_2}, and so on.

Then a *deterministic dynamical recognizer* ρ consists of a space $M = \mathbb{R}^d$, an alphabet A, a function f_a for each $a \in A$, an initial point x_0, and a subset

$H_{\text{yes}} \subset M$ called the *accepting subset*. The *language recognized by* ρ is $L_\rho = \{w | x_w \in H_{\text{yes}}\}$, the set of words for which applying the maps f_{w_i} on the initial point yields a point in the accepting set H_{yes}.

For example, suppose $M = \mathbb{R}$, $A = \{a, b\}$, $f_a(x) = x + 1$, $f_b(x) = x - 1$, $x_0 = 0$, and $H_{\text{yes}} = [0, \infty)$. Then if $\#_a(w)$ and $\#_b(w)$ are the number of a's and b's in w respectively, $x_w = \#_a(w) - \#_b(w)$, and $L(\rho)$ is the language of words with more a's than b's.

We can also define *non-deterministic* dynamical recognizers: for each $a \in A$, let there be several choices of function $f_a^{(1)}, f_a^{(2)}$ etc. Then we accept the word w if there exists a set of choices k_i that puts x_w in H_{yes}, i.e.

$$x_w^{(k)} = f_{w_{|w|}}^{(k_{|w|})} \circ \cdots \circ f_{w_2}^{(k_2)} \circ f_{w_1}^{(k_1)}(x_0) \in H_{\text{yes}} \text{ for some sequence } k.$$

Then for a given class $\mathcal{C}$ of functions, we define (by abuse of notation) the class $\mathcal{C}$ as the set of languages recognized by dynamical recognizers where:

1) f_a for all a is in $\mathcal{C}$.

2) H_{yes} is defined by a Boolean combination of a finite number of inequalities of the form $h(x) \geq 0$, with all h in $\mathcal{C}$.

We will indicate a non-deterministic class with an **N** in front. In particular:

Poly$_k$ and **NPoly**$_k$ are the language classes recognized by deterministic and non-deterministic polynomial recognizers of degree k.

Lin = **Poly**$_1$ and **NLin** = **NPoly**$_1$ are the deterministic and non-deterministic linear languages.

Poly = $\cup_k$**Poly**$_k$ and **NPoly** = $\cup_k$**NPoly**$_k$ are the deterministic and non-deterministic polynomial languages of any degree.

PieceLin and **NPieceLin** are the languages recognized by piecewise-linear recognizers with a finite number of components.

Elem and **NElem** are languages recognized by elementary functions, meaning compositions of algebraic, trigonometric, and exponential functions.

In this paper, we will further assume that the coefficients (and component boundaries in the piecewise-linear case) of all the f_a and h are rational. We study real coefficients in [15].

As a partial motivation for this work, we note that recurrent neural networks are being studied as models of language recognition [17] for regular [7], context-free [5], and context-sensitive [20] languages, as well as fragments of natural language [6], where grammars are represented dynamically rather than symbolically. The results herein then represent upper and lower limits on the grammatical capabilities of such networks in real time, with varying sorts of nonlinearities.

3   A language not in Poly or PieceLin, and its consequences

We will now put an upper bound on the recognition ability of deterministic piecewise-linear and polynomial maps of any degree.

Theorem 1 *The language*

$$L = \{w_1 \mathbb{c} w_2 \mathbb{c} \cdots \mathbb{c} w_m \$ v \mid v = w_i \text{ for some } i\}$$

where the w_i and v are in A^*, is in **NLin** *but not* **Poly** *or* **PieceLin**.

Proof.. Note that L can be "programmed" to recognize any finite language: if $u = w_1 \mathbb{c} w_2 \mathbb{c} \cdots \mathbb{c} w_m \$$ where $w_1, \ldots, w_m$ are all the words in a finite language L_u, then $uv \in L$ if and only if $v \in L_u$. Therefore, any recognizer ρ for L contains recognizers for all possible finite languages in its state space, i.e. it recognizes L_u if we use x_u instead of x_0 as our initial point. We will show that no polynomial recognizer of finite degree can have this property.

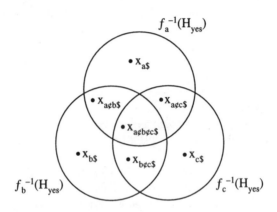

Fig. 1. The family of sets $f_v^{-1}(H_{\text{yes}})$ over all finite words v is independent.

A family of sets $S_1, \ldots, S_n$ is *independent* if all 2^n possible intersections of the S_i and their complements are non-empty; in other words, if the S_i overlap in a Venn diagram. But since $f_v(x_u) \in H_{\text{yes}}$ if and only if $v \in L_u$, x_u is in the following intersection of sets:

$$x_u \in \left(\bigcap_{v \in L_u} f_v^{-1}(H_{\text{yes}}) \right) \cap \left(\bigcap_{v \notin L_u} \overline{f_v^{-1}(H_{\text{yes}})} \right)$$

For instance, if $A = \{a, b, c\}$, then $x_{a\mathbb{c}b\$}$ is in $f_a^{-1}(H_{\text{yes}})$ and $f_b^{-1}(H_{\text{yes}})$, but not in $f_c^{-1}(H_{\text{yes}})$ as shown in figure 1. Since any such intersection is therefore non-empty, the family of sets $f_v^{-1}(H_{\text{yes}})$ is independent, where v ranges over any finite set of words.

Now a theorem of Warren [21] states that m polynomials of degree k can divide $\mathbb{R}^d$ into at most $(4emk/d)^d$ components if $m \geq d$. This number must be at least 2^m for all m of the sets defined by inequalities in these polynomials to be independent.

Suppose ρ is polynomial of degree k, has d dimensions, and has an alphabet with n symbols. Assume for the moment that H_{yes} is defined by a single polynomial inequality of degree k. Then $f_v^{-1}(H_{\text{yes}})$ is defined by a polynomial of degree $k^{|v|+1}$. Then for all n^l of the sets $f_v^{-1}(H_{\text{yes}})$ for words of length l to be independent, we need

$$\left(\frac{4en^l k^{l+1}}{d} \right)^d \geq 2^{n^l}$$

This is clearly false for sufficiently large l, since the right-hand side is doubly exponential in l while the left-hand side is only singly so.

If H_{yes} is defined by c inequalities instead of one, we simply replace n^l with cn^l on the left-hand side; the right-hand side remains the same, since we still need to create n^l independent sets.

Thus polynomial maps of a fixed degree, in a fixed number of dimensions, cannot contain in their state spaces recognizers for all finite languages; so L is not in **Poly**. A similar argument works for piecewise-linear maps, as long as their number of components is finite.

However, L is easily seen to be in **NLin**. With maps like $f_a(x) = nx + a$ where $0 \leq a < n$, we can read a word w into an integer variable $x = \overline{w}$; then just non-deterministically keep one of the $\overline{w}_i$ and ignore away the others, and check that $\overline{v} = \overline{w}_i$. □

Several corollaries follow easily from this example:

Corollary 2 PieceLin, Poly, *and* **Poly**$_k$ *for all k are properly contained in their non-deterministic counterparts* **NPieceLin, NPoly,** *and* **NPoly**$_k$.

We show in [15] that context-free languages can be recognized in **NPieceLin** and **NPoly**$_2$. The next corollary makes this inclusion sharper:

Corollary 3 *There are non-deterministic context-free languages not in* **Poly** *or* **PieceLin**.

Proof.. The *reversal* of a word w is $w^R = w_{|w|} \cdots w_2 w_1$. Let L' be a modified version of L in which $v^R = w_i$ for some i, instead of $v = w_i$. Then L' is context-free: it is accepted by a non-deterministic push-down automaton (PDA) that puts one of the w_i on the stack, ignores the others, and then compares v to it in reverse. However, L' is not in **Poly** or **PieceLin** by the same independent-set argument we used for L. □

Corollary 4 PieceLin, Poly, *and* **Poly**$_k$ *for $k \geq 2$ are not closed under reversal.*

Proof.. L^R, where the first word has to be equal to one that follows, is in **Poly**$_2$: let $y_0 \neq 0$ and $f_{\phi}(y) = y(\overline{v} - \overline{w}_i)$, and require that $y = 0$ or $\overline{v} = \overline{w}_1$ at the end. We can do the same thing with piecewise-linear maps. Since $L = (L^R)^R$, these classes can't be closed under reversal. □

Using arguments almost identical to those in [18] for the languages recognized by deterministic Turing machines in real time, we can also show that **PieceLin**, **Poly**, **Poly**$_k$ for all k are not closed under alphabetic homomorphism, Kleene star, or positive closure.

4   A unary version of L

Next, we will show that a unary version of L separates **Lin** from **PieceLin** and **Poly**$_2$:

Theorem 5 **Lin** *is properly contained in* **PieceLin** *and in* **Poly**$_2$.

Proof.. Consider a version of L where the w_i and v are over a one-symbol alphabet $\{a\}$:

$$L_{\text{unary}} = \{a^{p_1} \mathcal{c} a^{p_2} \mathcal{c} \cdots \mathcal{c} a^{p_m} \$ a^q \mid q = p_i \text{ for some } i\}$$

Suppose L_{unary} is in **Lin**. The composition of linear maps is linear, so if H_{yes} is described by c linear inequalities, then each of the sets $f_{a^i}^{-1}(H_{\text{yes}})$ is also. But these all have to be independent by the same argument as in theorem 1, so cl linear inequalities have to divide $\mathbb{R}^d$ into at least 2^l components. But for $k = 1$, Warren's inequality becomes

$$\left(\frac{4ecl}{d}\right)^d \geq 2^l$$

which is false for sufficiently large l. So L_{unary} is not in **Lin**.

However, L_{unary} is in **PieceLin**. Let $x_0 = y_0 = 0$ and $r_0 = 1$, with the following dynamics:

$$f_a(x, y, r) = (x, 2y \bmod 2, r/2)$$
$$f_{\mathcal{c}}(x, y, r) = \begin{cases} (x+r, x+r, 1) \text{ if } y \in [0, 1) \\ (x, x, 1) \text{ if } y \in [1, 2) \end{cases}$$
$$f_\$(x, y, r) = f_{\mathcal{c}}(x, y, r)$$

(We only need two 'pieces' of $2y \bmod 2$ since $y \in [0, 2)$ always.) The sequence $a^p \mathcal{c}$ adds 2^{-p} to x unless the 2^{-p} digit of x was already 1, i.e. unless $y \in [1, 2)$ where $y = 2^p x \bmod 2$. By the time we reach the $\$$, we have $x = \sum_i 2^{-p_i}$. Then with an additional variable w, let $f_\$(w) = x$, let $f_a(w) = 2w \bmod 2$, and let H_{yes} require that $w \in [1, 2)$, checking that the 2^{-q} digit of x is 1.

What about L_{unary}^R? It is in **Poly**$_2$ by the same construction as in corollary 4 above: just let $f_{\mathcal{c}}(y) = y(q - p_i)$ with $y_0 \neq 0$ and require that $y = 0$ or $q = p_1$. Similarly, it is in **PieceLin**. However, a similar independent-set argument [15] shows that it is not in **Lin**. □

It is an interesting open question whether L_{unary} is in $\textbf{Poly}_2$. For $k = 2$ and $n = 1$, Warren's inequality becomes

$$\left(\frac{4ecl\,2^{l+1}}{d}\right)^d \geq 2^l$$

and no longer yields a contradiction for large l.

The method of independent sets we use here is equivalent to the *Vapnik-Chervonenkis dimension*, which has been used in computational learning theory to quantify the difficulty of learning sets by example [2]. The VC dimension of a family of sets $\mathcal{F}$ is the size of the largest independent family $S \subset \mathcal{F}$. Then our arguments about independent sets can be re-stated in the following way: in $\mathbb{R}^d$ with d fixed, the VC dimension of the family

$$\left\{\, f_w^{-1}(H_{\text{yes}}) \,\middle|\, |w| \leq l \,\right\}$$

is $\mathcal{O}(l)$ for polynomial maps and $\mathcal{O}(1)$ for linear maps (this also follows from the results in [8]), while L and L_{unary} require a VC dimension of at least $\mathcal{O}(n^l)$ and $\mathcal{O}(l)$ respectively.

5   Higher classes: elementary functions

We now consider the classes **Elem** and **NElem**, where we allow exponential, trigonometric and polynomial functions, as well as their compositions. We will allow coefficients that are elementary functions of integers, such as rational or algebraic numbers.

Theorem 6 **Elem** *properly contains* **Poly**.

Proof.. We will show that L is in **Elem**. Recall its definition:

$$L = \{\, w_1 \phi w_2 \phi \cdots \phi w_m \$ v \mid v = w_i \text{ for some } i \,\}$$

By reading words into integers $\overline{w}$ as before, and letting $x_0 = 0$ and $f_\phi(x) = x + 2^{\overline{w}}$, by the time we reach the \$ we have $x = \sum_i 2^{\overline{w_i}}$. Then

$$x/2^{\overline{v}} \in \begin{cases} [2k+1, 2k+2] & \text{if } v = w_i \text{ for some } i \\ [2k, 2k+1] & \text{if } v \neq w_i \text{ for all } i \end{cases} \quad \text{for some integer } k$$

In other words, the $2^{\overline{v}}$ digit of x is 1 if $v = w_i$ and 0 otherwise. So let H_{yes} require that $\sin \pi(x/2^{\overline{v}}) < 0$ or $\cos \pi(x/2^{\overline{v}}) = -1$, i.e. $x/2^{\overline{v}} \in (2k+1, 2k+2)$ or $x = 2k + 1$.

Since L is in **Elem** but not **Poly**, the inclusion $\textbf{Poly} \subset \textbf{Elem}$ is proper.   □

Here we're using the fact that all the sets $S_j = \{x \mid \sin 2^j x < 0\}$ for $j = 0, 1, 2, \ldots$ are independent, i.e. the family $\{S_j\}$ has infinite VC-dimension.

6   Relationships with other models of analog computation.

There are several differences between Blum, Shub and Smale's (BSS) analog machines [1], Siegelmann and Sontag's neural networks (NN) [19], and dynamical recognizers.

First, BSS-machines can branch on polynomial inequalities during the course of the computation. Except for **PieceLin**, our recognizers have completely continuous dynamics except for the final measurement of H_{yes}. NN-machines are defined with piecewise-linear maps.

Secondly, BSS- and NN-machines are not restricted to real time, so that time complexity classes such as **P**, **EXPTIME** and so on can be defined for them.

Thirdly, BSS-machines can recognize "languages" whose symbols are real numbers, and can make real number guesses in their non-deterministic versions.

Fourthly, BSS-machines have unbounded dimensionality, with the number of variables growing linearly with time. Like ours, NNs have a fixed number of dimensions.

Finally, BSS-machines receive their entire input as part of their initial state. Therefore, they have at least n variables on input of length n. NN-machines, like ours, receive their input dynamically rather than as part of the initial state.

This last point seems entirely analogous to Turing machines. If we wish to consider sub-linear space bounds such as **LOGSPACE**, we need to use an *off-line* Turing machine which receives its input on a read-only tape separate from its worktape.

This suggests a unification of all three models. First of all, let **PiecePoly** and **NPiecePoly** be recognizer classes where the f_a are piecewise polynomials, with polynomial component boundaries.

Secondly, relax our real-time restriction by iterating an additional map f_{comp}, in the same class as the f_a, until x falls into some subset H_{halt}.

Thirdly, restrict BSS-machines to their Boolean part BP and to *digital non-determinism*, e.g. **DNP** [3].

And finally, define an *off-line BSS-machine* as one who receives its input dynamically in the first n steps, and which has a bound **VARS**$(f(n))$ on the number of variables it can use during the computation. (In [9] these are called *separated input and output* or SIO-BSS-machines.)

Then we can look at these classes in a unified way:

$$\textbf{PiecePoly}(\mathbb{R})\textbf{TIME}(\mathcal{O}(n^k))\textbf{VARS}(\mathcal{O}(n^k)) = \text{BP}(\textbf{P}_\mathbb{R}) \text{ (Blum, Shub and Smale [1])}$$

$$\textbf{NPiecePoly}(\mathbb{R})\textbf{TIME}(\mathcal{O}(n^k))\textbf{VARS}(\mathcal{O}(n^k)) = \text{BP}(\textbf{NDP}_\mathbb{R}) \text{ (Cucker and Matamala [3])}$$

$$\textbf{PieceLin}(\mathbb{R})\textbf{TIME}(\mathcal{O}(n^k))\textbf{VARS}(\mathcal{O}(n^k)) = \text{BP}(\textbf{P}_{\text{lin}}^<) \text{ (Meer [13] and Koiran [11])}$$

$$\textbf{PieceLin}(\mathbb{R})\textbf{TIME}(\mathcal{O}(n^k))\textbf{VARS}(\mathcal{O}(1)) = \textbf{NET} - \textbf{P} \text{ (Siegelmann and Sontag [19])}$$

$$\textbf{PiecePoly}_k(\mathbb{Z})\textbf{TIME}(n)\textbf{VARS}(\mathcal{O}(1)) = \textbf{PiecePoly}_k(\mathbb{Z}) \text{ (dynamical recognizers)}$$

and similarly for other complexity classes.

A number of proper inclusions among these classes follow immediately from the present work. For instance, just as for deterministic Turing machines [18], linear time is more powerful than real time, since linear time is closed under reversal [15] while real time is not.

7   Conclusion and directions for further work

In addition to the conjectures and open problems mentioned above, there are several directions in which one could extend this work.

1) We have seen that **PieceLin** and **NLin** are more powerful than **Lin**. Is **PieceLin** contained in any continuous class, and is **NLin** contained in any deterministic class? In other words, can branching and non-determinism be compensated for in real time by going to a more powerful class of functions such as **Elem**? We conjecture that these are fundamentally different computational resources, and that **NLin**, **PieceLin** and **Elem** are all incomparable.

2) In theorem 2, quadratic maps seem to be roughly equivalent to elementary maps when their input is given in unary. Is there any depth to this equivalence?

3) Can we exhibit a language not in **Elem** or **NLin**? We need methods other than VC-dimension for both of these.

4) Finally, we believe that classes in these models with sub-linear bounds on the number of variables, such as $\mathbf{PiecePoly}(\mathbb{Z})\mathbf{TIME}(\mathcal{O}(n))\mathbf{VARS}(\mathcal{O}(\log n))$, are very much worth studying. It ought to be possible to prove **VARS** hierarchy theorems analogous to the **SPACE** hierarchy for Turing machines (constant space is universal if unlimited time is allowed [14]).

Other suggestions and open problems are given in [15].

Acknowledgements. I thank Elizabeth Hunke and Mats Nordahl for careful readings of the manuscript; Mike Casey, Kiran Chilakamarri, Jeff Erickson, Michael Fischer, Jordan Pollack, Vicki Powers, and Hava Siegelmann for helpful communications; and Spootie the Cat for companionship.

References

1. L. Blum, M. Shub, and S. Smale, "On a theory of computation and complexity over the real numbers: NP-completeness, recursive functions and universal machines." *Bull. Amer. Math. Soc.* **21** (1989) 1-46.

2. A. Blumer, A. Ehrenfeucht, D. Haussler, and M.K. Warmuth, "Learning and the Vapnik-Chervonenkis dimension." *Journal of the ACM* **36(4)** (1989) 929-965.

3. F. Cucker and M. Matamala, "On digital nondeterminism." To appear in *Math. Systems Theory*.

4. F. Cucker and D. Grigoriev, "On the power of real Turing machines over binary inputs." NeuroCOLT Technical Report NC-TR-94-007 (1994).

5. S. Das, C.L.Giles, and G.Z.Sun, "Using prior knowledge in an NNPDA to learn languages." *Advances in Neural Information Processing Systems* **5** (1993) 65-72.

6. J. Elman, "Language as a dynamical system." In R.F. Port and T. van Gelder (Eds.), *Mind as Motion: Explorations in the Dynamics of Cognition*. MIT Press, 1995.

7. C.L. Giles, C.B. Miller, D. Chen, H.H. Chen, G.Z. Sun, and Y.C. Lee, "Learning and extracting finite-state automata with second-order recurrent networks." *Neural Computation* **2** (1992) 331-349

8. P.W. Goldberg and M.R. Jerrum, "Bounding the Vapnik-Chevonenkis dimension of concept classes parametrized by real numbers." *Machine Learning* **18** (1995) 131-148.

9. E. Grädel and K. Meer, "Descriptive complexity theory over the real numbers." NeuroCOLT Technical Report NC-TR-95-040 (1995), and in *Proc. of the 27th Symposium on the Theory of Computing* (1995) 315-324.

10. J.E. Hopcroft and J.D. Ullman, *Introduction to Automata Theory, Languages, and Computation.* Addison-Wesley, 1979.

11. P. Koiran, "Computing over the reals with addition and order." *Theoretical Computer Science* **133** (1994) 35-47.

12. P. Koiran, "A weak version of the Blum, Shub and Smale model." DIMACS Technical Report 94-10 (1994).

13. K. Meer, "Real Number Models under Various Sets of Operations." *Journal of Complexity* **9** (1993) 366-372.

14. C. Michaux, *Differential fields, machines over the real numbers and automata.* Ph.D. thesis, Université de Mons Hainaut, Faculte des Sciences (1991).

15. C. Moore, "Dynamical Recognizers: Real-time Language Recognition by Analog Computers." Santa Fe Institute Working Paper 96-05-023, submitted to *Theoretical Computer Science.*

16. C.H. Papadimitriou, *Computational Complexity.* Addison-Wesley, 1994.

17. J. Pollack, "The Induction of Dynamical Recognizers." *Machine Learning* **7** (1991) 227-252.

18. A.L. Rosenberg, "Real-time definable languages." *Journal of the ACM* **14** (1967) 645-662.

19. H. Siegelmann and E.D. Sontag, "Analog Computation Via Neural Networks." *Theoretical Computer Science* **131** (1994) 331-360.

20. M. Steijvers and P.D.G. Grünwald, "A recurrent network that performs a context-sensitive prediction task." NeuroCOLT Technical Report NC-TR-96-035 (1996), and in *Proceedings of the 18th Annual Conference of the Cognitive Science Society.* Erlbaum (in press).

21. H.E. Warren, "Lower bounds for approximation by nonlinear manifolds." *Trans. Amer. Math. Soc.* **133** (1968) 167-178.

Solving special polynomial systems by using structured matrices and algebraic residues

Bernard Mourrain, Victor Y. Pan

Abstract. We apply and extend some well-known and some recent techniques from algebraic residue theory in order to relate to each other two major subjects of algebraic and numerical computing, that is, computations with structured matrices and solving a system of polynomial equations. In the first part of our paper, we extend the Toeplitz and Hankel structures of matrices and some of their known properties to some new classes of structured (quasi-Hankel and quasi-Toeplitz) matrices, naturally associated to systems of multivariate polynomial equations. In the second part of the paper, we apply some results on computations with matrices of these new classes, together with some techniques from algebraic residues theory, in order to devise an algorithm for approximating a selected solution of a polynomial system of the form

$$\begin{cases} x_1^{d_1} - R_1(x_1, \ldots, x_n) = 0, \\ \vdots \\ x_n^{d_n} - R_n(x_1, \ldots, x_n) = 0, \end{cases}$$

where $\deg(R_i) < d_i$. The complexity of this algorithm is $\mathcal{O}(D^2 \log(D)^c)$, where $D = \prod_{i=1}^{n} d_i$ is the number of roots of the system.

1 Introduction

We apply and extend some well-known and some recent techniques from algebraic residue theory in order to relate to each other two major subjects of algebraic and numerical computing, that is, the computations with structured matrices and solving a system of polynomial equations. We also reveal some hidden correlations between these two subjects via the study of the associated operators of multivariate displacement. The latter operators naturally extend the univariate displacement operators, which define Toeplitz and/or Hankel structure of matrices (cf. [1]). In our multivariate case, we generalize such a matrix structure and arrive at the new classes of operators and structured matrices, which include operators and matrices associated to the polynomial systems of equations and which we call quasi-Hankel and quasi-Toeplitz operators and matrices since some well-known properties of Toeplitz and Hankel operators and matrices can be extended to them (see section 2). Due to high importance of computations with structured matrices (see e.g. [1]), our study of these matrix classes may be of independent technical interest. In section 3, we recall some basic definitions and facts about algebraic residues and extend them in order to apply, in section 4, to the solution of polynomial systems of n equations with n variables. For a special class of such systems (where the i-th equation has the form $P_i = x_i^{d_i} - R_i(x_1, \ldots, x_n)$ and where R_i has a total degree less than d_i), we reduce the solution to computing the associated residues, where we apply some results on computations with structured matrices from section 2. This enables us to compute (under some additional assumptions) a selected solution to the system, by using order of $D^2 \log(D)$ arithmetic operations, where $D = \prod_i^n d_i$. The latter result is a substantial improvement versus the previously known solutions requiring order of D^3 arithmetic operations. The result may also be of technical interest as the first example where combined application of structured matrices and algebraic residues leads to a substantial improvement of the known methods for solving polynomial systems of equations.

Next, we will state some definitions. $R = \mathbb{C}[x_1, \ldots, x_n]$ will denote the polynomial ring in variables $x_1, \ldots, x_n$ over the complex field $\mathbb{C}$, and $L = \mathbb{C}[x_1^{\pm 1}, \ldots, x_n^{\pm 1}]$ will denote the ring of Laurent's polynomials in the same variables. We will write $\mathbf{x} = (x_1, \ldots, x_n)$ and $\mathbf{x}^\alpha = x_1^{\alpha_1} \cdots x_n^{\alpha_n}$. For a vector $\alpha = (\alpha_1, \ldots, \alpha_n)$, we will write $|\alpha|$ to denote the 1-norm of this vector, $|\alpha| = \sum_{i=0}^n |\alpha_i|$. The total degree of a monomial $c\,\mathbf{x}^\alpha$, with a coefficient c, is $|\alpha|$. The total degree of a polynomial $\sum_\alpha c_\alpha \mathbf{x}^\alpha$, with coefficients c_α, is the highest total degree of its monomials, for which $c_\alpha \neq 0$. We will write $\lfloor S \rceil$ to denote the cardinality of a set S. ops will stand for "arithmetic operations". $\mathbf{e}_i$ will denote the i-th unit coordinate vector in $\mathbb{C}^n$.

Our study can be immediately extended from the complex field $\mathbb{C}$ to the case of any number field of constants having characteristic 0. Furthermore, with the exception of the results based on the interpolation techniques of [3] (cf. proposition 26), our study can be extended to the case of any field of constants.

2   Structured Matrices

In this section, we propose a generalization of the structure of Toeplitz and Hankel matrices to the case of matrices associated with multivariate polynomials having rows and columns indexed by monomials.

2.1   Quasi-Hankel and quasi-Toeplitz matrices, operators, and the associated generating polynomials: definitions and a correlation

DEFINITION 21 *Let E and F be two subsets of $\mathbb{Z}^n$ and let $M = (m_{\alpha,\beta})_{\alpha \in E, \beta \in F}$ be a matrix whose rows are indexed by the elements of E and columns by the elements of F.*

- M is an (E, F) quasi-Hankel *matrix iff, for all* $\alpha \in E, \beta \in F$, *the entries* $m_{\alpha,\beta} = h_{\alpha+\beta}$ *depend only on* $\alpha+\beta$, *that is, if for every* $i = 1, \ldots, n$, *we have* $m_{\alpha-e_i,\beta+e_i} = m_{\alpha,\beta}$ *provided that* $\alpha, \alpha - e_i \in E; \beta, \beta + e_i \in F$; *such a matrix* M *is associated with the Laurent polynomial* $H_M(\mathbf{x}) = \sum_{u \in E-F} h_u \mathbf{x}^{-u}$.
- M is an (E, F) quasi-Toeplitz *matrix iff, for all* $\alpha \in E, \beta \in F$, *the entries* $m_{\alpha,\beta} = t_{\alpha-\beta}$ *depend only on* $\alpha - \beta$, *that is, if for every* $i = 1, \ldots, n$, *we have* $m_{\alpha+e_i,\beta+e_i} = m_{\alpha,\beta}$, *provided that* $\alpha, \alpha + e_i \in E; \beta, \beta + e_i \in F$; *such a matrix* M *is associated with the polynomial* $T_M(\mathbf{x}) = \sum_{u \in E+F} t_u \mathbf{x}^u$.

For $E = [0, \cdots, m-1]$ and $F = [0, \ldots, n-1]$, definition 21 turns into the usual definition of Hankel (resp. Toeplitz) matrices [1].

DEFINITION 22 *Let* $\mathcal{P}_E : L \to L$ *be the projection map such that*

$$\mathcal{P}_E(\mathbf{x}^\alpha) = \mathbf{x}^\alpha$$

if $\alpha \in E$ *and* $\mathcal{P}_E(\mathbf{x}^\alpha) = 0$ *otherwise. Let* $\rho_E = Id - \mathcal{P}_E$, *where* Id *denotes the identity operator,* $Id(e) = e$ *for all* e. *For any element* Q *of* L, *let* $\mu_Q : L \to L$ *denote the operator of multiplication by* Q. *For any matrix* $M = (m_{\alpha,\beta})_{\alpha \in E, \beta \in F}$, *let* $\mathcal{M}$ *denote the linear map* $L \to L$ *such that*

$$\mathcal{M}(\mathbf{x}^\beta) = \sum_{\alpha \in E} m_{\alpha,\beta} \mathbf{x}^{-\alpha}$$

if $\beta \in F$ *and* $\mathcal{M}(\mathbf{x}^\beta) = 0$ *otherwise. The matrix of this linear operator coincides with the matrix* M *on* $(\mathbf{x}^{-\alpha}) \times (\mathbf{x}^\beta)$, *for* $\alpha \in E$, $\beta \in F$, *and is null elsewhere. We will call this operator an* (E, F) quasi-Hankel *(resp. an* (E, F) quasi-Toeplitz*) operator if the matrix* M *is an* (E, F) quasi-Hankel *(resp. an* (E, F) quasi-Toeplitz*) matrix.*

PROPOSITION 23 *If* M *is an* (E, F) quasi-Hankel *(resp. an* (E, F) quasi-Toeplitz*) matrix, then* $\mathcal{M} = \mathcal{P}_{-E} \circ \mu_{H_M} \circ \mathcal{P}_F$ *(resp.* $\mathcal{M} = \mathcal{P}_E \circ \mu_{T_M} \circ \mathcal{P}_F$*).*

To the end of this of section, we will assume that both sets E and F contain 0.

2.2 Multiplication of quasi-Hankel and quasi-Toeplitz matrices by vectors

Multiplication of an (E, F) quasi-Hankel *matrix by a vector* $\mathbf{v} = [v_\beta] \in \mathbb{C}^F$ *can be reduced to (Laurent) polynomial multiplication in the following way. Let* $M = (m_{\alpha,\beta})_{\alpha \in E, \beta \in F}$ *denote an* (E, F) quasi-Hankel *matrix, let* $H_M(\mathbf{x}) = \sum_{u \in E-F} h_u \mathbf{x}^{-u}$ *denote the associated Laurent polynomial, and let* $V(\mathbf{x}) =$

$\sum_{\beta \in F} v_\beta \mathbf{x}^\beta$. Then, we have

$$H_M(\mathbf{x})\, V(\mathbf{x}) = \sum_{u \in E-F, \beta \in F} \mathbf{x}^{-u+\beta}\, h_u\, v_\beta$$

$$= \sum_{\alpha = u - \beta \in E - 2\,F} \mathbf{x}^{-\alpha} \left(\sum_{\beta \in F} h_{\alpha+\beta}\, v_\beta \right)$$

where we assume that $v_\beta = 0$ if $u \notin E - F$, $h_u = 0$ if $u \notin E - F$. Therefore, for $\alpha \in E$, the coefficient of $\mathbf{x}^{-\alpha}$ equals

$$\sum_{\beta \in F} h_{\alpha+\beta}\, v_\beta = \sum_{\beta \in F} m_{\alpha,\beta}\, v_\beta$$

which is precisely the coefficient α of $M\,\mathbf{v}$.

A similar argument *reduces multiplication of an (E, F) quasi-Toeplitz matrix by a vector to multiplication of a pair of Laurent's polynomials.*

The stated reductions enable us to deduce the following result:

PROPOSITION 24 *An (E, F) quasi-Hankel (resp. an (E, F) quasi-Toeplitz) matrix M can be multiplied by a vector in $\mathcal{O}(N \log^2 N + N\, C_M)$ ops, where $N = \lfloor E - 2\,F \rceil$ (resp. $\lfloor E + 2F \rceil$) and where C_M bounds the cost of evaluating the polynomial H_M (resp. T_M).*

Proof. To obtain the latter complexity estimate, we reduce the problem to computing the product of the two polynomials $H_M(\mathbf{x})$ (resp. $T_M(\mathbf{x})$) and $V(\mathbf{x})$ and then apply a variant of the well-known techniques of evaluation and interpolation (cf. [1]). Namely, we first evaluate the polynomials $H_M(\mathbf{x})$ (resp. $T_M(\mathbf{x})$) and $V(\mathbf{x})$ on a fixed set of points, then pairwise multiply their values, to obtain the values of the product $H_M(\mathbf{x})\, V(\mathbf{x})$ (resp. $T_M(\mathbf{x})\, V(\mathbf{x})$), and finally, obtain the coefficients of this product by applying the known interpolation techniques for sparse polynomials (see e.g. section 1.9 of [1] or [10]).

In some special cases, we have better complexity estimates.

PROPOSITION 25 *In the case where $E = F = \{(\alpha_1, \ldots, \alpha_n) \in \mathbb{N}^n \; ; \; 0 \le \alpha_i \le d_i - 1\}$, an (E, F) quasi-Hankel (resp. an (E, F) quasi-Toeplitz) matrix can be multiplied by a vector in $\mathcal{O}(3^n\, D \log(3^n\, D))$ ops, where $D = \prod_i^n d_i$.*

Proof. At first, as in the proof of proposition 24, we reduce the problem to multiplication of a pair of the associated multivariate polynomials, $U(\mathbf{x})$ and $V(\mathbf{x})$, so that $U\,V$ is of degree at most $3d_i - 3$ in each variable x_i. The latter multiplication can be reduced to multiplication of a pair of univariate polynomials, by means of application of Kronecker's map,

$$x_1 = y, \; x_k = y^{D_k}, \; D_k = \prod_{i<k} (3d_i - 2), \; k = 1, \ldots, n,$$

which turns $U(\mathbf{x})\,V(\mathbf{x})$ into a univariate polynomial of degree at most

$$(3d_n - 3)D_n < \prod_{i=1}^{n}(3d_i - 2) < 3^n D.$$

By using fast Fourier transform (FFT), we may multiply such polynomials by using order of $3^n D \log(3^n D)$ ops [1].

REMARK 1 — *For practical implementation where D is very large, it is better to avoid using Kronecker's map so as to multiply two multivariate polynomials $U(\mathbf{x})$ and $V(\mathbf{x})$ of lower degrees, rather than two univariate polynomials of very high degrees. Then, the number of ops may grow a little, but FFT involves roots of 1 of a lower order.*

PROPOSITION 26 *In the case where $E = \{\alpha \in \mathbb{N}^n, |\alpha| \le k\}$, $F = \{\beta \in \mathbb{N}^n, |\beta| \le l\}$ and where the computations are over a field of constants containing the field of rational numbers, an (E, F) quasi-Toeplitz matrix can be multiplied by a vector in $\mathcal{O}(\sigma \log^3 \sigma)$ ops, where*

$$\sigma = \sigma_{2l+k,n} = \binom{2l + k + n}{n} = \mathcal{O}((e\,\frac{2l + k + n}{n})^n / \sqrt{n}),\ e = 2.718\ldots.$$

(The latter equation is implied by Stirling's formula.)

Proof. Proceed as in the proof of proposition 24 but use the interpolation technique of [3] for polynomials with bounded total degrees, instead of using the techniques of section 1.9 of [1] or [10].

2.3   Multivariate displacement operators and ranks (definition)

By convention, if $\mathcal{A} : L \to L$ is a linear operator, A will denote its matrix in a (sub)-basis of L. For all $\alpha, \beta \in \mathbb{Z}^n$, $[\mathcal{A}]_{\alpha,\beta} = A_{\alpha,\beta}$ is the coefficient of $\mathbf{x}^\alpha$ in $\mathcal{A}(\mathbf{x}^\beta)$.

DEFINITION 27 *For any subset E of $\mathbb{Z}^n$, we define the two following unit E-displacement matrices (operators):*

$$\mathcal{Z}_i^E = \mathcal{P}_E\,\mu_{x_i}\mathcal{P}_E$$

and

$$\mathcal{Z}_{-i}^E = \mathcal{P}_E\,\mu_{x_i^{-1}}\mathcal{P}_E.$$

In particular, for $E = [0, \ldots, n-1]$ and $i = 1$, we arrive at the well-known displacement matrix

$$Z_1^E = \begin{pmatrix} 0 & \cdots & \cdots & 0 \\ 1 & \ddots & & \vdots \\ 0 & \ddots & \ddots & \vdots \\ \vdots & \ddots & 1 & 0 \end{pmatrix}$$

and its transpose Z_{-1}^E (cf. [1]) .

DEFINITION 28 *Let E and F denote two subsets of $\mathbb{Z}^n$ and let $\mathcal{A}$ denote a linear operator $L \to L$. Then, the operators*

$$\mathcal{H}_i^+(\mathcal{A}) = \mathcal{A} - \mathcal{Z}_{-i}^{-E} \mathcal{A} \mathcal{Z}_i^F, \quad \mathcal{H}_i^-(\mathcal{A}) = \mathcal{A} - \mathcal{Z}_i^{-E} \mathcal{A} \mathcal{Z}_{-i}^F, \quad T_i^+(\mathcal{A}) = \mathcal{A} - \mathcal{Z}_{-i}^E \mathcal{A} \mathcal{Z}_i^F,$$

and $T_i^-(\mathcal{A}) = \mathcal{A} - \mathcal{Z}_i^E \mathcal{A} \mathcal{Z}_{-i}^F$ will be called the $(-, +, -E, F, i)$, $(+, -, -E, F, i)$, $(-, +, E, F, i)$, and $(+, -, E, F, i)$ displacements of $\mathcal{A}$, respectively). The ranks of these displacements will be called the $(-, +, -E, F, i)$, $(+, -, -E, F, i)$, $(-, +, E, F, i)$, and $(+, -, E, F, i)$ displacement ranks of $\mathcal{A}$, resp., and will be denoted $r_{-,+,-E,F,i}(\mathcal{A})$, $r_{+,-,-E,F,i}(\mathcal{A})$, $r_{-,+,E,F,i}(\mathcal{A})$, and $r_{+,-,E,F,i}(\mathcal{A})$, resp. The operators transforming $\mathcal{A}$ into the above displacements will be called the $(-, +, -E, F, i)$, $(+, -, -E, F, i)$, $(-, +, E, F, i)$, and $(+, -, E, F, i)$ displacement operators, resp.

2.4   Bounds on displacement ranks of quasi-Hankel and quasi-Toeplitz matrices

DEFINITION 29 *Hereafter, we write*

$$\delta_i(E) = \{\alpha : \alpha \in E \,;\, \alpha + e_i \notin E\}$$

(resp. $\delta_{-i}(E) = \{\alpha : \alpha \in E \,;\, \alpha - e_i \notin E\}$).

PROPOSITION 210 *For an $(E, F,)$ quasi-Hankel operator $\mathcal{M}$, we have the following bounds on its $(-, +, E, F, i)$ and $(+, -, E, F, i)$ displacement ranks:*

$$r_{-,+,-E,F,i}(\mathcal{M}) \le \lfloor \delta_i(-E) \rceil + \lfloor \delta_i(F) \rceil,$$

$$r_{+,-,-E,F,i}(\mathcal{M}) \le \lfloor \delta_{-i}(-E) \rceil + \lfloor \delta_{-i}(F) \rceil.$$

For an (E, F) quasi-Toeplitz operator $\mathcal{M}$, we have the following bounds on its $(-, +, E, F, i)$ and $(+, -, E, F, i)$ displacement ranks:

$$r_{-,+,E,F,i}(\mathcal{M}) \le \lfloor \delta_{-i}(E) \rceil + \lfloor \delta_i(F) \rceil,$$

$$r_{+,-,E,F,i}(\mathcal{M}) \le \lfloor \delta_i(E) \rceil + \lfloor \delta_{-i}(F) \rceil.$$

Proof. The proofs of all the four bounds of this proposition mimic each other, so we will only prove the first bound. According to proposition 23, we have

$$\mathcal{H}_i^+(\mathcal{M}) = \mathcal{P}_{-E} \left(\mu_U - \mu_{x_i^{-1}} \mathcal{P}_{-E}\, \mu_U \mathcal{P}_F \mu_{x_i} \right) \mathcal{P}_F.$$

where U is the polynomial associated with $\mathcal{M}$.

- If $\beta \notin F$ then $\mathcal{H}_i(\mathcal{M})(\mathbf{x}^\beta) = 0$.
- If β lies in $\delta_i(F)$, then $\mathcal{Z}_i^F(\mathbf{x}^\beta) = 0$ and $\mathcal{H}_i(\mathcal{M})(\mathbf{x}^\beta) = \mathcal{M}(\mathbf{x}^\beta) = \sum_{\alpha \in E} m_{\alpha,\beta}\, \mathbf{x}^\alpha$.
- If β lies in F but not in $\delta_i(F)$, then

$$\mu_{x_i^{-1}} \mathcal{P}_{-E}\, \mu_U \mathcal{P}_F \mu_{x_i}(\mathbf{x}^\beta) = x_i^{-1} \mathcal{P}_{-E}(U\, \mathbf{x}^\beta\, x_i)$$

and

$$\begin{aligned}
\mathcal{H}_i(\mathcal{M})(\mathbf{x}^\beta) &= \mathcal{P}_{-E} \left(U\, \mathbf{x}^\beta - x_i^{-1} \mathcal{P}_{-E}(U\, \mathbf{x}^\beta\, x_i) \right) \\
&= \mathcal{P}_{-E} \left(U\, \mathbf{x}^\beta - U\, \mathbf{x}^\beta + x_i^{-1} \rho_{-E}(U\, \mathbf{x}^\beta\, x_i) \right) \\
&= \mathcal{P}_{-E} \left(x_i^{-1} \rho_{-E}(U\, \mathbf{x}^\beta\, x_i) \right) \\
&= \sum_{\alpha \in \delta_i(-E)} m_{\alpha,\beta}\, \mathbf{x}^{-\alpha}.
\end{aligned}$$

Therefore, if $\alpha \notin \delta_i(-E)$ and $\beta \notin \delta_i(\mathcal{H}_i)$, then $[\mathcal{H}_i(\mathcal{M})]_{\alpha,\beta} = 0$. Thus, the rank of $\mathcal{H}_i(M)$ is at most $\lfloor \delta_i(-E) \rceil + \lfloor \delta_i(F) \rceil$.

In the particular case, where $E = F = \{(\alpha_1, \dots, \alpha_n) \in \mathbb{N}^n \;;\; 0 \le \alpha_i \le d_i - 1\}$, the displacement rank of $\mathcal{H}_i(\mathcal{M})$ is bounded by $2 \frac{D}{d_i} = 2 \prod_{j \neq i} d_j$.

2.5   Examples

Quasi-Toeplitz matrices: Let us consider the matrix associated with the linear map

$$\phi \; : \; V_0 \times \cdots \times V_n \to V,$$

$$(Q_0, \dots, Q_n) \mapsto \sum_{i=0}^n P_i\, Q_i,$$

where V_i is the vector space generated by the monomials $\mathbf{x}^\beta$ for $\beta \in F_i$, where E denotes the set of the exponents of the monomials of $(P_i Q_i)_{i=0,\dots,n}$, and where V is generated by the monomials $\mathbf{x}^\alpha$ for $\alpha \in E$. Such maps typically appear in the construction of resultant type matrices associated to the system $\{P_i = 0,\ i = 0, \dots, n\}$, of polynomial equations [8], [2].

Let M denote the matrix of this linear map in the monomial basis of $V_0 \times \cdots \times V_n$ and V. The rows of this matrix are indexed by the elements of the set E and the columns by the elements of the set $F_0 \sqcup \cdots \sqcup F_n$. Let x_0 be a new variable and let see $\mathbb{Z}^n$ as the subset of $\mathbb{Z}^{n+1}$ of elements of the form $(0, a_1, \ldots, a_n)$. Let e_0 denote the first canonical vector of $\mathbb{Z}^{n+1}$. Then the elements of the subset

$$F = \{i\, e_0 + \alpha \; ; \; 0 \le i \le n, \; \alpha \in F_{i+1}\}$$

index the rows of M.

Note that the $(\alpha, i\, e_0 + \beta)$-th entry of the matrix M is the coefficient of $\mathbf{x}^\alpha$ in $\mathbf{x}^\beta P_i$. It is also the coefficient of $\mathbf{x}^{\alpha-\beta}$ in P_i. Therefore, it depends only on $\alpha - \beta - i\, e_0$.

REMARK 2 — *Resultant type matrices and their transposes are quasi-Toeplitz matrices.*

REMARK 3 — *The polynomial associated to the Toeplitz operator ϕ is just*

$$T_\phi = \sum_{i=0}^n x_0^{-i}\, P_i.$$

Quasi-Hankel matrices: Let $\lambda \in \hat{L}$ be a linear form on L and consider the matrix

$$[\lambda(\mathbf{x}^{\alpha+\beta})]_{\alpha \in E, \beta \in F}.$$

This is an (E, F) quasi-Hankel matrix. As we will see in the next section, such matrices appear in algebraic residue theory. If B is a Gorenstein algebra (for instance, if B is a complete intersection) and has a finite dimension over $\mathbb{C}$, then any non-degenerating bilinear form q can be represented as $(a, b) \mapsto q(a, b) = \lambda(a\, b)$ where $\lambda \in \hat{L}$ is a linear form (see [6]) and any Gramm-Schmidt matrix, $(q(\mathbf{u}_i, \mathbf{u}_j))$, where $(\mathbf{u}_i)$ is a basis of B, is conjugated to a quasi-Hankel matrix.

3   Algebraic residues

In this section, we will recall some basic definitions from algebraic residue theory, referring the reader to [5], [6] for further details.

3.1   Definitions and basic facts

Let $R = \mathbb{C}[\mathbf{x}] = \mathbb{C}[x_1, \ldots, x_n]$ be the algebra of polynomials in x_i over the field $\mathbb{C}$. In addition to the vector (set) of variables $\mathbf{x}$, we consider the vectors $\mathbf{y} = (y_1, \ldots, y_n)$ and write $\mathbf{x}^{(0)} = \mathbf{x}$, $\mathbf{x}^{(1)} = (y_1, x_2, \ldots, x_n)$, $\ldots$, $\mathbf{x}^{(n)} = \mathbf{y}$. We define $\theta_i(P) = \frac{P(\mathbf{x}^{(i)}) - P(\mathbf{x}^{(i-1)})}{y_i - x_i}$, the *discrete differentiation* of P. For any sequence of $n+1$ polynomials $P_0, \ldots, P_n \in R$, let us construct the following polynomial in $\mathbf{x}$ and $\mathbf{y}$:

$$\Phi(P_0, P_1, \ldots, P_n) = \begin{vmatrix} P_0(\mathbf{x}) & \theta_1(P_0) & \cdots & \theta_n(P_0) \\ \vdots & \vdots & \vdots & \vdots \\ P_n(\mathbf{x}) & \theta_1(P_n) & \cdots & \theta_n(P_n) \end{vmatrix}, \tag{1}$$

and write $\Delta_{\mathbf{P}} = \Phi(1, P_1, \ldots, P_n) \in \mathbb{C}[\mathbf{x}, \mathbf{y}]$. Now, we can define the residue of $\mathbf{P} = (P_1, \ldots, P_n)$ as a unique linear form τ in the set of linear forms on R such that

1. τ vanishes on $(\mathbf{P})$,
2. $\Delta_{\mathbf{P}}(\tau) - 1 \in (\mathbf{P})$.

Hereafter, I will denote the ideal generated by the polynomials $P_1, \ldots, P_n$; $B = R/I$ will denote the quotient ring defined in R by I, and $\equiv$ will denote an equality in B.

If $(\mathbf{x}^\alpha)_{\alpha \in E}$ is a basis of B, then we have the following property:

$$\Delta_{\mathbf{P}} \equiv \sum_{\alpha \in E} \mathbf{x}^\alpha \, \mathbf{w}_\alpha(\mathbf{y}) \equiv \sum_{\alpha \in E} \mathbf{w}_\alpha(\mathbf{x}) \, \mathbf{y}^\alpha \mod (\mathbf{P}(\mathbf{x}), \mathbf{P}(\mathbf{y})).$$

Here, $(\mathbf{w}_\alpha)$ is the dual basis of $(\mathbf{x}^\alpha)$ for τ:

$$\tau(\mathbf{x}^\alpha \, \mathbf{w}_\beta) = \delta_{\alpha,\beta},$$

$\delta_{\alpha,\beta}$ is 1 if $\alpha = \beta$ and 0 otherwise. Thus, for any $b \in B$, we have the relations

$$b \equiv \sum_{\alpha \in E} \tau(b \, \mathbf{x}^\alpha) \, \mathbf{w}_\alpha \equiv \sum_{\alpha \in E} \tau(b \, \mathbf{w}_\alpha) \, \mathbf{x}^\alpha. \tag{2}$$

Moreover, for any polynomial $Q \in R$, we have

$$\Delta_{\mathbf{P}}(\mathbf{x}, \mathbf{y}) \, Q(\mathbf{x}) \equiv \Delta_{\mathbf{P}}(\mathbf{x}, \mathbf{y}) \, Q(\mathbf{y}) \mod (\mathbf{P}(\mathbf{x}), \mathbf{P}(\mathbf{y})). \tag{3}$$

Consequently, for any pair of distinct roots, ζ and η, of the polynomial system $\mathbf{P} = \mathbf{0}$, we have

$$\Delta_{\mathbf{P}}(\zeta, \eta) = 0. \tag{4}$$

3.2   Computation of the residues associated to systems of polynomial equations

We consider a system of polynomial equations of the special form

$$\begin{cases} P_1 = x_1^{d_1} - R_1(x_1, \ldots, x_n) = 0, \\ \vdots \\ P_n = x_n^{d_n} - R_n(x_1, \ldots, x_n) = 0, \end{cases}$$

with $\deg(R_i) < d_i$. Then, the vector space B has dimension $D = \prod_{i=1}^n d_i$, and a basis of B is the set of monomials $\mathbf{x}^\alpha$ with α in the set

$$E = \{(\alpha_1, \ldots, \alpha_n) \in \mathbb{N}^n \; ; \; 0 \le \alpha_i \le d_i - 1\}.$$

In this section, we compute the vector

$$[\tau(\mathbf{x}^\alpha)]_{\nu < |\alpha| \le \mu},$$

where $\nu = \sum_{i=1}^{n} d_i - n$ and $\mu > \nu$. Let $T_l = \{\alpha \in \mathbb{N}^n \text{ such that } |\alpha| \leq l\}$ and $T^{(l)} = \{\alpha \in \mathbb{N}^n \text{ such that } |\alpha| = l\}$. The set T_l contains $\sigma_l := \sigma_{l,n} := \lfloor T_l \rfloor = \binom{l+n}{n}$ elements. Due to Stirling's formula, $\sigma_l = \lfloor T_l \rfloor$ (the cardinality of T_l) is asymptotically equivalent to

$$\mathcal{O}((e\frac{l+n}{n})^n/\sqrt{n}), \ e = 2.718\dots.$$

Let $\mathcal{N}$ denote the set of all monomials $\mathbf{x}^\alpha$ for $\alpha \in T_\mu \backslash T_\nu = \{\alpha \in T_\mu; \alpha \notin T_\nu\}$. An element of $\mathcal{N}$ is divisible by one of the monomials $x_i^{d_i}$, thus $\mathcal{N}$ can be partitioned into n subsets as follows:

$$\mathcal{N} = x_1^{d_1}\mathcal{N}_1 \sqcup \cdots \sqcup x_n^{d_n}\mathcal{N}_n.$$

(for instance, $x_1^{d_1}\mathcal{N}_1$ is the subset of elements of $\mathcal{N}$ divisible by $x_1^{d_1}$, $x_2^{d_2}\mathcal{N}_2$ the subset of elements of $\mathcal{N}$ divisible by $x_2^{d_2}$ and not by $x_1^{d_1}$, $\dots$). Let V_i (resp. V) denote the vector space generated by the monomials in $\mathcal{N}_i$ (resp. $\mathcal{N}$). As in section 2.5, consider the map

$$\phi \ : \ V_1 \times \cdots \times V_n \to V,$$
$$(Q_1, \dots, Q_n) \mapsto \sum_{i=1}^{n} P_i\, Q_i.$$

For any set S in R, let $S^{(l)}$ (resp. $S^{(\leq l)}$) denote the subset of S formed by the elements that are homogeneous of a degree l (resp. of a degree at most l).

Due to the structure of the polynomials of the system $\mathbf{P} = \mathbf{0}$, the image of $(V^{(l-d_1)} \times \cdots \times V_n^{(l-d_n)})$ is in $V^{(\leq l)}$, and the matrix of this map [in the monomial basis of $(V^{(l-d_1)} \times \cdots \times V_n^{(l-d_n)})$ and $V^{(\leq l)}$] has the following form:

$$\begin{pmatrix} U_l \\ \\ I_{\sigma_l} \end{pmatrix} \left. \begin{array}{c} \\ \end{array} \right\} V^{(\leq l-1)} \ \left. \begin{array}{c} \\ \end{array} \right\} V^{(l)}$$

where U_l is a $(T_{l-1}, T^{(l)})$ quasi-Toeplitz matrix, $\sigma_l = \lfloor T_l \rfloor$, and I_{σ_l} is the $\sigma_l \times \sigma_l$ identity matrix. Thus, the complete matrix of ϕ in the monomial basis is a block upper triangular matrix of the size $\lfloor T_\mu \rfloor \times (\lfloor T_\mu \rfloor - \lfloor T_\nu \rfloor)$.

Let M be the submatrix of this matrix whose indices of rows and columns are in $\mathcal{N} = T_\mu \backslash T_\nu$ and let $\mathbf{v} = [v_\alpha]_{\alpha \in N}$ be the row corresponding to the monomial $x_1^{d_1-1} \cdots x_n^{d_n-1}$ in the matrix of ϕ. Then, we can compute the residues based on the next result:

PROPOSITION 31 *(see [6]).* *The following equations hold:*

$$\mathbf{t} = [\tau(\mathbf{x}^\alpha)]_{\alpha \in N} = -\mathbf{v}\, M^{-1}.$$

The matrix M is upper triangular, of size $(\lfloor T_\mu \rfloor - \lfloor T_\nu \rfloor) \times (\lfloor T_\mu \rfloor - \lfloor T_\nu \rfloor)$. Therefore, the vector $\mathbf{t}$ can be computed in $\mathcal{O}((\lfloor T_\mu \rfloor - \lfloor T_\nu \rfloor)^2)$ ops.

By using the quasi-Toeplitz structure of the matrix M, we can obtain a better complexity bound.

PROPOSITION 32 *The vector of the residues, $[\tau(\mathbf{x}^\alpha)]_{\alpha \in N}$, can be computed in $\mathcal{O}((\mu - \nu)\,\sigma \log^3 \sigma)$ ops, where*

$$\sigma = \sigma_{3\mu,n} = \binom{3\mu + n}{n} = \mathcal{O}((e\frac{3\mu + n}{n})^n / \sqrt{n}), \quad e = 2.718\ldots.$$

Proof. Computation of $[\tau(\mathbf{x}^\alpha)]_{\alpha \in N}$ is essentially reduced to solving the following linear system of equations:

$$[\mathbf{t}_{\nu+1}, \ldots, \mathbf{t}_\mu] = [\mathbf{v}_{\nu+1}, \ldots, \mathbf{v}_\mu] \begin{bmatrix} I_{\sigma_{\nu+1}} & U_{\nu+1,\nu+1} & \cdots & U_{\nu,\mu} \\ 0 & I_{\sigma_{\nu+2}} & \ddots & \vdots \\ \vdots & & \ddots & U_{\mu-1,\mu} \\ 0 & \cdots & 0 & I_{\sigma_\mu} \end{bmatrix}. \tag{5}$$

We immediately obtain from the latter system that $\mathbf{t}_{\nu+1} = \mathbf{v}_{\nu+1}$. Substituting this part of the solution vector into the system yields a subsystem in $[\mathbf{t}_{\nu+2}, \ldots, \mathbf{t}_\mu]$, having similar form. We will iterate this process in order to compute the entire vector $\mathbf{v}\,M^{-1}$, by using $\mu - \nu$ iteration steps. At the k-th iteration step, we multiply the vector $\mathbf{t}_{\nu+k}$ by the quasi-Toeplitz matrix $[U_{\nu+k,\nu+k}, \ldots, U_{\nu+k,\mu}]$ and also perform $\lfloor T_\mu \rfloor - \lfloor T_{k+\nu} \rfloor$ vector subtractions. By applying proposition 26, we multiply the vector $\mathbf{t}_{\nu+k}$ by the matrix $[U_{\nu+k,\nu+k}, \ldots, U_{\nu+k,\mu}]$ by using order of

$$\binom{2\nu + 2k + \mu + n}{n} \log^3 \left(\binom{2\nu + 2k + \mu + n}{n} \right)$$

ops, which dominates the computational cost of performing the k-th iteration step. Since $k \leq \mu - \nu$, the entire computation of the vector $(\mathbf{v}_{\nu+1}, \cdots, \mathbf{v}_\mu)$ only requires

$$\mathcal{O}((\mu - \nu)\,\sigma \log^3 \sigma)$$

ops, where $\sigma = \sigma_{3\,\mu,n} = \binom{3\mu + n}{n}$.

Hereafter, we will write $H_0 = [\tau(\mathbf{x}^{\alpha+\beta})]_{\alpha,\beta \in E}$ and $H_i = [\tau(x_i \mathbf{x}^{\alpha+\beta})]_{\alpha,\beta \in E}$. Clearly, H_i are quasi-Hankel matrices for all i. Since $(\mathbf{x}^\alpha)_{\alpha \in \mathbf{E}}$ is a basis in the quotient ring B, we will call H_0 a *basis residue matrix* of B.

PROPOSITION 33 *The matrices $H_i, i = 0, \ldots, n$ can be computed within*

$$\mathcal{O}\left(n^{3.5}\,(6\,e)^n \vartheta^{n+1} \log^3(\vartheta)\right) \text{ ops}$$

where $\vartheta = \frac{\sum_i d_i}{n}$

Proof. In order to compute H_i, we need to compute $\tau(\mathbf{x}^\alpha)$ up to $|\alpha| \leq \mu = 2\sum_{i=1}^n d_i - 2n + 1$. By Stirling formula, we have

$$\sigma_{3\,\mu,n} = \mathcal{O}\left((6\,e\frac{\sum_i d_i}{n})^n / \sqrt{n}\right), \quad e = 2.718\ldots,$$

which implies the previous bound

If all the degrees d_i are equal to each other, that is, if $d_i = d$, $i = 1,\cdots,n$, then, the complexity bound of the latter proposition turns into

$$\mathcal{O}(6^n\,e^n\,d^{n+1}\,n^{3.5}\log^3 d) = \mathcal{O}(\frac{1}{n^3}\,6^n\,e^n\,D^{1+\frac{1}{n}}\log(D))$$

ops, which is asymptotically better than D^2.

3.3   Linear solver with the basis residue matrix H_0

PROPOSITION 34 *The inverse of W_0 is the quasi-Hankel matrix H_0.*

Proof. We have the equations

$$\tau(\mathbf{x}^\alpha\,\mathbf{w}_\beta) = \sum_{\beta \in E} \tau(\mathbf{x}^{\alpha+\beta})\,w_{\beta,\alpha} = \delta_{\alpha,\beta}$$

and

$$H_0\,W_0 = I_D,$$

where I_D is the $D \times D$ identity matrix. Therefore, the inverse of H_0 is the matrix of the coefficients of the dual basis of $(\mathbf{x}^\alpha)$ in this basis.

PROPOSITION 35 *The solution vector $\mathbf{z}$ to the system $H_0\,\mathbf{z} = \mathbf{w}$ can be computed in $\mathcal{O}(3^n\,D^2\log(3^n D))$ ops.*

Proof. We first compute the values $s_k = \mathbf{v}^T\,\tilde{H}_0^k\,\mathbf{u}$, where $\mathbf{u}$ and $\mathbf{v}$ are two random vectors, $\tilde{H}_0 = TH_0$, T is a random square Toeplitz matrix, and $k = 0,\ldots,2D + 1$. We reduce multiplication of the matrix $\tilde{H}_0$ by a vector to multiplication of H_0 by a vector, followed by multiplication of T by the resulting vector. By applying proposition 25, we multiply H_0 by a vector by using $\mathcal{O}(3^n D\log(3^n D))$ ops. Multiplication of the Toeplitz matrix T by a vector takes $\mathcal{O}(D\log D)$ ops (cf. [1]). Therefore, we may evaluate $\tilde{H}_0^k\,\mathbf{u}$, for $k = 0,\ldots,2D + 1$, by using $\mathcal{O}(3^n D^2\log(3^n D))$ ops. Then, the values $s_0,\ldots,s_{2D+1}$ can be computed by using $\mathcal{O}(D^2)$ ops.

Due to our random choice of Toeplitz matrix T, the characteristic and minimal polynomials of $\tilde{H}_0$ coincide with each other, $c_{\tilde{H}_0}(x) = m_{\tilde{H}_0}(x)$, with a sufficiently high probability (cf. [7] and regularization 2.13.3 on page 206 of [1]). Now, assuming that they do coincide and having the values $s_0, \ldots, s_{2D+1}$ available, we will follow [1] and [7] and will compute the coefficients of the characteristic polynomial $c_{\widetilde{H}_0}(\lambda) = \sum_{i=0}^{D} c_i \lambda^i$ of $\widetilde{H}_0$. This computation is reduced to the solution (by using $\mathcal{O}(D \log^2 D)$ ops, cf. [1]) of a Toeplitz linear system of D equations. The latter system is non-singular, with a high probability, due to the equality, with a high probability, of the minimal polynomial of $\widetilde{H}_0$ to the characteristic polynomial of $\widetilde{H}_0$ and to the random choice of the vectors $\mathbf{u}$ and $\mathbf{v}$ (cf. [1] or [7]). Having the coefficients c_i, $i = 0, \ldots, D$, of the characteristic polynomial of the matrix $\widetilde{H}_0$ available and applying the Cayley-Hamilton theorem, we obtain the expression

$$\widetilde{H}_0^{-1} = -\sum_{i=1}^{D} \frac{c_i}{c_0} \widetilde{H}_0^{i-1},$$

where $c_0 = (-1)^D \det(\widetilde{H}_0)$ [$c_0 \neq 0$, with a high probability, since so are $\det(T)$ and $\det(H_0)$]. Thus, computing $H_0^{-1}\mathbf{w}$ requires $D - 1$ multiplications of $\widetilde{H}_0$ by vectors and D additions of vectors. According to proposition 25, these operations can be performed in $\mathcal{O}(3^n D^2 \log(3^n D))$ ops.

3.4 Multiplication in $B = R/I$

Let us write $\mathbf{w}_\alpha = \sum_{\beta \in E} w_{\beta,\alpha} \mathbf{x}^\alpha$ and $W_0 = (w_{\alpha,\beta})_{\alpha, \beta \in E}$.

Let μ_i denote the operator of multiplication by x_i in B and let $M_i = (m_{\alpha,\beta}^{(i)})_{\alpha, \beta \in E}$ denote its matrix in the basis $(\mathbf{x}^\alpha)$. Then, we have

$$x_i \mathbf{x}^\alpha \equiv \sum_{\gamma \in E} m_{\gamma,\alpha}^{(i)} \mathbf{x}^\gamma$$

and

$$\tau(x_i \mathbf{x}^{\alpha+\beta}) = \sum_{\gamma \in E} \tau(\mathbf{x}^{\beta+\gamma}) \, m_{\gamma,\alpha}^{(i)}.$$

In other words, we have

$$H_i = H_0 \, M_i. \qquad (6)$$

Note that the matrix $(n_{\alpha,\beta}^{(i)})$ of multiplication by x_i in the basis $(\mathbf{w}_\alpha)$ is M_i^T, for we have

$$m_{\alpha,\beta}^{(i)} = \langle x_i \mathbf{x}^\beta | \mathbf{w}_\alpha \rangle = \tau(x_i \, \mathbf{x}^\beta \, \mathbf{w}_\alpha)$$

and

$$n_{\alpha,\beta}^{(i)} = \langle x_i \mathbf{w}_\beta | \mathbf{x}^\alpha \rangle = \tau(x_i \, \mathbf{w}_\beta \, \mathbf{x}^\alpha).$$

PROPOSITION 36 *After a precomputation of*

$$\mathcal{O}\left(n^{3.5} \, (9\,e)^n \, \vartheta^{n+1} \log^3(\vartheta)\right) \; ops$$

where $\vartheta = \frac{\sum_i d_i}{n}$, *two elements of B can be multiplied in*

$$\mathcal{O}\left(3^n D^2 \log(3^n D)\right) + 4^n D \log^2(4^n D))$$

ops.

Proof. We want to compute $f g$ in B where

$$f := \sum_{\alpha \in E} f_\alpha \mathbf{x}^\alpha,$$
$$g := \sum_{\alpha \in E} g_\alpha \mathbf{x}^\alpha.$$

According to the equation (2), we have

$$f g = \sum_{\alpha \in E} \tau(f g \mathbf{x}^\alpha) \mathbf{w}_\alpha$$
$$= \sum_{\alpha \in E, \beta \in E, \gamma \in E} \tau(\mathbf{x}^{\alpha+\beta+\gamma}) f_\beta g_\gamma \, \mathbf{w}_\alpha.$$

Let $S := \sum_{u \in 3\,E} \tau(\mathbf{x}^u)\mathbf{x}^{-u} = \sum_{u \in 3\,E} \tau_u \mathbf{x}^{-u}$, with the convention that $\tau_u = 0$ if $u \notin 3\,E$. According to (32), this polynomial can be computed within

$$\mathcal{O}\left(n^{3.5}\,(9\,e)^n\,\vartheta^{n+1}\log^3(\vartheta)\right)$$

ops. This computation has to be done once. Then,

$$S g = \sum_{u \in 3\,E, \gamma \in E} \mathbf{x}^{-u+\gamma} \tau_u g_\gamma$$
$$= \sum_{v = u - \gamma \in 3\,E - E} \mathbf{x}^{-v} \sum_{\gamma \in E,\ v+\gamma \in 3\,E} \tau_{v+\gamma} g_\gamma.$$

The support of this polynomial is in $3\,E - E$, therefore by applying the known interpolation techniques for sparse polynomials ([1], [10]), the product of these two polynomials can be computed within $\mathcal{O}(4^n D \log^2(4^n D))$. Similarly,

$$S f g = \sum_{v \in 3\,E, \beta \in E} \mathbf{x}^{-v+\beta} \left(\sum_{\gamma \in E, v+\gamma \in 3\,E - E} \tau_{v+\gamma} g_\gamma \right) f_\beta$$
$$= \sum_{\alpha = v - \beta \in 3\,E - 2\,E} \mathbf{x}^{-\alpha} \sum_{\beta \in E, \gamma \in E} \tau_{\alpha+\beta+\gamma} f_\beta g_\gamma.$$

Therefore, the coefficients of $\mathbf{x}^{-\alpha}$ in $S f g$ for $\alpha \in E$ are precisely the coefficients of $f g$ in the dual basis $(\mathbf{w}_\alpha)$ of B. Note that these coefficients only involve the coefficients of $\mathbf{x}^{-v}$ in $S g$ for which $v \in 2\,E$, therefore the cost of such a computation is $\mathcal{O}(3^n D \log^2(3^n D))$.

In order to obtain the coefficients of $f\,g$ in the basis $(\mathbf{x}^\alpha)$, we multiply the vector

$$\mathbf{t} = [\sum_{\beta\in E, \gamma\in E} \tau_{\alpha+\beta+\gamma}\, f_\beta\, g_\gamma]_{\alpha\in E}$$

by the matrix $W_0 = [\mathbf{w}_\alpha]_{\alpha\in E} = H_0^{-1}$, that is, we solve the linear system of equations $H_0\,\mathbf{s} = \mathbf{t}$. According to proposition (35), this computation can be done within $\mathcal{O}(3^n\,D^2 \log(3^n D))$ ops.

4   Application to solving polynomial systems of equations

4.1   Computing selected roots of a polynomial system

Let Z denote the set of all common roots of the system $\mathbf{P} = 0$. We will assume that *they are all distinct*. Let J be the Jacobian of $\mathbf{P}$. For any $\zeta \in Z$, we have $J(\zeta) \neq 0$.

PROPOSITION 41   *If the roots of* $\mathbf{P}$ *are simple,*

$$e_\zeta = \frac{1}{J(\zeta)}\Delta_{\mathbf{P}}(\mathbf{x},\zeta),\ \zeta \in Z$$

is a linear basis of orthogonal idempotent of B, *of sum* 1.

Proof. According to the equation (3), for any $Q \in R$ and for any $\zeta \in Z$, we have

$$\Delta_{\mathbf{P}}(\mathbf{x},\zeta)\,Q(\mathbf{x}) \equiv \Delta_{\mathbf{P}}(\mathbf{x},\zeta)\,Q(\zeta)$$

in the quotient ring B. Therefore,

$$\Delta_{\mathbf{P}}(\mathbf{x},\zeta)\,\Delta_{\mathbf{P}}(\mathbf{x},\zeta) \equiv J(\zeta)\Delta_{\mathbf{P}}(\mathbf{x},\zeta),$$

and $e_\zeta = \frac{1}{J(\zeta)}\Delta_{\mathbf{P}}(\mathbf{x},\zeta)$ is an idempotent ($J(\zeta) \neq 0$, assuming all roots of the system $\mathbf{P} = 0$ are distinct). Moreover, according to (4), we have

$$\Delta_{\mathbf{P}}(\mathbf{x},\zeta)\,\Delta_{\mathbf{P}}(\mathbf{x},\eta) \equiv \Delta_{\mathbf{P}}(\mathbf{x},\zeta)\Delta_{\mathbf{P}}(\zeta,\eta) \equiv 0,$$

for any pair of distinct roots $\zeta, \eta \in Z$, which shows that $e_\zeta\, e_\eta \equiv 0$ unless $\zeta = \eta$. Now, we recall from the definition of the residue τ and from the the Euler-Jacobi identity (cf. [6]) that

$$\Delta_{\mathbf{P}}(\tau) = 1 \text{ (by definition)}$$
$$= \sum_{\zeta\in Z} \frac{1}{J(\zeta)}\,\Delta_{\mathbf{P}}(\mathbf{x},\zeta) = \sum_{\zeta\in Z} e_\zeta \text{ (by the Euler}-\text{Jacobi identity).}$$

By decomposing any element h of B in the basis e_ζ, we obtain that

$$h(\mathbf{x}) = \sum_{\zeta \in Z} h(\mathbf{x}) \, e_\zeta \equiv \sum_{\zeta \in Z} h(\zeta) \, e_\zeta.$$

Here, the second equation follows since $e_\zeta h(x) \equiv e_\zeta h(\zeta)$. Squaring h in the quotient ring B gives us

$$h^2 \equiv \sum_{\zeta \in Z} h(\zeta)^2 \, e_\zeta.$$

Here and hereafter, for any element $b \in B$, $[b]$ denotes the vector of the coefficients of b in the basis $(\mathbf{x}^\alpha)_{\alpha \in E}$. In particular, $[1] = (1, 0, \cdots, 0)$ if the basis starts with the monomial 1. Let denote by $|| \cdot ||$ a norm in $\mathbb{C}^D$ [say, for the Euclidean norm,

$$||\mathbf{v}|| = (\mathbf{v}, \mathbf{v}) = \left(\sum_{i=1}^{D} |v_i|^2 \right)^{1/2}, \mathbf{v} = (v_i), \, i = 1, \ldots, D].$$

By abuse of notation, for any element $b \in B$, $||b||$ will denote $||[b]||$. Let $h \in R$ and assume that there is a unique root $\zeta \in Z$, for which the norm of $h(\zeta)$ is maximum, so that

$$|h(\zeta)|/|h(\eta)| - 1 \geq \rho \qquad (7)$$

for some fixed positive ρ and for any $\eta \in Z$ distinct from ζ. (Since all the roots in Z are assumed to be distinct, we may, in principle, ensure the latter relation with a high probability, by means of random linear substitution of the vector of the variables $\mathbf{x}$.) Then, by iteratively computing and normalizing the squares,

$$h_0 = h, \; h_{i+1} \equiv h_i^2/||h_i^2||, \; i = 0, 1, \ldots, k - 1,$$

so that we have

$$\epsilon_k := \left|\left| \frac{h_k}{||h_k||} - \frac{e_\zeta}{||e_\zeta||} \right|\right| \leq \frac{c}{(1 + \rho)^{2^k}}$$

and $\epsilon_k \leq 2^{-b}$ in $k = k(\rho, b) = \mathcal{O}(\log(b/\rho))$ recursive steps. Therefore, squaring and normalizing in B, we will converge to a multiple of the element e_ζ.

From the previous section, we have the bound of $\mathcal{O}(3^n D \log(3^n D) + D^2 \log D)$ ops on the computational cost of squaring in B, which means that

$$\mathcal{O}\left(3^n D^2 \log(3^n D) + 4^n D \log^2(4^n D) \right)$$

ops suffice in order to approximate the element $e_\zeta/||e_\zeta||$ within the error norm bound 2^{-b}, assuming the equation (7).

We refer the reader to [9] and [4] on a similar approach in the univariate case.

4.2   Transition from e_ζ to ζ

By definition, we have

$$e_\zeta = \frac{1}{J(\zeta)}\Delta_\mathbf{P}(\mathbf{x},\zeta) = \frac{1}{J(\zeta)}\sum_{\alpha\in E}\mathbf{w}_\alpha(\mathbf{x})\,\zeta^\alpha.$$

This can be rewritten as

$$[e_\zeta] = \frac{1}{J(\zeta)}W_0\,[\zeta^\alpha]_{\alpha\in E}.$$

Here, W_0 denotes the matrix $(\mathbf{w}_\alpha)_{\alpha\in E}$ in the monomial basis $(\mathbf{x}^\alpha)_{\alpha\in E}$. Then, we obtain that

$$[\zeta^\alpha]_{\alpha\in E} = J(\zeta)\,H_0\,[e_\zeta]. \tag{8}$$

Let $f = H_0 e_\zeta$. Then according to (8), we have $f_0 = 1/J(\zeta)$ and the ith coordinate of ζ is $\frac{f_{x_i}}{f_0}$. According to proposition (35), we arrive at the following result:

PROPOSITION 42   *The transition from e_ζ to the root ζ of the system $\mathbf{P} = 0$ can be performed by using $\mathcal{O}(3^n D^2 \log(3^n D))$ ops.*

4.3   The closest root

Suppose that we seek a root of the system $\mathbf{P} = 0$ for which x_1 is the closest to a given value $u \in \mathbb{C}$. Let us assume that u is not a projection of any root of the system $\mathbf{P} = 0$ and that $x_1 - u$ has reciprocal in B. Let $\rho_1(\mathbf{x})$ denote such a reciprocal. We have $\rho_1(\mathbf{x})(x_1 - u) \equiv 1$ and $\rho_1(\zeta) = \frac{1}{\zeta_1 - u}$. Therefore, a root for which x_1 is the closest to u_1 is a root for which $|\rho_1(\zeta)|$ is the largest. Consequently, iterative squaring of $\rho_1 = \rho_1(\zeta)$ shall converge to this root.

   The polynomial ρ_1 can be computed in the following way. Let μ_1 denote multiplication by x_1 in B. Then $\rho_1 = (\mu_1 - u)^{-1}[1]$. According to the matrix equation (6), we have

$$[\rho_1] = H_0\,(H_1 - u\,H_0)^{-1}[1],$$

which can be computed within $\mathcal{O}(3^n D^2 \log(3^n D))$ ops.

   One may compute several roots of the polynomial system, by applying the latter computation (successively or concurrently) to several initial values u.

5   Conclusion

In this paper, we extend the structure of Toeplitz and Hankel matrices to a new class of structured matrices, which include matrices associated to the polynomial systems of equations and which we call quasi-Hankel and quasi-Toeplitz operators. Exploiting the fact that multiplication of such matrices by vectors is "fast", we devise an algorithm, based on residues computations, for approximating a selected solution of a polynomial system of the form $P_i = x_i^{d_i} - R_i(x_1, \ldots, x_n)$ where R_i has a total degree less than d_i.

The key ingredients of this algorithm are a) the inversion of basis residue matrices b) fast multiplication of two elements of $B = R/I$, c) iterative method to select one root. The complexity analysis leads to a substantial improvement of the known methods for solving polynomial systems of equations. Though we consider a special class of polynomial systems, we expect to extend such an approach to general systems $R_i = 0, i = 1, \ldots, n$, by means of the known techniques of homotopic deformation and the local continuity of the residue.

References

1. D. Bini and V. Y. Pan. *Polynomial and matrix computations, Vol 1 : Fundamental Algorithms*. Birkaüser, Boston, 1994.
2. J. Canny and I. Emiris. An efficient algorithm for the sparse mixed resultant. In G. Cohen, T. Mora, and O. Moreno, editors, *Proc. Intern. Symp. Applied Algebra, Algebraic Algor. and Error-Corr. Codes, Springer's Lecture Notes in Computer Science, vol. 263*, pages 89–104, 1993.
3. J. Canny, E. Kaltofen, and Y. Lakshman. Solving systems of non-linear polynomial equations faster, *Proc. ACM-SIGZAM Intern. Symp. on Symb. and Alg. Comput. (ISSAC'89)*, pages 121-128, ACM Press, New York, 1989.
4. J. P. Cardinal. On two iterative methods for approximating the roots of a polynomial. In J. Renegar, M. Shub, and S. Smale, editors, *Proc. AMS-SIAM Summer Seminar on Math. of Numerical Analysis*, Park City, Utah, July 1995. *Lectures in Applied Math.*, 1996.
5. J. P. Cardinal and B. Mourrain. Algebraic approach of residues and applications. In J. Renegar, M. Shub, and S. Smale, editors, *Proc. AMS-SIAM Summer Seminar on Math. of Numerical Analysis*, Park City, Utah, July 1995. *Lectures in Applied Math.*, 1996.
6. M. Elkadi and B. Mourrain. Approche Effective des Risidus Algébriques. Rapport de Recherche 2884, INRIA, 1996.
7. E. Kaltofen and V.Y. Pan. Processor efficient parallel solution of linear systems over an abstract field. In *Proc. 3rd Ann. ACM Symp. on Parallel Algorithms and Architectures*, pages 180–191. ACM Press, New York, 1991.
8. F. S. Macaulay. Some formulae in elimination. *Proc. London Math. Soc.*, 1(33):3–27, 1902.
9. J. Sebastiao e Sylva. Sur une methode d'approximation semblable a celle de Graeffe. *Portugal. Math.*, 2:271-279, 1941.
10. R. Zippel. Interpolating polynomials from their values. *J. Symbolic Computation*, 9:375–403, 1990.

Numerical Integration of Differential Equations on Homogeneous Manifolds*

Hans Munthe-Kaas and Antonella Zanna

Abstract. We present an overview of intrinsic integration schemes for differential equations evolving on manifolds, paying particular attention to homogeneous spaces. Various examples of applications are introduced, showing the generality of the methods. Finally we discuss abstract Runge–Kutta methods. We argue that homogeneous spaces are the natural structures for the study and the analysis of these methods.

1 Introduction

In the last few years there has been an increasing interest in the design of numerical integration techniques which preserve important qualitative properties of differential equations. This includes the recent work on preserving the symplectic structure of Hamiltonian systems [15], orthogonality [5], isospectrality [2] and on designing numerical methods which stay on a prescribed manifold [3, 4, 6, 7, 11, 12, 13].

In this paper we are concerned with the latter problem, in particular numerical integration of differential equations evolving on homogeneous spaces. The structure of a homogeneous space is both specific enough to allow the definition and the analysis of numerical integration methods, and, at the same time, general enough to include most domains in mathematical modeling. Some examples are:

- Any Lie group; e.g. $SO(n)$ associated with orthogonal flows and $SL(n)$ related to volume preserving flows.
- The classical manifolds: $\mathbb{R}^n$, the n-sphere S^n, projective spaces $RP(n)$, the Stiefel and Grassman manifolds.
- Any group of diffeomorphisms from a manifold $\mathcal{M}$ onto itself, acting transitively on $\mathcal{M}$.

There are two main approaches for integrating differential equations on manifolds, associated with *embedded* and *intrinsic* methods. The first consists of methods where the manifold is embedded in a vector space and a classical integration scheme is employed. The work of Calvo, Iserles and Zanna [3, 7] shows

---

* This work is sponsored by NFR under contract no. 111038/410, through the SYNODE project. WWW: **http://www.imf.unit.no/num/synode**

that except for a few special, yet important cases, it is impossible to devise classical integration schemes such that the numerical solution will stay on the correct manifold. Intrinsic methods are based on natural movements on the manifold, therefore they are independent of the particular embedding and remain on the correct manifold by construction. The main price we pay for this is that we need to compute exponential mappings, either as matrix exponentials or by computing flows of special vectorfields.

In this paper we present a brief overview of the intrinsic methods introduced in [4, 6, 13, 17]. We will focus on the main structures and ideas rather than the technical details, and present several applications.

Finally we will address the question of abstract Runge–Kutta (RK) methods; what are the basic operations needed to formulate RK methods in an abstract setting? This is an important topic in object-oriented program design and a fundamental issue in understanding the foundations of the methods themselves. We will argue that homogeneous spaces are the natural structure to formulate RK methods, and unless we have this structure, it seems difficult to define and analyze RK-type integration schemes.

2   Numerical Integration on Lie Groups

A Lie group is defined as a manifold G equipped with a binary operation $\cdot$: $G \times G \to G$ satisfying the axioms of a group, such that the mappings $(a, b) \mapsto a \cdot b$ and $a \mapsto a^{-1}$ are smooth. We define right multiplications in the group as:

$$R_a(b) = b \cdot a, \quad \text{for } a, b \in G.$$

Examples of Lie groups are matrix Lie groups, where G consist of a set of non-singular matrices and the product is the matrix product. We let $GL(n)$ denote the set of all non-singular $n \times n$ matrices. Another example is given by $\text{Diff}(\mathcal{M})$, the set of diffeomorphisms on a manifold $\mathcal{M}$, where the product is the composition of diffeomorphisms.

The Lie algebra of the group G is defined as its tangent space at the identity, $\mathfrak{g} = T G|_e$, and it has the structure of a (real or complex) linear space equipped with a bilinear skew-symmetric form $[\_, \_] : \mathfrak{g} \times \mathfrak{g} \to \mathfrak{g}$, called the *Lie bracket*. If $G = GL(n)$ then $\mathfrak{g}$ is the space of $n \times n$ matrices and the Lie bracket is the matrix commutator $[u, v] = uv - vu$. If $G = \text{Diff}(\mathcal{M})$, then $\mathfrak{g} = \mathfrak{X}(\mathcal{M})$ is the set of tangent vectorfields on $\mathcal{M}$ and the Lie bracket is given as the Lie–Jacobi bracket of vectorfields, which, given $F, G \in \mathfrak{X}(\mathcal{M})$, yields

$$[F, G]^i = \sum_j \left(F^j \frac{\partial G^i}{\partial x^j} - G^j \frac{\partial F^i}{\partial x^j} \right).$$

The exponential function is a mapping $\exp : \mathfrak{g} \to G$, such that $t \mapsto \exp(tv)$ is a one-parameter subgroup of G. In the case of matrix Lie groups, it reduces to the matrix exponential

$$\exp(tv) = \sum_{j=0}^{\infty} \frac{(tv)^j}{j!},$$

while on $\mathrm{Diff}(\mathcal{M})$ it returns the flow of a given vector field. This can be expressed as

$$\frac{\partial}{\partial t} \exp(tF)(y)\bigg|_{t=0} = F(y).$$

2.1   The Crouch–Grossman (CG) Methods

The starting point for the methods of Crouch and Grossman, as presented in [4, 11], is the introduction of a *frame* on the manifold $\mathcal{M}$, i.e. a set of smooth vectorfields $\{E_1, \ldots, E_d\} \subset \mathfrak{X}(\mathcal{M})$, which at each point $p \in \mathcal{M}$ span the tangent space $T\mathcal{M}|_p$. It follows that an ordinary differential equation on $\mathcal{M}$ can be written as

$$y' = F(y) = \sum_{i=1}^{d} E_i(y) f_i(y),$$

where $f_i : \mathcal{M} \to \mathbb{R}$ are smooth scalar functions. We denote by F_p the vectorfield F with coefficients frozen at p relative to the frame:

$$F_p(y) = \sum_{i=1}^{d} E_i(y) f_i(p). \tag{1}$$

The explicit s-stage Crouch–Grossman method with stepsize h is given as:

Algorithm 1 (RK–CG):

$p = y_k$
$\nu_1 = p$
for $r = 2, \ldots, s$
$\quad \nu_r = \exp(ha_{r,r-1}F_{\nu_{r-1}}) \circ \exp(ha_{r,r-2}F_{\nu_{r-2}}) \circ \cdots \circ \exp(ha_{r,1}F_{\nu_1})(p)$
end
$y_{k+1} = \exp(hb_sF_{\nu_s}) \circ \exp(hb_{s-1}F_{\nu_{s-1}}) \circ \cdots \circ \exp(hb_1F_{\nu_1})(p)$

where the exponentiation of a vectorfield denotes its flow.

If $\mathcal{M} = G$ is a matrix Lie group and $E_i(y) = v_iy$ are right invariant vectorfields generated by a basis $\{v_1, \ldots, v_d\}$ spanning $\mathfrak{g}$, we can compute the flows via the matrix exponential

$$\exp\left(\sum_{i=1}^{d} \alpha_i E_i\right)(y) = \exp\left(\sum_{i=1}^{d} \alpha_i v_i\right) y.$$

Crouch and Grossman construct methods of order three. Owren and Marthinsen [11] develop a systematic order theory for this type of methods. They present a fourth-order method and prove that five stages are required to obtain order four. Due to non-commutative effects, the number of order conditions for this class of methods grows faster than for classical Runge–Kutta schemes. It is therefore difficult to construct higher order methods of this type, and it is at the moment unknown if methods of order five exist.

2.2   The Munthe-Kaas (MK) Methods

The Runge–Kutta-type methods of Munthe-Kaas were introduced in [12] and further developed in [13]. The main difference between the CG and MK methods is that whereas CG perform the approximations in the Lie group, MK does them in the Lie algebra. In other words, CG advance by a composed product of exponentials, while MK first combines elements in the Lie algebra, and then advances by a single exponential mapping. It turns out that the order theory for the MK approach is much simpler than for the CG approach due to the fact that, whereas operations in the Lie group are nonlinear, the Lie algebra is a linear space.

A general initial value problem on a Lie group can be written as

$$y'_t = R'_{y_t}(f(y_t)), \quad y_0 = p, \tag{2}$$

where $f : G \to \mathfrak{g}$. If G is a matrix Lie group, (2) becomes

$$y'_t = f(y_t)y_t, \quad y_0 = p. \tag{3}$$

The general s-stage MK methods with stepsize h for the initaial value problems (2),(3) is:

Algorithm 2 (RK–MK):

$y_0 = p$
for $n = 0, 1, \ldots$
    for $i = 1, 2, \ldots, s$
        $u_i = h \sum_{j=1}^s a_{i,j} k_j$
        $\tilde{u}_i = \zeta_i(u_i, k_1, k_2, \ldots, k_s)$
        $k_i = f(\exp(\tilde{u}_i) \cdot y_n)$
    end
    $v = h \sum_{j=1}^s b_j k_j$
    $\tilde{v} = \zeta(v, k_1, k_2, \ldots, k_s)$
    $y_{n+1} = \exp(\tilde{v}) \cdot y_n$
end

where $u_i, \tilde{u}_i, k_i, v, \tilde{v} \in \mathfrak{g}$ and $y_i \in G$. The real constants $a_{i,j}, b_j$ and the correction functions $\zeta_i()$ and $\zeta()$ determine a particular scheme.

In [13] the order theory for such methods is developed in terms of a *Lie-Butcher* series for the numerical and the analytical solution. The starting point of the analysis is the identification of the tangent space $TG|_p$ at an arbitrary point p with $\mathfrak{g}$ by means of the right multiplication. (In other words, TG and $\mathfrak{g}$ are related by the canonical right-invariant Maurer–Cartan form on G). In this respect, any curve on G may be lifted to a curve on $\mathfrak{g}$, and the order conditions arise by comparing terms in the Lie–Butcher series for the analytical and the numerical solution on $\mathfrak{g}$. The main result of [13] is the following:

Theorem 1 *If the coefficients $a_{i,j}$ and b_j satisfy the classical Runge–Kutta order conditions up to order q, and the correction functions $\zeta_i()$ and $\zeta()$ satisfy certain conditions given in [13], then Algorithm 2 defines a q'th order method on an arbitrary Lie group.*

An important consequence of the above result is that the task of computing the correction functions is independent of the task of finding the coefficients $a_{i,j}$ and b_j, thus when the correction functions are constructed to order q, any classical RK method of the same order can be turned into a MK-type method of order q on an arbitrary Lie group. In Algorithm 3, presented in Sect. 3, we give the correction functions for order four.

2.3 The Method of Iterated Commutators

The method of iterated commutators, introduced by Iserles in [6], is given a lengthy treatment in another paper in this volume [18] as well as in [17]. We will therefore not repeat the algorithm here, but just state its basic properties. In its original form it is presented as an algorithm for solving linear matrix differential equations with variable coefficients. It is naturally generalized to Lie-group differential equations of the form

$$y_t' = R_{y_t}'(f(t)), \quad y_0 = p,$$

where $f : \mathbb{R} \to \mathfrak{g}$ depends on time but not on space. Such equations are called *equations of Lie type*. The method advances by a product of exponential mappings, and in that respect it somehow resembles the RK–CG algorithm. Its theoretical explanation is, however, very different from both RK–CG and RK–MK schemes. The method can be explained as a numerical implementation of the technique of Lie reduction. In the case where G is solvable, the Lie reduction technique can be used to express the solution in terms of quadratures. The numerical technique is based on numerical quadratures and it works also for non-solvable groups. Zanna [17] has recently generalized this approach to the general case when $f : G \to \mathcal{M}$.

3   Homogeneous Spaces

A *left action* of a Lie group G on a manifold $\mathcal{M}$ is a smooth map $\lambda : G \times \mathcal{M} \to \mathcal{M}$ such that

$$\lambda(e, m) = m,$$
$$\lambda(g_1 \cdot g_2, m) = \lambda(g_1, \lambda(g_2, m)), \quad g_1, g_2 \in G, \quad m \in \mathcal{M},$$

where e denotes the identity element in G. We henceforth assume that the action is *transitive*, which means that for any two points $m_1, m_2 \in \mathcal{M}$ there exist a $g \in G$ such that $\lambda(g, m_1) = m_2$. A manifold with a transitive Lie group action is called a *homogeneous space*.

For a given fixed point $p \in \mathcal{M}$, let G_p denote the *stabilizer* of p,

$$G_p = \{\, g \in G \mid \lambda(g, p) = p \,\}.$$

The set G_p is a subgroup of G. A fundamental result [1] states that G/G_p and $\mathcal{M}$ are naturally diffeomorphic, where G/G_p denotes the left cosets of G_p in G. Hence, we may equivalently define a homogeneous space as the quotient of two Lie groups.

As for Lie groups we used the right multiplication to identify $\mathfrak{g}$ with $TG|_p$, we now use λ to map $\mathfrak{g}$ onto $T\mathcal{M}|_p$. Let $\lambda_p : G \to \mathcal{M}$ be defined as $\lambda_p(g) = \lambda(g, p)$. Then,

$$\lambda_p{}' : \mathfrak{g} \to T\mathcal{M}|_p, \qquad v \mapsto \lambda_p{}'(v) = \left. \frac{\partial}{\partial t} \lambda(\exp(tv), p) \right|_{t=0}. \tag{4}$$

This mapping is surjective since λ is transitive. It is, however, not injective. If $\mathfrak{g}_p \subset \mathfrak{g}$ denotes the Lie algebra of G_p one may verify that $T\mathcal{M}|_p \simeq \mathfrak{g}/\mathfrak{g}_p$. Thus two vectors $v, w \in \mathfrak{g}$ correspond to the same direction in $T\mathcal{M}|_p$ if and only if $v - w \in \mathfrak{g}_p$.

We now have all the necessary tools to define the order for methods on $\mathcal{M}$, and to derive the order conditions. It is a tedious work, yet straightforward, to check that all definitions and results which in [13] are stated for integration on G, do hold for integration on $\mathcal{M}$. The correspondence is obtained by replacing the right multiplication on G, $R_p(g) = g{\cdot}p$, where both $g, p \in G$, with the action $\lambda_p(g) = \lambda(g, p)$, where $g \in G, p \in \mathcal{M}$. The general initial value problem on $\mathcal{M}$ can be written as:

$$y_t' = \lambda_{y_t}{}'(f(y_t)), \qquad y_0 = p, \tag{5}$$

where $f : \mathcal{M} \to \mathfrak{g}$.

Algorithm 3 is a fourth-order integration scheme for such equations. The correction functions ζ_i and ζ are given for order four.

Algorithm 3 Explicit fourth-order RK–MK on homogeneous spaces:

Choose the coefficients $a_{i,j}$ and b_j of a classical s-stage, fourth-order explicit RK method. Let $c_i = \sum_{j=1}^{s} a_{i,j}$, and $d_i = \sum_{j=1}^{s} a_{i,j} c_j$. Compute the coefficients (m_1, m_2, m_3) by solving the linear system:

$$(m_1 \ m_2 \ m_3) \begin{pmatrix} c_2 & c_2^2 & 2d_2 \\ c_3 & c_3^2 & 2d_3 \\ c_4 & c_4^2 & 2d_4 \end{pmatrix} = (1 \ 0 \ 0).$$

Then the fourth-order RK–MK algorithm for (5) is given by

$y_0 = p$
for $n = 0, 1, 2, \ldots$
$\quad I_1 = k_1 = f(y_n)$
$\quad$ for $i = 2, \ldots, s$
$\quad\quad u_i = h \sum_{j=1}^{i-1} a_{i,j} k_j$
$\quad\quad \tilde{u}_i = u_i - \frac{c_i h}{6} [I_1, u_i]$
$\quad\quad k_i = f(\lambda(\exp(\tilde{u}_i), y_n))$
$\quad$ end
$\quad I_2 = (m_1(k_2 - I_1) + m_2(k_3 - I_1) + m_3(k_4 - I_1))/h$
$\quad v = h \sum_{j=1}^{s} b_j k_j$
$\quad \tilde{v} = v - \frac{h}{4}[I_1, v] - \frac{h^2}{24}[I_2, v]$
$\quad y_{n+1} = \lambda(\exp(\tilde{v}), y_n)$
end

We can for instance base the algorithm on the classical four-stage fourth-order RK scheme whose coefficients are

$$a_{2,1} = \tfrac{1}{2}, \ a_{3,2} = \tfrac{1}{2}, \ a_{4,3} = 1, \text{ all other } a_{i,j} = 0,$$
$$b_1 = \tfrac{1}{6}, \quad b_2 = \tfrac{1}{3}, \quad b_3 = \tfrac{1}{3}, b_4 = \tfrac{1}{6},$$

to find

$$c_1 = 0, \ c_2 = \tfrac{1}{2}, \ c_3 = \tfrac{1}{2}, \ c_4 = 1,$$
$$d_1 = 0, \ d_2 = 0, \ d_3 = \tfrac{1}{4}, \ d_4 = \tfrac{1}{2},$$
$$m_1 = 2, m_2 = 2, m_3 = -1.$$

The RK–MK algorithms are abstractly formulated, hence admit several different interpretations. Some of them will be presented in the next section.

4   Examples

4.1   The Classical Case $\mathbb{R}^n$

The real vector space $\mathbb{R}^n$ is a Lie group where the group product is the sum of vectors. The Lie algebra is also $\mathbb{R}^n$, the Lie bracket is given as $[u, v] = 0$ and the exponential as $\exp(u) = u$. The initial value problem (2) becomes

$$y'_t = f(y_t), \quad \text{for } f : \mathbb{R}^n \to \mathbb{R}^n.$$

We may verify that the RK–CG and RK–MK schemes both reduce to their classical RK counterpart and that the method of iterated commutators in this case becomes solution of $y' = f(t)$ by a single quadrature.

4.2   The n-Sphere

Let $G = SO(n+1)$ be the Lie group of real orthogonal $(n+1) \times (n+1)$ matrices with determinant equal to one. Its Lie algebra $\mathfrak{g} = \mathfrak{so}(n+1)$ is the set of skew-symmetric matrices. The n-sphere $\mathcal{M} = S^n = \{ y \in \mathbb{R}^{n+1} \mid y^T y = 1 \}$ is the homogeneous space related to the action given by the matrix-vector product $\lambda(A, y) = Ay$. Thus, $\lambda_y'(V) = Vy$, and the general initial value problem on the sphere is

$$y_t' = f(y_t)y_t, \quad y_0 = p,$$

where $f : S^n \to \mathfrak{so}(n+1)$.

4.3   Isospectral Flows

This topic is covered thoroughly in [2]. Let m_0 be a symmetric $n \times n$ matrix[2], $m_0 \in \mathrm{Sym}(n)$. The matrix differential equation

$$y' = [f(y), y] = f(y)y - yf(y), \quad y_0 = m_0, \tag{6}$$

where $f(y) \in \mathfrak{so}(n)$, is an example of an isospectral flow evolving in $\mathrm{Sym}(n)$. The adjective 'isospectral' reflects the fact that the eigenvalues of y are invariant under the flow. We define the following submanifold of $\mathrm{Sym}(n)$, $\mathcal{M} = \{ y = am_0a^T \mid a \in SO(n) \}$, and define the action $\lambda : SO(n) \times \mathcal{M} \to \mathcal{M}$ as

$$\lambda(a, y) = aya^T.$$

The action λ is transitive. Furthermore, from (4), using the fact that in $a^T = a^{-1}$, we get

$$\lambda_y'(v) = \frac{\partial}{\partial t} \exp(tv)y\exp(-tv)\Big|_{t=0} = vy - yv = [v, y],$$

thus (6) is also a special case of (5), and Algorithm 3 can be used also here.

4.4   Integration Methods Based on Frames

In this section we intend to show that the methods based on frames fit in the formalism of homogeneous spaces introduced in Sect. 3.

Let $\mathfrak{g}$ be the Lie algebra generated by the frame $\{E_1, \ldots, E_d\} \subset \mathfrak{X}(\mathcal{M})$ and let $G \subset \mathrm{Diff}(\mathcal{M})$ be the collection of flows on $\mathcal{M}$ generated by exponentiating $\mathfrak{g}$. We define the action $\lambda : G \times \mathcal{M} \to \mathcal{M}$ as

$$\lambda(\phi, y) = \phi(y).$$

We wish to solve

$$y' = \sum_{i=1}^{d} E_i f_i(y) = F_y(y), \tag{7}$$

---

[2] The set $\mathrm{Sym}(n)$ is a manifold, but not a Lie group.

where $F_y \in \mathfrak{g}$ is the vectorfield of frozen coefficients in (1). The map

$$F_p : \mathcal{M} \to \mathfrak{g}, \qquad p \mapsto F_p,$$

is of the same form as f in (5). Since

$$\lambda_y'(F_p) = \frac{\partial}{\partial t} \exp\left(tF_p\right)(y)\Big|_{t=0} = F_p(y),$$

it follows that (7) can be written as

$$y' = \lambda_y'\left(F_y\right),$$

which is of the general form (5). This shows that Algorithm 3 can also be interpreted in this context. The central equations become

$$k_i = f\left(\lambda(\exp(\tilde{u}_i), y_n)\right) = F_{\exp(\tilde{u}_i)(y_n)}$$
$$y_{n+1} = \lambda\left(\exp(\tilde{v}), y_n\right) = \exp(\tilde{v})(y_n),$$

and the brackets such as $[I_1, u_i]$ are Lie–Jacobi brackets between vectorfields.

4.5 Partial Differential Equations of Evolution

Several important partial differential equations of evolution arising in Physics can be written as a differential equation on (a subgroup of) the Lie group $\text{Diff}(\mathcal{M})$. Given $\psi, \phi \in \text{Diff}(\mathcal{M})$, we let $\lambda(\psi, \phi) = \psi \circ \phi$. Therefore,

$$\lambda_\phi'(f) = \frac{\partial}{\partial t} \exp\left(tf\right)(y)\Big|_{t=0} \circ \phi = f \circ \phi,$$

where $f \in \mathfrak{X}(\mathcal{M})$. Thus (5) becomes

$$\phi_t' = f(\phi_t) \circ \phi_t.$$

The wave equation in elasticity, where ϕ is the configuration of the body (departure from initial position), the Poisson–Vlasov equations in plasma physics and the Euler flows in fluid dynamics are examples that fit naturally into the above description. The Euler flows evolve in the subgroup of volume-preserving diffeomorphisms on $\mathcal{M}$. A thorough treatment of these examples can be found in [9, 10]. It is possible to interpret Algorithm 3 in this context, although the numerical consequences of this interpretation is much more subtle. Since we cannot represent an infinite dimensional Lie group exactly, both the bracket and the exponential map have to be approximated. One procedure is, for instance, to represent $\mathfrak{X}(\mathcal{M})$ on a grid, and in this light the Lie-group approach becomes a version of the well known *method of lines*, where the spatial discretization is hidden for the time integration routine.

Comparing the Lie-group approach and the method of lines, we may gain important insight into strategies for discretizing partial differential equations, in such a way that important qualitative features are preserved (e.g. volume preservation for Euler flows). Secondly, it follows that the time integration can be performed independently of the spatial discretization. This suggests an object-oriented approach to integration of PDEs, whereby we can hide specific spatial discretization techniques within some appropriate classes.

5   Abstract RK Methods

One of our initial motivations for studying integration on manifolds, was an investigation of object-oriented programming techniques applied to the field of numerical computing. The main issue in object-oriented programming is the separation between 'what' and 'how', in other words, between *specification* and *implementation*. Formal specifications are a tool for finding a modularization of the program into classes. Ideally, the specification of a class should express a mathematical structure independently of a particular implementation. The implementation itself should be hidden within the class so that it can be later replaced with another implementation without doing modifications to the other classes. A numerical discretization is a matter of choosing an implementation, and it belongs to the 'how'-part of the programming. If we further follow this line of thought, we may conclude that also the choice of coordinate systems for a differential equation is a matter of 'how' rather than 'what'. A consequence of this philosophy is that it is important to search for *coordinate-free numerical techniques* [14], i.e. numerical algorithms that can be expressed independently of particular coordinate systems.

The examples we have presented in this paper display that RK-type methods may be expressed in an abstract coordinate-free manner involving the following basic structures:

- A Lie algebra defined as a vector space with a Lie bracket $[\_, \_]$.
- A Lie group G with a multiplication rule.
- An exponential map $\exp : \mathfrak{g} \to G$.
- A homogeneous space $\mathcal{M}$ and an action $\lambda : G \times \mathcal{M} \to \mathcal{M}$.
- A differential equation (5) defined by a function $f : \mathcal{M} \to \mathfrak{g}$.

Using modern computer languages (for instance C++), it is possible to program the time integration process in this abstract manner, such that particular implementations of these objects are subject to change.

A question is whether all the elements of this abstraction are necessary for expressing RK-type time integration schemes. The question is, at present, without a precise answer. However the principal component of the abstraction is that we have a domain $\mathcal{M}$ equipped with some basic continuous movements that can take us in all possible directions from every point. It seems difficult to give meaning to RK-type methods if we do not have such a structure.

6   Concluding Remarks

It is our feeling that this problem of integrationg differential equations on manifolds is becoming more (and better) understood. We believe that the Lie-group approach is indeed a big steep in this direction. Yet the basic ideas that we have presented are still in early stage and devote further analysis. Many other issues, like stability, and error control etc. need to be addressed in the future work.

References

1. Bryant, R.L.: An introduction to Lie groups and symplectic geometry. In Geometry and Quantum Field Theory, (D.S. Freed and K.K. Uhlenbeck eds) AMS IAS/Park City Math. Series, vol. 1, 1995.
2. Calvo, M.P., Iserles, A., Zanna, A.: Numerical solution of isospectral flows, DAMTP Techn. Rep. NA03/95, University of Cambridge, 1995.
3. Calvo, M.P., Iserles, A., Zanna, A.: Runge–Kutta methods on manifolds. In A.R. Mitchell 75th Birthday Volume, (D.F. Griffiths & G.A. Watson eds), World Scientific, Singapore, 57–70, 1996.
4. Crouch, P.E., Grossman, R.: Numerical integration of ordinary differential equations on manifolds, J. Nonlinear Sci., **3** (1993), 1-33.
5. Dieci L., Russel D.R., van Vleck E.S.: Unitary integrators and applications to continuous orthonormalization techniques, SIAM J. Numer. Anal., **31** (1994), 261–281.
6. Iserles, A.: Solving ordinary differential equations by exponentials of iterated commutators, Numer. Math., **5** (1984), 183–199.
7. Iserles, A.: Numerical methods on (and off) manifolds. This volume.
8. Iserles, A., Zanna, A.: Qualitative numerical analysis of ordinary differential equations, DAMTP Tech. Rep. 1995/NA05, to appear in Lectures in Applied Mathematics (J. Renegar, M. Shub & S. Smale eds), AMS, Providence RI, 1995.
9. Marsden J.E., Hughes J.R.: Mathematical Foundations of Elasticity, Prentice-Hall, 1983.
10. Marsden J.E., Ratiu T.S.: Introduction to Mechanics and Symmetry, Springer-Verlag, New-York, 1994.
11. Marthinsen, A., Owren, B.: Order conditions for integration methods based on rigid frames, to appear.
12. Munthe-Kaas, H.: Lie–Butcher Theory for Runge–Kutta Methods, BIT, **35** (1995), 572–587.
13. Munthe-Kaas, H.: Runge–Kutta methods on Lie groups. Submitted to BIT.
14. Munthe-Kaas H., Haveraaen M.: Coordinate Free Numerics – Closing the gap between 'Pure' and 'Applied' mathematics? In proc. of ICIAM -95, Zeitschrift fuer Angewandte Mathematik und Mechanik, Akademie Verlag, Berlin, 1996.
15. Sanz–Serna J.M., Calvo M.P.: Numerical Hamiltonian Problems, Chapman & Hall, London, 1994.
16. Varadarajan, V.S.: Lie Groups, Lie Algebras, and Their Representations, Springer-Verlag, New York, 1984.
17. Zanna, A.: The method of iterated commutators for ordinary differential equations on Lie groups, DAMTP Techn. Rep. NA12/96, University of Cambridge, 1996.
18. Zanna, A., Munthe-Kaas, H.: Iterated Commutators, Lie's Reduction Method and Ordinary Differential Equations on Matrix Lie Groups. This volume.

A Convergence proof of an Iterative Subspace Method for Eigenvalues Problems [*]

Suely Oliveira

Abstract. The generalized Davidson algorithm can be seen as a method which uses preconditioned residuals to create a subspace where it is easier to find the smallest eigenvalue and its eigenvector. In this paper theoretical results proving convergence rates are shown. In addition, we investigate the use of multigrid as a preconditioner for this method and describe a new algorithm for calculating some other eigenvalue–eigenvector pairs as well, while avoiding problems of misconvergence. The advantages of implicit restarts are also investigated.

1   Generalized Davidson Algorithm

The Davidson algorithm [6] and its variants provide a means to calculate eigenvalues of large, sparse, or structured symmetric matrices. While it has been known to, and used by, quantum chemists, this algorithm has only recently come to the attention of mathematicians through the work of Saad [14] and others. In this paper the term "Davidson algorithm" is used to mean the generalized Davidson algorithm which allows any matrix M_λ as a preconditioner, rather than restricting the choice of preconditioner to the diagonal preconditioner in [6]. The Generalized Davidson algorithm is shown in Figure 1.

Quantum chemists have frequently studied matrices that are strongly diagonally dominant, and so the "diagonal preconditioner" $M_\lambda = (D - \lambda I)^{-1}$ with $D = \text{diag}(A)$ gives extremely rapid convergence. For more general cases we use other preconditioners. Numerical results with a multigrid preconditioner applied to a PDE eigenproblem are shown in the last Section.

In the Davidson method, if the preconditioner used is $M_\lambda = \alpha I$, $\alpha \neq 0$ the method is equivalent to the Lanczos algorithm in exact arithmatic. In fact, with this preconditioner, the Davidson algorithm is essentially an expensive implementation of the Lanczos algorithm with complete reorthogonalization [7, §9.2.3]. Both the Lanczos and Davidson algorithms involve the computation of an orthonormal basis for a subspace from a given starting vector x_0. For the Davidson algorithm the subspace also depends on the preconditioner used. The

[*] This research is supported by NSF grant ASC-9528912 and a Texas A&M University Interdisciplinary Research Initiative Award.

Lanczos method generates successive bases for Krylov subspaces

$$\text{span}\{\, q_0, q_1, \ldots, q_{k-1} \,\} = K_k(x_0, A) := \text{span}\{\, x_0, Ax_0, \ldots, A^{k-1}x_0 \,\}.$$

On the other hand, the Davidson algorithm generates an orthonormal basis $\{\, v_0, v_1, \ldots, v_{k-1} \,\}$ for subspaces which depend on A and x_0 and the M_λ matrices. From these orthonormal bases, "approximating matrices" are computed by both methods: $T_k = Q_k^T A Q_k$ with $Q_k = [q_0, q_1, \ldots, q_{k-1}]$ for the Lanczos algorithm, and $S_k = V_k^T A V_k$ with $V_k = [v_0, v_1, \ldots, v_{k-1}]$ for the Davidson algorithm.

Algorithm 1 – Generalized Davidson Algorithm
Given an initial vector $x_1 \neq 0$ and iteration limit m, eigenvalue number p, and convergence tolerance ϵ, compute V_k, S_k, $\widehat{\lambda}_k$ and u_k.

> *Set $v_1 \leftarrow x_1/\|x_1\|_2$.*
> *$V_1 \leftarrow [v_1]$; $W_0 \leftarrow [\,]$.*
> **for** $k = 1, \ldots, m$ **do**. . .
> > *$w_k \leftarrow Av_k$.*
> > *$W_k = [W_{k-1}, w_k]$.*
> > *Compute $V_k^T w_k$ and make it the last column & row of*
> > *$S_k = V_k^T A V_k = V_k^T W_k$.*
> > *$\ell \leftarrow \min(k, p)$*
> > *Compute the ℓth smallest eigenpair $\widehat{\lambda}_k$, y_k of S_k: $S_k y_k = \widehat{\lambda}_k y_k$, $y_k \neq 0$.*
> > *(This can be done using the QR algorithm.)*
> > *$u_k \leftarrow V_k y_k$ (Ritz vector).*
> > *$r_k \leftarrow Au_k - \widehat{\lambda}_k u_k = W_k y_k - \widehat{\lambda}_k V_k y_k$ (residual vector).*
> > *If $\|r_k\|_2 < \epsilon$ then exit loop.*
> > *If $k < m$*
> > > *$t_k \leftarrow M_{\widehat{\lambda}_k} r_k$.*
> > > *$v_{k+1} \leftarrow mgs(V_k, t_k)$ where $mgs(V_k, t_k)$ is the result of applying the modified Gram–Schmidt process to orthonormalize t against the columns of V_k.*
> > > *$V_{k+1} \leftarrow [V_k, v_{k+1}]$*

By the nature of Krylov subspaces and the symmetry of A, it turns out that T_k is a tridiagonal matrix, while for the Davidson algorithm the S_k matrices are usually dense, with little structure beyond symmetry.

Previously convergence results have been obtained for the Davidson algorithm (e.g., [5, 14]) but without proving any convergence *rate*. In this work we give a bound on the rate of convergence for Davidson type algorithms. This bound depends on the preconditioner M and the original matrix A.

The connection between the quality of the preconditioner and the convergence rate of the generalized Davidson algorithm has not been shown elsehwere. Crouzeix, Philippe and Sadkane [5] gave a convergence proof for the generalized Davidson method which does not show formal results on the rate of convergence.

They assume that the preconditioners are uniformly bounded and uniformly positive definite and give qualitative advice on properties for a good preconditioner for the Davidson algorithm.

In Section 3 we develop an algorithm which is a variant of the Davidson algorithm for computing the pth eigenvalue of A (starting from the most negative eigenvalue) for small to modest p. Rather than use the pth eigenvalue of S_k for $p \le k$, eigenvalue and eigenvector pairs are found *in order*. The method used to derive the rate of convergence for the smallest eigenvector in the Davidson algorithm can be used to obtain the rate of convergence for this variant as well. This new algorithm avoids problems of misconvergence, i.e. convergence to an eigenvalue other than the one currently sought, that can occur with the Davidson algorithm and related algorithms such as inverse iteration.

In Section 3 we also incorporate the use of implicit restarts for this class or algorithms. While this increases the number of iterations needed for a given number of eigenvalues at a given accuracy, the numerical results indicate that this increase is very small, and that implicitly restarted methods are quite efficient.

2　Proof of rate of convergence

Let A be the given matrix whose eigenvalues and eigenvectors are wanted. The preconditioner M is given for one step, and Davidson's algorithm is used with u_k being the current computed approximate eigenvector. The current eigenvalue estimate is the Rayleigh quotient $\widehat{\lambda}_k = \rho_A(u_k) = (u_k^T A u_k)/(u_k^T u_k)$. Let the exact eigenvector with the smallest eigenvalue of A be u, and

$$Au = \lambda u.$$

(If λ is a repeated eigenvalue of A, then we can let u be the normalized projection of u_k onto this eigenspace.)

Theorem 1 *Let P be the orthogonal projection onto $ker(A - \lambda I)^{\perp}$. Suppose that A and M are symmetric positive definite. If*

$$\|P - PMP(A - \lambda I)\|_2 \le \sigma < 1,$$

then for almost any starting value x_1, the convergence of the eigenvalue estimates $\widehat{\lambda}_k$ converge to λ ultimately geometrically with convergence factor σ^2, and the angle between the computed eigenvector and the exact eigenspace goes to zero ultimately geometrically with convergence factor σ.

Proof.. The proof of convergence (but without a rate estimate) is given in Crouzeix, Philippe and Sadkane [5] which also applies under these conditions. Therefore we assume that k is sufficiently large and that $u_k = \alpha_k u + \beta_k s_k$ with β_k "sufficiently small". Set $\|u_k\|_2 = \|u\|_2 = \|s_k\|_2 = 1$ and $s_k \perp u$, so $\alpha_k^2 + \beta_k^2 = 1$.

The current estimate of λ is $\widehat{\lambda}_k = \rho_A(u_k) = (u_k^T A u_k)/(u_k^T u_k)$. Then

$$\begin{aligned} \widehat{\lambda}_k &= \lambda \alpha_k^2 + 2\alpha_k \beta_k \lambda u^T s_k + \beta_k^2 s_k^T A s_k = \lambda \alpha_k^2 + \beta_k^2 s_k^T A s_k \\ &= \lambda + \beta_k^2 (s_k^T A s_k - \lambda). \end{aligned} \qquad (1)$$

The residual vector is

$$r_k = Au_k - \widehat{\lambda}_k u_k = \beta_k(As_k - \lambda s_k) + \alpha_k \beta_k^2(\lambda - s_k^T As_k)u + \beta_k^3(\lambda - s_k^T As_k)s_k$$
$$= \beta_k(A - \lambda I)s_k + O(\beta_k^2).$$

Therefore the vector added to the Davidson subspace is

$$t_k = Mr_k = \beta_k M(A - \lambda I)s_k + O(\beta_k^2).$$

The new eigenvalue and eigenvector estimates are the smallest Ritz value and its Ritz vector, respectively, for the expanded subspace. Thus the new eigenvalue estimate $\widehat{\lambda}_{k+1} \leq \rho_A(\widehat{u}_k)$ where $\widehat{u}_k$ is a linear combination of u_k and t_k. The vector $\widehat{u}_k$ is chosen in such a way that the convergence rate of the method is more easily demonstrated. The remaniner of the proof is shown in [13]. There we show that the convergence of the eigenvalue estimates is geometric with convergence factor σ^2:

$$\frac{\widehat{\lambda}_{k+1} - \lambda}{\widehat{\lambda}_k - \lambda} \leq \sigma^2 + O(\beta_k). \tag{2}$$

If $\lambda_2 > \lambda$ is the second *distinct* eigenvalue, using 1 we get $\rho_A(u_k) \geq \lambda + (\lambda_2 - \lambda)\beta_k^2$, or $\beta_k^2 \leq (\widehat{\lambda}_k - \lambda)/(\lambda_2 - \lambda)$. Thus

$$|\beta_{k+1}| \leq \sqrt{\frac{\widehat{\lambda}_{k+1} - \lambda}{\lambda_2 - \lambda}} \leq (\sigma + O(\beta_k))\sqrt{\frac{\widehat{\lambda}_k - \lambda}{\lambda_2 - \lambda}}$$

which shows that $\beta_k \rightarrow 0$ geometrically with convergence factor σ. As $\beta_k = \sin \angle(u_k, u)$, it follows that the angle between the exact and computed eigenvectors converges geometrically as desired. $\square$

Note that the performance of the Davidson algorithm is independent of the scaling of M; that is, replacing M by θM for any $\theta \neq 0$ does not change the computed eigenvalues or eigenvectors (except possibly for the sign of the eigenvectors). We restrict $\theta > 0$ so that θM will be positive definite. Then we can replace $\sigma = \|P - PMP(A - \lambda I)\|_2$ with

$$\sigma = \min \theta > 0\|P - \theta PMP(A - \lambda I)\|_2.$$

3   Restarted Davidson for Multiple Eigenvalues and its Convergence

The Davidson algorithm calculates the smallest or the largest eigenvalue of a symmetric matrix. This can be done by using the Lanczos or Arnoldi algorithms as well. In this Section we show our algorithm which calculates the first p smallest (or biggest) eigenvalues of a matrix.

Our algorithm starts by first using the Davidson algorithm to compute the smallest eigenvalue and eigenvector to within a certain error tolerance for the residual vector. Then the algorithm continues, extending the S_k and V_k matrices, until the second eigenvalue is computed. When that eigenvalue is computed to sufficient accuracy, the process continues extending the S_k and V_k matrices, computing the third eigenvalue, and so on. This is effectively a form of deflation [13], without locking in errors from previous eigenvectors.

Algorithm 2 – Restarted Davidson for Multiple Eigenvalues *Given a matrix A, an initial vector x_1, iteration limit m, number of eigenpairs maxp, restart index q, and convergence tolerance ϵ, compute approximations $\widehat{\lambda}k$ and u_k, $k = 1, \ldots, maxp$.*

> *Set $v_1 \leftarrow x_1/\|x_1\|_2$.   (initial guess)*
> *$V_1 \leftarrow [v_1]$; $S \leftarrow [\ ]$;*
> *$iter \leftarrow 0$.   (iteration number)*
> *For $p = 1, \ldots, maxp$ do   (approx for p^{th} pair)*
> > *For $k = 1, \ldots, m$ do*
> > > *$iter \leftarrow iter + 1$.*
> > > *$w_k \leftarrow Av_k$.*
> > > *$W_k = [W_{k-1}, w_k]$.*
> > > *Compute $V_k^T w_k$ and make it the last column and row of $S_k = V_k^T A V_k$*
> > > *$dimS \leftarrow dimS + 1$*
> > > *If $dimS \geq (m + q)$ (restart S)*
> > > > *Compute $S_k = Y_k \Lambda_k Y_k^T$, the complete symmetric eigendecomposition of S_k, where Λ_k is a diagonal matrix with its entries ordered in increasing order.*
> > > > *$dimS = m$*
> > > > *$S_k \leftarrow (\Lambda_k)_{(m \times m)}$.*
> > > > *$Y_k \leftarrow$ first m columns of Y_k.*
> > > > *$V_k \leftarrow V_k Y_k$.*
> > > *$l \leftarrow min(dimS, p)$.*
> > > *Compute the lth smallest eigenpair $\widehat{\lambda}_k$, y_k of S_k.*
> > > *$u_k \leftarrow V_k y_k$ (calculating the eigenvector of A).*
> > > *$r_k \leftarrow Au_k - \widehat{\lambda}_k V_k u_k = W_k - \widehat{\lambda}_k y_k$.   (residual vector)*
> > > *If $\|r_k\|_2 < \epsilon$ then   (exit inner loop)*
> > > *$t_k \leftarrow Mr_k$.   (preconditioning)*
> > > *$v_{k+1} \leftarrow mgs(V_k, t_k)$.   (apply modified Gram Schmidt)*
> > > *$V_{k+1} \leftarrow [V_k, v_{k+1}]$.*
> *Stop.*

A geometric convergence rate can be found for this method (which obtains eigenvalues beyond the smallest (or largest) eigenvalue) by modifying Theorem 1. In the following theorem assume that

$$\sigma' = \|P' - P'MP'\,P'(A - \lambda_p I)P'\|_2$$

where P' is the orthogonal projection onto the orthogonal complement of the span of the first $p - 1$ eigenvectors. Then we can shown, in a similar way to

Theorem 1 that the convergence factor for the new algorith is $(\sigma')^2$ To prove Theorem 2 we use the fact that $P's_k = s_k$, as s_k is orthogonal to the bottom p eigenvectors. and that although $(A - \lambda_p I)$ is no longer positive semi-definite, $P'(A - \lambda_p I)P'$ is.

Theorem 2 *Suppose that A and M are symmetric positive definite and that the first $p - 1$ eigenvectors have been found exactly. Let P' be the orthogonal projection onto the orthogonal complement of the span of the first $p - 1$ eigenvectors of A. If*

$$\|P' - P'MP'(A - \lambda_p I)P'\|_2 \leq \sigma' < 1,$$

then for almost any starting value x_1, the eigenvalue estimates $\widehat{\lambda}_k$ obtained by our modified Davidson algorithm for multiple eigenvalues converge to λ_p ultimately geometrically with convergence factor $(\sigma')^2$, and the angle between the exact and computed eigenvector goes to zero ultimately geometrically with convergence factor σ'.

As with Theorem 1, the preconditioner M can be scaled by any factor $\theta > 0$ giving the improved estimate:

$$\sigma' = \min \theta > 0 \|P' - \theta P'MP'(A - \lambda_p I)P'\|_2.$$

Another feature of algorithm 2 is the use of implicit restarts. This form of restart is suggested in [15]; however, computational results are not shown there. Note that implicit restarts do not affect the convergence proof. The following approach was used here: fix m and q; when the dimension of the subspace used is $m + q$, perform a complete symmetric eigendecomposition of S_k:

$$S_k = Y_k \Lambda_k Y_k^T.$$

Here Y_k is an orthogonal $(m + q) \times (m + q)$ matrix. Have the eigenvalues in Λ_k ordered so that $(\Lambda_k)_{11} \leq (\Lambda_k)_{22} \leq \cdots \leq (\Lambda_k)_{m+q,m+q}$. Let $Y_k = [Y_{k,1}, Y_{k,2}]$ where $Y_{k,1}$ is $(m + q) \times m$ and $Y_{k,2}$ is $(m + q) \times q$. If only the first m smallest eigenvalues of A are sought, the q largest eigenvalues of S_k can be dropped: $S_k^{(+)} = Y_{k,1}^T S_k Y_{k,1} = \Lambda_{k,1}$, and $V_k^{(+)} = V_k Y_{k,1}$. The Davidson algorithm can then be continued with $S_k^{(+)}$ and $V_k^{(+)}$ instead of S_k and V_k. This can be done using only $2n(m + q) + O((m + q)^2)$ floating point locations, by performing the update $V_k^{(+)} = V_k Y_{k,1}$ in place, using Householder operations.

This addition to the algorithm will periodically reduce the dimension of the subspace used. As the dimension of matrix S_k increases, the amount of work needed per iteration of the Davidson algorithm increases as $O(dimS\, n + dimS^3)$. Clearly it is desirable to keep the size of matrix S_k small, consistent with quickly obtaining accurate approximations for eigenvalues and eigenvectors. The restart idea is not new or particular to this algorithm [4] but our results will show that our new algorithm can particular benefit from this idea. Using implicit restarts can increase the number of iterations needed. Nevertheless, as we will see from the numerical results, this is more efficient than if it is ignored, especially if many eigenvalues are sought. Also, for many problems, memory is the limiting factor, which makes the use of implicit restarts essential.

4 Numerical Experiments

In this section we show results for the restarted and non-restarted versions of the new algorithm. The test case here is the Laplacian operator which has multiple eigenvalues and is a good test to illustrate how our methods handle misconvergence (convergence to another eigenvector other than the pth) . Our algorithm uses a multigrid preconditioner. The multigrid is a $(2,1)$ FMV (Full Multigrid V-cycle) method [11]. The red-black Gauss–Seidel method was used as the smoother. Note that the preconditioner is not symmetric and has not been modified to use the eigenvalue estimate.

An eigenvalue/eigenvector pair is considered to have converged if the relative residual norm $\|Ax - \lambda x\|_2/\|x\|_2 < 10^{-7}$. Note that the Davidson algorithm is a particular case of ours and is identical until the first eigenvalue is calculated.

The residual behavior for our algorithm is shown in Table 1. After calculating the first algorithm ours continues calculating eigenvalues eigenvectors pairs until the $10^t h$ eigenvalue/eigenvector pair. One version was restarted so as to keep the dimension of the subspace between 15 and 17 inclusive, the other had no restart in it. The results are identical for the first 16 iterations, thus in Table 1 we show the relative residuals after iteration 16.

iter'n #	16	17*	18	19*	20
no restart	1.24×10^{-6}	3.07×10^{-7}	6.755×10^{-8}	1.26×10^0	5.45×10^{-1}
restart	1.24×10^{-6}	3.07×10^{-7}	6.757×10^{-8}	1.23×10^0	5.11×10^{-1}
iter'n #	21*	22	23*	25	25*
no restart	4.54×10^{-2}	1.38×10^{-2}	$2.48 \times 10^{+2}$	$1.32 \times 10^{+1}$	1.44×10^0
restart	9.48×10^{-2}	1.55×10^{-2}	$2.61 \times 10^{+2}$	$2.49 \times 10^{+1}$	1.67×10^0
iter'n #	26	27*	28	29*	30
no restart	6.41×10^{-2}	3.26×10^{-3}	1.89×10^{-4}	2.75×10^{-6}	6.80×10^{-7}
restart	1.56×10^{-1}	7.11×10^{-3}	4.03×10^{-4}	2.23×10^{-5}	9.99×10^{-7}

Table 1. Comparison of $\|Ax - \lambda x\|_2/\|x\|_2$

The implicit restarts were performed after the iterations marked by an asterisk. As can be seen from Table 1, the non-restarted method converges faster, although the implicitly restarted method is close behind. This shows that the performance degradation due to implicit restarting in terms of the number of iterations is small. Thus implicit restart gives an efficient procedure for our algorithm since the cost of an iteration is much smaller than with the non-restarted method.

An important issue about our methods is how the convergence rate varies with the size of the problem. In Figure 1 results are shown for grid sizes: 31×31, 63×63, and 127×127 (961, 3969, and 16129 unknowns respectively). In this Figure the logarithm of the residual norm $\log_{10}(\|Ax - \lambda x\|_2/\|x\|_2)$ is plotted.

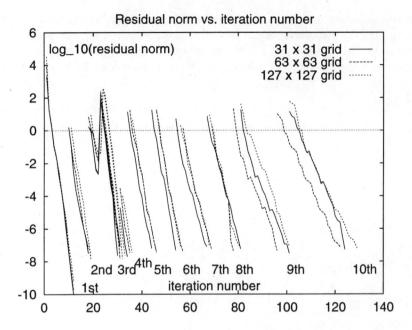

Fig. 1. Comparison of methods for three different 2-D grids

In Table 2 we compare for compare convergence factors across different grids for the restarted method. The convergence factors are computed by taking the ith root of the ratio of the first and last residual norms for computing the eigenvalue and eigenvector pair, where i is the number of steps between the first and last residual norms. The test cases for the convergence factors in Figure 1 and Table 2 were for subspace dimensions ranging from 13 to 15.

grid	\multicolumn{10}{c}{*eigenvalue number*}
	1	2	3	4	5	6	7	8	9	10
31×31	0.065	0.109	0.275	0.101	0.147	0.139	0.250	0.264	0.364	0.430
63×63	0.058	0.120	0.220	0.110	0.138	0.067	0.161	0.183	0.240	0.270
127×127	0.058	0.105	0.250	0.138	0.155	0.147	0.245	0.240	0.367	0.479

Table 2. Convergence factors for different grids

Figure 2 plots $\log_{10}(\|Ax - \lambda x\|_2/\|x\|_2)$ against the iteration number. The three different cases are for subspace dimensions 13 to 15, 15 to 17, and 17 to 19. In this Figure we notice that the convergence factor while searching for the 10th eigenvalue, for example, is worse than for previous eigenvalues. The degradation in the convergence factors for the last few eigenvalues is mainly due to the limit of the subspace dimension in the restarted algorithm. To illustrate this better, runs with the two-dimensional problem were performed for a 63×63 grid using the restarted method using different minimum and maximum dimensions were made and the convergence factors obtained are shown in Table 3.

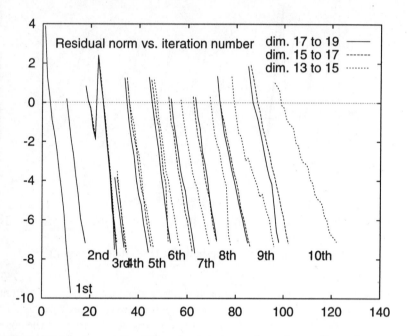

Fig. 2. Residual norm for different minimum and maximum dimensions

Convergence factors for the residual norms show the degradation as the number of the eigenvalue approaches the minimum dimension m used in the restarted algorithm. The minimum and maximum dimension values used here are fairly close to each other, and so the method is restarted quite frequently. In practice q (and m) can be much larger, especially if parallel architectures are considered. In a later paper parallel performance results for the new algorithms will be investigated.

subspace dimensions	eigenvalue number									
	1	2	3	4	5	6	7	8	9	10
13 to 15	0.058	0.120	0.220	0.110	0.138	0.067	0.161	0.183	0.240	0.270
17 to 19	0.058	0.120	0.201	0.128	0.129	0.115	0.163	0.154	0.222	0.202

Table 3. Convergence factors for different
minimum and maximum subspace dimensions

References

1. J. H. Bramble. *Multigrid Methods*, volume 294 of *Pitman Research Notes in Mathematical Sciences*. Longman Scientific & Technical, Essex, England, 1993.
2. A. Brandt, S. McCormick, and J. Ruge. Multigrid methods for differential eigenproblems. *SIAM J. Sci. Stat. Comp.*, 4(2):244–260, 1983.
3. A. Brandt and S. Ta'asan. Multigrid methods for nearly singular and slightly indefinite problems. In W. Hackbusch and U. Trottenberg, editors, *Multigrid Methods II*, number 1228 in Lecture Notes in Mathematics, pages 100–122. Springer-Verlag, 1985.
4. D. Calvetti, L. Reichel, and D.C. Sorenson. An implicitly restarted Lanczos method for large symmetric eigenvalue problems. *Electronic Trans. Numer. Anal.*, 2:1–21, March 1994.
5. M. Crouzeix, B. Philippe, and M. Sadkane. The Davidson method. *SIAM J. Sci. Comput.*, 15(1):62–76, 1994.
6. Ernest R. Davidson. The iterative calculation of a few of the lowest eigenvalues and corresponding eigenvectors of large real-symmetric matrices. *J. Comp. Physics*, 17:87–94, 1975. Note.
7. G. Golub and C. Van Loan. *Matrix Computations*. Johns Hopkins Press, 2nd edition, 1989. 1st edition published 1983 by North Oxford Academic.
8. M. Gu and S. C. Eisenstat. A stable and efficient algorithm for the rank-one modification of the symmetric eigenproblem. *SIAM J. Matrix Anal. Appl.*, 14(4):1266–1276, 1994.
9. R.A. Horne and C.A. Johnson. *Matrix Analysis*. Cambridge Uni. Press, Cambridge, 1985.
10. R. B. Lehoucq, D. C. Sorensen, and P. Vu. *ARPACK: An implementation of implicitly re-started Arnoldi iteration that computes some of the eigenvalues and eigenvectors of a large sparse matrix*. Comput. Appl. Math. Dept., Rice University, 1996. Software available from `netlib`.
11. S.F. McCormick, editor. *Multigrid Methods*. Frontiers in Applied Math. SIAM, Philadelphia, PA, 1987.
12. R. B. Morgan. Computing interior eigenvalues of large matrices. *Linear Alg. Appln.*, 154-156:289–309, 1991.
13. S. Oliveira. A preconditioned iterative subspace algorithm for computing eigenvalues and eigenvectors. *Numerical Linear Algebra and its Applications*, 1996. submitted.
14. Y. Saad. *Numerical Methods for Large Eigenvalue Problems*. Algorithms and Architectures for Advanced Scientific Computing Series. Manchester University Press and Halsted Press (an imprint of J. Wiley & Sons), New York, Brisbane, Toronto, 1992.
15. G. L. G. Sleijpen and H. A. van der Vorst. A Jacobi–Davidson iternation method for linear eigenvalue problems. *SIAM J. Matrix Anal.*, 1996. To appear. Also available as Preprint 856, Dept. Mathematics, University Utrecht, Utrecht, 1995.
16. D.C. Sorenson. The k-step Arnoldi process. In T.F. Coleman and Y. Li, editors, *Large-Scale Numerical Optimization*, pages 228–237. SIAM, Philadelphia, 1990.
17. H. Yserentant. Old and new convergence proofs for multigrid methods. In A. Iserles, editor, *Acta Numerica*, chapter Chap. 6, pages 285–326. Cambridge University Press, 1993.

Regularity of Minimizers
of the Mumford-Shah Functional

Diego Pallara

Abstract. A variational formulation for the image segmentation problem due to D. Mumford and J. Shah has risen a number of mathematical problems, which are interesting both from the point of view of applications and in their own right. In this paper we report on some results concerning the existence and the regularity of the solutions of a class of variational problems whose prototype is the Mumford-Shah problem.

1   Introduction

In the seminal paper [32] D. Mumford and J. Shah proposed a variational formulation for the image segmentation problem in computer vision. The datum is a function g defined in a rectangle R of the plane with values in $[0, 1]$ representing the grey level of a picture, and one wants to solve a joint smoothing and edge-detection problem, i.e. to obtain a piecewise smoothed image u outside a set of contours Γ (which is expected to be a system of smooth curves) coming from the sharp discontinuities of g. This output is achieved by the minimization of the functional

$$G(\Gamma, u) = \int_{R \setminus \Gamma} \left[|\nabla u|^2 + \alpha (u - g)^2 \right] \, dx + \beta \, \text{length}(\Gamma) \qquad (1)$$

among all the rectifiable curves Γ and all the functions $u \in C^1(R \setminus \Gamma)$, where α and β are positive scale and contrast parameters. We refer also to [31] for a wide discussion of varius variational models of image segmentation.

The solution of the variational problem posed in [32] would be a positive answer to the following

Mumford-Shah conjecture For every continuous function g there exists a minimizer in the set of all pairs (Γ, u) such that Γ is a finite set of points joined by a finite set of C^1-arcs, and $u \in C^1(R \setminus \Gamma)$.

The Mumford-Shah functional (1) has been generalized to n-dimensional spaces as follows:

$$G(\Gamma, u) = \int_{\Omega \setminus \Gamma} \left[|\nabla u|^2 + \alpha (u - g)^2 \right] \, dx + \beta \mathcal{H}^{n-1}(\Gamma \cap \Omega), \qquad (2)$$

where Ω is an open set in $\mathbf{R}^n$, $g \in L^\infty \cap L^2(\Omega)$, $\mathcal{H}^{n-1}$ is the $(n-1)$-dimensional Hausdorff measure and (Γ, u) is an admissible pair if $\Gamma \subset \mathbf{R}^n$ is a closed set and $u \in C^1(\Omega \setminus \Gamma)$. The minimization of the functional G has become the prototype of a class of variational problems, called free discontinuity problems after [22], i.e. problems where the unknown is a pair (Γ, u) with Γ a lower dimensional closed set and u a function smooth outside Γ. Variational models of this kind suitably describe problems where both volume and surface energies are present; we refer to [7] for an up-to-date report on the main developments of the theory of free discontinuity problems. Even though the case $n = 1$ is not trivial and interesting in applications (e.g. for segmentation problems in acoustics) in this paper we assume $n \geq 2$ because the 1-dimensional methods are obviously different.

The aim of the present survey is the presentation of some research work coming from the Mumford-Shah functional, with emphasis on that part which has been developed in general space dimensions. We shall concentrate our attention on the existence of minimizers of G in the class of admissible pairs, as proved in [24] through a suitable weak formulation (Section 2), and mainly on the regularity theory for the minimizers which has been developed in [10], [8], [9] along these lines (Section 3). Further properties of minimizers, applications and open problems are presented in Sections 4, 5.

We end this Introduction by discussing some properties of the functional (2) which have led to the already mentioned weak formulation. The first remark is that one can reduce the variables Γ and u to only one, in the following sense: if Γ is given then u is determined as the solution of the variational problem in the Sobolev space $W^{1,2}$

$$\min \left\{ \int_{\Omega \setminus \Gamma} \left[|\nabla u|^2 + \alpha(u - g)^2 \right] \, dx, \ u \in W^{1,2}(\Omega \setminus \Gamma) \right\} ; \qquad (3)$$

on the other hand, one can think of u as defined in the whole of Ω and allow u to be discontinuous along $(n-1)$-dimensional sets: calling $\Gamma = \Gamma_u$ the closure of the discontinuity set and requiring $u \in W^{1,2}(\Omega \setminus \Gamma)$, the functional turns out to depend only on u.

Notice also that minimizing G by the direct method of the calculus of variations is not easy because there is no topology on the closed sets Γ which ensures compactness of minimizing sequences and lower semicontinuity of the Hausdorff measure.

It follows from these remarks that a suitable space of possibly discontinuous functions could help to solve the problem. The classical candidate for this is the space BV of the functions with bounded variation, i.e. functions whose distributional gradient is a measure; accordingly, if u is taken in BV, ∇u in (2) must be understood as the absolutely continuous part of the gradient of u, see §2 for more details. Unfortunately, $BV(\Omega)$ turns out to be too large: in fact, the set of BV continuous functions with $\nabla u = 0$ a.e. is dense in $L^2(\Omega)$. As a consequence, $G(\Gamma_u, u)$ reduces to $\alpha \int_\Omega |u - g|^2$ for these functions, and can be arbitrarily small, no matter what g is.

A good subspace of BV which has proved to be suitable for this problem (and many other free discontinuity problems) is the space of special functions of bounded variation introduced by E. De Giorgi and L. Ambrosio in [23]. Its definition, together with its main properties, is given in the next section.

2   Preliminaries and the Existence Theory

In this section, after recalling some notions from geometric measure theory and about the structure of BV functions, we sketch the proof of the existence theorem for minimizers of the Mumford-Shah functional.

For complete information on the basic notions recalled here we refer to [27], [28], [30], [33], [36], [38]. We shall follow [24] for the proof of the existence of minimizers of G. A different proof, valid only in $\mathbf{R}^2$, is given in [19], see also [31].

We shall denote by $|E|$ the Lebesgue measure of E, and recall the definition of the Hausdorff measures: if $k \in]0, n]$ and $E \subset \mathbf{R}^n$, the k-dimensional Hausdorff measure of E is defined as follows:

$$\mathcal{H}^k(E) = \frac{\omega_k}{2^k} \lim_{\delta \to 0} \inf \left\{ \sum_{h=1}^{\infty} (\mathrm{diam} E_h)^k , \ E \subset \bigcup_{h=1}^{\infty} E_h, \ \mathrm{diam} E_h < \delta \right\} ,$$

where $\omega_k = \Gamma(1/2)^k/\Gamma(k/2+1)$ is a normalizing factor which gives the measure of the unit ball in $\mathbf{R}^k$ for integer values of k; $\mathcal{H}^0$ is the counting measure. In the following, we will be mainly concerned with $(n-1)$-dimensional measure, hence we put $m = n - 1$ in order to simplify the notation.

Another important notion is that of rectifiability; we say that $E \subset \mathbf{R}^n$ is countably $(\mathcal{H}^k, k)$-rectifiable if $\mathcal{H}^k$ almost all E can be covered by a sequence of k-dimensional C^1 manifolds Γ_h, i.e.

$$\mathcal{H}^k \left(E \setminus \bigcup_{h=1}^{\infty} \Gamma_h \right) = 0 .$$

Countably rectifiable sets E with locally finite measure have approximate tangent space $\mathcal{H}^k$ a.e., in the sense that for $\mathcal{H}^k$ a.e. $x \in E$ there exists $T_x \in G(n,k)$ (the Grassmann manifold of k-dimensional subspaces of $\mathbf{R}^n$) such that setting $E_\varrho = \varrho^{-1}(E - x)$, the equality

$$\lim_{\varrho \to 0} \int_{E_\varrho} \phi(y) \, d\mathcal{H}^k(y) = \int_{T_x} \phi(y) \, d\mathcal{H}^k(y)$$

holds for every $\phi \in C_0^1(\mathbf{R}^n)$. We shall identify any $S \in G(n,k)$ with the $n \times n$ matrix representing the orthogonal projection onto S, and denote by $\|S-T\|$ the euclidean norm induced by $\mathbf{R}^{n^2}$. We denote by $\mathcal{A}_k$ the set of affine k-dimensional subspaces of $\mathbf{R}^n$.

The theory of SBV functions relies on some fine properties of measurable functions; let $u : \Omega \subset \mathbf{R}^n \to \mathbf{R}$ be measurable and $x \in \Omega$; we say that $z \in \mathbf{R} \cup \{\pm\infty\}$ is the approximate limit of u at x (and we set $z = \operatorname{ap} \lim\limits_{y \to x} u(y)$) if

$$\lim_{\varrho \to 0} \varrho^{-n} \int_{B_\varrho(x)} |u(y) - z| \, dy = 0$$

(where $B_\varrho(x) = \{y \in \mathbf{R}^n : |x - y| < \varrho\}$). We say that u is approximately continuous at x if the approximate limit exists and is finite; in this case, we set $z = \tilde{u}(x)$ and we call jump set of u (denoted S_u) the Lebesgue negligible set where u is not approximately continuous. We say that u is approximately differentiable at x if u is approximately continuous at x and there exists a vector $\nabla u(x) \in \mathbf{R}^n$ such that

$$\operatorname{ap} \lim_{y \to x} \frac{u(y) - \tilde{u}(x) - \langle \nabla u(x), y \rangle}{|y - x|} = 0 \ .$$

If u is approximately differentiable at x then its approximate gradient $\nabla u(x)$ is unique and of course if u is differentiable then ∇u is the classical gradient.

Let now be $u \in BV(\Omega)$, i.e. assume that $u \in L^1(\Omega)$ and that Radon measures $D_i u$ $(i = 1, \ldots, n)$ exist such that

$$\int_\Omega \phi D_i u = - \int_\Omega u \frac{\partial \phi}{\partial x_i} \, dx$$

for every $\phi \in C_0^1(\Omega)$; we denote by $|Du|$ the total variation of the distributional derivative $Du = (D_1 u, \ldots, D_n u)$ and we decompose Du into its absolutely continuous part $D^a u$ and its singular part $D^s u$. The following properties hold:

(i) S_u is countably $(\mathcal{H}^m, m)$-rectifiable, hence $\mathcal{H}^m$ for a.a. $x \in S_u$ there exists a measure theoretic normal unit vector ν_x, orthogonal to the approximate tangent space T_x to S_u at x.

(ii) For $\mathcal{H}^m$ a.a. $x \in S_u$ the traces $u^+(x)$, $u^-(x)$ exist and are identified by

$$\lim_{\varrho \to 0} \varrho^{-n} \int_{B_\varrho^+(x)} |u(y) - u^+(x)| \, dy = 0$$

and

$$\lim_{\varrho \to 0} \varrho^{-n} \int_{B_\varrho^-(x)} |u(y) - u^-(x)| \, dy = 0 \ ,$$

where $B_\varrho^+(x) = \{y \in B_\varrho(x) : \langle y, \nu_x \rangle > 0\}$, $B_\varrho^-(x) = \{y \in B_\varrho(x) : \langle y, \nu_x \rangle < 0\}$; of course, since ν_x is determined up to the sign, changing ν_x by $-\nu_x$ entails that the traces are in turn exchanged.

(iii) u is approximately differentiable a.e. in $\Omega \setminus S_u$ and the density of $D^a u$ gives the approximate gradient ∇u.

(iv) The singular part $D^s u$ of Du is further decomposed as $D^J u + D^C u$, where $D^J u(B) = D^s u(B \cap S_u)$ and $D^C u(B) = D^s u(B \setminus S_u)$; the measure $D^J u$ is given by

$$D^J u(B) = \int_{B \cap S_u} \left(u^+(x) - u^-(x)\right) \nu_x \, d\mathcal{H}^m(x)$$

and moreover $\mathcal{H}^m(B) < +\infty$ implies $D^C u(B) = 0$ for every Borel set B.

The measure $D^C u$ is also called the "Cantor part" of Du because if u is the classical Cantor-Vitali function then $Du = D^C u$; notice that the functions u such that $Du = D^C u$ are precisely those BV functions mentioned in the Introduction which make the infimum of the Mumford-Shah functional vanish on BV. By (iv) this phenomenon appears to be tied to the fact that the derivative of these functions is concentrated on a set of intermediate dimension between $n-1$ and n (like Cantor's middle third set in the real line). At the same time, functions whose distributional derivative contains only volume and surface terms are well suited to study functionals with volume and surface terms. These remarks led to the introduction of the space SBV of special functions with bounded variation: $u \in SBV(\Omega)$ if $u \in BV(\Omega)$ and $D^C u = 0$, i.e.

$$Du(B) = \int_B \nabla u(x) \, dx + \int_{B \cap S_u} \left(u^+(x) - u^-(x)\right) \nu_x \, d\mathcal{H}^m(x)$$

for any Borel set B. As usual, $u \in SBV_{\text{loc}}(\Omega)$ means that $u \in SBV(A)$ for any open A with $\overline{A} \subset \Omega$ compact.

The space SBV enjoys very useful compactness and lower semicontinuity properties in connection with functionals like G.

Theorem 1 *Let $A \subset \mathbf{R}^n$ be a bounded open set and let $\phi : [0, +\infty) \to [0, +\infty)$ be a convex function such that $\phi(t)/t \to +\infty$ as $t \to +\infty$; if $(u_h) \subset SBV(A)$ is a sequence satisfying*

$$\int_A \phi(|\nabla u_h|) \, dx + \mathcal{H}^m(S_{u_h}) + \|u_h\|_\infty \leq C < +\infty$$

then, (u_h) has a subsequence (still denoted (u_h)) converging in $L^1(A)$ to $u \in SBV(A)$. Moreover, ∇u_h weakly converges to ∇u in $L^1(A; \mathbf{R}^n)$ and

$$\int_A \phi(|\nabla u|) \, dx \leq \liminf_{h \to \infty} \int_A \phi(|\nabla u_h|) \, dx \ ,$$

$$\mathcal{H}^m(S_u) \leq \liminf_{h \to \infty} \mathcal{H}^m(S_{u_h}) \ .$$

This is a particular case of the main theorems proved in [4], [1] (see also [6]) where more general surface integrals are considered.

We are now in a position to define the weak form of G for $u \in SBV(\Omega)$: it is formally obtained by substituting the variable Γ with S_u:

$$\bar{G}(u) = \int_\Omega \left[|\nabla u|^2 + \alpha(u-g)^2\right] \, dx + \beta \mathcal{H}^{n-1}(S_u) \ . \tag{4}$$

In this way, as announced in the Introduction, we have obtained a functional which depends only on the function u; of course, in general S_u is not a closed set (it may be even dense in Ω).

Notice that the infimum of $\bar{G}$ does not increase by assuming $\|u\|_\infty \leq \|g\|_\infty$; in fact, given $u \in SBV(\Omega)$, if v is the function obtained by truncation at levels $\pm\|g\|_\infty$, then $v \in SBV(\Omega)$ and $\bar{G}(v) \leq \bar{G}(u)$. As a consequence, by Theorem 1 the existence of minimizers of $\bar{G}$ in $SBV(\Omega)$ follows. Moreover, it is not difficult to prove (see [24, Lemma 2.3]) that if (Γ, u) is an admissible pair for G and $u \in L^\infty(\Omega)$ (as above, it is not restrictive to assume $\|u\|_\infty \leq \|g\|_\infty$) then u is (equivalent to) a SBV function with $S_u \subset \Gamma$; as a consequence,

$$\min\{\bar{G}(u) : u \in SBV(\Omega)\} \leq \inf\{G(\Gamma, u) : (\Gamma, u) \text{ admissible pair}\} . \tag{5}$$

At this point, the question is whether, for u minimizer of $\bar{G}$, the pair $(\overline{S}_u, u)$ is an admissible one for the functional G and which is the link between $\bar{G}(u)$ and $G(\overline{S}_u, u)$: by using (3) it is not difficult to see that $u \in C^1(\Omega \setminus \overline{S}_u)$ (see Section 4 for some details), but the the theory of SBV functions does not give any control on the measure of $\overline{S}_u$, which in general can be much greater than the measure of S_u, even infinite. So, the main step consists in estimating $\mathcal{H}^m(\overline{S}_u \setminus S_u)$. The following result, proved in [24], is the first step towards the proof of the existence of minimizers of G and the regularization of the weak solutions.

Lemma 2 (density lower bound) *There exist a positive constant θ depending only on the space dimension, such that if u minimizes $\bar{G}$ in $SBV(\Omega)$ then*

$$\mathcal{H}^m(S_u \cap B_\varrho(x)) \geq \theta \varrho^m \tag{6}$$

for any $B_\varrho(x) \subset \Omega$, with $x \in \overline{S}_u$ and ϱ satisfying $\alpha\|g\|_\infty^2 \varrho \leq \beta$.

The above estimate entails the following consequence:

Theorem 3 *Let $u \in SBV_{\text{loc}}(\Omega)$ be a minimizer of $\bar{G}$; then*

$$\mathcal{H}^m(\overline{S}_u \cap \Omega \setminus S_u) = 0 . \tag{7}$$

This result implies that $G(\overline{S}_u, u) = \bar{G}(u)$, hence the equality holds in (5) and the minimum of G is achieved.

In the following, we will always refer to "essential minimizers" and we will call them *optimal segmentations*, i.e. pairs (Γ, u) such that no admissible pair (Γ', u') exists such that Γ' is strictly contained in Γ, $u' \equiv u$ in $\Omega \setminus \Gamma$ and $G(\Gamma', u') \leq G(\Gamma, u)$. Observe that, even in this restricted sense, minimizers are not unique in general.

A further very interesting consequence of Lemma 2 is the possibility of approximating, in a variational sense, the functional G by a sequence of elliptic integral functionals, see [12], [13].

As mentioned in the Introduction, the singular set Γ of an optimal segmentation is expected to be a piecewise smooth hypersurface (curve in the two-dimensional case), whereas the existence theory ensures only its rectifiability (in fact, Γ is – essentially – the singular set of a BV function) and the density lower bound of Lemma 2. This last result can be improved in two dimensions, giving a "uniform rectifiability" property of Γ, see [21], and also the above mentioned alternative proof of the existence of minimizers (see [19], [31]).

3   Regularity Results for Quasi Minimizers

In this section we present the main ideas underlying the regularity theory for the jump set of an optimal segmentation (Γ, u), as developed in general space dimensions by L. Ambrosio, N. Fusco and the Author in the last few years. The theory has been developed for a wider class of hypersurfaces Γ, including those coming from optimal segmentations; in fact, the leading term in $\bar{G}$ is the functional

$$F(u, \Omega) = \int_\Omega |\nabla u|^2\, dx + \mathcal{H}^{n-1}(S_u) \ ,$$

where a scaling argument allows us to fix $\beta = 1$ for simplicity, hence we look at G as a perturbation of F. Accordingly, we give the following definitions, where the localization in relatively compact subdomains of Ω is needed because it can happen that the energy of a local minimizer is infinite on the whole of Ω (e.g. if $\Omega = \mathbf{R}^n$, a case which will be considered in Section 5). As usual, $E \subset\subset A$ means that the closure of E is a compact set contained in the interior of A.

Definition 4 (local minimizers) We say that $u \in SBV_{\mathrm{loc}}(\Omega)$ is a local minimizer in Ω if

$$\int_A |\nabla u|^2\, dx + \mathcal{H}^m(S_u \cap A) < +\infty \qquad\qquad \forall A \subset\subset \Omega \qquad (8)$$

and

$$\int_A |\nabla u|^2\, dx + \mathcal{H}^m(S_u \cap A) \le \int_A |\nabla v|^2\, dx + \mathcal{H}^m(S_v \cap A)$$

whenever $v \in SBV_{\mathrm{loc}}(\Omega)$ and $\{v \ne u\} \subset\subset A \subset\subset \Omega$.

Definition 5 (quasi minimizers) We will call deviation from minimality, denoted $\mathrm{Dev}(u, \Omega)$, of a function $u \in SBV_{\mathrm{loc}}(\Omega)$ satisfying (8) the smallest $\lambda \in [0, +\infty]$ such that

$$\int_A |\nabla u|^2\, dx + \mathcal{H}^m(S_u \cap A) \le \int_A |\nabla v|^2\, dx + \mathcal{H}^m(S_v \cap A) + \lambda$$

for any $v \in SBV_{\mathrm{loc}}(\Omega)$ such that $\{v \ne u\} \subset\subset A \subset\subset \Omega$. Clearly, $\mathrm{Dev}(u, \Omega) = 0$ if and only if u is a local minimizer in Ω. Moreover, we say that u is a quasi minimizer in Ω if there exists a nondecreasing function $\omega(t) : (0, +\infty) \to [0, +\infty)$ such that $\omega(t) \downarrow 0$ as $t \downarrow 0$ and for any ball $B_\varrho(x) \subset \Omega$

$$\mathrm{Dev}(u, B_\varrho(x)) \le \varrho^m \omega(\varrho) \ . \qquad (9)$$

We denote by $\mathcal{M}_\omega(\Omega)$ the class of functions satisfying (9).

Remark. It is easy to see (cf [10, Remark 2.6]) that minimizers of $\bar{G}$ are quasi minimizers with $\omega(\varrho) = 4\alpha\|g\|_\infty^2 \omega_n \varrho$. Moreover (see [10, Theorem 2.7 and Proposition 2.8]), it is also possible to prove that the density lower bound (6) holds for quasi minimizers as well, with ϱ depending on ω, hence also (7) holds for any quasi minimizer u; for this reason, we shall use the notation $K(u)$ for $\overline{S}_u$ when u is a quasi minimizer.

The density lower bound is the starting point to investigate the regularity of $K(u)$ along the lines of the regularity theory of varifolds as developed by W. K. Allard ([2], see also [33]) and K. Brakke, see [15], which is in turn reminiscent of the regularity techniques for minimal boundaries, see [25]. Let us recall that for $k \in \{1,\ldots,n\}$ an integer k-varifold in $\mathbf{R}^n$ is a $(\mathcal{H}^k, k)$ rectifiable set E with locally finite $\mathcal{H}^k$ measure, endowed with a $\mathcal{H}^k$-measurable integer valued multiplicity function. These objects have been introduced and deeply studied in connection with various formulations of the minimal area problem; we do not enter into the details, but assume unit multiplicity for simplicity and limit ourselves to recall the first variation of the k-dimensional area, which leads to the notion of generalized mean curvature, and the "tilt lemma", which is the starting point of Allard's regularity theorem. Given $\eta \in [C_0^1(\mathbf{R}^n)]^n$, the first variation formula along $\tau_\varepsilon(x) = x + \varepsilon\eta(x)$, setting $E_\varepsilon = \{\tau_\varepsilon(x), x \in E\}$, reads

$$\frac{d}{d\varepsilon}\mathcal{H}^k(E_\varepsilon)\Big|_{\varepsilon=0} = \int_E \sum_{i=1}^n \delta_i \eta_i \, d\mathcal{H}^k = \int_E \mathrm{div}^\tau \eta \, d\mathcal{H}^k$$

where $(\delta_1,\ldots,\delta_n)$ is the projection of the gradient operator on the approximate tangent space of E and $\mathrm{div}^\tau \eta$ has the meaning of tangential divergence of η. If a $\mathcal{H}^k$ locally summable vector field H exists on E such that

$$\int_E \mathrm{div}^\tau \eta \, d\mathcal{H}^k = -\int_E \langle H, \eta\rangle \, d\mathcal{H}^k$$

for any regular vector field η as above, then H is called the generalized mean curvature of E. Allard's regularity theorem ensures that if E has generalized mean curvature $H \in L^p(E, d\mathcal{H}^k)$ for some $p > k$ then, up to a relatively closed set S with $\mathcal{H}^k(S) = 0$, E is locally a k-dimensional manifold of class $C^{1,\alpha}$ with $\alpha = 1 - k/p$.

Notice that the $C^{1,\alpha}$ regularity of E is equivalent to the $C^{0,\alpha}$ regularity of the map $x \mapsto T_x$; this leads to look for a suitable control of the quantity

$$\mathbf{T}(x,\varrho) = \min_{T \in G(n,k)} \varrho^{-k} \int_{B_\varrho(x) \cap E} \|T_y - T\|^2 \, d\mathcal{H}^k(y) \ ,$$

which gives a measure of the oscillation of the approximate tangent space T_y: if $\mathbf{T}(x,\varrho)$ decays fast enough as $\varrho \to 0$, then T_x is $C^{0,\alpha}$ and E is $C^{1,\alpha}$. A key step in Allard's proof is that $\mathbf{T}(x,\varrho)$ can be controlled by more convenient quantities, namely the sum of the integral of the distance from the minimizing k-plane T_0 and the scaled L^p norm of H:

$$\varrho^{-(k+2)} \int_{B_\varrho(x) \cap E} \text{dist}^2(y - x, T_0)\, d\mathcal{H}^k(y) + \varrho^{1-k/p} \left(\int_{B_\varrho(x) \cap E} |H|^p\, d\mathcal{H}^k \right)^{1/p} ;$$

this estimate is analogous to the Caccioppoli's inequality which is well-known in the regularity theory for elliptic partial differential equations.

To see how such a scheme can be used in our context, we use the same deformation field to compute the first variation of F; following [17], for $u \in SBV(\Omega)$ verifying (8) and $\eta \in [C_0^1(\mathbf{R}^n)]^n$ as above, we set

$$\tau_\varepsilon(x) = x + \varepsilon\eta(x), \qquad u_\varepsilon(\tau_\varepsilon(x)) = u(x) \tag{10}$$

and we obtain:

$$\frac{d}{d\varepsilon}F(u_\varepsilon)\Big|_{\varepsilon=0} = \int_\Omega \left[|\nabla u|^2 \text{div}\,\eta - 2\langle \nabla u, \nabla u \nabla \eta \rangle \right] dx + \int_{S_u} \text{div}^\tau \eta\, d\mathcal{H}^m. \tag{11}$$

Notice that in (11) the surface term is the first variation of the area of S_u, whereas the volume term can be controlled by the Dirichlet energy so that (11) allows us to control the mean curvature of $K(u)$ by the Dirichlet energy (obviously, a correction is needed for *quasi* minimizers, see Lemma 6 below). These computations, together with the preceding discussion on Allard's theorem, suggest that the relevant quantities to approach the partial regularity of $K(u)$ are the *tilt*, the *flatness* and the scaled Dirichlet energy, defined respectively by:

$$\mathbf{T}(x, \varrho) = \min_{T \in G(n,m)} \varrho^{-m} \int_{B_\varrho(x) \cap K(u)} \|T_y - T\|^2\, d\mathcal{H}^m(y)$$

$$= \min_{\nu \in \partial B_1} \varrho^{-m} \int_{B_\varrho(x) \cap K(u)} \|\nu_y - \nu\|^2\, d\mathcal{H}^m(y)$$

$$\mathbf{A}(x, \varrho) = \min_{A \in \mathcal{A}_m} \varrho^{-(m+2)} \int_{B_\varrho(x) \cap K(u)} \text{dist}^2(y, A)\, d\mathcal{H}^m(y)$$

$$\mathbf{D}(x, \varrho) = \varrho^{-m} \int_{B_\varrho(x)} |\nabla u|^2\, dy.$$

The above quantities have the right scaling properties in connection with quasi minimality: in fact, they do not change by setting

$$u_\varrho(y) = \frac{1}{\sqrt{\varrho}} u(x_0 + \varrho y), \tag{12}$$

and noticing that $u \mapsto u_\varrho$ maps $\mathcal{M}_\omega(B_\varrho(x_0))$ into $\mathcal{M}_\omega(B_1(0))$.

The tilt estimate in our case can be stated as follows (see [10, Lemma 6.1]).

Lemma 6 (Tilt estimate) *Let $u \in \mathcal{M}_\omega(\Omega)$. For any $\beta \in (0,1)$ there exist $c(\beta) > 0$, $\varrho_\beta > 0$ such that*

$$\mathbf{T}(x, \beta\varrho) \le c(\beta) \left[\mathbf{D}(x, \varrho) + \mathbf{A}(x, \varrho) + \sqrt{\omega(\varrho)} \right]$$

provided $\varrho < \varrho_\beta$.

From the tilt estimate the following regularity criterion follows: the way to obtain it consists essentially in proving first that if $\mathbf{A}(x, \varrho)$ is small enough in a ball, then (in a smaller ball) a large portion of of $K(u)$ can be covered by the graph of a lipschitz function, and hence that the procedure which gives the lipschitz continuous approximation yields in fact a function whose graph covers entirely the set $K(u)$ (in a still smaller ball); the tilt estimate gives the further regularity of f and, as a consequence, of $K(u)$.

Theorem 7 *Let $u \in \mathcal{M}_\omega(\Omega)$ and assume that, for some constants $C > 0$, $\gamma > 0$, $\omega(t) \leq Ct^{2\gamma}$ and*

$$\mathbf{D}(x, \varrho) + \mathbf{A}(x, \varrho) \leq C\varrho^\gamma \tag{13}$$

for any ball $B_\varrho(x) \subset \Omega$ and any $x \in K(u)$. Then $\Omega \cap K(u)$ is a $C^{1,\gamma/(m+2)}$- hypersurface.

In [8] it is proved that any point $x \in K(u)$ where $\mathbf{T}(x, \varrho) + \mathbf{D}(x, \varrho)$ is less than a critical treshold $\varepsilon_0 > 0$ (which depends on the space dimension) for a sufficiently small ϱ is a regular point of K. This gives the following result on the regular points of $K(u)$.

Theorem 8 *Let $\mathcal{M}_\omega(\Omega)$ be defined in Definition 4, $u \in \mathcal{M}_\omega(\Omega)$, and assume that*

$$\omega(\varrho) \leq c_0\varrho^{2\gamma} \tag{14}$$

for some constants $c_0 \geq 0$, $\gamma > 0$ and let $\alpha = \min\{1/4, \gamma\}/(m+2)$. Then, there is a positive constant $\varepsilon_0(c_0, \gamma)$ such that for any $x \in K(u)$ and any ball $B_\varrho(x) \subset \Omega$ with $\varrho < 1$, the condition

$$\mathbf{A}(x, \varrho) + \mathbf{D}(x, \varrho) < \varepsilon_0$$

implies that $B_{\varrho/2}(x) \cap K(u)$ is a $C^{1,\alpha}$ hypersurface.

It is interesting to note that this approach leads to a constructive characterization of the singular set S:

$$x \in S \quad \Longleftrightarrow \quad \limsup_{\varrho \to 0^+} [\mathbf{A}(x, \varrho) + \mathbf{D}(x, \varrho)] \geq \varepsilon_0 . \tag{15}$$

An easy consequence (see [8, Remark 3.2]) of the above remark is that S is closed and $\mathcal{H}^m(S) = 0$. Thus, we can state the main regularity theorem as follows. Notice that by the Remark after Definition 5, minimizers of the Mumford-Shah functional verify (14).

Theorem 9 *Let $u \in \mathcal{M}_\omega(\Omega)$, with ω satisfying (14). Then, up to a closed $\mathcal{H}^m$ negligible set, $K(u)$ is locally a $C^{1,\alpha}$ regular hypersurface.*

We refer to the original papers for the complete proof of these results, and also to [11] for a brief sketch where the main points are described, limiting ourselves to a few comments. The main idea is that, like in many regularity results, if $\mathbf{A}(x, \varrho) + \mathbf{D}(x, \varrho) < \varepsilon_0$ (with ε_0 small enough) then (13) holds in the ball $B_{\varrho/2}(x)$, provided that *both* **D** and **A** decay faster on smaller balls contained in $B_\varrho(x)$: so, the strategy is based on two decay properties. The first one concerns **D**: if **A** is much smaller than **D** then the decay of **D** improves on smaller radii (this is related to classical elliptic regularity theory); the second one concerns **A**, it is true if **D** is comparable with **A** and is related to the classical flatness improvement theorems of De Giorgi, Allard, Almgren. Of course this double blow-up reflects the different behaviour of the two terms in the functional F. Finally, if $\mathbf{D}(x,\varrho)+\mathbf{A}(x,\varrho)$ is small enough a joint iteration can be used, leading to (13) in $B_{\varrho/2}(x)$ and to the regularity near x.

We shall come back to consider more specifically quasi minimizers arising as optimal segmentations in the next section: in this case Theorem 8 can in fact be strengthened.

4   Further Properties of Optimal Segmentations and Applications

In this section we consider optimal segmentations, i.e. pairs $(K(u), u)$ minimizing the Mumford-Shah functional G and show how the regularity results obtained for the jump set of general quasi minimizers can be improved. Moreover, we shall consider the possibility of using as minimizing sequences for the functional G only admissible pairs (Γ, u) where the hypersurface Γ is piecewise smooth and $u \in C^\infty(\Omega \setminus \Gamma)$. Our regularity results will be applied in order to see that the infimum of G over these admissible pairs is equal to the absolute minimum; as a consequence, approximate solutions of the minimum problem can be computed in this restricted class. Using a standard terminology, this result can be stated by saying that the Lavrentieff phenomenon does not occur. Our results are close to those of [18], who uses piecewise affine hypersurfaces in order to obtain discrete approximations of optimal segmentations, and [26].

Let $(K(u), u)$ be an optimal segmentation: as we have already observed, the function u is the solution of the variational problem (3), so that the computation of the first variation along $u + \varepsilon\phi$, with $\phi \in C_0^\infty(\Omega \setminus K(u))$, shows that u is a weak solution of the following elliptic problem:

$$\begin{cases} \Delta u = \alpha(u - g) & \text{in } \Omega \setminus K(u) \\ \dfrac{\partial u}{\partial \nu} = 0 & \text{on } \partial\Omega \cup K(u) \end{cases} \tag{16}$$

It follows by using standard elliptic equations theory (see e.g. [16]) that the hypothesis $g \in L^\infty(\Omega)$ ensures that $u \in C^1(\Omega \setminus K(u))$ (a result which has already been mentioned). As far as the boundary condition on $K(u)$ is concerned, we remark that the $C^{1,\alpha}$ regularity of $K(u)$ outside the singular set S does not yet ensure that it holds in a classical sense. In [9, Section 2] it is shown that ∇u has a locally Hölder continuous extension on both sides of $K(u) \setminus S$, a result which will be useful in order to improve the regularity of $K(u) \setminus S$ itself. To see how a suitable bootstrap argument can be used to this aim, let us argue as in

Section 3: computing the first variation of G along the directions u_ε given by (10), we obtain the following formula, analogous to (11):

$$\frac{d}{d\varepsilon}G(u_\varepsilon)\Big|_{\varepsilon=0} = \int_\Omega \left(|\nabla u|^2 + \alpha(u-g)^2\right) \operatorname{div} \eta \, dx \qquad (17)$$

$$-2\int_\Omega \left[\alpha(u-g)\langle \nabla g, \eta\rangle + \langle \nabla u, \nabla u \nabla \eta\rangle\right] dx + \int_{K(u)} \operatorname{div}^\tau \eta \, d\mathcal{H}^m .$$

Let us draw a first consequence from (11) and (17): let $A \subset \Omega$ be an open set, and suppose that $K(u) \cap A$ is the graph of a $C^{1,\alpha}$ function ϕ of $n-1$ variables; then, it is possible to prove (see [9, Section 3]) that ϕ is the distributional solution of an equation of the form

$$-\operatorname{div}\left(\frac{\nabla \phi}{\sqrt{1+|\nabla \phi|^2}}\right) = H \qquad (18)$$

for a suitable bounded function H. This amounts to say that $K(u) \cap A$ has bounded mean curvature, since the left hand side of (18) is the mean curvature of $K(u)$. Heuristically, the above formula can be explained as follows: by choosing a variation η in (17) (resp. (11)) normal to $K(u)$ and integrating by parts in $\Omega \setminus K(u)$, one obtains that the mean curvature of $K(u)$ coming from the surface terms equals the jump of $|\nabla u|^2 + \alpha(u-g)^2$ (resp. $|\nabla u|^2$) across $K(u)$, which formally comes from the integration by parts.

The next step is to prove that the solutions of (18) are in fact more regular. This argument gives the first statement in Theorem 10 below. Assuming further regularity on the datum g, the formal computation mentioned above can be justified, and this in turn allows to gain further regularity on $K(u)$, using again equation (18) with a more regular right hand side. Let us state these results in a precise way in the following theorem.

Theorem 10 *Let $(K(u), u)$ be an optimal segmentation; then, outside a closed and $\mathcal{H}^m$ negligible singular set, $K(u)$ is (locally) a hypersurface of class $C^{1,\alpha}$ for every $\alpha < 1$ in general space dimension, and $C^{1,1}$ in two space dimensions.*

Moreover, if $A \subset \Omega$ is an open set where $K(u)$ is regular and $g \in C^{k,\beta}_{\mathrm{loc}}(A)$ for some integer $k \geq 1$ and some $\beta \in (0,1]$, then there exists $s \in (0,1]$ depending on the space dimension n and on β such that $K(u)$ is $C^{k+2,s}$ regular.

The two dimensional case of Theorem 10 has been independently obtained, via techniques from harmonic analysis and conformal mapping in [20], [21] and also in [14].

We turn now our attention to the non occurrence of the Lavrentieff phenomenon and to the approximation of the minimum of G by pairs with piecewise smooth jump set.

Let us define the class of piecewise smooth hypersurfaces we will use.

Definition 11 Let $K \subset \mathbf{R}^n$ be compact. We say that K is piecewise smooth if there exist finitely many C^∞ hypersurfaces $\Gamma_1, \ldots, \Gamma_N$ with C^∞ boundaries $\partial\Gamma_1, \ldots, \partial\Gamma_N$ such that $K = \bigcup_{i=1}^{N} \Gamma_i$ and

$$(\Gamma_i \setminus \partial\Gamma_i) \cap (\Gamma_j \setminus \partial\Gamma_j) = \emptyset$$

whenever $i \neq j$. We denote by $\mathcal{K}$ the class of these sets.

The main result can be stated as follows, and is an easy consequence of a density property of SBV functions with smooth singular set in the whole of SBV; it is proved in [9, Section 5].

Theorem 12 *Let $\Omega \subset \mathbf{R}^n$ be a bounded open set with Lipschitz boundary, α, β positive, $g \in L^\infty(\Omega)$ and G, $\bar{G}$ given by (2), (4); then*

$$\inf\{G(\Gamma, u): \ \Gamma \in \mathcal{K}, \ u \in C^\infty(\Omega \setminus K)\} = \min\{\bar{G}(u): \ u \in SBV(\Omega)\} \ .$$

The idea of the proof is not difficult to explain. Let u_0 be a minimizer of $\bar{G}$, and extend u_0 to an open set $A \supset \overline{\Omega}$ in such a way that $\mathcal{H}^m(S_{u_0} \cap \partial\Omega) = 0$ and $F(u_0, A) < +\infty$ (in this extension the regularity of $\partial\Omega$ is used). Let (w_h) be a sequence of $C^\infty(A)$ functions such that $h\|w_h - u_0\|_2 \to 0$; let further u_h be the solution of the minimum problem

$$\min_{u \in SBV} \left\{ \int_A \left[|\nabla u|^2 + h|u - w_h|^2\right] dx + \beta\mathcal{H}^m(S_u) \right\}. \tag{19}$$

By Theorem 10, the singular set $K(u_h)$ is C^∞ outside a closed negligible set S_h. Fix $\varepsilon > 0$; covering S_h by a finite number of sufficiently small balls and using a truncation argument, $K(u_h)$ can be modified into a set $K_h \in \mathcal{K}$ in such a way that the functional in (19) does not increase more that ε. The fast convergence of w_h to u_0 implies that also $u_h \to u_0$ in L^2; the final step is to prove that $\nabla u_h \to \nabla u$ in L^2 norm, and that $\mathcal{H}^m(K_h) \to \mathcal{H}^m(K(u_0))$. This ends the proof.

5   Local Minimizers and Open Problems

In this section we shall deepen the analysis of the necessary minimality conditions for the functional F and of the properties of local minimizers. The interest in doing this, in connection with the properties of optimal segmentations, relies on the already exploited "quasi minimality", but even in the asymptotic properties of optimal segmentations under blow-up. The blow-up technique is by now classical in the study of the regularity of solutions of variational problems, and has been introduced in connection with the regularity of minimal surfaces, see [25]. In minimal surfaces theory the blow-up of a minimal surface produces (thanks to a monotonicity formula) a minimal cone, i.e. a still minimizing surface which is also invariant under further rescaling. Other monotonicity formulas have been heavily exploited in different variational problems, and in particular in other geometric variational problems (see e.g. [34]) and in many free boundary problems, those which seem to be the closest to free discontinuity problems, see e.g. the book [29].

In the same vein, the blow-up technique, applied to optimal segmentations, leads to compare the blow-up limits with local minimizers in $\mathbf{R}^n$, i.e. SBV functions minimizing the functional F with respect to arbitrary compactly supported perturbations. At a first look, candidates for being local minimizers in this situations (beside the constant functions) could seem to be either functions which are harmonic in the whole of $\mathbf{R}^n$ (remember that if u is a local minimizer in $\mathbf{R}^n$ then u is harmonic in $\mathbf{R}^n \setminus K(u)$, cf (16)), or functions with two values separated by a hyperplane or more generally functions with a finite number of values separated by a minimal partition of $\mathbf{R}^n$: the first case corresponds to $K(u) = \emptyset$, i.e. to a concentration of the energy on the volume integral, the other ones to $\nabla u = 0$ a.e., namely a concentration of the energy on the surface term. We shall see that none of the above situations occur, even though functions assuming only two values separated by a hyperplane occur as blow-up limits at every regular point of $K(u)$. We follow [17, Section 5] and give some details, because the original paper has never been published in widely circulated journals. Our results can be usefully compared with those of [14], where the blow-up technique is successfully applied to two dimensional optimal segmentations.

Let us first recall that if u is a local minimizer then the value of $F(u)$ on a ball B_ϱ can be easily estimated.

In fact, let u be a quasi minimizer in Ω and let v be a function constant in $B_\varrho \subset\subset \Omega$ and agreeing with u outside: the jump set of v is contained in $S_u \cup \partial B_\varrho$ and $0 \leq F(v, \Omega) - F(u, \Omega) \leq \mathcal{H}^m(\partial B_\varrho) - F(u, B_\varrho)$, whence

$$F(u, B_\varrho) \leq \mathcal{H}^m(\partial B_\varrho) = n\omega_n \varrho^m \ . \tag{20}$$

We now prove an estimate similar to the classical Caccioppoli inequality (see [17, Lemma 5.1]).

Lemma 13 *If u is a local minimizer on $\mathbf{R}^n$, then for every real λ and for every ball $B_{2\varrho}$ the following estimate holds:*

$$\int_{B_\varrho} |\nabla u|^2 \, dx \leq \frac{c}{\varrho^2} \int_{B_{2\varrho} \setminus B_\varrho} |u - \lambda|^2 \, dx \ , \tag{21}$$

where c does not depend either on ϱ or on n.

Proof. It is enough to prove the thesis only for $\lambda = 0$. Let be $\phi \in C_0^\infty$ with $0 \leq \phi \leq 1$, $\phi = 1$ in B_ϱ and $\nabla\phi \leq C/\varrho$. Set $u_\varepsilon = u + \varepsilon\phi^2 u$; since $K(u) = K(u_\varepsilon)$, the inequality $F(u) \leq F(u_\varepsilon)$ holds, and it implies

$$\int_{B_{2\varrho}} \left(\phi^2 |\nabla u|^2 + u\langle \nabla u, \nabla\phi^2 \rangle \right) dx = 0 \ .$$

Then

$$\int_{B_{2\varrho}} \phi^2 |\nabla u|^2 \, dx = -2 \int_{B_{2\varrho}} \phi u \langle \nabla u, \nabla \phi \rangle \, dx \le 2 \int_{B_{2\varrho} \setminus B_\varrho} |u \nabla \phi| \, |\phi \nabla u| \, dx$$

$$\le 2 \left(\int_{B_{2\varrho} \setminus B_\varrho} |u \nabla \phi|^2 \, dx \right)^{1/2} \left(\int_{B_{2\varrho}} |\phi \nabla u|^2 \, dx \right)^{1/2}$$

$$\le \frac{2C}{\varrho} \left(\int_{B_{2\varrho} \setminus B_\varrho} |u|^2 \, dx \right)^{1/2} \left(\int_{B_{2\varrho}} \phi^2 |\nabla u|^2 \, dx \right)^{1/2} ,$$

and the thesis follows.   □

Using (20) and Lemma 13 the following result can be easily obtained.

Proposition 14 *Let u be a local minimizer in $\mathbf{R}^n$. Then, if either $S_u = \emptyset$ or S_u is bounded and $\int |\nabla u|^2 dx$ is finite, then u is constant.*

Proof. Assume first that $S_u = \emptyset$. Then u is harmonic in the whole of $\mathbf{R}^n$, hence $|\nabla u|^2$ is subharmonic and the mean value inequality, together with (20), implies:

$$|\nabla u(\xi)|^2 \le \frac{1}{\omega_n \varrho^n} \int_{B_\varrho(\xi)} |\nabla u|^2 \, dx \le \frac{n}{\varrho}$$

for any $\xi \in \mathbf{R}^n$ and $\varrho > 0$, and u must be constant.

Let now S_u be bounded, e.g. $S_u \subset B_\varrho$, and set

$$\lambda = \frac{1}{|B_{2\varrho} \setminus B_\varrho|} \int_{B_{2\varrho} \setminus B_\varrho} u(x) \, dx .$$

Since $u \in H^1(B_{2\varrho} \setminus B_\varrho)$, by Poincaré inequality and Lemma 13 we get

$$\int_{B_\varrho} |\nabla u|^2 \, dx \le \frac{c}{\varrho^2} \int_{B_{2\varrho} \setminus B_\varrho} |u - \lambda|^2 \, dx \le c' \int_{B_{2\varrho} \setminus B_\varrho} |\nabla u|^2 \, dx .$$

Now filling the hole, i.e. adding c' times the left-hand side, we obtain

$$\int_{B_\varrho} |\nabla u|^2 \, dx \le \tau \int_{B_{2\varrho}} |\nabla u|^2 \, dx ,$$

where $\tau = c'/(1 + c') < 1$. It immediately follows:

$$\int_{B_\varrho} |\nabla u|^2 \, dx \le \tau^k \int_{B_{2^k \varrho}} |\nabla u|^2 \, dx \to 0 \qquad \text{as } k \to +\infty$$

so $\int_{B_\varrho} |\nabla u|^2 dx = 0$ for every ball containing S_u and again u must be constant.
                                                                         □

Another result which can be easily proved is that functions assuming only two values with a hyperplane as jump set are not local minimizers in $\mathbf{R}^n$.

Proposition 15 *For every $c \neq 0$ the function $u : \mathbf{R}^n \to \mathbf{R}$ defined by*

$$u(x_1, \ldots, x_n) = \begin{cases} c & \text{for } x_n > 0 \\ -c & \text{for } x_n < 0 \end{cases}$$

is not a local minimizer in $\mathbf{R}^n$.

Proof. Let $\varrho > 2c^2$ and $L > \dfrac{4(n-1)\varrho^2}{\varrho - 2c^2}$ and define the comparison function

$$v(x_1, \ldots, x_n) = \begin{cases} \dfrac{c}{\varrho} x_n & \text{for } (x_1, \ldots, x_n) \in [0, L]^{n-1} \times [-\varrho, \varrho] \\[2mm] u(x) & \text{otherwise} \end{cases}.$$

Then it is easy to check that for every open set A containing $[0, L]^{n-1} \times [-\varrho, \varrho]$ the inequality $F(v, A) < F(u, A)$ holds. □

An obvious consequence is that a piecewise constant function with jump set containing e.g. a half hyperplane cannot be a local minimizer in $\mathbf{R}^n$; in particular, in $\mathbf{R}^2$ a function with three different constant values in three congruent angles is not a local minimizer. This example is interesting because in two dimensions the singular points of the jump set $K(u)$ of an optimal segmentation are expected to be either terminal points (crack tips) or triple junctions of regular curves forming equal angles, and this last situation is asymptotically that one described before. We shall add some comments on this point at the end of this section.

The above remarks show that it is not easy to guess how a local minimizer should look; nevertheless in two space dimensions there are strong reasons to believe that the function (in polar coordinates)

$$u_0(\varrho, \vartheta) = \sqrt{\frac{2}{\pi}} \varrho^{1/2} \sin \frac{\vartheta}{2}$$

is a local minimizer in the whole of $\mathbf{R}^2$, even though no proof of this is known at the present time. A stronger conjecture (see [22]) is that u_0 is the *unique* nontrivial local minimizer in $\mathbf{R}^2$, up to rigid motions. Notice that u_0 is invariant under scaling (12), hence it is in some sense the analogous of the minimal cones in the theory of minimal surfaces, and moreover it has the nice property of "equipartition of energy", i.e.

$$\int_{B_\varrho} |\nabla u_0|^2 \, dx = \mathcal{H}^1(S_{u_0} \cap B_\varrho)$$

for every ball centered at 0 (the "crack tip" of the jump set). Moreover, u_0 satisfies all the necessary properties already presented, and even that one stated in the next lemma (see [17, Propositions 3.9 and 4.1] and [32]).

Proposition 16 *Let u be a local minimizer in $B_1 \subset \mathbf{R}^2$, and assume that its jump set S_u is given in polar coordinates as follows:*

$$S_u = \{(\varrho, \vartheta) : \ \vartheta = \psi(\varrho)\} \ ,$$

with $\psi \in C^{1,1}\big([0,1[\big) \cap C^2\big(]0,1[\big)$ and (without loss of generality) $\psi(0) = \pi$. Then u has the form

$$u(\varrho, \vartheta) = c \pm u_0(\varrho, \vartheta) + u_R(\varrho, \vartheta) \ ,$$

with $u_R(\varrho, \vartheta) = o(\varrho)$ and the angular coordinate ϑ is defined in such a way that it is continuous in $B_1 \setminus S_u$ and vanishes on the positive x axis; moreover

$$\lim_{\varrho \to 0} H(\varrho, \psi(\varrho)) = 0 \ ,$$

where $H(\varrho, \psi(\varrho))$ is the mean curvature of S_u.

Proof. We shall identify $\mathbf{R}^2$ with $\mathbf{C}$ as usual; by [37, Satz 7], $B_1 \setminus S_u$ can be conformally mapped onto the half disk $D = \{re^{i\varphi}, \ 0 < r < 1, \ |\varphi| < \pi/2\}$ through a transformation T whose asymptotic behaviour near the origin is $T(\varrho e^{i\vartheta}) = \varrho^{1/2} e^{i\vartheta/2} + o(\varrho^{1/2})$. Localizing in a half disk D_λ concentric to D with a small radius λ, the transformed function $v = u \circ T^{-1}$ solves the following mixed problem:

$$\begin{cases} \Delta v = 0 & \text{in } D_\lambda \\[2mm] v(\lambda, \varphi) = u \circ T^{-1}(\lambda, \varphi) & \text{for } -\pi/2 < \varphi < \pi/2 \\[2mm] \dfrac{\partial v}{\partial \nu}\left(r, \pm\dfrac{\pi}{2}\right) = 0 & \text{for } 0 \leq r < \lambda \ . \end{cases}$$

The solution can be computed by separation of the variables and coming back to u the expression

$$u(\varrho, \vartheta) = c + a_1 \varrho^{1/2} \sin \frac{\vartheta}{2} + a_2 \varrho \cos \vartheta + o(\varrho) \tag{22}$$

is found. Using the equality between the jump of $|\nabla u|^2$ across S_u and the mean curvature of S_u we deduce that (up to the sign) the mean curvature of S_u is given by $H(\varrho, \psi(\varrho)) = 2a_1 a_2 \varrho^{-1/2} \sin(\psi(\varrho)/2) + o(1)$; since by hypothesis H must stay bounded as $\varrho \to 0$, we get $a_1 a_2 = 0$. It remains to prove that $|a_1| = \sqrt{2/\pi}$, so that $a_2 = 0$. To this aim, insert in (11) a variation $\eta = (\eta_1, \eta_2)$ tangencial to S_u and use the divergence theorem on S_u (see [33, p. 43]), (11) and again the divergence theorem on the disk B_ε to obtain

$$\eta_1(0,0) = -\int_{S_u} \mathrm{div}^\tau \eta \, d\mathcal{H}^1 = \int_{B_1} \big[|\nabla u|^2 \mathrm{div}\, \eta - 2\langle \nabla u, \nabla u \nabla \eta\rangle\big] \, dx$$

$$= \lim_{\varepsilon \to 0} \int_{B_1 \setminus B_\varepsilon} \big[|\nabla u|^2 \mathrm{div}\, \eta - 2\langle \nabla u, \nabla u \nabla \eta\rangle\big] \, dx$$

$$= \lim_{\varepsilon \to 0} \int_{\partial B_\varepsilon} \big[|\nabla u|^2 \langle \eta, \nu_\varepsilon\rangle - 2\langle \nabla u, \eta\rangle \langle \nabla u, \nu_\varepsilon\rangle\big] \, d\mathcal{H}^1 \ ,$$

where ν_ε is the inward pointing unit normal to ∂B_ε. Using the above relation and (22) we deduce $a_1^2 = \pi/2$ and the proof is complete.

$\square$

Notice that the proof of the above proposition in particular shows that the function $u(\varrho, \vartheta) = c\varrho^{1/2} \sin\vartheta/2$ can be a local minimizer only if $c^2 = 2/\pi$.

Coming back to the Mumford-Shah conjecture, what is still missing is a deeper analysis of the singular set S: first of all, up to now, S is known only to be closed and $\mathcal{H}^1$ negligible, whereas it is expected to be (locally) finite. Moreover, the points of S are expected to be either points where different regular arcs meet or points where a regular arc ends; referring to (15), in the first situation $\mathbf{D}(x, \varrho) \to 0$ as $\varrho \to 0$, whereas $\mathbf{A}(x, \varrho) \not\to 0$, and in the second one $\mathbf{A}(x, \varrho) \to 0$, whereas $\mathbf{D}(x, \varrho) \not\to 0$. By (20) it readily follows that not more than three arcs may meet, and by first variation they must make congruent angles. At "crack tips" an asymptotic behaviour like that one described in Proposition 16 is expected, with a definite tangent in the limit (see [32, Section 3]). However, these results are not yet been proved without further regularity assumptions; for example, the "triple junction" property is proved in [32] under the hypothesis that $K(u)$ is C^2 up to the singular point. Further results in this direction have been proved, under a priori topological assumptions on $K(u)$, in [14].

In general space dimension n the singular set S is closed and $\mathcal{H}^{n-1}$ negligible, and it is expected (see [22]) to have locally finite $\mathcal{H}^{n-2}$ measure (remember that $\mathcal{H}^0$ is the counting measure), but the situation is much more obscure: it is likely true that different regular portions of $K(u)$ meeting together form (almost everywhere) regular surfaces of lower dimension and behave asymptotically as a minimal partition of $\mathbf{R}^n$ in a suitable sense, see [3]. This situation would be analogous to the triple junction in the plane, but a classification of minimal partitions of the space is known only in three dimensions, see [35].

As mentioned before, the study of the dimension and the structure of the singular set seems to be tied to a better understanding of the behaviour of blow-up limits of optimal segmentations, and of local minimizers in the whole space. Unfortunately, even in dimension 2, no (nontrivial) local minimizer in the whole space is known, and this is another open problem: to develop a method suitable to check that a given SBV function is a local minimizer.

References

1. Alberti, G. and Mantegazza, C.: A note on the theory of SBV functions, Boll. Un. Mat. Ital., to appear.
2. Allard, W. K.: On the first variation of a varifold, Ann. of Math. **95**, 1972, 417–491.
3. Almgren, F. J.: Existence and regularity almost everywhere of solutions to elliptic variational problems with constraints, Mem. Am. Math. Soc., **165**, 1976.
4. Ambrosio, L.: A compactness theorem for a new class of functions of bounded variation, Boll. Un. Mat. Ital. **3-B** (1989), 857-881.
5. Ambrosio, L.: Existence theory for a new class of variational problems, Arch. Rat. Mech. Anal. **111** (1990), 291-322.

6. Ambrosio, L.: A new proof of *SBV* compactness theorem, Calc. Var., **3** (1995), 127–137.

7. Ambrosio, L.: Free discontinuity problems and special function with bounded variation, Proc. 2nd european congress of Mathematics, Budapest, July 21-27, 1996, to appear.

8. Ambrosio, L., Fusco N. and Pallara, D.: Partial regularity of free discontinuity sets II, (1995), to appear.

9. Ambrosio, L., Fusco N. and Pallara, D.: Higher regularity of solutions of free discontinuity problems, (1996), to appear.

10. Ambrosio, L. and Pallara, D.: Partial regularity of free discontinuity sets I, (1994), to appear.

11. Ambrosio, L. and Pallara, D.: Partial regularity in free discontinuity problems, in: Progress in partial differential equations: The Metz surveys IV, (M. Chipot and I. Shafrir eds.), Pitman Res. Notes in Math. 345, 1996, 3-17.

12. Ambrosio, L. and Tortorelli, V. M.: Approximation of functionals depending on jumps by elliptic functionals via Γ-convergence, Comm. Pure Appl. Math. **43** (1990), 999–1036.

13. Ambrosio, L. and Tortorelli, V. M.: On the approximation of free discontinuity problems, Boll. Un. Mat. Ital. **6-B** (1992), 105–123.

14. Bonnet, A.: On the regularity of edges in image segmentation, to appear.

15. Brakke, K.: The motion of a surface by its mean curvature, Princeton U. P., 1978.

16. Brézis, H.: Analyse fonctionnelle, théorie et applications, Masson, 1982.

17. Carriero, M., Leaci, A., Pallara D. and Pascali, E.: Euler conditions for a minimum problem with free discontinuity set, Preprint Dip. di Matematica, **8** Lecce, 1988.

18. Chambolle, A.: Image segmentation by variational methods: Mumford and Shah functional and the discrete approximations, Siam J. Appl. Math. **55** (1995), 827-863.

19. Dal Maso, G., Morel, J. M. and Solimini, S.: A variational method in image segmentation: existence and approximation results, Acta Math., **168** (1992), 89–151.

20. David, G.: C^1-arcs for minimizers of the Mumford-Shah functional, to appear.

21. David, G. and Semmes, S.: On the singular set of minimizers of the Mumford-Shah functional, to appear.

22. De Giorgi, E.: Free discontinuity problems in calculus of variations, in: Frontiers in pure and applied mathematics, a collection of papers dedicated to J.L.Lions on the occasion of his 60th birthday, R. Dautray ed., North Holland, 1991.

23. De Giorgi, E. and Ambrosio, L.: Un nuovo tipo di funzionale del calcolo delle variazioni, Atti Accad. Naz. Lincei Cl. Sci. Fis. Mat. Nat. **(8) 82** (1988), 199-210.

24. De Giorgi, E., Carriero M. and Leaci, A.: Existence theorem for a minimum problem with free discontinuity set, Arch. for Rational Mech. Anal. **108** (1989), 195-218.

25. De Giorgi, E., Colombini F.and Piccinini, L. C.: Frontiere orientate di misura minima e questioni collegate, Quaderni Sc. Norm. Sup., Pisa, 1972.

26. Dibos, F. and Séré, E.: An approximation result for minimizers of the Mumford-Shah functional, preprint.

27. Evans, L. C. and Gariepy, R.: Measure theory and fine properties of functions, CRC Press, 1992.

28. Federer, H.: Geometric measure theory, Springer, 1969.

29. Friedman, A.: Variational principles and free boundary problems, Wiley, 1982.

30. Giusti, E.: Minimal surfaces and functions with bounded variation, Birkhäuser, 1984.

31. Morel, J. M. and Solimini, S.: Variational models in image segmentation, Birkhäuser, 1994.
32. Mumford, D. and Shah, J.: Optimal approximation by piecewise smooth functions and associated variational problems, Comm. on Pure and Appl. Math. **17** (1989) 577-685.
33. Simon, L.: Lectures on geometric measure theory, Proceedings of the Centre for Mathematical Analysis, Australian National University, Canberra 1983.
34. Simon, L.: Theorems on regularity and singularity of energy minimizing maps, Birkhäuser, 1996.
35. Taylor, J. E.: The structure of singularities in soap-bubble-like and soap-film-like minimal surfaces, Ann. of Math., **103**, (1976), 489-524.
36. Vol'pert, A. I.: Spaces BV and quasi-linear equations, Math. USSR Sb. **17** (1967), 225-267.
37. Warschawski, S.: Über das Randwerthalten der Ableitung der Abbildungsfunktion bei konformer Abbildung, Math. Z. **35** (1932), 321-456.
38. Ziemer, W. P.: Weakly differentiable functions, Springer, 1989.

Tests and Constructions of Irreducible Polynomials over Finite Fields

Shuhong Gao and Daniel Panario

Abstract. In this paper we focus on tests and constructions of irreducible polynomials over finite fields. We revisit Rabin's (1980) algorithm providing a variant of it that improves Rabin's cost estimate by a $\log n$ factor. We give a precise analysis of the probability that a random polynomial of degree n contains no irreducible factors of degree less than $O(\log n)$. This probability is naturally related to Ben-Or's (1981) algorithm for testing irreducibility of polynomials over finite fields. We also compute the probability of a polynomial being irreducible when it has no irreducible factors of low degree. This probability is useful in the analysis of various algorithms for factoring polynomials over finite fields. We present an experimental comparison of these irreducibility methods when testing random polynomials.

1 Motivation and results

For a prime power q and an integer $n \geq 2$, let $\mathbb{F}_q$ be a finite field with q elements, and $\mathbb{F}_{q^n}$ be its extension of degree n. Extensions of finite fields are important in implementing cryptosystems and error correcting codes. One way of constructing extensions of finite fields is via an irreducible polynomial over the ground field with degree equal to the degree of the extension. Therefore, finding irreducible polynomials and testing the irreducibility of polynomials are fundamental problems in finite fields.

A probabilistic algorithm for finding irreducible polynomials that works well in practice is presented in [26]. The central idea is to take polynomials at random and test them for irreducibility. Let I_n be the number of irreducible polynomials of degree n over a finite field $\mathbb{F}_q$. It is well-known (see [21], p. 142, Ex. 3.26 & 3.27) that

$$\frac{q^n}{n} - \frac{q(q^{n/2} - 1)}{(q-1)n} \leq I_n \leq \frac{q^n - q}{n}. \tag{1}$$

This means that a fraction $1/n$ of the polynomials of degree n are irreducible, and so we find on average one irreducible polynomial of degree n after n tries. In order to transform this idea into an algorithm one has to consider irreducibility tests. In sections 2 and 3, we focus on tests for irreducibility. Let $f \in \mathbb{F}_q[x]$, $\deg f = n$, be a polynomial to be tested for irreducibility. Assume that $p_1, \ldots, p_k$ are the distinct prime divisors of n. In practice, there are two general approaches for this problem:

- Butler (1954): f is irreducible if and only if $\dim \ker(\Phi - I) = 1$, where Φ is the Frobenius map on $\mathbb{F}_q[x]/(f)$ that sends $h \in \mathbb{F}_q[x]/(f)$ to $h^q \in \mathbb{F}_q[x]/(f)$, and I is the identity map on $\mathbb{F}_q[x]/(f)$ (see [4]);
- Rabin (1980): f is irreducible if and only if $\gcd(f, x^{q^{n/p_i}} - x) = 1$ for all $1 \leq i \leq k$, and $x^{q^n} - x \equiv 0 \bmod f$ (see [26]).

Other irreducibility tests can be found in [14], [31], and [12].

In this paper, we concentrate on Rabin's test, and a variant presented in [1]. In section 2, we review Rabin's and Ben-Or's irreducibility algorithms. We state a variant of Rabin's algorithm that allows a $\log n$ factor saving. In section 3, we focus on Ben-Or's algorithm. This leads us to study the behavior of *rough* polynomials, i.e., polynomials without irreducible factors of low degrees. The analysis is expressed as an asymptotic form in n, the degree of the polynomial to be tested for irreducibility. First, we fix a finite field $\mathbb{F}_q$, and then we study asymptotics on q. As was noted in [7], probabilistic properties of polynomials over finite fields frequently have a shape that resembles corresponding properties of the cycle decomposition of permutations to which they reduce when the size of the field goes to infinity. An instance of this is derived for the probability that a polynomial of degree n over $\mathbb{F}_q$ contains no factors of degree m, $1 \leq m \leq O(\log n)$, when $q \to +\infty$. This probability relates naturally with Ben-Or's algorithm. The probability of a polynomial being irreducible when it has no irreducible factors of low degree provides useful information for factoring polynomials over finite fields (see for instance [12], §6). We provide the probability of a polynomial being irreducible when it has no irreducible factors of degree at most $O(\log n)$.

In section 4, we give an experimental comparison on the algorithms discussed in section 2. We provide tables of running times of the algorithms for various fields and polynomial degrees. These results suggest that Ben-Or's algorithm has a much better average time behavior than others, even though its worst-case complexity is the worst.

Very sparse irreducible polynomials are useful for several applications: pseudorandom number generators using feedback shift registers ([15]), discrete logarithm over $\mathbb{F}_{2^n}$ ([6], [23]), and efficient arithmetic in finite fields (Shoup private communication, 1994). However, few results are known about these polynomials beyond binomials and trinomials (see [22], Chapter 3, and the references there). In section 5, we present a construction of irreducible polynomials over $\mathbb{F}_q$ of degree n with up to $O(1)$ nonzero terms (not necessarily the lowest coefficients), for infinitely many degrees n.

We assume that arithmetic in $\mathbb{F}_q$ is given. The cost measure of an algorithm will be the number of operations in $\mathbb{F}_q$. The algorithms in this paper use basic polynomial operations like products and gcds. We consider in this paper exclusively FFT based arithmetic; similar results hold for classical arithmetic. Let $M(n) = n \log n \log \log n$. For constants τ_1 and τ_2, the cost of multiplying two polynomials of degree at most n using "fast" arithmetic ([28], [27], [5]) can be taken as $\tau_1 M(n)$, and the cost of a gcd between two polynomials of degree at most n can be taken as $\tau_2 \log n M(n)$ operations in $\mathbb{F}_q$. The number of multiplications needed to compute $h^q \bmod f$ by means of the classical *repeated squaring* method (see [20], p. 441–442), where h is a polynomial over $\mathbb{F}_q$ of degree less than n, is $C_q = \lfloor \log_2 q \rfloor + \nu(q)$, with $\nu(q)$ the number of ones in the binary representation of q. Therefore, the cost of computing $h^q \bmod f$ by this method is $\tau_1 C_q M(n)$ operations in $\mathbb{F}_q$ using FFT based methods.

2   Irreducibility tests

In this section, we review Rabin's and Ben-Or's tests, and we present a variant of Rabin's method.

2.1   Rabin irreducibility test and an improvement

Algorithm: Rabin Irreducibility Test
Input: A monic polynomial $f \in \mathbb{F}_q[x]$ of degree n,
                and $p_1, \ldots, p_k$ all the distinct prime divisors of n .
Output: Either "f is irreducible" or "f is reducible".

```
for j := 1 to k do
    n_j := n/p_j;
for i := 1 to k do
    g := gcd(f, x^{q^{n_i}} − x mod f);
    if g ≠ 1, then 'f is reducible' and STOP;
endfor;
g := x^{q^n} − x mod f;
if g = 0, then 'f is irreducible'
          else 'f is reducible';
```

The correctness of Rabin's algorithm is based on the following fact (see [26], p. 275, Lemma 1).

Fact 2.1 *Let $p_1, \ldots, p_k$ be all the prime divisors of n, and denote $n/p_i = n_i$, for $1 \le i \le k$. A polynomial $f \in \mathbb{F}_q[x]$ of degree n is irreducible in $\mathbb{F}_q[x]$ if and only if $\gcd\left(f, x^{q^{n_i}} - x \bmod f\right) = 1$, for $1 \le i \le k$, and f divides $x^{q^n} - x$.*

The basic idea of this algorithm is to compute $x^{q^{n_i}} \bmod f$ independently for each value $n_1, \ldots, n_k$ by repeated squaring, and then to take the correspondent gcd. The worst-case analysis given in [26] is $O\left(nM(n) \log n \log q\right)$ operations in $\mathbb{F}_q$. However, it can be shown that $O\left(nM(n) \log \log n \log q\right)$ is an upper bound of the number of operations in $\mathbb{F}_q$ for this algorithm. Indeed, first note that the number of distinct prime factors of n is at most $\log n$. The cost of k exponentiations is

$$\sum_{i=1}^{k} \frac{n}{p_i} \log q \, M(n) \le nM(n) \log q \sum_{i=1}^{\log n} \frac{1}{p_i} \le nM(n) \log q \, H_{\log n},$$

where $H_m = \sum_{k=1}^{m} 1/k$ is the harmonic sum. Using the well-known approximation of the harmonic sum ([16], p. 452), $H_{\log n} = \log \log n + \gamma + O(\frac{1}{\log n})$, we obtain $O(nM(n) \log \log n \log q)$, which dominates the cost $O(M(n) \log^2 n)$ of computing k gcd's. Therefore, the total cost of Rabin's algorithm is $O(nM(n) \log \log n \log q)$.

As an improvement, we propose the following variant for the computation of $x^{q^{n_i}} - x \bmod f$, for $1 \le i \le k$.

Algorithm: Variant of Rabin Irreducibility Test
Input: A monic polynomial $f \in \mathbb{F}_q[x]$ of degree n,
           and $p_1, \dots, p_k$ all the distinct prime divisors of n .
Output: Either "f is irreducible" or "f is reducible".

```
n₀ := 0;  h₀ := x;
for j := 1 to k do
     nⱼ := n/pⱼ;
sort(n₁,...,nₖ);   (* Assume n₁ < n₂ < ... < nₖ. *)
for i := 1 to k do
     hᵢ := h_{i-1}^{q^{nᵢ-n_{i-1}}}  mod f;
     g := gcd(f, hᵢ - x);
     if g ≠ 1, then 'f is reducible' and STOP;
endfor;
g := hₖ^{q^{n-nₖ}} - x mod f;
if g = 0, then 'f is irreducible'
          else 'f is reducible';
```

Theorem 2.2 *The above variant of Rabin's algorithm correctly tests for polynomial irreducibility, and uses $O(nM(n) \log q)$ operations in $\mathbb{F}_q$.*

Proof. The correctness of the algorithm follows from the correctness of the power computations. We prove that $h_i = x^{q^{n_i}} \bmod f$, $1 \le i \le k$, by induction on k. Basis: when $k = 1$,

$$h_1 \equiv h_0^{q^{n_1-n_0}} \equiv \left(x^{q^{n_0}}\right)^{q^{n_1-n_0}} \equiv x^{q^{n_0} \cdot q^{n_1}} \equiv x^{q^{n_1}} \bmod f.$$

Inductive step: for some k, $h_i = x^{q^{n_i}} \bmod f$, $1 \le i \le k$. Then,

$$h_{k+1} \equiv h_k^{q^{n_{k+1}-n_k}} \equiv \left(x^{q^{n_k}}\right)^{q^{n_{k+1}-n_k}} \equiv x^{q^{n_k} \cdot q^{n_{k+1}-n_k}} \equiv x^{q^{n_{k+1}}} \bmod f.$$

With this variant, in the worst-case, the number of polynomial multiplications in Rabin's algorithm to compute all powers using repeated squaring is

$$n_1 \log q + (n_2 - n_1) \log q + \cdots + (n - n_k) \log q = n \log q.$$

Hence the cost of k exponentiations is $O(nM(n)\log q)$. Since the number of distinct prime factors of n is at most $\log n$, the cost of taking all the gcd's in the algorithm is $O(M(n)\log^2 n)$. Therefore the total cost of this variant is $O(nM(n)\log q)$. $\qquad\qquad\qquad\qquad\qquad\qquad\qquad\qquad\qquad\qquad\qquad\qquad\qquad$ $\square$

2.2   Ben-Or irreducibility test

Algorithm: Ben-Or Irreducibility Test
Input: A monic polynomial $f \in \mathbb{F}_q[x]$ of degree n.
Output: Either "f is irreducible" or "f is reducible".

```
for i := 1 to n/2 do
    begin
        g := gcd(f, x^(q^i) - x mod f);
        if g ≠ 1, then 'f is reducible' and STOP;
    end;
'f is irreducible';
```

The correctness of Ben-Or's procedure is based on the following fact (see [21], p. 91, Theorem 3.20).

Fact 2.3 *For $i \geq 1$, the polynomial $x^{q^i} - x \in \mathbb{F}_q[x]$ is the product of all monic irreducible polynomials in $\mathbb{F}_q[x]$ whose degree divides i.*

Indeed, Ben-Or's algorithm computes $x^{q^i} \bmod f$, and $\gcd(f, x^{q^i} - x)$ for $i = 1, \ldots, \frac{n}{2}$. The polynomial is reducible if and only if one of the gcd's is different from 1.

In the worst case, this algorithm computes $\frac{n}{2}$ times a qth power and a gcd of polynomials of degree at most n. Recalling the cost of these operations from section 1, the worst-case behavior of Ben-Or's algorithm is $O(nM(n)\log(qn))$ using FFT based multiplication algorithms, and therefore, it is worse than our Rabin's variant. However, as can be seen from our theoretical and experimental results in Sections 3 and 4, Ben-Or's algorithm is very efficient. The main reason for the efficiency of this algorithm is that random polynomials of large degree are very likely to have an irreducible factor of small degree, and Ben-Or's algorithm quickly discards these polynomials (see also [19]).

We should point out that the average cases of Rabin and our variant are not known. Ben-Or's average-case analysis is only known when q goes to infinity in the following sense. Let $s(f)$ be the expected value of the smallest degree among the irreducible factors of f; then the expected cost of Ben-Or's algorithm is $O(s(f)M(n)\log(qn))$. Ben-Or ([1], Theorem 2) derives an $O(\log n)$ estimate for $s(f)$. In fact, he relates the factorial decomposition of polynomials with the cyclic decomposition of permutations. The result follows from the study of the expected length of the shortest cycle in a random permutation ([30]). However, this relation between irreducible factors of polynomials and cycles of permutations just holds when the size of the field is large, as it was observed in [7].

3   Distribution of rough polynomials

Polynomials without irreducible factors of low degree make Ben-Or's irreducibility test to execute a large number of iterations. The probability that a random polynomial of degree n contains no factors of low degree gives meaningful information on the behavior of Ben-Or's algorithm. We call a polynomial m-*rough* if it has no irreducible factors of degrees $\leq m$. In this section we are interested in the distribution of rough polynomials.

The following theorem is proved in [2] when m is fixed.

Theorem 3.1 *Denote by $P_q(n, m)$ the probability of a random monic polynomial of degree n over $\mathbb{F}_q$ being m-rough. Then when $n \to \infty$,*

$$P_q(n, m) = \prod_{k=1}^{m} \left(1 - \frac{1}{q^k}\right)^{I_k} (1 + o(1)),$$

uniformly for q and $1 \leq m \leq O(\ln n)$.

Proof. Let $\mathcal{I}$ be the collection of all monic irreducible polynomials in $\mathbb{F}_q$. Formally, the summation of all monic polynomials with all irreducible factors with degree $> m$ is

$$P = \prod_{\omega \in \mathcal{I},\ |\omega| > m} (1 + \omega + \omega^2 + \cdots) = \prod_{\omega \in \mathcal{I},\ |\omega| > m} (1 - \omega)^{-1}.$$

Let z be a formal variable, and $|\omega|$ the degree of $\omega \in \mathcal{I}$. The substitution $\omega \mapsto z^{|\omega|}$ produces the generating function $P_m(z)$ of polynomials with all irreducible factors having degrees $> m$

$$P_m(z) = \prod_{\omega \in \mathcal{I},\ |\omega| > m} \left(1 - z^{|\omega|}\right)^{-1} = \prod_{k > m} (1 - z^k)^{-I_k} = \frac{1}{1 - qz} \prod_{k=1}^{m} (1 - z^k)^{I_k}.$$

Note that m may vary when $n \to +\infty$, and thus we can not apply the transfer lemmas in [8, 24].

As usual, $[z^n] P_m(z)$ represents the coefficient of z^n in $P_m(z)$, and observe that $P_q(n, m) = [z^n] P_m(z)/q^n$. In order to estimate $P_q(n, m)$, we apply Theorem 10.8 in [24]. $P_m(z)$ presents a pole of order 1 at $z = \frac{1}{q}$ with residue

$$-\frac{1}{q} \prod_{k=1}^{m} (1 - q^{-k})^{I_k}.$$

Denote by $g_q(m)$ the product $\prod_{k=1}^{m}(1-q^{-k})^{I_k}$. Suppose that $m \leq c\ln n$ for some constant $c > 0$. Let b be a constant such that $1 < b < e^{1/c}$, and take $r = \frac{b}{q} > \frac{1}{q}$. By Odlyzko ([24], Theorem 10.8),

$$\left|[z^n]P_m(z) + \left(-\frac{1}{q}g_q(m)\right)\left(\frac{1}{q}\right)^{-n-1}\right| \leq w\,r^{-n} + \left(r - \frac{1}{q}\right)^{-1}r^{-n}\frac{1}{q}g_q(m),$$

where $w = \max_{|z|=r}|P_m(z)|$. Therefore,

$$|P_q(n,m) - g_q(m)| \leq \left(w + \left(r - \frac{1}{q}\right)^{-1}\frac{1}{q}g_q(m)\right)(rq)^{-n}$$

$$= \left(w + \frac{1}{b-1}g_q(m)\right)/b^n \leq \left(w + \frac{1}{b-1}\right)/b^n,$$

as $0 \leq g_q(m) \leq 1$. Since $b > 1$ is a constant independent of m, q and n, we only need to estimate w in term of n. When $|z| = r = \frac{b}{q}$, $|1 - qz| \geq b - 1$, and $|1 - z^k| \leq 1 + r^k$. Considering that $I_k\,r^k \leq q^k\,r^k = b^k$, we obtain

$$|P_m(z)| = \frac{1}{|1-qz|}\prod_{k=1}^{m}|1 - z^k|^{I_k} \leq \frac{1}{b-1}\prod_{k=1}^{m}(1 + r^k)^{I_k}$$

$$= \frac{1}{b-1}\prod_{k=1}^{m}\exp\left(I_k\log(1 + r^k)\right) \leq \frac{1}{b-1}\prod_{k=1}^{m}\exp(I_k\,r^k)$$

$$\leq \frac{1}{b-1}\prod_{k=1}^{m}\exp(b^k) = \frac{1}{b-1}\exp\left(\sum_{k=1}^{m}b^k\right) \leq \frac{1}{b-1}\exp\left(\frac{b^{m+1}}{b-1}\right)$$

$$\leq \frac{1}{b-1}\exp\left(\frac{b}{b-1}b^{c\ln n}\right) = \frac{1}{b-1}\exp\left(\frac{b}{b-1}n^{c\ln b}\right).$$

Hence, $w \leq \frac{1}{b-1}\exp\left(\frac{b}{b-1}n^{c\ln b}\right)$, and

$$\left(w + \frac{1}{b-1}\right)/b^n \leq \frac{2}{b-1}\exp\left(\frac{b}{b-1}n^{c\ln b} - (\ln b)\,n\right).$$

By the next theorem,

$$g_q(m) \geq \frac{1}{em} \geq \frac{1}{ec\log n}.$$

Therefore,

$$\left|\frac{P_q(n,m)}{g_q(m)} - 1\right| \leq \frac{2}{b-1}ec\log n \exp\left(\frac{b}{b-1}n^{c\ln b} - (\ln b)\,n\right). \qquad (2)$$

As $\ln b > 0$ and $c\ln b < 1$, the right-hand of (2) approaches to 0 as $n \to \infty$. Since the quantity on the right-hand of (2) is independent of q and m, we see that $P_q(n,m)/g_q(m)$ approaches to 1 uniformly for q and $m \leq c\lg n$ when $n \to \infty$. This completes the proof.   □

In the next theorem we estimate the function $g_q(m) = \prod_{k=1}^m \left(1 - \frac{1}{q^k}\right)^{I_k}$.

Theorem 3.2 *For any prime power q and positive integer m, we have*

$$\frac{1}{em} \leq \exp\left(-\sum_{k=1}^m \frac{1}{k}\right) \leq \prod_{k=1}^m (1 - q^{-k})^{I_k} \leq \left(1 - \frac{1}{\sqrt{q}}\right)^{-\frac{q}{q-1}} \cdot \exp\left(-\sum_{k=1}^m \frac{1}{k}\right).$$

When $q \to \infty$, we have

$$g_q(m) = \prod_{k=1}^m \left(1 - \frac{1}{q^k}\right)^{I_k} \to e^{-H_m} \sim \frac{e^{-\gamma}}{m},$$

where γ is the Euler's constant, and $e^{-\gamma} = 0.56416\ldots$

Proof. Note that $\log(1+x) \leq x$ for $x > -1$, and $\sum_{k=1}^\infty \frac{1}{kq^{k/2}} = -\log\left(1 - \frac{1}{\sqrt{q}}\right)$. By (1), we have

$$g_q(m) = \prod_{k=1}^m \exp\left(I_k \log\left(1 - \frac{1}{q^k}\right)\right)$$

$$\leq \prod_{k=1}^m \exp\left(-\frac{I_k}{q^k}\right) = \exp\left(-\sum_{k=1}^m \frac{I_k}{q^k}\right) \leq \exp\left(-\sum_{k=1}^m \frac{\frac{q^k}{k} - \frac{q(q^{k/2}-1)}{(q-1)k}}{q^k}\right)$$

$$\leq \exp\left(-\sum_{k=1}^m \frac{1}{k}\right) \cdot \exp\left(\frac{q}{q-1} \sum_{k=1}^m \frac{q^{k/2} - 1}{kq^k}\right)$$

$$\leq \exp\left(-\sum_{k=1}^m \frac{1}{k}\right) \cdot \left(\exp\left(\sum_{k=1}^\infty \frac{1}{kq^{k/2}}\right)\right)^{\frac{q}{q-1}}$$

$$= \left(1 - \frac{1}{\sqrt{q}}\right)^{-\frac{q}{q-1}} \cdot \exp\left(-\sum_{k=1}^m \frac{1}{k}\right).$$

We have derived the upper bound for $g_q(m)$. The lower bound was derived by the authors when studying lower bounds for the Euler totient function for polynomials and for the density of normal elements (see [11]). Since it is simple, we reproduce it here. As $I_k \leq (q^k - 1)/k$ and $0 < 1 - 1/q^k < 1$, we have

$$g_q(m) = \prod_{k=1}^m \left(1 - \frac{1}{q^k}\right)^{I_k} \geq \prod_{k=1}^m \left(1 - \frac{1}{q^k}\right)^{\frac{q^k-1}{k}}$$

$$= \prod_{k=1}^m \left(\left(1 + \frac{1}{q^k - 1}\right)^{q^k-1}\right)^{-\frac{1}{k}} \geq \prod_{k=1}^m \exp\left(-\frac{1}{k}\right) = \exp\left(-\sum_{k=1}^m \frac{1}{k}\right).$$

Let H_m be the harmonic sum, i.e., $H_m = \sum_{k=1}^{m} 1/k$. Then $H_m \leq 1 + \int_1^m 1/x\,dx = 1 + \log m$, and thus $e^{-H_m} \geq 1/(em)$. When $q \to \infty$

$$g_q(m) = \prod_{k=1}^{m} \left(1 - \frac{1}{q^k}\right)^{I_k} \sim e^{-H_m}.$$

Using the well-known approximation of the harmonic sum $H_m = \log m + \gamma + O(\frac{1}{m})$, we have

$$e^{-H_m} \sim \frac{e^{-\gamma}}{m}, \quad m \to \infty. \tag{3}$$

Finally, this result is in accordance with the correspondent one of permutations with no cycles of length m or less (see [29]).   □

We provide in Table 1 below the values of $g_q(m)$, $P_q(n, m)$ and their ratio, when $m = \log n$ and $q = 2$ for several $n < 1000$. This shows that the convergence of $g_q(m)$ to $P_q(n, m)$ is very fast. Moreover, as n grows, $P_q(n, m)$ quickly decreases. For instance, for a random polynomial of degree 900, there is a probability of more than 0.9 of having a factor of degree at most 9. This is another explanation for the efficiency of Ben-Or's algorithm. Indeed, it is enough to search for irreducible polynomials of degree at most $O(\log n)$ in order to have a high probability of finding a factor.

For the remaining of this section we concentrate on the probability that a polynomial be irreducible if it has no irreducible factors of low degree. Many algorithms for factoring polynomials over finite fields comprise the following three stages: *squarefree factorization* (replace a polynomial by a squarefree one which contains all the irreducible factors of the original polynomial with exponents reduced to 1); *distinct-degree factorization* (split a squarefree polynomial into a product of polynomials whose irreducible factors have all the same degree); and *equal-degree factorization* (factor a polynomial whose irreducible factors have the same degree).

As things now stand, distinct-degree factorization is the bottleneck of the polynomial factorization problem (see [14], [18], [13]). This step of the factorization process works as follows: at any point k, all the irreducible factors of degree up to k have been found, and all the irreducible factors of degree greater than k remain to be determined from a factor g.

A natural way of improving the distinct-degree factorization step is by testing the irreducibility of the remaining factor g. Unfortunately, in the worst-case asymptotic scenario, the cost of the irreducibility test is about the same as the distinct-degree factorization algorithm. An alternative to overcome this problem is given in [12] (§6). The central idea is to run the irreducibility test and the distinct-degree factorization algorithm in parallel, feeding the former with partial information obtained by the latter (see the details in [12]).

It is clear that the probability of a monic polynomial being irreducible when it has no irreducible factors of low degree provides useful information in the above process. In the following, we derive an asymptotic formula for this probability.

n	m	$P_q(n,m)$	$g_q(m)$	P/g
2	1	.25000000000000000000000000	.25000000000000000000000000	1.0000
3	1	.25000000000000000000000000	.25000000000000000000000000	1.0000
4	2	.18750000000000000000000000	.18750000000000000000000000	1.0000
5	2	.18750000000000000000000000	.18750000000000000000000000	1.0000
6	2	.18750000000000000000000000	.18750000000000000000000000	1.0000
7	2	.18750000000000000000000000	.18750000000000000000000000	1.0000
8	3	.14062500000000000000000000	.14355468750000000000000000	.97963
9	3	.14453125000000000000000000	.14355468750000000000000000	1.0068
10	3	.14355468750000000000000000	.14355468750000000000000000	.99997
20	4	.11828613281250000000000000	.11828541755676269531250000	1.0000
30	4	.11828541755676269531250000	.11828541755676269531250000	1.0000
40	5	.09776907367631793022155762	.09776907366723652792472876	1.0000
50	5	.09776907366723719405854354	.09776907366723652792472876	1.0000
60	5	.09776907366723652792472876	.09776907366723652792472876	1.0000
70	6	.08484899050039278356888779	.08484899050039278175814854	1.0000
80	6	.08484899050039278175860054	.08484899050039278175814854	1.0000
90	6	.08484899050039278175814857	.08484899050039278175814854	1.0000
100	6	.08484899050039278175814854	.08484899050039278175814854	1.0000
200	7	.07367738498865927351164168	.07367738498865927351164168	.99999
300	8	.06551498664534958936373162	.06551498664534958936373162	1.0000
400	8	.06551498664534958936373162	.06551498664534958936373162	1.0000
500	8	.06551498664534958936373162	.06551498664534958936373162	1.0000
600	9	.05872097388539275926570230	.05872097388539275926570230	1.0000
700	9	.05872097388539275926570230	.05872097388539275926570230	1.0000
800	9	.05872097388539275926570230	.05872097388539275926570230	1.0000
900	9	.05872097388539275926570230	.05872097388539275926570230	1.0000

Table 1. Values of $P_q(n,m)$ and $g_q(m)$, with $m = \log n$ and $q = 2$.

Theorem 3.3 *Let $P_q^I(n,m)$ be the probability that a polynomial of degree n over $\mathbb{F}_q$ be irreducible if it has no factors of degree less than or equal to m, $1 \le m \le O(\ln n)$. Then, as n, m and q approach to infinity,*

$$P_q^I(n,m) \sim e^\gamma \frac{m}{n},$$

where γ is the Euler's constant.

Proof. This probability can be estimated considering the subset of irreducible polynomials of degree n over $\mathbb{F}_q$ inside the set of polynomials of degree n over $\mathbb{F}_q$ without irreducible factors of degree less than or equal to m, $1 \le m \le O(\ln n)$. Using (1), Theorems 3.1 and 3.2, when n, m and q approach to infinity, we obtain

$$P_q^I(n,m) = \frac{I_n}{q^n\, P_q(n,m)} \sim \frac{\frac{1}{n}}{\frac{e^{-\gamma}}{m}} = e^\gamma \frac{m}{n}.$$

4   Experimental results

In this section, we describe an implementation of the algorithms discussed in Section 2. We provide a running time comparison of the algorithms for random polynomials on a Sun Sparc 20 computer. The algorithms were implemented on a C++ software due to Shoup. This package contains classes for finite fields and polynomials over finite fields with implementations for basic operations such as multiplication, taking gcd, and so on (for a description of the software see [32]).

We tested all algorithms with the same random polynomials. A summary table for the average time in seconds of CPU in the case $\mathbb{F}_2$ is presented in Table 2 and Table 3. The degrees n in Table 2 and in Table 3 were chosen such that n has many prime divisors and few prime divisors, respectively. The number of polynomials tested was $10 * n$ for Table 2 and $5 * n$ for Table 3, where n is the degree of the polynomials being tested for irreducibility. It can be seen from the Tables that even in the case of n with few prime divisors Ben-Or has the best behavior among the three algorithms. The worst-case scenario for these algorithms happens when testing irreducible polynomials. We include a column with the number of irreducible polynomials that were tested for each degree.

n	Rabin	Rabin's Variant	Ben-Or	Number of Irreducible
105	0.7990	0.6133	0.2000	9
210	2.7652	2.6942	0.9938	14
330	10.5672	17.4147	2.3443	8
420	10.5702	4.3971	1.5189	7

Table 2. Average time in seconds for testing $10 * n$ polynomials over $\mathbb{F}_2$ of degree n with many prime divisors.

n	Rabin	Rabin's Variant	Ben-Or	Number of Irreducible
101	4.1564	8.9188	0.5901	4
256	20.8451	46.9867	2.7950	5
331	32.7156	71.8685	2.7719	9

Table 3. Average time in seconds for testing $5 * n$ polynomials over $\mathbb{F}_2$ of degree n with few prime divisors.

Similar results for testing irreducibility of polynomials over the finite field $\mathbb{F}_{1021}$ are presented in Table 4 below.

n	Rabin	Rabin's Variant	Ben-Or	Number of Irreducible
101	186.1383	280.2633	13.0198	4
105	78.2952	61.8381	11.2629	5
210	227.5800	174.7400	23.0000	3

Table 4. Average time in seconds for testing $5*n$ polynomials of degree n over $\mathbb{F}_{1021}$.

We also tested the case of very large fields. For instance, the average time in seconds of CPU for testing 315 polynomials of degree 105 over $\mathbb{F}_p$, p a prime with 100 bits, was:

Rabin	774.054
Rabin's variant	613.267
Ben-Or	137.565

In this case, 5 irreducible polynomials were found.

These timings suggest that Ben-Or's algorithm has a much better average time behavior than others, even though its worst-case complexity is the worst among them. A variant of the above algorithms that economizes gcd's computations can be given using Ben-Or's ideas up to $O(\log n)$ iterations and our Rabin's variant after that point. For this algorithm together with more experimental results, see [25].

5   Construction of sparse irreducible polynomials

When implementing the irreducibility tests in Section 2, we wanted to experiment our programs on various polynomials of large degrees. For reducible polynomials, they are likely to have small factors, as shown in Section 3, and the programs terminate almost immediately. However, when testing an irreducible polynomial, we do not have a priori any idea of how long it will take to complete the task. It is desirable to have some simple polynomials which we know in advance are irreducible so that we can test the correctness of our programs and know the approximate time our computer needs on various degrees. This would also help us in deciding the range of degrees to compare the tests. By "simple", we mean polynomials that can either be constructed easily (without testing for irreducibility) or have only a few nonzero terms. The problem of constructing sparse irreducible polynomials is also of independent interest.

A well-known open problem is to construct irreducible polynomials over $\mathbb{F}_2$ of degree n with at most $O(\log n)$ nonzero terms in its lowest coefficients. These polynomials are useful in the discrete logarithm problem ([6], [23]). Shoup (private communication, 1994) points out that if $\mathbb{F}_{2^n} = \mathbb{F}_2[x]/(f)$ with $f = x^n + g$ irreducible and $g \in \mathbb{F}_2[x]$ of small degree, say $\deg g \leq 2 \log n$, then exponentiation in $\mathbb{F}_{2^n}$ can be achieved with $O(n^2 \log\log n)$ operations in $\mathbb{F}_2$ and using storage for $O(n/\log n)$ elements from $\mathbb{F}_{2^n}$ (see also [9]). Experimental results show that such polynomial f exists for $n \leq 1000$ taking $\deg g \leq 2 + \log_2 n$.

Shparlinski ([33]) gives a construction of irreducible polynomials with degrees of the form $4 \cdot 3^k \cdot 5^\ell$ over $\mathbb{F}_2$, $4 \cdot 2^k \cdot 5^\ell$ over $\mathbb{F}_3$, and $2 \cdot 2^k \cdot 3^\ell$ over $\mathbb{F}_p$, for any prime $p > 3$, and k, l nonnegative integers. In the following, we first generalize his construction and then construct explicitly several infinite families of irreducible polynomials.

Theorem 5.1 *Let $p_1, \ldots, p_k$ be any fixed primes and $n = m p_1^{e_1} \ldots p_k^{e_k}$ where m is the multiplicative order of q modulo $p_1 \cdots p_k$ and $e_1, \ldots, e_k$ are arbitrary nonnegative integers. Then, over any finite field $\mathbb{F}_q$ whose characteristic is distinct from $p_1, \ldots, p_k$, there is an irreducible polynomial of degree n with at most $2m + 1 \leq 2(p_1 - 1) \cdots (p_k - 1) + 1$ nonzero terms.*

Proof. Let $\ell = m$ if all p_i are odd. If one of p_i, say p_1, is 2 and $q \equiv 3 \bmod 4$, let $\ell = \mathrm{lcm}(2, m)$. Then, $p_i | (q^\ell - 1)$ for $1 \leq i \leq \ell$, and $4 | (q^\ell - 1)$ if p_1 is even.

Let β be an element in $\mathbb{F}_{q^\ell}$ that is not a p_ith power in $\mathbb{F}_{q^\ell}$ for $1 \leq i \leq k$. The minimal polynomial $\zeta(x)$ of β over $\mathbb{F}_q$ has degree ℓ. By Lidl and Niederreiter ([21], p. 124, Theorem 3.75),

$$x^{p_1^{e_1} p_2^{e_2} \cdots p_k^{e_k}} - \beta$$

is irreducible over $\mathbb{F}_{q^\ell}$ for all nonnegative integers $e_1, \ldots, e_k$. Then

$$\zeta\left(x^{p_1^{e_1} p_2^{e_2} \cdots p_k^{e_k}}\right) \tag{4}$$

is irreducible over $\mathbb{F}_q$ of degree $\ell p_1^{e_1} p_2^{e_2} \cdots p_k^{e_k}$. The polynomial (4) has at most $\ell + 1 \leq 2m + 1$ nonzero terms.

Irreducible polynomials from the construction of Theorem 5.1 are usually very sparse. In fact, the number of nonzero terms depends only on the prime factors of n, and if we fix them and let the exponents grow arbitrarily then these polynomials have only $O(1)$ nonzero terms (even though not necessarily the lowest coefficients). This can be seen from the following examples.

Example 5.2 *$q \equiv 1 \bmod 4$ and $n = 2^k$. Take $a \in \mathbb{F}_q$ to be any quadratic nonresidue. Then $x^{2^k} - a$ is irreducible over $\mathbb{F}_q$ for all $k \geq 0$. For instance,*

i. if $q = p \equiv \pm 3 \bmod 8$ is a prime, then take $a = 2$;
ii. if $q = p \equiv \pm 5 \bmod 12$ is a prime, then take $a = 3$;
iii. if $q = p \equiv \pm 2 \bmod 5$ is a prime, then take $a = 5$.

These results are from [17] (Proposition 5.1.3 for the first two, and Theorem 2, p. 54, for the third).

Example 5.3 $q \equiv 3 \bmod 4$ *and* $n = 2^k$. *In this case,* $\ell = 2$ *in the proof of Theorem 5.1. We need to find a quadratic nonresidue in* $\mathbb{F}_{q^2}$ *and compute its minimal polynomial. The following elegant solution is from [3]. Suppose that* $q = p^m$ *where m is odd and $p \equiv 3 \bmod 4$ is a prime. Let* $2^v \mid (p+1)$, $2^{v+1} \nmid (p+1)$. *Then* $v \geq 2$. *Construct* $u \in \mathbb{F}_p$ *iteratively as follows:*

$$u_1 = 0,$$

$$u_i = \pm \left(\frac{u_{i-1} + 1}{2} \right)^{\frac{p+1}{4}} \pmod{p}, \quad for \ 1 < i < v,$$

$$u_v = \pm \left(\frac{u_{v-1} - 1}{2} \right)^{\frac{p+1}{4}} \pmod{p},$$

where, at each step, one can take any of the signs arbitrarily. Let $u = u_v$. *Then* $x^2 - 2ux - 1$ *is the minimal polynomial of some quadratic nonresidue in* $\mathbb{F}_{p^2}$. *Therefore*

$$x^{2^k} - 2ux^{2^{k-1}} - 1$$

is irreducible over $\mathbb{F}_p$, *and over* $\mathbb{F}_q$ *as well, for all* $k \geq 1$.

Example 5.4 $q = 2$. *Theorem 5.1 yields the following families of irreducible polynomials over* $\mathbb{F}_2$ *for all* $k, \ell, m, n \geq 0$:

$$x^{2 \cdot 3^k} + x^{3^k} + 1$$
$$x^{3 \cdot 7^k} + x^{7^k} + 1$$
$$x^{4 \cdot 3^k \cdot 5^\ell} + x^{3^k \cdot 5^\ell} + 1$$
$$x^{6 \cdot 3^k \cdot 7^\ell} + x^{3^k \cdot 7^\ell} + 1$$
$$x^{10 \cdot 3^k \cdot 11^\ell \cdot 31^m} + x^{3 \cdot 3^k \cdot 11^\ell \cdot 31^m} + 1$$
$$x^{12 \cdot 3^k \cdot 5^\ell \cdot 7^m \cdot 13^n} + x^{8 \cdot 3^k \cdot 5^\ell \cdot 7^m \cdot 13^n} + x^{2 \cdot 3^k \cdot 5^\ell \cdot 7^m \cdot 13^n} + x^{3^k \cdot 5^\ell \cdot 7^m \cdot 13^n} + 1.$$

For other explicit constructions of irreducible polynomials, see [22] (Chapter 3), and [10].

Acknowledgement. We thank Bruce Richmond for helpful discussions and Victor Shoup for the use of his package. Part of this work was done by the first author while visiting the University of Waterloo whose hospitality and support are gratefully acknowledged.

References

1. BEN-OR, M. Probabilistic algorithms in finite fields. In *Proc. 22nd IEEE Symp. Foundations Computer Science* (1981), pp. 394–398.
2. BLAKE, I., GAO, S., AND LAMBERT, R. Constructive problems for irreducible polynomials over finite fields. In *Information Theory and Applications*, A. Gulliver and N. Secord, Eds., vol. 793 of *Lecture Notes in Computer Science*. Springer-Verlag, 1994, pp. 1–23.

3. BLAKE, I., GAO, S., AND MULLIN, R. Explicit factorization of $x^{2^k} + 1$ over $\mathbb{F}_p$ with prime $p \equiv 3 \pmod 4$. *Appl. Alg. Eng. Comm. Comp.* 4 (1993), 89–94.

4. BUTLER, M. On the reducibility of polynomials over a finite field. *Quart. J. Math. Oxford 5* (1954), 102–107.

5. CANTOR, D., AND KALTOFEN, E. On fast multiplication of polynomials over arbitrary algebras. *Acta. Inform. 28* (1991), 693–701.

6. COPPERSMITH, D. Fast evaluation of logarithms in fields of characteristic two. *IEEE Trans. Info. Theory 30* (1984), 587–594.

7. FLAJOLET, P., GOURDON, X., AND PANARIO, D. Random polynomials and polynomial factorization. In *Proc. 23rd ICALP Symp.* (1996), vol. 1099 of *Lecture Notes in Computer Science*, Springer-Verlag, pp. 232–243.

8. FLAJOLET, P., AND ODLYZKO, A. Singularity analysis of generating functions. *SIAM Journal on Discrete Mathematics 3 2* (1990), 216–240.

9. GAO, S., VON ZUR GATHEN, J., AND PANARIO, D. Gauss periods and efficient arithmetic in finite fields. Submitted to *Journal of Symbolic Computation* (extended abstract in *Proc. LATIN'95*, vol. 911 of *Lecture Notes in Computer Science*, 311–322), 1995.

10. GAO, S., AND MULLEN, G. Dickson polynomials and irreducible polynomials over finite fields. *J. Number Theory 49* (1994), 118–132.

11. GAO, S., AND PANARIO, D. Density of normal elements. Submitted to *Finite Fields and their Applications* (abstract in *AMS Abstracts* **102** Fall 1995, # 904-68-227, p. 798), 1995.

12. VON ZUR GATHEN, J., AND GERHARD, J. Arithmetic and factorization of polynomials over $\mathbb{F}_2$. In *Proc. ISSAC'96, Zürich, Switzerland* (1996), L. Y.N., Ed., ACM press, pp. 1–9.

13. VON ZUR GATHEN, J., AND PANARIO, D. A survey on factoring polynomials over finite fields. Submitted to the special issue of the MAGMA conference in *J. Symb. Comp.*, 1996.

14. VON ZUR GATHEN, J., AND SHOUP, V. Computing Frobenius maps and factoring polynomials. *Comput complexity 2* (1992), 187–224.

15. GOLOMB, S. W. *Shift register sequences.* Aegean Park Press, Laguna Hills, California, 1982.

16. GRAHAM, R., KNUTH, D., AND PATASHNIK, O. *Concrete Mathematics*, 2nd ed. Addison-Wesley, Reading, MA, 1994.

17. IRELAND, K., AND ROSEN, M. *A Classical Introduction to Modern Number Theory*, 2nd ed. Springer-Verlag, Berlin, 1990.

18. KALTOFEN, E., AND SHOUP, V. Subquadratic-time factoring of polynomials over finite fields. In *Proc. 27th ACM Symp. Theory of Computing* (1995), pp. 398–406.

19. KNOPFMACHER, J., AND KNOPFMACHER, A. Counting irreducible factors of polynomials over a finite field. *SIAM Journal on Discrete Mathematics 112* (1993), 103–118.

20. KNUTH, D. *The art of computer programming, vol.2: seminumerical algorithms*, 2nd ed. Addison-Wesley, Reading MA, 1981.

21. LIDL, R., AND NIEDERREITER, H. *Finite fields*, vol. 20 of *Encyclopedia of Mathematics and its Applications*. Addison-Wesley, Reading MA, 1983.

22. MENEZES, A., BLAKE, I., GAO, X., MULLIN, R., VANSTONE, S., AND YAGHOOBIAN, T. *Applications of Finite Fields.* Kluwer Academic Publishers, Boston, Dordrecht, Lancaster, 1993.

23. ODLYZKO, A. Discrete logarithms and their cryptographic significance. In *Advances in Cryptology, Proceedings of Eurocrypt 1984* (1985), vol. 209 of *Lecture Notes in Computer Science*, Springer-Verlag, pp. 224–314.

24. ODLYZKO, A. Asymptotic enumeration methods. In *Handbook of Combinatorics*, R. Graham, M. Grötschel, and L. Lovász, Eds. Elsevier, 1996.

25. PANARIO, D. Combinatorial and algebraic aspects of polynomials over finite fields. PhD Thesis, in preparation, 1996.

26. RABIN, M. O. Probabilistic algorithms in finite fields. *SIAM J. Comp. 9* (1980), 273–280.

27. SCHÖNHAGE, A. Schnelle Multiplikation von Polynomen über Körpern der Charakteristik 2. *Acta Inf. 7* (1977), 395–398.

28. SCHÖNHAGE, A., AND STRASSEN, V. Schnelle Multiplikation großer Zahlen. *Computing 7* (1971), 281–292.

29. SEDGEWICK, R., AND FLAJOLET, P. *An Introduction to the Analysis of Algorithms*. Addison-Wesley, Reading MA, 1996.

30. SHEPP, L., AND LLOYD, S. Ordered cycle lengths in a random permutation. *Trans. Amer. Math. Soc. 121* (1966), 340–357.

31. SHOUP, V. Fast construction of irreducible polynomials over finite fields. *J. Symb. Comp. 17* (1995), 371–391.

32. SHOUP, V. A new polynomial factorization algorithm and its implementation. *J. Symb. Comp. 20* (1996), 363–397.

33. SHPARLINSKI, I. Finding irreducible and primitive polynomials. *Appl. Alg. Eng. Comm. Comp. 4* (1993), 263–268.

Numerical Linear Algebra in Optical Imaging

Robert J. Plemmons

Abstract. This work involves two-stage approaches to enhancing the quality of images taken through the atmosphere. First, a matrix control problem arising in *adaptive-optics* is discussed. The problem involves optimal real-time control of very fast-acting deformable mirrors designed to compensate for atmospheric turbulence and other image degradation factors, such as wind-induced telescope vibration (windshake). The surface shapes of the mirrors must change rapidly to correct for time-varying optical distortions. The second stage of compensating for the effects of atmospheric turbulence generally occurs off-line, and consists of the post-processing step of *image restoration*. Here, the work involves large-scale computations, using either a simultaneous image of a natural guide star or a large ensemble of images corresponding to different atmospheric realizations, to deconvolve the blurring effects of atmospheric turbulence.

1 Atmospheric Imaging

Researchers in science and engineering are actively seeking to overcome the degradation of astronomical image quality caused by the effects of atmospheric turbulence and other image degradation processes. Atmospheric turbulence has especially frustrated astronomers since telescopes were invented. The effects are in part due to the mixing of warm and cold air layers, resulting in changes in air density. This causes parts of the light waveforms to be slowed by different amounts, distorting the image. The resulting twinkling of the stars and other effects are the main limitations of ground-based imaging. Modern astro-imaging methods are now giving ground-based, air-to-air and air-to-ground imaging vital tools. Science and engineering challenges come in the real-time measurement of the atmospheric turbulence and the use of those measurements to actuate the control systems of the deformable mirrors, and to perform large-scale image post-processing computations. As pointed out in recent articles in IEEE Spectrum [1], Scientific American [4], Popular Science [9], and National Geographic Magazine [10], exciting technological breakthroughs are rapidly coming to the aid of scientists attempting to de-blur atmospheric images.

The improvement in ground-based image quality can be attempted in two stages. The first stage occurs as the observed image is initially formed. Specially designed *deformable mirrors* operating in a closed-loop adaptive-optics system can partially compensate for the effects of atmospheric turbulence. Optical systems detect the distortions using either a natural guide star or a guide star artificially generated using range-gated laser backscatter. A wavefront sensor measures the optical distortions which can then be partially nullified by deforming a flexible mirror in the telescope. To be effective, these corrections have to be performed at real-time speed. Adaptive-optics control systems form the subject of much of our recent research, see, e.g. [2, 3]. A study is made of a non–smooth matrix optimization problem arising in adaptive–optics, which involves the optimal real–time control of a very fast–acting deformable mirror designed to compensate for atmospheric turbulence and other image degradation factors, such as wind–induced telescope vibration (windshake). The surface shape of this mirror must change rapidly to correct for time–varying optical distortions caused by these sources of image degradation. One formulation of this problem yields a functional

$$f(U) = \sum_{i=1}^{n} \max_{j}\{(U^T M_j U)_{ii}\}$$

over orthogonal matrices U, where U defines the basis of deformable mirror control modes to be used, and the collection of $n \times n$ symmetric matrices M_j characterize adaptive–optics performance over multiple control bandwidths. Selecting the orthogonal matrix U which maximizes $f(U)$ corresponds to defining a basis of deformable mirror control modes and their associated control bandwidths. We consider first the situation which can arise in practical applications where the matrices M_j are "nearly" pairwise commutative. Besides providing useful bounds, these results lead to a simple corollary providing a closed–form solution for maximizing f in the special case where M_j are simultaneously diagonalizable. The general optimization problem is approached using a heuristic Jacobi–like algorithm, and results in a previous paper on this approach by the authors are extended. Numerical tests using the algorithm indicate that in the presence of windshake, the performance of a closed–loop adaptive–optics system can be improved by the selection of distinct and independently optimized control bandwidths for separate modes of the wave–front distortion profile.

In some situations, windshake vibration (noise) on the telescope system must also be taken into account as the image is initially acquired. For example, in the Gemini 8-meter telescope adaptive-optics systems currently being installed on mountains in Chile and Hawaii, windshake noise, in addition to atmospheric turbulence, is an important factor influencing our work [3].

The second stage of enhancing the quality of optical images generally occurs off–line, and consists of the postprocessing step of image restoration. Image restoration is an ill–posed inverse problem which involves the removal or minimization of degradations caused by noise and blur in an image, resulting from, in this case, imaging through a medium. Letting $\mathcal{H}$ denote the blurring operator and η the noise function, then the image restoration problem with additive noise can be expressed as a linear operator equation $g = \mathcal{H}f + \eta$, where g and the unknown f denote functions containing the information of the recorded and original images, respectively. A classical approach employed is that of penalized least squares, which is also called Tikhonov regularization in the inverse problems literature. This requires minimization of the functional

$$\|\mathcal{H}f - g\|^2 + \alpha J(f),$$

where α is a positive (regularization) parameter and the functional $J(f)$ serves the purpose of stabilizing the least squares problem and penalizing certain undesirable artifacts like spurious oscillations in the computed f. Our work is surveyed on the topic of space–varying regularization approaches, and associated techniques for accelerating the convergence of iterative image postprocessing computations. Here, large-scale computations, again using either a natural guide star or a guide star artificially generated using range-gated laser backscatter, are used to deconvolve the blurring effects of atmospheric turbulence, e. g., [5, 7, 8].

We consider first, however, adaptive closed-loop control methods [2, 3] proposed for *stage one* of the reconstruction process: a *multiple bandwidth* method for optimizing the real-time control of deformable mirrors. Closed loop adaptive optics systems must compensate for time varying wave front distortions on the basis of noisy sensor measurements. Time varying distortions are most accurately corrected by a system with a high control bandwidth, but the noise in wave front sensor measurements is best rejected by reducing the control bandwidth and temporally smoothing the wave front correction to be applied. These trends lead to a performance tradeoff between the residual wave front distortions due to sensor noise and servo lag, and the optimum control bandwidth which minimizes the sum of these two effects will depend upon the wave front sensor noise level and the temporal characteristics of the wave front errors to be corrected. Because wave front sensor noise statistics and the temporal dynamics of the wave front distortion to be corrected vary as a function of spatial frequency, it is possible to improve adaptive optics performance by using this more sophisticated multiple bandwidth control approach [2, 3].

In addition to applications to ground-based astronomy, multiple control bandwidth techniques and fast image restoration methods considered in this paper may be useful, more generally, whenever (a) the spatial-temporal characteristics of the phase errors to be corrected are reasonably well understood, and (b) the temporal statistics of the errors are considerably different at different spatial frequencies. Potential applications in medicine, communications, and laser technology might include:

- Retinoscopy. In this case the idea is to eliminate the aberrations of the eye and fluid in the eye.
- Endoscopy. Images are obtained through fluids, and optical image enhancement can often improve the image quality.
- Microscopy. This involves confocal, fluorescence, and scanning microscopy, often in 3 dimensions, producing images which often must be compensated by adaptive-optics and image restoration.
- Surveillance. This can involve, for example, satellite imaging of the earth for military or civilian purposes.

- Underwater Imaging. Again, this technology is known to be limited by turbulence and turbidity.
- Lasers. This involves intra- and extra- cavity beam correction and aiming, particularly for high power lasers used in military and in industrial material processing applications.

2   Two Inverse Problems

An analysis, both theoretical and practical, has been given in our paper [3] for a matrix trace maximization problem and some of its generalizations. This paper extends earlier work on an optimization method from [2] for the real-time control of deformable mirrors with the use of multiple bandwidths for adaptive-optics applications.

Numerical tests using a Jacobi-like algorithm on a 400 node IBM SP2 computer indicate that the performance of a closed-loop adaptive-optics system which must compensate for the effects of both atmospheric turbulence and telescope windshake may be *greatly improved* by the selection of distinct and independently optimized control bandwidths for separate modes of the wave-front-distortion profile.

Although the simulation tests reported in [3] are for problems of modest dimensions, future tests will involve adaptive optics systems utilizing deformable mirrors with many more degrees of freedom. The adaptive-optics model assumed for the calculations in our current work is based upon the current Gemini design and includes a Shack-Hartmann wave front sensor with 8 by 8 subapertures and a continuous facesheet deformable mirror with 9 by 9 actuators. Segmented mirror systems with very large numbers of degrees of freedom also are of interest. For example, the SELENE system [6] with segmented rather than continuous facesheet mirrors, is envisioned to have about 250,000 subapertures. Designing an adaptive optics system with a very large number of degrees of freedom taxes many design areas. Because of the bandwidth required for atmospheric compensation, a most fundamental question is whether the optimal control algorithms can be implemented quickly enough, perhaps on a special purpose parallel signal processor. Also, Milman and Fijany [6] have recently developed adaptive optics algorithms for segmented mirrors, based on fast Fourier transformation methods for Poisson solvers, that offer a significant degree of parallelism on conventional parallel architectures. Our plans are to to bring experience in using FFTs on conventional massively parallel machines such as the SP2 [7, 8] to bear on these deformable mirror optimal control problems.

The results given in our work may be relevant for the adaptive-optics system to be installed on the Gemini-North 8-meter telescope to be located on the mountain of Mauna Kea in Hawaii, where both windshake and atmospheric turbulence are significant sources of image degradation. The multiple bandwidth approach can have a significant performance advantage over the single bandwidth algorithm when the effect of windshake on image quality is included. This is especially true when the direction of the wind is a priori known and can be included in the optimization of the multiple bandwidth algorithm.

The inclusion of windshake has interesting practical applications. It would be best to be able to update the algorithm in real-time as the wind direction and wind strength changes without interrupting or interfering with normal telescope operations.

The second stage of enhancing the quality of optical images occurs off-line, and consists of the post-processing step of *image restoration*. Image restoration is an ill-posed inverse problem which involves the removal or minimization of degradations caused by noise and blur in an image, resulting from, in this case, imaging through a medium. Our work in [8] concerns a new space-varying regularization approach, and associated techniques for accelerating the convergence of iterative image post-processing computations. In [8] we have presented a new segmentation–based preconditioning scheme, which we refer to as *SVR*, leading to space–varying regularized iterative restoration of optical images. Denoising methods, including total variation minimization, followed by segmentation–based preconditioning methods for minimum residual conjugate gradient iterations, were investigated. Numerical tests were made on both simulated and practical atmospheric imaging problems. The SVR preconditioning method appears to have certain advantages:

- Even if a stopping criteria does not halt the algorithm precisely at the appropriate solution (either a few iterations before or after), it appears that the SVR method will still provide a restoration that is close to optimal. We have observed that the SVR method rapidly reduces the relative error, and then stabilizes at a level near that of a near optimal solution for several iterations. Examining the norms of the difference of successive iterations, in conjunction with one of the standard stopping criteria, leads to a robust iterative method.
- The SVR preconditioning scheme can easily be implemented in parallel. In this case, the total computational cost of our scheme is essentially the same as for fast preconditioned restoration methods.
- The appropriate segmentation procedure used, as well as the number of segments chosen, is problem dependent. More studies and comparisons are needed in this regard. But analyzing the image in a well–chosen segmented form appears to yield better overall separation of the signal and noise subspaces for the purpose of choosing the space–varying preconditioners, and the resulting iterative restoration using SVR can effectively suppress noise contamination of the iterates. Over smoothing the image is avoided, and singular image features such as sharp edges and oscillatory textures can be restored without resorting to a computationally expensive nonlinear optimization method such as total variation restoration.

Acknowledgment. Research sponsored by the U.S. Air Force Office of Scientific Research under grant F49620-94-1-0261, and by the National Science Foundation under grant CCR-96-23356.

References

1. T. E. Bell, *Electronics and the stars*, IEEE Spectrum **32, no. 8**, 16–24 (1995).
2. B. L. Ellerbroek, C. Van Loan, N. P. Pitsianis and R. J. Plemmons, *Optimizing closed-loop adaptive-optics performance using multiple control bandwidths*, J. Optical Soc. Amer. A **11**, 2871–2886 (1994).
3. B. L. Ellerbroek, C. Van Loan, N. P. Pitsianis and R. J. Plemmons, *Multiple control bandwidth computations in adaptive-optics*, Preprint, March (1996).
4. J. W. Hardy, *Adaptive Optics*, Scientific American **270, no. 6**, 60–65 (1994).
5. R. L. Lagendijk and J. Biedmond, *Iterative Identification and Restoration of Images*, Kluwer Press, Boston, 1991.
6. M. Milman and A. Fijany, *Wave-front control algorithms and analysis for a dense adaptive optics system with segmented deformable mirror*, J. Optical Soc. Amer. A **13**, 365–378 (1996).
7. J. Nagy, R. Plemmons and T. Torgersen, *Iterative image restoration using approximate inverse preconditioning*, IEEE Trans. on Image Processing, **5, no. 7**, 1151–1162 (1996).
8. J. Nagy, P. Pauca, R. Plemmons and T. Torgersen, *Space-varying restoration of optical images*, Preprint, August (1996).
9. J. Nelson, *Reinventing the telescope*, Popular Science, **85**, 57–59, (1995).
10. R. A. Smith, *Seeing the universe through new eyes*, National Geographic Magazine **185, no. 1**, 2–41 (1994).
11. R. K. Tyson, *Principles of Adaptive Optics*, Academic Press, N. Y., (1991).

Explicit symplectic integration of rod dynamics

Sebastian Reich

In this paper, we derive an explicit and symplectic integration scheme for elastic rods. To do so, we first discuss various spatial discretizations that yield an appropriate finite-dimensional truncation of the *Simo-Marsden-Krishnaprasad* formulation of rod dynamics [1]. Then, similar to rigid body dynamics [2], we formulate an explicit time-integrator by splitting the Hamiltonian in an appropriate way. The resulting scheme is 2nd order, explicit, symplectic, and preserves the underlying symmetries of rod dynamics. We also consider the case of an unshearable and unextensible rod. It is demonstrated that this leads to a Hamiltonian formulation with holonomic constraints that can be integrated by an appropriate modification of the constrained schemes considered in [3]. Application of our scheme also includes continuum models of DNA [4].

References

1. Simo, J.C., Marsden, J.E., and Krishnaprasad, P.S., The Hamiltonian Structure of Nonlinear Elasticity: The Material and Convective Representations of Solids, Rods and Plates, *Arch. Rational Mech. Anal.* **104**(1988), 125–183.
2. Reich, S., Symplectic integrators for systems of rigid bodies, *Fields Institute Communications*, to appear.
3. Reich, S., Symplectic Integration of Constrained Hamiltonian Systems by Composition Methods, *SIAM J. Numer. Anal.* **33**(1996), 475–491.
4. Dichmann, D.J., Li, Y., and Maddocks, J.H., Hamiltonian Formulations and Symmetries in Rod Mechanics, to appear in: *IMA Volumes in Mathematics and its Applications*, eds. Mesirov, Schulten, Sumners, 1995.

Toric Laminations, Sparse Generalized Characteristic Polynomials, and a Refinement of Hilbert's Tenth Problem

J. Maurice Rojas*

Abstract. This paper reexamines univariate reduction from a toric geometric point of view. We begin by constructing a binomial variant of the u-resultant and then retailor the generalized characteristic polynomial to sparse polynomial systems. We thus obtain a fast new algorithm for univariate reduction and a better understanding of the underlying projections. As a corollary, we show that a refinement of Hilbert's Tenth Problem is decidable in single-exponential time. We also obtain interesting new algebraic identities for the sparse resultant and certain multisymmetric functions.

1 Introduction

We give a new approach for combatting degeneracy problems which occur when reducing large polynomial systems to univariate polynomials. Taking a positive attitude, we will actually make use of these degeneracies to better understand polynomial system solving. We do this by applying new observations involving toric varieties to establish faster, more reliable algorithms for univariate reduction. Our techniques provide an intrinsic geometric setting for such reductions and fully exploit the sparsity of any polynomial system specified by its monomial term structure.

Algebraically reducing the problem of solving a polynomial system to solving a single univariate polynomial has been known for over a century and has been applied with increasing efficiency in computational algebra over the last two decades. (For example, see [Laz81, KL92, PK95, GHMP95] and the references therein.) However, there are difficulties with this method which *still* have not been completely addressed. Three such problems are the following:

A: Since univariate reduction corresponds to projecting the solution space onto a projective line, the fiber over a point with finite coordinate might contain roots at infinity.

---

* This research was completed at City University of Hong Kong and was funded by an N.S.F. Mathematical Sciences Postdoctoral Fellowship.

B: The univariate polynomial one reduces to might be identically zero, due to (1) degeneracies in the underlying projection, or (2) the presence of infinitely many roots.

C: Univariate reduction is still too slow for many problems of interest.

We circumvent A by the technique of *toric laminations*. By constructing a generalization of Canny's *generalized characteristic polynomial* (GCP) [Can90, Shu93] — the *toric* (or *sparse*) GCP — we can dispose of B$_2$. Our toric geometric approach also helps us better understand B$_1$. These two new techniques are contained in theorems 1–3 below.

Although problem B$_2$ was solved earlier in [Ren87, Can90], the toric GCP greatly improves the complexity bounds given there [Roj96b]. In particular, we address problem C by using the *sparse resultant* [GKZ90, EC95] throughout our development. Fast methods for polynomial system solving were derived via the *sparse u-resultant* in [Emi94, EC95], but problem B was not addressed there. Our approach to univariate reduction thus unifies the GCP with the the sparse u-resultant. Furthermore, we generalize both techniques to certain *toric varieties* [KKMS73, Ful93, Stu95].

One advantage of a toric variety setting is reducing the problem of *exact root counting* for polynomial systems, almost completely, to convex geometry. Generically sharp *upper bounds* on the number of (complex) isolated roots, in terms of mixed volume, first appeared just over two decades ago [Kus75]. These bounds have since been generalized to algebraically closed fields [Dan78, Roj96a] and to various subsets of affine space (other than the algebraic torus) [Kho78, HS96, Roj96a]. However, the following algorithm appears to be the first example of a convex geometric approach to counting the exact number of roots.

Theorem 1 *Let $F(x_1, \ldots, x_n)$ be an $n \times n$ polynomial system with support E and assume that our polynomials and roots are considered over an algebraically closed ground field K. Pick $a \in \mathbb{Z}^n$ such that the segment $[\mathbf{O}, a]$ is not perpendicular to any facet of any Newton polytope of F. Define $\pi_a(u_+, u_-) := \mathrm{Res}_{E'}(F, u_+ + u_- x^a)$ where $\{u_\pm\}$ is a pair of algebraically independent indeterminates and $E' := (E, \{\mathbf{O}, a\})$. Finally, let $\varepsilon_\pm$ be the lowest exponent of $u_\pm$ occuring in π_a. Then F has exactly $\mathcal{N} := \mathcal{M}(E) - \varepsilon_+ - \varepsilon_-$ isolated roots (counting multiplicities) in $(K^*)^n$, provided $\mathcal{N} < \infty$.*

In the above, $\mathrm{Res}_{E'}(\cdot)$ and $\mathcal{M}(\cdot)$ respectively denote the sparse resultant and mixed volume [EC95, DGH96, Roj96a]. The key contribution of the above theorem (which is proved in section 5) is to make explicit use of roots at toric infinity when doing univariate reduction. Furthermore, by instead calculating a particular *coefficient* of π_a, we can actually compute (up to sign) certain multisymmetric functions of the roots of F and thus do more than just count the number of isolated roots. Admittedly, such a computation can be quite difficult, but recent advances in interpolation techniques, e.g., [Zip93], are making this approach increasingly feasible. More to the point, we will see in section 5 that finding a root of π_a (when $\pi_a \not\equiv 0$) is equivalent to calculating an expression of the form $-\zeta_1^{a_1} \cdots \zeta_n^{a_n}$ where $(\zeta_1, \ldots, \zeta_n) \in (K^*)^n$ is a root of F.

We may also calculate the number of *distinct* roots in a similar way.

Corollary 1 *Following the notation of theorem 1, let $\pi'_a(u_+, u_-)$ be the square-free part of $\pi_a(u_+, u_-)$. Also let $\varepsilon'_\pm$ be the lowest exponent of $u_\pm$ occurring in π'_a. Then F has exactly $\mathcal{N}' := \deg(\pi'_a) - \varepsilon'_+ - \varepsilon'_-$ roots in $(K^*)^n$, provided $\mathcal{N}' < \infty$ and the projection of roots $\{\zeta \in (K^*)^n \mid F(\zeta) = 0\} \longrightarrow \{\zeta^a\}$ is injective.*

Recall that a *ridge* is a polytope face of codimension 2 [Zie95]. Although the "bad" case $\mathcal{N} = \infty$ is an annoyance, it is a small annoyance and can actually be made use of in certain cases.

Theorem 2 *Following the notation of theorem 1, let P be the polytope $\sum \text{Conv}(E_i)$. Define the* ambiguity locus, *$L(E, a)$, to be the subvariety of $\mathcal{T}_P$ corresponding to the union of the ridges of P perpendicular to $[\mathbf{O}, a]$. Then $\mathcal{N} = \infty \iff F$ has a root lying in $L(E, a)$ or infinitely many roots in $(K^*)^n$. Also, $L(E, a)$ has codimension 2 and, within the space of all systems with support contained in E, the F with $\mathcal{N} = \infty$ form a subvariety of codimension ≥ 2.*

The toric variety (over K), $\mathcal{T}_P$, corresponding to a polytope P is detailed in [KSZ92, Ful93, GKZ94, Stu95, Roj96a], and polynomial roots in a toric compactification are described (in a manner closest to our present framework) in [Roj96a]. By replacing the binomial $u_+ + u_- x^a$ in our preceding results with another specially chosen indeterminate polynomial, we can completely avoid ambiguity loci. This is detailed further in [Roj96b] and, to a certain extent, in our final main theorem below. However it is worth noting that the roots of F *always* avoid the ambiguity locus when $n \leq 2$ (cf. lemma 1).

Remark 1 *The existence of infinitely many roots of F at toric infinity and the existence of a root of F within the ambiguity locus are independent in general [Roj96b]. Thus it is worthwhile to know something about $L(E, a)$.*

Remark 2 *A correspondence useful for visualizing theorems 1 and 2 is to identify toric infinity $(\mathcal{T}_P \backslash (K^*)^n)$ with the boundary of the polytope P (cf. definition 1). Computing π_a then amounts to splitting ∂P into two signed halves, much like cracking an egg-shell. In particular, the half with sign $\pm$ consists precisely of those faces of P having an inner normal with $\pm$ inner product with a. The ambiguity locus is then precisely the common boundary of these two halves. We will see later that π_a can be viewed as a coordinatization of a particular rational map $\mathcal{T}_P \dashrightarrow \mathbb{P}^1_K$. Thus, provided no roots lie in the ambiguity locus, $\varepsilon_\pm$ is the sum of the intersection numbers of the roots of F which lie in the $\pm$ half.*

A general way to deal with the existence of *infinitely* many roots of F within $\mathcal{T}_P$ is the following.

Theorem 3 *[Roj96b] Following the notation of theorem 2, suppose $g(x) := \sum_{e \in A} u_e x^e$ where the u_e are algebraically independent indeterminates and $A \subset \mathbb{Z}^n$ is nonempty and finite. Further suppose that $\mathcal{P} := (P_1, \ldots, P_n)$ is an n-tuple of integral polytopes in $\mathbb{R}^n$ containing the support of F and let D be an irreducible fill of $\mathcal{P}$. Setting $F^* := (\sum_{e \in D_i} x^e \mid i \in [1..n])$, define $\mathcal{H}(s; u) := \text{Res}_{E'}(F + sF^*, g)$,* *where s is a new indeterminate and $E' := (E, A)$. Finally, considering $\mathcal{H}$ as a polynomial in s with coefficients in $K[u]$, let $\mathcal{F}_A(u)$ be the coefficient of the lowest power term of $\mathcal{H}$. Then $\mathcal{F}_A$ is a homogeneous polynomial with the following properties:*

1. $\mathcal{F}_A$ *is divisible by* $g(\zeta)$ *for any isolated root* $\zeta \in (K^*)^n$ *of* F.
2. *If* $\text{Conv}(A)$ *is compatible with* $\mathcal{P}$, *then* $\mathcal{F}_A$ *has degree* $\mathcal{M}(\mathcal{P})$ *and to every irreducible factor of* $\mathcal{F}_A$ *there naturally corresponds a root of* F *in* $\mathcal{T}_P$.

We call $\mathcal{H}$ a toric (or sparse) generalized characteristic polynomial for (F, A).

□

The combinatorial portion of the above theorem can be explained as follows: We say that D *fills* $\mathcal{P}$ iff $D := (D_1, \ldots, D_n)$ satisfies $D_i \subseteq P_i$ for all $i \in [1..n]$ and $\mathcal{M}(D) = \mathcal{M}(\mathcal{P})$ [Roj94, RW96]. An *irreducible* fill is then simply a fill which is minimal with respect to n-tuple containment. Finding such a D amounts to a combinatorial preprocessing step which need only be done once for a given set of problems, provided E is fixed. This preprocessing step (along with the assertion in (2)) is explained in section 6, and compatibility is defined in section 5. We also outline the proof of the above theorem in section 6. Combined with the toric variety version of Bernshtein's theorem [Roj96a], we thus see that the above theorem gives us a way to work with the isolated roots of a sparse polynomial system, even in the presence of infinitely many roots. Furthermore, the roots defined by $\mathcal{F}_A$ which do *not* correspond to any isolated root of F can be interpreted as contributions from the excess components.

We apply our main theorems to some simple examples in section 3. Our results are also extremely useful for root counting over $\mathbb{R}$, $\mathbb{Q}$, and even $\mathbb{Z}$. For example, one can combine standard univariate techniques such as Sturm sequences [GLRR94, Roy95] with our preceding results to quickly count the number of real roots. Better still, we can go to an even smaller ring: In section 4 we will give a short proof (under some mild hypotheses) of the following refinement of Hilbert's Tenth Problem.

Theorem 4 *[Roj96b] Consider the problem of determining all integer solutions of the following class of Diophantine systems:*
Hilb(d): *Multivariate polynomial systems with integer coefficients and d-dimensional complex solution set.*
Then there is an algorithm which, given any instance of Hilb(0)*, finds all integer solutions, or certifies that there are none, in single-exponential time.*   □

All of the above results stem from an observation on the vanishing of the sparse resultant. In particular, one basic property of the sparse resultant is that (F, f_{n+1}) has a root in $(K^*)^n \implies \text{Res}_{E'}(F, f_{n+1}) = 0$ (provided $E' = (E, E_{n+1})$ and f_{n+1} is a polynomial over K with support contained in $E_{n+1} \subset \mathbb{Z}^n$). However, the converse assertion is **not** always true. (The polynomial system $(x+y+1, x+y+2, x+y+3)$ with $E' = \{\mathbf{0}, \hat{e}_1, \hat{e}_2\}^3$ gives a simple counter-example.) The correct statement in which *both* implications hold seems to be known only folklorically, while the *unmixed* case $E_1 = \cdots = E_{n+1} = \mathcal{A}$ (over $\mathbb{C}$) is contained in various recent works, e.g., [KSZ92, GKZ94]. So we will make use of the following more general theorem.

Vanishing Theorem for Resultants *[Roj96b] Let $P_i := \mathrm{Conv}(E_i)$ for all $i \in [1..n+1]$ and set $P := \sum_{i=1}^{n+1} P_i$. Then $\mathrm{Res}_{E'}(F, f_{n+1}) = 0 \iff \bigcap_{i=1}^{n+1} \mathcal{D}_P(f_i, P_i) \neq \emptyset$.* $\qquad\qquad\square$

In the above, $\mathcal{D}_P(f_i, P_i)$ is the *toric effective divisor* of $\mathcal{T}_P$ corresponding to (f_i, P_i) [Kho77, Roj96a]. This result provides a geometric analogue, over a *general* algebraically closed field, of the product formula for the sparse resultant [PS93].

We close this introduction with an important note:

Our approach to univariate reduction provides a first step toward an intrinsic method for root finding in the algebraic torus $(K^*)^n$. Although one could in principle reduce to various coordinate subspaces to tackle the problem of geometrically intrinsic elimination in K^n, such an approach runs into complications with intersection multiplicities on the coordinate hyperplanes and is really *not* intrinsic to K^n. We propose a more intrinsic approach to *affine elimination theory* by defining a new resultant operator — the *affine* sparse resultant — in [Roj96c]. Following this route, *all* of our main theorems generalize easily to root finding in affine space minus an arbitrary union of coordinate hyperplanes. This is pursued in greater depth in [Roj96a, Roj96c] and the complexity bound of theorem 4 is further refined in the latter work.

2 Related Work

We point out a few important recent approaches to elimination theory which can profit from our results.

From an applied angle, our observations on degeneracies and handling polynomial systems with infinitely many roots nicely complement the work of Emiris and Canny [EC95]. In particular, their sparse resultant based algorithms for polynomial system solving can now be made to work even when problem B occurs. Also, an added benefit of working torically (as opposed to the classical approach of working in projective space) is the increased efficiency of the sparse resultant: the resulting matrix calculations (for polynomial system solving) are much smaller and faster. In particular, whereas it was remarked in [Can90] that Gröbner basis methods are likely to be faster than the GCP for sparse polynomial systems, the toric GCP appears to be far more competitive in such a comparison.

From a more theoretical point of view, our focus on univariate reduction is a useful addition to Sturmfels' foundational work on sparse elimination theory in the large [Stu93]. (We also point out that this paper provides a wonderfully clear introduction to some of the toric variety techniques we refer to.)

Going further into the intersection of algebraic geometry and complexity theory, there is a beautiful new approach to elimination theory founded by the school of Heintz et. al. [PK95, GHMP95]. Our results fill in some toric geometry missing from their more algebraic framework. In particular, we propose that toric laminations, i.e., "fibering" $\mathcal{T}_P$ by a toric projection (instead of fibering $(K^*)^n$ by coordinate projections), can usefully complement their methods. For example, although their formulation in terms of straight-line programs is extremely general, it appears that their complexity bounds can be significantly improved with our techniques in the case of polynomial systems which are "sparse" in the specified supports sense we use here. This is explored further in [Roj96b].

3   Examples

We will give a 2×2 example of our first two main theorems.

3.1   Root Counting with Sparse Resultants

Let $n = 2$ and consider the bivariate polynomial system $F := (x^3 + y^4 - 1, x^4 + y^5 - 1)$. Also let E_1 and E_2 respectively be the supports $\{\mathbf{O}, 3\hat{e}_1, 4\hat{e}_2\}$ and $\{\mathbf{O}, 4\hat{e}_1, 5\hat{e}_2\}$, and set $P_i := \mathrm{Conv}(E_i)$ for all i. Clearly then the Newton polygons are two triangles and the segment $[\mathbf{O}, (1, 1)]$ is not perpendicular to any edge of any P_i. So we can set $a := (1, 1)$ and apply theorem 1 to count the roots of F.

Using the Maple sparse resultant software package written by John Canny and Paul Pedersen (available on-line at ftp://robotics.eecs.berkeley.edu/pub/SparseResultant) we can calculate a matrix whose determinant is a nonzero multiple of the sparse resultant $\mathrm{Res}_{E'}(F, u_{\pm} + u_{-}x^a)$, where $E' := (E_1, E_2, \{\mathbf{O}, a\})$ and the coefficients of F have been replaced by algebraically independent indeterminates. Due to an element of randomness in their code (the output matrices can vary in size on different runs) we got a matrix of size 35×35. This computation took about 44 seconds on a Sun 4m Computeserver and computing the determinant of the resulting (sparse) indeterminate matrix took another 8 seconds.

From [Stu94] we know that the sparse resultant itself should have degree $\mathcal{M}(E_1, E_2) + \mathcal{M}(E_1, \{\mathbf{O}, a\}) + \mathcal{M}(E_2, \{\mathbf{O}, a\}) = 16 + 7 + 9 = 32$. (This calculation is easily done by hand via the general formula $\mathcal{M}(E_1, E_2) = \mathrm{Area}(\mathrm{Conv}(E_1 + E_2)) - \mathrm{Area}(\mathrm{Conv}(E_1)) - \mathrm{Area}(\mathrm{Conv}(E_2))$.) So by factoring in Maple (which takes another 2 seconds) we can isolate the sparse resultant and specialize coefficients to obtain

$$\pi_a(u_+, u_-) = \mathrm{Res}_{E'}(x^3 + y^4 - 1, x^4 + y^5 - 1, u_+ + u_- xy) =$$

$$-8u_-^{12}u_+^4 - 6u_-^{11}u_+^5 + 20u_-^9 u_+^7 + 19u_-^8 u_+^8 - 2u_-^7 u_+^9 - 6u_-^6 u_+^{10}$$

$$-14u_-^5 u_+^{11} - 22u_-^4 u_+^{12} - 7u_-^3 u_+^{13} - 9u_-^2 u_+^{14} - u_- u_+^{15} - u_+^{16}$$

So $\varepsilon_+ = 4$, $\varepsilon_- = 0$, and theorem 1 tells us that F has exactly $16 - 4 = 12$ roots, counting multiplicities, in $(K^*)^2$. Furthermore, after factoring our resulting π_a, corollary 1 tells us that the multiplicity of every root of F in $(K^*)^2$ is 1. As for the meaning of the coefficients of π_a, the standard Newton identities imply, for example, that the multisymmetric functions $\sum_{F(\zeta_1, \zeta_2) = 0} \zeta_1 \zeta_2$ and $\sum \mu_1 \mu_2 \zeta_1 \zeta_2 \omega_1 \omega_2$

are, respectively, ±1 and ±7 (the latter sum ranging over all sets $\{\mu, \zeta, \omega\}$ of 3 distinct roots of F).

The roots of F can also be counted by localization as suggested in [Math250B], but the toric geometry in this example is rather nice to observe. In particular, letting P be the quadrilateral $P_1 + P_2$, the coordinate cross $\{xy = 0\}$ is naturally embedded in $\mathcal{T}_P$ and corresponds precisely to the two lower-left edges of P (cf. remark 2). Also, the ambiguity locus consists of two distinct points; and $\varepsilon_+ = 4 \implies F$ has precisely 4 roots, counting multiplicities, on the coordinate cross. The latter fact is also easily checked via Gröbner basis methods.

We have included timing information solely for illustrative purposes. It is also not hard to see that our calculations can be sped up tremendously with suitably specialized code.

4 A Refinement of Hilbert's Tenth Problem

We will present a concise proof of theorem 4 under the following three hypotheses:

1. The instances of Hilb(0) considered are $n \times n$ polynomial systems.
2. The complex roots of an instance must have all coordinates nonzero.
3. No instance can have a complex root at toric infinity.

These hypotheses are present mainly for technical reasons and are removed in [Roj96b]. They may also be removed with other recent techniques [BPR94, PK95, GHMP95], but through our toric techniques it seems possible to obtain a general complexity bound which is at worst quadratic in the mixed volume. This would be the best asymptotic bound to date for solving Diophantine systems.

Proof of Theorem 4: First we outline the following algorithm:

Step 1: Following the notation of theorem 3, let A be the vertices of a standard n-simplex in $\mathbb{R}^n$ and $\mathcal{P}$ the n-tuple of Newton polytopes of F. Form the polynomial $\mathcal{F}_A$.

Step 2: Using the Lenstra-Lenstra-Lovasz algorithm [LLL82], factor $\mathcal{F}_A$ over $\mathbb{Z}$.

Step 3: Any integral root $(z_1, \ldots, z_n) \in \mathbb{Z}^n$ of F corresponds precisely to a linear, integral factor of $\mathcal{F}_A$ of the form $\pm(u_{\mathbf{O}} + z_1 u_{\hat{e}_1} + \cdots + z_n u_{\hat{e}_n})$.

Step 1 runs in time quadratic in the mixed volume, applying hypotheses (1) and (3), and the results of [Stu93, Emi96]. In fact, we are merely computing a sparse u-resultant in Step 1. To prove this (and the fact that this sparse u-resultant is not identically 0) one uses slightly more general versions of theorems 1 and 2 [Roj96b]. One also needs the fact that hypothesis (3) implies that the complex zero set of F in $(K^*)^n$ is zero-dimensional, but this is an immediate consequence of corollary 3 of [Roj96a]. We also point out that working more generally in terms of $\mathcal{F}_A$ aids in later removing hypothesis (3) [Roj96b].

To see that Step 2 runs in single-exponential time in the input size, one need only observe that (a) the coefficient growth in the u-resultant calculation can be reasonably bounded, and (b) the LLL-algorithm runs within time single-exponential in n and polynomial in the bit-sizes of the coefficients of F. Part (a) is detailed in [Roj96b] and part (b) was already proved in [LLL82].

That we can terminate as described in Step 3 follows immediately from hypothesis (2) and the Vanishing Theorem for Resultants.   □

We conclude this section with some brief notes on related work.

First, we point out the rather surprising fact that restructuring Hilbert's Tenth Problem as the union $\bigcup_{d=0}^{\infty} \text{Hilb}(d)$ seems to be new. (In particular, Hilbert's Tenth Problem was originally stated as deciding the existence of a *single* integral root for *one* polynomial in several variables.) Although [Mat93, pg. 69] gives two references implying that Hilb(8) is undecidable, nothing seems to be known about Hilb(d) for $0 < d < 9$. We find this shocking, considering the vast effort within arithmetic geometry to apply complex geometric invariants to Diophantine problems, e.g., Falting's theorem and the deep conjectures of Lang and Vojta [CS86].

We also point out that Hilb(0) being a single-exponential problem may not be so new: Teresa Krick and Luis-Miguel Pardo-Vasallo have pointed out to the author that theorem 4 can also be readily derived from [PK95, GHMP95]. (Indeed, such an argument could have already been derived directly from [Ren87, LLL82] in 1987.) Pardo-Vasallo has also pointed out that Marie-Françoise Roy lectured in 1995 on a similar result, presumably derived from [BPR94]. Nevertheless, optimal bounds are open territory; not to mention the 7-long gap in the status of Hilb(d).

5 Toric Varieties and Sparse Resultants

Our notation is a slight variation of what is used in [Ful93], and is described at greater length in [Roj96a]. However, we will at least list our cast of main characters:

Definition 1 *[Roj96a] Given any $w \in \mathbb{R}^n$, we will use the following notation:*

$T =$ *The algebraic torus* $(K^*)^n$.

$U_w =$ *The affine chart of $\mathcal{T}_P$ corresponding to the cone σ_w of* Fan(P).

$P^w =$ *The face of P with inner normal w.*

$L_w =$ *The* dim(P^w)*-dimensional subspace of $\mathbb{R}^n$ parallel to the face P^w of P.*

$x_w =$ *The point in U_w corresponding to the semigroup homomorphism $\sigma_w^\vee \cap \mathbb{Z}^n \longrightarrow \{0,1\}$ mapping $p \mapsto \delta_{w \cdot p, 0}$, where δ_{ij} denotes the Kronecker delta.*

$O_w =$ *The T-orbit of x_w.*

$V_w =$ *The closure of O_w in $\mathcal{T}_P$.*

$p_w =$ *The first integral point not equal to $\mathbf{O}$ met along the ray generated by w. (When $w \in \mathbb{Q}^n \setminus \{\mathbf{O}\}$.)*

$\mathcal{E}_P(Q) =$ *The T-invariant of $\mathcal{T}_P$ corresponding to a polytope Q with which P is compatible.*

$\text{Div}(f) =$ *The Weil divisor of $\mathcal{T}_P$ defined by a rational function f on $(K^*)^n$.*

$\mathcal{D}_P(f, Q) = $ *The toric effective divisor of $\mathcal{T}_P$ corresponding to (f, Q)* $= \mathrm{Div}(f) + \mathcal{E}_P(Q)$.

$\mathcal{D}_P(F, \mathcal{P}) = $ *The (nonnegative) cycle in the Chow ring of $\mathcal{T}_P$ defined by $\bigcap_{i=1}^n \mathcal{D}_P(f_i, P_i)$ (following the notation of theorem 3).*

We will assume the reader to be familiar with normal fans of polytopes and the construction of a toric variety from a fan [Ful93, GKZ94]. For example, we say that a polytope P is *compatible* with Q iff every cone of $\mathrm{Fan}(Q)$ is a union of cones of $\mathrm{Fan}(P)$. We will also make frequent use of the natural correspondence between the face interiors $\{\mathrm{RelInt}P^w\}$ and the T-orbits $\{O_w\}$ [KSZ92, Ful93, GKZ94]. The following lemma gives a more explicit algebraic analogy between the vertices of P and the maximal affine charts of $\mathcal{T}_P$.

Lemma 1 *[Roj96a] Following the notation of definition 1, the defining ideal in $K[x^e \mid e \in \sigma_w^\vee \cap \mathbb{Z}^n]$ of $U_w \cap \mathcal{D}_P(F, \mathcal{P})$ is $\langle x^{b_1} f_1, \ldots, x^{b_n} f_n \rangle$, for any $b_1, \ldots, b_n \in \mathbb{Z}^n$ such that $b_i + P_i^w \subseteq L_w$ for all $i \in [1..n]$.* $\square$

The following lemma, giving a projective description of $\mathcal{T}_P$, is more or less a straightforward consequence of [Ful93, sec. 3.4] and [GKZ94, ch. 5–6].

Lemma 2 *Let $A \subset \mathbb{Z}^n$ be nonempty and finite, and define $\varphi_A : (K^*)^n \longrightarrow \mathbb{P}_K^{|A|-1}$ to be the rational map defined by $(t_1, \ldots, t_n) \mapsto [\cdots : t^e : \cdots]_{e \in A}$. Then $\mathrm{Conv}(A)$ compatible with $P \implies \varphi_A$ extends to a proper morphism $\overline{\varphi}_A$ from $\mathcal{T}_P$ onto a subvariety of $\mathbb{P}_K^{|A|-1}$. Furthermore, if $A = (\alpha P') \cap \mathbb{Z}^n$ for some P' compatible with P and some $\alpha \in \mathbb{N}$, then $\overline{\varphi}_A$ is an embedding for α sufficiently large.* $\square$

Theorems 1 and 2 then follow easily from the Vanishing Theorem for Resultants and lemma 1. Lemma 2 will be important in the proof of theorem 3.

Proof of Theorem 2: By lemma 1, it immediately follows that the closure of the zero set in $\mathcal{T}_P$ of any binomial of the form $x^a - c$ (with $c \in K^*$) must intersect $L(E, a)$. So by the Vanishing Theorem for Resultants, π_a must be divisible by infinitely many distinct binomials of the form $u_+ + cu_-$. So the existence of a root of F within the ambiguity locus implies that $\mathcal{N} = \infty$. Similarly, the existence of infinitely many roots of F within $(K^*)^n$ implies that $\mathcal{N} = \infty$ as well.

Conversely, suppose that F has no roots within $L(E, a)$ and only finitely many roots within $(K^*)^n$. Then it follows from lemma 1 and the Vanishing Theorem that u_+ (resp. u_-) divides π_a iff F has a root in some torus orbit O_w with $w \cdot a > 0$ (resp. $w \cdot a < 0$). More to the point, it follows similarly that $u_+ + cu_-$ divides π_a iff some root of F lies in the closure of the hypersurface $\{x^a = c\}$ in $\mathcal{T}_P$. Thus we are done with the first portion of the theorem.

That $L(E, a)$ has codimension 2 is more or less a direct consequence of definition 1. As for the codimension of the space of "degenerate" F being ≥ 2, it suffices to show that these F are contained in a union of irreducible codimension 2 subvarieties. The latter fact follows easily by checking the initial term systems $\mathrm{in}_w(F)$ corresponding to the edges of P, the Antipodality Theorem of [Roj96a], and theorem 1.3 of [Stu94]. Note, however, that the results of [Stu94] are stated over the complex numbers, but this is only a minor technicality: in this case, the results of [Roj96a] immediately imply that we can apply theorem 1.3 over *any* algebraically closed field.   □

So in summary, computing π_a amounts to laminating $\mathcal{T}_P$ with hypersurfaces of the form $\overline{\{c_+ x^a = c_-\}}$. These laminae all intersect at the ambiguity locus, and toric infinity is precisely the union of two degenerate laminae.

Proof of Theorem 1 and Corollary 1: This theorem follows upon making the above paragraph a bit more rigourous. In particular, it is clear from our preceding proof that the map $\varphi : (K^*)^n \longrightarrow K^*$ defined by $x \mapsto x^a$ extends naturally to a proper morphism $\overline{\varphi} : \mathcal{T}_P \setminus L(E, a) \longrightarrow \mathbb{P}^1_K$. The theorem then follows immediately from the fact that when $\mathcal{D}_P(F, \mathcal{P})$ avoids the ambiguity locus, π_a is precisely the Chow form of the subscheme $\overline{\varphi}(\mathcal{D}_P(F, \mathcal{P}))$ of $\mathbb{P}^1_K$. This last fact follows from the basic functorial properties of the sparse resultant and Chow forms [GKZ94, DS95].

In particular, when $\mathcal{N} < \infty$, the multiplicity of a factor $c_+ u_+ + c_- u_-$ of π_a is precisely the sum of the intersection numbers of the components of $\mathcal{D}_P(F, \mathcal{P})$ lying in the fiber $\overline{\varphi}^{-1}([c_- : c_+])$. So $\mathcal{N}$ (resp. $\mathcal{N}'$) is indeed the exact number of roots, counting multiplicities (resp. *not* counting multiplicities), of F in $(K^*)^n$.

□

Note that it is actually possible for a positive-dimensional component of $\mathcal{D}_P(F, \mathcal{P})$ lying at toric infinity to make a a finite contribution to $\mathcal{N}$ (or $\mathcal{N}'$) [Roj96b]. This can be interpreted as an alternative way of associating a positive finite intersection number to an excess component.

6   Toric Generalized Characteristic Polynomials

First we comment on assertion (2): the association of roots of F in $\mathcal{T}_P$ to factors of $\mathcal{F}_A$ follows naturally from lemma 2.

As for the combinatorial preprocessing step, the following example illustrates a simple fundamental case.

Example 1 (The "Dense" Case) *Suppose $\mathcal{P}$ is the n-tuple $(d_1 \Delta, \ldots, d_n \Delta)$ where $\Delta \subset \mathbb{R}^n$ is the standard n-simplex and $d_i \in \mathbb{N}$ for all i. It is then easily verified that the n-tuple $D := (\{\mathbf{O}, d_1 \hat{e}_1\}, \ldots, \{\mathbf{O}, d_n \hat{e}_n\})$ is an irreducible fill of $\mathcal{P}$. Letting A be the vertices of Δ, theorem 3 then implies that $\mathcal{H}$ is a variant (over a general algebraically closed field) of the original GCP applied to an $n \times n$ system of equations with degrees $d_1, \ldots, d_n$ [Can90]. In particular, our construction here has an extra s-monomial in the first n equations. Note also that Conv(A) and P are homothetic and $\mathcal{T}_P \cong \mathbb{P}^n_K$. Neglecting the extra s-monomials, setting $d_i = 1$ for all i, and suitably specializing the coefficients of g, we can then recover the usual characteristic polynomial of a matrix.*

We point out, however, that the computational coomplexity of finding an irreducible fill is an open question. However, the connection between fills and polynomial system solving appears to be new and, we hope, provides added incentive to investigate filling. Also, even if finding a fill is difficult, this step need only be done *once* for a given family of problems, provided E remains fixed. The situation where the monomial term structure of a polynomial system is fixed once and for all (and the coefficients may vary thousands of times) actually occurs frequently in many practical contexts, such as robot control or computational geometry.

In any event, the toric GCP is an algebraic perturbation method and irreducible fills provide a combinatorial means of inducing "general position" into the roots of F. For example, the following lemma implies that F^* is sufficiently generic in a useful sense.

Lemma 3 *Following the notation of theorem 3, for any point v lying in some D_i, there exists a $w \in \mathbb{R}^n \backslash \{\mathbf{O}\}$ such that $\{i\}$ is the unique essential subset of $\mathcal{P}^w$ and $v = D_i \cap P_i^w$. In particular, F^* has exactly $\mathcal{M}(P)$ roots (counting multiplicities) in $(K^*)^n$.* $\square$

This lemma follows easily from the techniques of [Roj94], particularly section 2.5.

Given that F^* has the generic number of roots in $(K^*)^n$, theorem 3 then follows from easily from the Vanishing Theorem for Resultants and an algebraic homotopy argument [Roj96b]. A similar homotopy technique appears in [RW96, Roj96a], and the "dense" case (with $K = \mathbb{C}$) is covered in [Can90, Shu93]. (We point out, however, that toric variety language was not used in the last two works.) Thus, the usual GCP is (almost) the special case of the toric GCP where $\mathcal{T}_P$ is complex projective space.

However, a caveat which must be pointed out is that our present toric GCP is primarily suited for $(K^*)^n$, while the original GCP is mainly suited for affine space. This accounts for the extra s-monomials in our toric GCP compared to the older GCP in the "dense" case. To *completely* generalize and improve the GCP, it is necessary to use the *affine* sparse resultant and this is pursued further in [Roj96c]. For instance, by replacing the sparse resultant with the affine sparse resultant, we can indeed recover Canny's GCP in the dense case.

7 Acknowledgements

The main results of this paper were derived and presented during a visit of the author at City University of Hong Kong. The author thanks Steve Smale for his invitation, warm hospitality, and even warmer beach hikes. He also thanks his roommates (and colleagues) Jean-Pierre Dedieu and Gregorio Malajovich for their input, but most of all for not getting too mad at the pile of dishes he consistently left in the sink.

References

[BPR94] Basu, S., Pollack, R., and Roy, M.-F., *"On the Combinatorial and Algebraic Complexity of Quantifier Elimination,"* In Proc. IEEE Symp. Foundations of Comp. Sci., Santa Fe, New Mexico, 1994, to appear.

[Can90] Canny, John F., *"Generalised Characteristic Polynomials,"* J. Symbolic Computation (1990) 9, pp. 241–250.

[CS86] Cornell, Gary and Silverman, Joseph H., *Arithmetic Geometry,"* with contributions by M. Artin, et. al., Springer-Verlag, 1986.

[Dan78] Danilov, V. I., *"The Geometry of Toric Varieties,"* Russian Mathematical Surveys, 33 (2), pp. 97–154, 1978.

[DGH96] Dyer, M., Gritzmann, P., and Hufnagel, A., *"On the Complexity of Computing Mixed Volumes,"* SIAM J. Comput., to appear (1996).

[DS95] Dalbec, John, and Sturmfels, Bernd, *"Introduction to Chow Forms,"* Invariant Methods in Discrete and Computational Geometry (Curaçao, 1994), pp. 37–58, Kluwer Academic Publishers, Dordrecht, 1995.

[EC95] Emiris, Ioannis Z. and Canny, John F., *"Efficient Incremental Algorithms for the Sparse Resultant and the Mixed Volume,"* Journal of Symbolic Computation, vol. 20 (1995), pp. 117–149.

[Emi94] Emiris, Ioannis Z., *"Sparse Elimination and Applications in Kinematics,"* Ph.D. dissertation, Computer Science Division, U. C. Berkeley (December, 1994), available on-line at http://www.inria.fr/safir/SAFIR/Ioannis.html.

[Emi96] Emiris, Ioannis Z., *"On the Complexity of Sparse Elimination,"* manuscript, INRIA, March 1996, available on-line at http://www.inria.fr/safir/SAFIR/Ioannis.html.

[Ful93] Fulton, William, *Introduction to Toric Varieties*, Annals of Mathematics Studies, no. 131, Princeton University Press, Princeton, New Jersey, 1993.

[GHMP95] Giusti, M., Heintz, J., Morais, J. E., Pardo, L. M., *"When Polynomial Systems Can Be 'Solved' Fast?,"* Proc. 11th International Symposium, AAECC-11, Paris, France, July 17-22, 1995, G. Cohen, M. Giusti and T. Mora, eds., Springer LNCS 948 (1995) pp. 205–231.

[GKZ90] Gel'fand, I. M., Kapranov, M. M., and Zelevinsky, A. V., *"Discriminants of Polynomials in Several Variables and Triangulations of Newton Polytopes"* Algebra and Analysis (translated from Russian) 2, pp. 1–62, 1990.

[GKZ94] Gel'fand, I. M., Kapranov, M. M., and Zelevinsky, A. V., *Discriminants, Resultants and Multidimensional Determinants*, Birkhäuser, Boston, 1994.

[GLRR94] Gonzalez, L., Lombardi, H., Recio, T., Roy, M.-F., *"Sturm-Habicht Sequence, Determinants and Real Roots of Univariate Polynomials,"* Quantifier Elimination and Cylindrical Algebraic Decomposition, Texts and Monographs in Symbolic Computation, B. Caviness and J. Johnson, Eds., Springer-Verlag, Wien, New York, 1994, to appear.

[HS96] Huber, Birkett and Sturmfels, Bernd, *"Bernshtein's Theorem in Affine Space,"* Discrete and Computational Geometry, to appear, 1996.

[Kho77] Khovanskii, A. G., *"Newton Polyhedra and Toroidal Varieties,"* Functional Anal. Appl., 11 (1977), pp. 289–296.

[Kho78] Khovanskii, A. G., *"Newton Polyhedra and the Genus of Complete Intersections,"* Functional Analysis (translated from Russian), Vol. 12, No. 1, January–March (1978), pp. 51–61.

[KL92] Kapur, Deepak and Lakshman, Yagati N., *"Elimination Methods: an Introduction,"* Symbolic and Numerical Computation for Artificial Intelligence, B. Donald et. al. (eds.), Academic Press, 1992.

[KKMS73] Kempf, G., Knudsen, F., Mumford, D., Saint-Donat, B., *Toroidal Embeddings I*, Lecture Notes in Mathematics 339, Springer-Verlag, 1973.

[KSZ92] Kapranov, M. M., Sturmfels, B., and Zelevinsky, A. V., *"Chow Polytopes and General Resultants,"* Duke Mathematical Journal, Vol. 67, No. 1, July, 1992, pp. 189–218.

[Kus75] Kushnirenko, A. G., *"A Newton Polytope and the Number of Solutions of a System of k Equations in k Unknowns,"* Usp. Matem. Nauk., 30, No. 2, pp. 266–267 (1975).

[Laz81] Lazard, D. *"Résolution des Systèmes d'équations Algébriques,"* Theor. Comp. Sci. 15, pp. 146–156.

[LLL82] Lenstra, A. K., Lenstra, H. W., and Lovász, L. *"Factoring Polynomials with Rational Coefficients,"* Math. Ann. 261 (1982), pp. 515-534.

[Mat93] Matiyasevich, Yuri V. *Hilbert's Tenth Problem,* MIT Press, MIT, Cambridge, Massachusetts, 1993.

[Math250B] Sturmfels, Bernd, *Math 250B (Commutative Algebra),* graduate course, U. C. Berkeley, fall 1995, homework #2, problem #3.

[PK95] Pardo-Vasallo, Luis-Miguel and Krick, Teresa, *"A Computational Method for Diophantine Approximation,"* to appear in Proc. MEGA '94, Birkhäuser, Progress in Math. (1995).

[PS93] Pedersen, P. and Sturmfels, B., *"Product Formulas for Sparse Resultants and Chow Forms,"* Mathematische Zeitschrift, 214: 377–396, 1993.

[Ren87] Renegar, Jim, *"On the Worst Case Arithmetic Complexity of Approximating Zeros of Systems of Polynomials,"* Technical Report, School of Operations Research and Industrial Engineering, Cornell University.

[Roj94] Rojas, J. Maurice, *"A Convex Geometric Approach to Counting the Roots of a Polynomial System,"* Theoretical Computer Science (1994), vol. 133 (1), pp. 105–140.

[Roj96a] Rojas, J. Maurice *"Toric Intersection Theory for Affine Root Counting,"* submitted to the Journal of Pure and Applied Algebra, also available on-line at http://www-math.mit.edu/~rojas.

[Roj96b] Rojas, J. Maurice *"When do Resultants Really Vanish?,"* manuscript, MIT.

[Roj96c] Rojas, J. Maurice *"Affine Elimination Theory and Solving Certain Diophantine Systems Quickly,"* manuscript, MIT.

[Roy95] Roy, Marie-Françoise *"Basic Algorithms in Real Algebraic Geometry and their Complexity: from Sturm Theorem to the Existential Theory of Reals,"* manuscript, IRMAR, Rennes, France, 1995.

[RW96] Rojas, J. M., and Wang, Xiaoshen, *"Counting Affine Roots of Polynomial Systems Via Pointed Newton Polytopes,"* Journal of Complexity, vol. 12, June (1996), pp. 116–133, also available on-line at http://www-math.mit.edu/~rojas.

[Shu93] Shub, Mike, *"Some Remarks on Bézout's Theorem and Complexity Theory,* From Topology to Computation: Proceedings of the Smalefest, pp. 443–455, Springer-Verlag, 1993.

[Stu93] Sturmfels, Bernd, *"Sparse Elimination Theory,"* In D. Eisenbud and L. Robbiano, editors, Proc. Computat. Algebraic Geom. and Commut. Algebra 1991, pages 377–396, Cortona, Italy, 1993, Cambridge Univ. Press.

[Stu94] Sturmfels, Bernd, *"On the Newton Polytope of the Resultant,"* Journal of Algebraic Combinatorics, 3: 207–236, 1994.

[Stu95] Sturmfels, Bernd, *Gröbner Bases and Convex Polytopes,* Lectures presented at the Holiday Symposium at New Mexico State University, December 27–31, 1994.

[Zie95] Ziegler, Gunter M., *Lectures on Polytopes,* Graduate Texts in Mathematics 152, Springer-Verlag, New York, 1995.

[Zip93] Zippel, R., *Effective Polynomial Computation,* Kluwer Academic Publishers, Boston, 1993.

Finite-Dimensional Feedback Control of a Scalar Reaction-Diffusion Equation via Inertial Manifold Theory

Ricardo Rosa and Roger Temam

Abstract. Using inertial manifold theory, a finite-dimensional feedback controller is constructed for a nonlinear parabolic partial differential equation so that the dynamics of the closed-loop system is close in a weighted C^1-metric for the vector fields to a finite dimensional dynamics prescribed in advance. As a consequence, some structurally stable dynamics, for instance, can be imposed on the controlled system.

1 Introduction

Consider the semilinear open-loop system

$$\begin{cases} \frac{du}{dt} + Au = f(u) + Bg, \\ y = Cu, \end{cases} \tag{1}$$

where u is the state-variable living in an infinite-dimensional space, y is the finite-dimensional control output, and g is the finite-dimensional control input. We assume that B and C are bounded linear operators, A generates a strongly continuous semigroup of bounded linear operators, and f is a given nonlinear term. We would like to construct g as a function of y so that the closed-loop system behaves in a certain desired way.

Note that from a pratical point of view, it is very important that we look for g as a function of y rather than u so that the feedback law is finite-dimensional, suitable for pratical implementations. Usual works on control theory construct g as a function of u so that some further approximation is necessary for pratical implementations. This further approximation might lead to instabilities and the so-called spillover problem, which is exactly what we are trying to solve here.

A long list of works is available on the related linear problem when the stabilization to the origin is desired; a short list reads [1, 2, 3, 13, 14, 16, 17]. From those results, one can easily obtain local (near fixed points) stabilization results for the nonlinear problem (see M. J. Balas [4], for instance). Some works on the nonlinear problem are also available and use ideas borrowed from inertial manifold theory: The works of Y. You [20] and M. Taboada and Y. You [21] present global stabilization results for specific equations on nonlinear elasticity and make use of some nice properties of the nonlinear term (it commutes with the spectral projectors of the linear term A) so that a spectral gap condition (not satisfied in this case) is not necessary. The work of H. Sano and N. Kunimatsu [18] borrows ideas from the linear case and use the so-called dynamic observers, but it is still a local result since it requires a smallnest condition on the global Lipschitz constant of the nonlinear term. Finally, the work of P. Brunovský [5] is in some sense a global result, but it assumes essentially that the control and observation operators B and C are spectral projectors of the linear operator A, which is not true in most applications. Our work differs from those above since the nonlinear term is not assumed to commute with the spectral projectors of A, and neither are the operators B and C such spectral projectors. Also, the result we obtain is not localized to a small neighborhood of a fixed point. Moreover, as in the work of P. Brunovský, we are interested not only in the stabilization problem but also on imposing a more complicated dynamics on the system under control.

Our aim is to construct a finite-dimensional feedback control for a specific concrete open-loop problem so that the closed-loop system behaves in a desired way given a priori by a finite-dimensional system. We follow the idea of P. Brunovský, but we do not assume B and C to be spectral projectors of A, which introduces major difficulties to the problem. One major consequence of this fact is that the control is obtained only in an approximate sense. We use inertial manifold theory to reduce the closed-loop system to a finite-dimensional system and we show that the vector field of this finite-dimensional system is close in a weighted C^1-metric to some finite-dimensional vector field given a priori. Hence, if the given finite-dimensional system is structurally stable [12, 11, 19] in a suitable sense (see Section 5, Theorem 2), the same dynamics will be seen in the closed loop system.

It is worth noting that the idea is general and works for problems other than the specific example studied here as long as the necessary spectral gap condition holds. Some further extensions of this result to the specific case of the stabilization towards a fixed point are also possible when the spectral gap condition does not hold; results in this direction will be presented elsewhere.

This article is organized as follows: Section 2 presents the open-loop system and sets the framework and notation to be used in the sequel. Section 3 establishes some properties of the operators B and C related to the existence of left or right inverses. The feedback law $g = g(y)$ is introduced in Section 4. Then, Section 5 studies the closed-loop system and establishes the main result.

2   The Open-Loop System

The open-loop system we consider is the following nonlinear scalar reaction-diffusion equation with zero Dirichlet boundary condition and distributed observation and control:

$$\begin{cases} u_t(t,x) = u_{xx}(t,x) + f(u(t,x)) + \sum_{i=1}^{I-1} g_i(t)\psi_i(x), & x \in (0,\pi),\ t > 0, \\ y(t) = (y_j(t))_{j=1}^{J-1} = (u(t,x_j))_{j=1}^{J-1}, & t \geq 0, \\ u(t,0) = 0 = u(t,\pi),\ t > 0, \qquad u(0,x) = u_0(x),\ x \in [0,\pi], \end{cases} \tag{2}$$

where u is the state variable, y is the observation, $g = (g_i)_i$ is the control, f is a nonlinear term, and $I, J \in \mathbb{N}$. The functions ψ_i are called the actuators and are assumed to lye in $H_0^1(0, \pi)$, while the points x_j are distinct points in $(0, \pi)$ called the obervation points and assumed to increase with j. We further assume that we are given another set of points $\{\tilde{x}_i\}_{i=0}^{I}$, with $0 = \tilde{x}_0 < \ldots < \tilde{x}_i < \tilde{x}_{i+1} < \ldots < \tilde{x}_I = \pi$, and that ψ_i, for $i = 1, \ldots, I - 1$, is given more precisely by

$$\psi_i(x) = \begin{cases} (x - \tilde{x}_{i-1})/\tilde{h}_i, & x \in [\tilde{x}_{i-1}, \tilde{x}_i), \\ (\tilde{x}_{i+1} - x)/\tilde{h}_{i+1}, & x \in [\tilde{x}_i, \tilde{x}_{i+1}), \\ 0, & \text{otherwise}, \end{cases} \tag{3}$$

where $\tilde{h}_i = \tilde{x}_i - \tilde{x}_{i-1}$. Set also $h_j = x_j - x_{j-1}$, $h = \max_j h_j$, and $\tilde{h} = \max_i \tilde{h}_i$.

We consider this equation in the phase space $E = H_0^1(0, \pi)$ endowed with the norm $\|u\| = |Du|$, for $u \in E$, where $|\cdot|$ denotes the usual L^2-norm on $(0, \pi)$ and Du denotes the derivative of u. We also denote by $((\cdot, \cdot))$ and $(\cdot, \cdot)$ the corresponding inner-products in E and $L^2(0, \pi)$, respectively.

For the nonlinear term, we assume more precisely that $f(s) = \lambda \sin s$, for $s \in \mathbb{R}$, where $\lambda > 0$, or any other function that can be regarded as a function $f \in C^1(E)$ with

$$\|f(u) - f(v)\| \leq M_1 \|u - v\|, \qquad \forall u, v \in E, \tag{4}$$
$$\|Df(u) - Df(v)\|_{\mathcal{L}(E)} \leq M_2 \|u - v\|^\nu, \qquad \forall u, v \in E, \tag{5}$$

for $M_1, M_2 > 0$ and $0 < \nu \leq 1$, where Df denotes the differential of f in E, and $\|\cdot\|_{\mathcal{L}(E)}$ denotes the operator norm for Df.

We can write (2) in the abstract form

$$\begin{cases} \dfrac{du}{dt} + Au = f(u) + Bg, \\ y = Cu, \end{cases} \tag{6}$$

by setting $A = -\partial_{xx}$ as an operator in E and by defining operators $B : X \to E$ and $C : E \to Y$, with $X \cong \mathbb{R}^{I-1}$ and $Y \cong \mathbb{R}^{J-1}$, as

$$Bg = \sum_{i=1}^{I-1} g_i \psi_i(x), \qquad \text{for} \quad g = (g_i)_{i=1}^{I-1} \in X, \tag{7}$$

$$Cu = ((Cu)_j)_{j=1}^{J-1} = (u(x_j))_{j=1}^{J-1}, \qquad \text{for} \quad u = u(x) \in E. \tag{8}$$

We endow X and Y with the norms given by

$$\|g\|_X^2 = \sum_{i=1}^{I-1} \left| \frac{g_i - g_{i-1}}{\tilde{h}_i} \right|^2 \tilde{h}_i, \ g \in X, \text{ and } \|y\|_Y^2 = \sum_{j=1}^{J-1} \left| \frac{y_j - y_{j-1}}{h_j} \right|^2 h_j, \ y \in Y. \tag{9}$$

It is not very difficult to show that

$$\|B\|_{\mathcal{L}(X,E)} = 1, \qquad \|C\|_{\mathcal{L}(E,Y)} = 1. \tag{10}$$

The operator A is a self-adjoint operator with eigenvalues given by $\{\lambda_n = n^2\}_{n \in \mathbb{N}}$ and eigenvectors $\{\varphi_n = \sin nx\}_{n \in \mathbb{N}}$. We let P_n denote the spectral projector associated with the first n eigenvalues of A. Clearly, P_n is orthogonal in E, as well as in $L^2(0, \pi)$. We also set $Q_n = I - P_n$, where I is the identity operator.

3 The Input and Output Control Operators B and C

In this section, we shall show that with a large number of properly located actuators and observation points, the operators B and C have respectively right and left inverses on appropriate spectral spaces of the operator A.

For $u \in C_0^2((0, \pi))$, one can show that

$$u'(x) = \frac{u(x_j) - u(x_{j-1})}{h_j} - \frac{1}{h_j} \int_{x_{j-1}}^{x_j} \int_x^s u''(r) \, dr \, ds, \quad \text{for } x \in [x_{j-1}, x_j],$$

and then that

$$u'(x)^2 \leq 2 \left| \frac{u(x_j) - u(x_{j-1})}{h_j} \right|^2 + 2h_j \int_{x_{j-1}}^{x_j} u''(r)^2 \, dr, \quad \text{for } x \in [x_{j-1}, x_j],$$

and eventually that $\|u\|^2 \leq 2\|Cu\|_Y^2 + 2h^2|u''|^2$. Then, for $u \in P_m E$, we have $|u''|^2 \leq \lambda_m |u'|^2 = \lambda_m \|u\|^2$, so that for h small enough,

$$\|u\|^2 \leq \frac{2}{1 - 2h^2\lambda_m} \|Cu\|^2, \qquad \forall u \in P_m E. \tag{11}$$

Hence, for h small enough depending on m, the map $CP_m : P_m E \to CP_m E \subset Y$ is one-to-one and admits an inverse defined on its image $CP_m E$ and denoted by $L : CP_m E \to P_m E$. Since $CP_m E$ is a linear subspace of Y and Y is a Hilbert space, we can define an orthogonal projector $P_{Y,m}$ in Y onto $CP_m E$. Then, we can define the operator $(CP_m)_\ell^{-1} = LP_{Y,m} : Y \to P_m E \subset E$, which is clearly a left inverse for CP_m. Moreover, from (11) and the fact that $P_{Y,m}$ is orthogonal it follows that

$$\| (CP_m)_\ell^{-1} \|_{\mathcal{L}(Y,E)} \leq \sqrt{\frac{2}{1 - 2h^2\lambda_m}}. \tag{12}$$

Now, for the operator B, consider the operator $\tilde{C} : E \to X$ given by $\tilde{C}u = (u(\tilde{x}_i))_{i=1}^{I-1}$. As for C, we have $\|\tilde{C}\|_{\mathcal{L}(E,X)} = 1$. Consider $u \in C_0^\infty((0, \pi))$ and set

$$e \equiv u - B\tilde{C}u = u(x) - \sum_{i=1}^{I-1} u(\tilde{x}_i)\psi_i(x).$$

Note since $\psi_i(\tilde{x}_i) = 1$ that $e(\tilde{x}_i) = 0$. Also, since $B\tilde{C}u$ is linear on each $[\tilde{x}_{i-1}, \tilde{x}_i]$ we find

$$e \in C_0^\infty([\tilde{x}_{i-1}, \tilde{x}_i]), \quad \text{with } e'' = u'' \text{ on } (\tilde{x}_{i-1}, \tilde{x}_i).$$

Hence,

$$\|u - B\tilde{C}u\|^2 = \int_0^\pi e'(x)^2 \, dx = \sum_{i=1}^I \int_{\tilde{x}_{i-1}}^{\tilde{x}_i} e'(x)^2 \, dx \leq \sum_{i=1}^I \tilde{h}_i^2 \int_{\tilde{x}_{i-1}}^{\tilde{x}_i} e''(x)^2 \, dx$$

$$\leq 4\tilde{h}^2 \sum_{i=1}^I \int_{\tilde{x}_{i-1}}^{\tilde{x}_i} u''(x)^2 \, dx = 4\tilde{h}^2 |u''|^2,$$

where we used the Poincaré inequality for e', which has zero average on each of the subintervals $(\tilde{x}_{i-1}, \tilde{x}_i)$. If $u \in P_n E$, then $|u''|^2 \leq \lambda_n \|u\|^2$, so that

$$\|u - B\tilde{C}u\|^2 \leq 4\lambda_n \tilde{h}^2 \|u\|^2, \qquad \forall u \in P_n E.$$

Then, since P_n is orthogonal, $\|u - P_n B\tilde{C}u\|^2 = \|P_n(u - B\tilde{C}u)\|^2 \leq \|u - B\tilde{C}u\|^2$. Hence,

$$\|u - P_n B\tilde{C}u\|^2 \leq 4\lambda_n \tilde{h}^2 \|u\|^2, \qquad \forall u \in P_n E. \tag{13}$$

Then,

$$\|u\|^2 = \|u - P_n B\tilde{C}u\|^2 + \|P_n B\tilde{C}u\|^2 \leq 4\lambda_n \tilde{h}^2 \|u\|^2 + \|P_n B\tilde{C}u\|^2,$$

so that, if $\tilde{h}$ is small enough,

$$\|u\|^2 \leq \frac{1}{1 - 4\tilde{h}^2 \lambda_n} \|P_n B\tilde{C}u\|^2, \qquad \forall u \in P_n E. \tag{14}$$

Define $K = P_n B\tilde{C}|_{P_n E} : P_n E \to P_n E$. Then, it is straightforward to see that

$$\|K\|_{\mathcal{L}(P_n E)} \leq \|P_n\|_{\mathcal{L}(E)} \|B\|_{\mathcal{L}(X,E)} \|\tilde{C}\|_{\mathcal{L}(E,X)} = 1 \tag{15}$$

and

$$\|u\| \leq \sqrt{\frac{1}{1 - 4\tilde{h}^2 \lambda_n}} \|Ku\|, \qquad \forall u \in P_n E, \tag{16}$$

Thus, K is one-to-one and, hence, it is also surjective since it maps $P_n E$ into itself and $P_n E$ is of finite dimension. Therefore, K is bijective with

$$\|K^{-1}\|_{\mathcal{L}(P_n E)} \leq \sqrt{\frac{1}{1 - 4\tilde{h}^2 \lambda_n}}. \tag{17}$$

Set then $(P_n B)_r^{-1} = \tilde{C} K^{-1} : P_n E \to X$. It is easy to see that $(P_n B)_r^{-1}$ is then a right inverse for $P_n B$ with (since $\tilde{C}$ has norm 1)

$$\|(P_n B)_r^{-1}\|_{\mathcal{L}(P_n E, X)} \leq \sqrt{\frac{1}{1 - 4\tilde{h}^2 \lambda_n}}. \tag{18}$$

4   The Finite-Dimensional Controller

Let now $n_0 \in \mathbb{N}$ be given and let $N : P_{n_0}E \to P_{n_0}E$ be a nonlinear function satisfying

$$\|N(u) - N(v)\| \le M_1'\|u - v\|, \qquad \forall u, v \in P_{n_0}E, \qquad (19)$$

$$\|DN(u) - DN(v)\|_{\mathcal{L}(E)} \le M_2'\|u - v\|^\nu, \qquad \forall u, v \in P_{n_0}E, \qquad (20)$$

for some $M_1', M_2' > 0$, and for ν as in (5), so that the desired dynamics we want to achieve for (2) is that given by the following n_0-dimensional system on $P_{n_0}E$:

$$\frac{dz}{dt} = N(z). \qquad (21)$$

Set $L = 4(\lambda_{n_0} + M_1 + M_1')$ and consider m and n arbitrary such that

$$m \ge n > \frac{6(M_1 + L) - 1}{2}. \qquad (22)$$

Choose then the $x_j{}'s$ and the $\tilde{x}_i{}'s$ such that

$$\tilde{h} \le \frac{\sqrt{3}}{4\lambda_{n_0}^{1/2}} \qquad \text{and} \qquad h \le \frac{1}{2\lambda_m^{1/2}}, \qquad (23)$$

so that

$$\sqrt{\frac{1}{1 - 4\tilde{h}\lambda_n}} \le 2 \qquad \text{and} \qquad \sqrt{\frac{2}{1 - 2h\lambda_m}} \le 2. \qquad (24)$$

We can then define $g : Y \to X$ by

$$g(y) = (P_n B)_r^{-1} \left[A P_{n_0} (CP_m)_\ell^{-1} y + N(P_{n_0} (CP_m)_\ell^{-1} y) \right. $$
$$\left. - P_n f((CP_m)_\ell^{-1} y) \right], \qquad \forall y \in Y. \qquad (25)$$

Note then that g is globally Lipschitz continuous with

$$\text{Lip } g \le \| (P_n B)_r^{-1} \|_{\mathcal{L}(P_n E, X)} \left[\|AP_{n_0}\|_{\mathcal{L}(E)} + \text{Lip } N + \text{Lip } f \right] \| (CP_m)_\ell^{-1} \|_{\mathcal{L}(Y, E)}$$

$$\le \sqrt{\frac{1}{1 - 4\tilde{h}\lambda_n}} \sqrt{\frac{2}{1 - 2h\lambda_m}} [\lambda_{n_0} + M_1' + M_1]$$

$$\le 4(\lambda_{n_0} + M_1 + M_1').$$

Thus

$$\text{Lip } g \le L. \qquad (26)$$

As opposed to the work of P. Brunovský [5], we do not obtain a fixed point map for g. This is because we do not assume that C is a spectral projector, hence the image through C of the inertial manifold for the closed-loop equation depends on g and is not simply $P_n E$, so that we were not able to find a left inverse (defined on all Y) for C restricted to the inertial manifold. The remedy was to replace the inertial manifold by the much bigger, but independent of g, flat manifold $P_m E$. For larger m the inertial manifold is closer to $P_m E$. If we could imbed the inertial manifold into some $P_m E$ or if we could find a left inverse as mentioned above, the final result (see Section 5) would be exact.

5   The Closed-Loop System

With g given by (25) we can write (6) in the closed-loop form

$$\frac{du}{dt} + Au = f(u) + Bg(Cu). \tag{27}$$

We shall also consider the following auxiliary equation:

$$\frac{dv}{dt} + Av = P_m f(P_m v) + P_m Bg(CP_m v). \tag{28}$$

The idea is to show that both equations have inertial manifolds and that the inertial form of the latter equation has the desired dynamics given by (21), while the inertial form of the closed-loop equation is close to that of the auxiliary equation in the proper metric.

Note that the right hand side of both the equations above are globally Lipschitz with Lipschitz constant less than or equal to $M_1 + L$. Thus with the choice of n given by (22) it is easy to see that

$$\lambda_{n+1} - \lambda_n > 6(M_1 + L), \tag{29}$$

which is the spectral gap condition in this case necessary for the existence of an inertial manifold as a graph over $P_n E$. Because of lack of space we simply refer the reader to the works [8, 9, 6, 10, 7, 15] on the existence of inertial manifolds. The exact form of the spectral gap condition that we use here appears in R. Rosa and R. Temam [15, Theorem 2.1].

Hence, there are inertial manifolds

$$\mathcal{M} = \text{graph } \Phi, \qquad \Phi : P_n E \to Q_n E,$$
$$\mathcal{N} = \text{graph } \Psi, \qquad \Psi : P_n E \to Q_n E,$$

respectively for (27) and (28). This means that $\mathcal{M}$ (and similarly for $\mathcal{N}$) is an invariant manifold for (27) with Φ Lipschitz continuous so that the restriction of (27) to $\mathcal{M}$ is a finite dimensional differential equation called the inertial form. Moreover, the asymptotic completeness holds, i.e., for each $u_0 \in E$, there exists $\tilde{u}_0 \in \mathcal{M}$ such that if $u = u(t)$ and $\tilde{u} = \tilde{u}(t)$ denote the solutions of (27) with $u(0) = u_0$ and $\tilde{u}(0) = \tilde{u}_0$, then

$$\|u(t) - \tilde{u}(t)\| \leq Ke^{-\kappa t}, \qquad \forall t \geq 0, \tag{30}$$

where $K = (\|u_0\|) > 0$ depends on the norm of u_0 in E, and $\kappa > 0$ is a constant. This implies that the long-time dynamics of (27) is essentially that of the inertial form (note that $\tilde{u}$ belongs to $\mathcal{M}$ since $\mathcal{M}$ is invariant). The corresponding inertial forms on $P_n E$ are

$$\frac{dp}{dt} + Ap = P_n f(p + \Phi(p)) + P_n Bg(C(p + \Phi(p))), \tag{31}$$

and

$$\frac{d\rho}{dt} + A\rho = P_n f(P_m(\rho + \Psi(\rho))) + P_n Bg(CP_m(\rho + \Psi(\rho))). \tag{32}$$

Note that

$$\frac{d\rho}{dt} + A(P_n - P_{n_0})\rho = -AP_{n_0}\rho + P_n f\left(P_m(\rho + \Psi(\rho))\right)$$
$$+ P_n B (P_n B)_r^{-1} \Big[AP_{n_0} (CP_m)_\ell^{-1} CP_m(\rho + \Psi(\rho))$$
$$+ N\left(P_{n_0} (CP_m)_\ell^{-1} CP_m(\rho + \Psi(\rho))\right)$$
$$- P_n f\left((CP_m)_\ell^{-1} CP_m(\rho + \Psi(\rho))\right)\Big]$$
$$= -AP_{n_0}\rho + P_n f\left(P_m(\rho + \Psi(\rho))\right) + [AP_{n_0} P_m(\rho + \Psi(\rho))$$
$$+ N\left(P_{n_0} P_m(\rho + \Psi(\rho))\right) - P_n f\left(P_m(\rho + \Psi(\rho))\right)]$$
$$= N(P_{n_0}\rho).$$

Thus, the inertial form for (28) reads

$$\frac{d\rho}{dt} + A(P_n - P_{n_0})\rho = N(P_{n_0}\rho), \tag{33}$$

and can be split for $\rho = \rho_1 + \rho_2$, $\rho_1 \in P_{n_0} E$, $\rho_2 \in (P_n - P_{n_0})E$ as

$$\begin{cases} \dfrac{d\rho_1}{dt} = N(\rho_1), \\ \dfrac{d\rho_2}{dt} + A(P_n - P_{n_0})\rho_2 = 0. \end{cases} \tag{34}$$

The system (34) above is now decoupled with

$$\rho_2(t) = e^{-tA(P_n - P_{n_0})}\rho_2(0) = \mathcal{O}(e^{-(n_0+1)^2 t}), \qquad \text{as } t \to +\infty.$$

Hence, the long-time dynamics of the inertial form and, hence, of the auxiliary equation (28) is given by the system $\dot{\rho}_1 = N(\rho_1)$.

Concerning the inertial form (31), we can write it as

$$\frac{dp}{dt} + A(P_n - P_{n_0})p = N(P_{n_0}p) + E(p), \tag{35}$$

where $E(p)$ is regarded as an error term given by

$$E(p) = P_n f(p + \Phi(p)) + P_n Bg\left(C(p + \Phi(p))\right) \tag{36}$$
$$- P_n f(p + P_m \Psi(p)) - P_n Bg\left(C(p + P_m \Psi(p))\right).$$

Note that

$$\|E(p)\| \le (M_1 + L)\|\Phi(p) - P_m \Psi(p)\| \tag{37}$$
$$= (M_1 + L)\|\Phi(p) - \Psi(p)\|,$$

where the equality follows because $\Psi(p)$ already lies in $P_m E$, which is not difficult to see.

Now, a rather extensive computation, which would double the size of this work and which is then omitted, is necessary in order to estimate the distance from Ψ to Φ and between their derivatives. One can do that using the techniques appearing in [15] for finding the inertial manifolds as fixed points of a certain map. By properly analysing this map and the corresponding map used for finding the differential of the function whose graph is the inertial manifold, one can obtain the following estimates:

$$\|E(p)\| \le \frac{1}{\lambda_m^{1/2}} \left(\alpha_1 + \alpha_2 \|p\|\right), \qquad \forall p \in P_n E, \tag{38}$$

$$\|DE(p)\|_{\mathcal{L}(P_n E)} \le \frac{\alpha_3}{\lambda_m^{1/2}} + \frac{\alpha_4}{\lambda_m^{\nu/2}}, \qquad \forall p \in P_n E, \tag{39}$$

where the $\alpha_i's$ are constant depending on $n_0, n, M_0, M_0', M_1, M_1', M_2, M_2'$ and ν, where $M_0 = \|f(0)\|$ and $M_0' = \|N(0)\|$, or, more precisely,

$$\alpha_i = \alpha_i(n_0, n, M_0, M_0', M_1, M_1') \qquad \text{for } i = 1, 2, 3,$$
$$\alpha_4 = \alpha_4(n_0, n, M_0, M_0', M_1, M_1', M_2, M_2', \nu).$$

We can now summarize in the form of a theorem what we obtained above:

Theorem 1 *Consider the open-loop system (2) with the hypotheses given in Section 2, and let a finite-dimensional system (21) be given with $n_0 \in \mathbb{N}$ and N satisfying (19) and (20). Let n, m, h and $\tilde{h}$ be such that (22) and (23) hold.*

If a feedback law $g = g(y)$ is given by (25), then the corresponding closed-loop equation (27) possesses an inertial manifold as a graph over $P_n E$ whose inertial form (35) is C^1-close to (33) (which has essentially the same dynamics as (21)) in a weighted metric for the vector fields as estimated in (38) and (39).

As a consequence of Theorem 1 above, certain structurally stable dynamics can be imposed on the closed-loop system. For instance, it is straighforward to see that the folowing result holds:

Theorem 2 *Assume the hypotheses in Theorem 1 hold. Assume also that N is such that for some $r_0 > 0$, we have $((N(z), z)) \le -\beta \|z\|$, for all $\|z\| \ge r_0$, for some $\beta > 0$, and that the flow induced by $\dot{z} = N(z)$ for z restricted to the ball $B_{r_0}^{n_0} = \{z \in P_{n_0}; \|z\| \le r_0\}$ is structurally stable.*

If $g = g(y)$ is given by (25) with m chosen large enough, then the long-time dynamics of the inertial form (35) of the closed-loop equation (27) is contained in the ball $B_{r_0}^n = \{p \in P_n E; \|p\| \le r_0\}$ and the corresponding flow restricted to this ball $B_{r_0}^n$ is topologically equivalent to the flow given by (33), so that the dynamics of the closed-loop system is essentially that of $\dot{z} = N(z)$.

References

1. Balas, M. J.: Modal control of certain flexible dynamic systems. SIAM J. Control Optim. **16** (1978), no. 3, 450-462
2. Balas, M. J.: Trends in large space structure control theory: Fondest hopes, wildest dreams. IEEE Trans. Automat. Control, Vol. **AC-27** (1982), no. 3, 522-533
3. Balas, M. J.: Finite-dimensional controllers for linear distributed parameter systems: Exponential stability using residual mode filters. J. Math. Anal. Appl. **133** (1988), 283-296

4. Balas, M. J.: Nonlinear finite-dimensional control of a class of nonlinear distributed parameter systems using residual mode filters: A proof of local exponential stability. J. Math. Anal. Appl. **162** (1991) 63-70
5. Brunovský, P.: Controlling the dynamics of scalar reaction diffusion equations by finite dimensional controllers. In "Modelling and Inverse Problems of Control for Distributed Parameter Systems" (A. Kurzhanski and I. Lasiecka, Eds.), pp 22-27, Lecture Notes in Control and Information Sciences, Vol. 154, Springer-Verlag, Berlin, 1991
6. Chow, S-N., Lu, K.: Invariant manifolds for flows in Banach spaces. J. Diff. Eqs. **74** (1988), 285-317
7. Debussche, A., Temam, R.: Inertial Manifolds and their dimension. In "Dynamical Systems, Theory and Applications" (S.I. Anderson, A.E. Anderson and O. Ottason, Eds), World Scientific Publishing Co., 1993
8. Foias, C., Sell, G.R., Temam, R.: Variétés inertielles des équations différentielles dissipatives. C.R. Acad. Sci. Paris, Série I, **301** (1988), 139-142
9. Foias, C., Sell, G.R., Temam, R.: Inertial manifolds for nonlinear evolutionary equations. J. Diff. Eqs. **73** (1988), 309-353
10. Foias, C., Sell, G.R., Titi, E.: Exponential Tracking and approximation of inertial manifolds for dissipative nonlinear equations. J. Dyn. Diff. Eqs. **1** (1989), 199-243
11. Guckenheimer, J., Holmes, P.J.: Nonlinear Oscillations, Dynamical Systems, and Bifurcations of Vector Fields. Springer-Verlag, New York, Heidelber, Berlin. 1983
12. Palis, J., deMelo, W.: Geometric Theory of Dynamical Systems: An Introduction. Springer-Verlag, New York, Heidelber, Berlin. 1982
13. Nambu, T.: Feedback stabilization of diffusion equations by a functional observer. J. Diff. Eqs. **43** (1982), 257-280
14. Nambu, T.: On stabilization of partial differential equations of parabolic type: Boundary observation and feedback. Funkcial. Ekvac. **28** (1985), 267-293
15. Rosa, R., Temam., R.: Inertial manifolds and normal hyperbolicity. ACTA Applicandae Mathematicae (to appear).
16. Sakawa, Y.: Controllability for partial differential equations of parabolic type. SIAM J. Control **12** (1974), no. 3, 389-400
17. Sakawa, Y.: Feedback stabilization of linear diffusion systems. SIAM J. Control Optim. **21** (1983), no. 5, 667-676
18. Sano, H., Kunimatsu, N.: An application of inertial manifold theory to boundary stabilization of semilinear diffusion systems. J. Math. Anal. Appl. **196** (1995) 18-42
19. Wiggins, S.: Introduction to Applied Nonlinear Dynamical Systems and Chaos. Texts in Applied Mathematics 2. Springer-Verlag, New York, Heidelberg, Berlin. 1990
20. You, Y.: Inertial manifolds stabilization in nonlinear elastic systems with structural damping. In "Differential Equations with Applications to Mathematical Physics", pp 335-346, Math. Sci. Engrg., Vol. 192, Academic Press, Boston, 1993
21. Taboada, M., You, Y.: Global dynamics and control of a nonlinear beam equation. In "Qualitative Aspects and Applications of Nonlinear Evolution Equations" (Trieste, 1993), pp 109-122, World Sci. Publishing, River Edge, NJ, 1994

Computational aspects of jacobian matrices*

Aron Simis**

1 Introduction

The underlying theme of this talk is the jacobian matrix of a (finite) set of multivariate polynomials $\mathbf{f} = \{f_1, \ldots, f_m\} \subset R = k[\mathbf{X}] = k[X_1, \ldots, X_d]$. Here k stands for a field, eventually of characteristic zero. Jacobian matrices are classical, useful in almost all branches of mathematics and there is a famous conjecture that bears the name! Is this enough reason to make them a central theme in a talk? Perhaps. The full justification here would go far beyond our modest purpose which is, namely, to convey their theoretical and computational usefulness in everyday commutative algebra and algebraic geometry.

Such a project, as it turns out, is rather too ambitious and will necessarily reflect anyone's personal taste. I have chosen to expound on jacobian topics that nourish effectiveness, by which I mean, more precisely, that they involve a theorem in commutative algebra whose statement involves a jacobian matrix in a non-trivial fashion and the conditions - hypothesis and result - are effectively computable (though may never actually be computed!).

We now shortly describe the contents of the sections - a more detailed explanation is to be found in the body of the very sections and subsections.

We start with a brief digression on the classical jacobian criterion for singularities. A full account on this venerable topic can be found in very many excellent books (cf. [2], [4], [12] to mention a few). This short section is meant as a warm-up to the subject and, simultaneously, as guide to some recent computational openings that stem directly from a rewording or a revitalization, if one can say so, of the jacobian criterion.

The second topic gets harder, therefore I chose to subdivide it into three short subsections. It builds a suggestive jacobian theory for quadratic forms of bidegree $(1,1)$ in a given bipartition of the variables. The main results have been proved in [7], but I keep going back to those ideas so often that its inclusion

---

AMS 1990 Mathematics Subject Classification. Primary 13H10, 13A30; Secondary 13C15, 13C40, 13H05, 13H10, 13N05, 13P10, 14B07, 14C17, 14E22, 14E25, 14M05, 14M25.
** Partially supported by CNPq, Brazil.

here was unavoidable. Moreover, I will emphasize the computational side of the matter more than it was done in [*loc. cit.*] so as to tie it up with the subsequent subsections. The presentation here also varies slightly from the original one.

The next section is truly computational, so it may better justify the inclusion of this talk in the *Proceedings* of the Conference. It deals with a device to iteratively compute the generators of the symbolic powers of an ideal in a polynomial ring. The proofs will appear in [5]. The method supersedes, to our view, the previously existing ones which, incidentally, were mostly *ad hoc* methods lacking a good iterative sequence so as to allow an effective implementation. Two *Macaulay* scripts based on this method were kindly written by D. Eisenbud (more to be said in the talk in general about the use of scripts and other computational routines). However computational the method, the result has some unexpected consequences such as a systematic ("differential") embedding of the symbolic normal cone algebra of an affine ring A into a polynomial ring over A.

ACKNOWLEDGEMENTS. This talk reflects conversations held with W. Vasconcelos for a good many years, for which I am deeply indebted. His account [13] on computational methods in commutative algebra and algebraic geometry stands alone as a must in the subject.

2   The jacobian criterion revisited

In order to avoid tedious repetitions of "dynamic" hypotheses on the characteristic of the ground field k, we assume in this section that char $k = 0$ - note, however, that many arguments will go through by just assuming that k is a perfect field. Let $R = k[\mathbf{X}] = k[X_1, \ldots, X_d]$ be a polynomial ring in d variables over k and et $I = (\mathbf{f}) = (f_1, \ldots, f_n) \subset R$ be an ideal. The codimension of I will be denoted ht I (for "height", a widely accepted alternative terminology). Let $\Theta = \Theta(\mathbf{f}) = (\partial f_j / \partial X_i)_{i,j}$ denote the jacobian matrix of the generators of I. Set $A = R/I$. One knows that the A-module coker$(\Theta) \otimes_R A$ can be identified with the module $\Omega_{A|k}$ of Kähler k-differentials of A.

Definition 2.1 The *jacobian ideal of order* r of $A = R/I$ (or simply, the rth *jacobian ideal* of A) is the rth Fitting ideal of the A-module $\Omega_{A|k}$. The jacobian ideal of order dim A is called the *jacobian ideal* of A.

Since the Fitting ideals of a module are invariants of the module, the jacobian ideals of A will be independent of the chosen generators $\mathbf{f}$. Thus, the rth *jacobian ideal* of A will be denoted simply by $J_r(A)$ and, if no confusion arises, we denote the jacobian ideal of A by $J(A)$. One says that A satisfies condition (R_s) (à la Serre) if the local ring A_P is regular for every prime ideal $P \subset A$ of codimension $\leq$ s, in which case one leisurely says that A is nonsingular in codimension s. Clearly, by definition, $(R_s) \Rightarrow (R_{s-1})$, hence to say that A is locally regular everywhere means that it satisfies (R_s) with $s = \dim A$. Moreover, if I is homogeneous (so that A is naturally a graded, standard k-algebra) then in order to check condition (R_s) it is enough to consider homogeneous prime ideals $P \subset A$. To say that the associated $\operatorname{Proj} A$ is a nonsingular variety will mean that A satisfies (R_s) with $s = \dim A - 1$. Here is a restatement of the classical jacobian criterion in terms of these conditions.

Proposition 2.2 *Let $r = \dim A$. Then, for $s \geq 0$, A satisfies condition (R_s) if and only if the codimension of the ideal $J(A)$ is at least $s + 1$.*

In particular, A is locally regular everywhere if and only $J(A)$ has codimension $\dim A + 1$, i.e., if and only if $J(A) = A$ (In the homogeneous case, the condition would read: $J(A)$ is an A_+-primary ideal). Stated in this way, the criterion reveals its effectiveness. Indeed, $J(A)$ can be written as the quotient ideal $(I_g(\Theta) + I)/I$, where $g = \operatorname{ht} I$ and $I_g(\Theta)$ denotes the ideal of $R = k[\mathbf{X}]$ generated by the $g \times g$ minors of the jacobian matrix of the generators of I. Thus, the criterion asks whether $\operatorname{ht}(I_g(\Theta) + I) \geq s + 1$, which is effective and computable in principle.

Now, why is $J(A)$ so much more important that the remaining jacobian ideals of A? The reason is this: if $\Omega_{A|k}$ has a well-defined rank (e.g., if A is puredimensional and generically a complete intersection) then its rank is necessarily $\dim A$, hence all jacobian ideals $J_r = (0)$ for $r < \dim A$. That is, the lower jacobian ideals - sometimes called "higher" jacobian ideals when thinking of the sizes of the corresponding minors - are all trivial.

This is as far as the classical theory goes. Try and go modern requires more than a talk. We will mention a few directions in which the criterion has evolved, some basically subsumed in the old theory, some quite subtler. As it will appear, the higher jacobian ideals will play a role as will the lifting of the lower ones to R (the Fitting ideals of the so-called *jacobian module*, to be considered in one of the subsections below).

Start with the meaning of condition (R_0). If I has no embedded primes then this condition simply says that I is a radical ideal (i.e., that the ring $A = R/I$ is reduced). So, for an ideal I without embedded primes, the jacobian criterion says that it is radical if and only if $\operatorname{ht}(I_g(\Theta) + I) \geq 1$. Now, this condition is equivalent, at least if A is equidimensional, to saying that $\operatorname{ht} I_g(\Theta) \geq \operatorname{ht} I + 1$, hence $I_g(\Theta)$ could not be contained in any associated prime of A, so it must contain an element which is not a zero-divisor on A.

Summing up, for an equidimensional $A = R/I$ with no embedded prime ideals (sometimes one says that the ideal I is *unmixed*), one has that A is reduced if and only if there is an element $h \in I_g(\Theta)$ such that $I : (h) = I$ (cf. [12, Proposition 10.4.14]).

Next on the row would be condition (R_1). By the jacobian criterion, this means that $\operatorname{ht} J(A) \geq 2$. Now, let's assume "slightly" more about the jacobian ideal, namely, that $\operatorname{depth} J(A) \geq 2$, i.e., that one can find a sequence of two elements $h_1, h_2 \in J(A)$ forming a regular sequence in the ring A - recall that (R_0) allows to find the first of these, but not necessarily one more. Also, assuming that A is a domain, this condition can be nicely restated in terms of the divisorial equality $A = A : J$. It implies not only that A is regular in codimension one but also in "depth one" - a condition that Serre baptized as (S_2). Summing it up, if A is a domain then it is normal if and only if the equality $A = A : J$ holds.

We will briefly discuss in the lecture the question of the effectiveness of the latter criterion of normality.

We conclude with an application of the higher order jacobian ideals (i.e., lower order size minors!) for a preliminary question to the (difficult) question of (effectively) finding a primary decomposition of an ideal $I \subset R$.

Proposition 2.3 ([12, Theorem 10.3.10]) (char $k = 0$) *Let* $g = \operatorname{ht} I$. *If* $r \geq \dim A$ *is such that* $\dim A / J_{r+1}(A) < \dim A$ *then*

$$\{P \in V(I) \mid \dim R/P = \dim A\} = \{P \in V(I : I_{d-r}(\Theta)) \mid \dim R/P = \dim A\}.$$

Moreover, if one can choose $r = \dim A$ *then the primary components of* $I : I_{d-r}(\Theta)$ *of dimension* $\dim A$ *are prime ideals.*

What this means is that one can approach some of the primary components of I by steps and eventually decide whether they are prime.

3   Quadrics and jacobian modules

One finds by experiment that jacobian matrices of forms of degree 2 are easier to deal with because the corresponding entries are linear forms - beware, however, since the underlying geometry of matrices whose entries are forms of degree one is a deep and fascinating subject altogether, with many recent openings. At any rate, our objective in this section is to show a particular behaviour of quadrics in regard to jacobian matrices (here and in the sequel, the word *quadric* will mean a form of degree 2 rather than the hypersurface it defines).

3.1   Exchangeable modules and quadrics of bidegree $(1, 1)$

Classical algebraic geometry carried a whole lot of information stemming from the so-called *correspondences*. The word had once a more limited meaning, but in the general sense a correpondence is a closed subvariety of a product $\mathbb{P}^{d-1} \times \mathbb{P}^{n-1}$ of projective spaces. As such, its defining ideal is generated by bihomogeneous polynomials in the bigraded polynomial ring $k[\mathbf{X}, \mathbf{T}] = k[X_1, \ldots, X_d; T_1, \ldots, T_n]$, in which, X_i (resp. T_j) has bidegree $(1, 0)$ (resp. $(0, 1)$). Among the bihomogeneous polynomials, the ones of bidegree $(1, 1)$ play an outstanding role, for at least two reasons: first, they define subvarieties whose singular locus shows a rather moderate complexity; second, their constituent terms are squarefree monomials - an important feature from the combinatorial viewpoint, as these terms may be set in bijection with the edges of a simple bipartite graph.

Algebraically, there is yet another way of setting up these data so as to bring in well-known constructions from ring and module theory. Thus, let $\mathcal{J}$ stand for an ideal of the bigraded polynomial ring $k[\mathbf{X}, \mathbf{T}]$ generated by bihomogeneous polynomials $G_1, \ldots G_m$ of bidegree $(1, 1)$. Then, one can consider two modules, one over the ring $R = k[\mathbf{X}]$ and the other over the ring $S = k[\mathbf{T}]$, both defined in terms of a free representation as follows:

$$R^m \xrightarrow{\varphi} R^n \longrightarrow E \to 0, \quad S^m \xrightarrow{\psi} S^d \longrightarrow F \longrightarrow 0,$$

where φ (resp. ψ) is the jacobian matrix of $G_1, \dots G_m$ with respect to the variables $\mathbf{T}$ (resp. $\mathbf{X}$).

Definition 3.1 We say that the above jacobian matrices (or the modules they present) are *exchangeable* (or *in correspondence*, if only to fill our yearning for some classical terminology).

Remark 3.2 The previous terminology employed in [7] was *jacobian duals*. We will refrain from using it here in order to avoid potential confusion with a special case of the construction to be mentioned in the next subsection.

It follows that the bigraded residue ring $k[\mathbf{X}, \mathbf{T}]/\mathcal{J}$ is isomorphic as k-algebra to both $\mathcal{S}_R(E)$ and $\mathcal{S}_S(F)$, the respective symmetric algebras of E and F.

Conversely, given a finitely generated R-module E defined as the cokernel of a matrix φ whose entries are liner forms, one can present the symmetric algebra $\mathcal{S}_R(E) \simeq k[\mathbf{X}, \mathbf{T}]/\mathcal{J}(\varphi)$, where $\mathcal{J}(\varphi)$ is generated by the entries of the product $(\mathbf{T})\varphi$. This gives a way of moving back and forth between such modules and certain bigraded ideals of $k[\mathbf{X}, \mathbf{T}]$.

Now, the symmetric algebra of a module can easily turn out to be an intractable ring (for instance, if E is a module of rank 0). In this case, one should rather study the properties of the various annihilators involved - we give a feeling for it in the next subsections. In this subsection, we will be trying to look at "good" modules, such as torsionfree or reflexive ones and even modules of sections of homogeneous bundles on projective space. Alas, we are more interested in the arithmetic properties involved than in the underlying set theoretic geometry. Thus, far fetched designations such as *Cohen–Macaulay, Gorenstein, factorial*, (S_2), *homological dimension*, etc, will be in our aim.

As it turns out, there are quite a number of cases where one can say more about the two exchangeable jacobian matrices. Here is one:

Proposition 3.3 *With the preceding notation, assume moreover that* $(m = d$ *and) the rows of* φ *are relations of the variables* $\mathbf{X}$ *(written as a column vector). Then the exchanged matrix* ψ *is skew-symmetric.*

Strange as it may appear, this condition is met by quite a few interesting correspondences, including some that yield sections of bundles (such as the Tango–Trautmann–Vetter bundle). To see this, one has first to develop some more theory.

As before, let E be a finitely generated R-module which is the cokernel of a matrix φ whose entries are liner forms. We assume, moreover, that $\mathcal{S}(E)$ is a torsionfree R-algebra, in which case one can actually see that it must be an integral domain. We let $\nu(E)$ denote the minimal number of generators of E. The following proposition accomodates the basic results in this setup (cf. [7]).

Proposition 3.4 *Let* ψ *denote the exchanged jacobian matrix over the ring* $S = k[\mathbf{T}]$ *and let* F *be its cokernel.*

(i) $\nu(E) \leq \dim R + \operatorname{rank} E - 1$.

(ii) $\mathcal{S}(F)$ *is a domain (in particular, F is a torsionfree S-module).*

Assume, moreover, that equality holds in (i). *Then:*

(iii) *Let ψ' denote (any arbitrarily chosen) $d \times (d-1)$ submatrix of ψ of rank $d-1$. Then the exchanged jacobian module F is isomorphic to the ideal $I \subset S$ generated by the maximal minors of ψ' divided by their greatest common divisor.*

(iv) *(Jacobian criterion) The ideal $I \subset S$ in* (iii) *has codimension one less than that of the locus $Sing(\mathcal{S}(E)) \cap V(\mathbf{X})$.*

(v) *Let $N_S(I) = S[It, t^{-1}]/t^{-1}S[It, t^{-1}]$ be the normal cone algebra of I. Then $N_S(I)$ is integral if and only if $\mathcal{S}(E)$ is a normal ring and $\nu(E_P) \le \mathrm{ht}\, P + \mathrm{rank}\, E - 2$ holds for every $P \in \mathrm{Spec}\,(R) \setminus \{(\mathbf{X})\}$ not lying in the free locus of E.*

(vi) *If the conditions of* (v) *hold then $\mathcal{S}(E)^{**} \simeq S[It, t^{-1}]$, where $\mathcal{S}(E)^{**}$ denotes the* factorial closure *of $\mathcal{S}(E)$.*

Remark 3.5 For the precise notion of factorial closure, we refer to the survey [11]. For details on the dimension and integrality of symmetric algebras, as well as the above condition on the local number of generators of E, see [9] and [10] - in these references one explains how this seemingly ineffective condition is actually effectively computable by means of the codimensions of the ideals generated by minors of the matrix φ. Since it is usually rather hard to computationally decide if the normal cone algebra $N_S(I)$ is integral (it actually amounts to look at *all* symbolic powers of the ideal I - see next section), one is now given a choice to mechanically compute Gröbner bases of ideals generated by determinants of matrices with linear entries. As for the normality of $\mathcal{S}(E)$, the part that checks the non-singularity in codimension one boils down to the computation of (the relative codimension of) the jacobian ideal of the bigraded ideal $\mathcal{J}$; often, this can be read off the codimension of the exchanged ideal I. Usually, the most difficult part is to check that $\mathcal{S}(E)$ satisfies the depth condition (S_2) of Serre. In practice, one verifies first if $\mathcal{S}(E)$ is a Cohen–macaulay ring. Alas, this is a formidable job! Alternatively, there is an interesting effective method devised by Vasconcelos [12, Chapter] that computes the so-called S_2-ification of a ring.

An application to describing arithmetic properties of certain bundles is as follows. Let $d \ge 4$ and E be an R-module with a free resolution of the form

$$0 \longrightarrow R \xrightarrow{\eta} R^d \xrightarrow{\varphi} R^{2d-3} \longrightarrow E \longrightarrow 0,$$

with φ a linear matrix and $\eta = \begin{bmatrix} x_1 \\ \vdots \\ x_d \end{bmatrix}$. For a suitable φ, Vetter [14] has shown how to obtain the module E of global sections of a (homogeneous) indecomposable vector bundle of rank $d-2$ on $\mathbb{P}^{d-1}$. The bundle extends to any dimension the known Tango bundle and explains a construction of Trautmann, hence we call them the *Tango–Trautmann–Vetter* modules (or bundles).

Corollary 3.6 *Let E be a Tango–Trautmann–Vetter module over $k[x_1, \ldots, x_d]$.*

(a) *If d is odd, then both $\mathcal{S}(E)$ and $S[It, t^{-1}]$ are normal Cohen–Macaulay integral domains and there is an isomorphism $\mathcal{S}(E)^{**} \simeq S[It, t^{-1}]$.*

(b) *If d is even, then $k[\mathbf{X}, \mathbf{T}]/(\mathbf{X} \cdot \psi, \mathrm{Pfaff}(\psi))$ is a Cohen–Macaulay domain and is isomorphic to $\mathcal{S}(E)^{**}$, where $\mathrm{Pfaff}(\psi)$ denotes the pfaffian of the $d \times d$ (square, symmetric) matrix ψ.*

The confront between the module of sections E and the exchanged ideal I is fruitful. In the talk we will display their computational background for low values of d.

3.2   Secant varieties and jacobian modules

Let $V \subset \mathbb{P}^{n-1}$ be a variety. The set theoretic definition of the secant variety to V in $\mathbb{P}^{n-1}$ is the closure of the union of (true) secant lines to V. A more precise definition consists in blowing up the product $V \times V$ along its diagonal and taking the projective variety in $\mathbb{P}^{n-1}$ whose affine cone is the so called fiber cone. To get the algebraic "picture", let $I \subset k[\mathbf{X}]$ stand for the (homogeneous) defining ideal of V and let $k[\mathbf{x}, \mathbf{y}] = k[\mathbf{X}, \mathbf{Y}]/(I, I(\mathbf{Y}))$, where $\mathbf{Y}$ is a duplicate of the $\mathbf{X}$-variables and $I(\mathbf{Y})$ is the ideal I read in the $\mathbf{Y}$-variables. Then the subalgebra $k[x_1 - y_1, \ldots, x_n - y_n] \subset k[\mathbf{x}, \mathbf{y}]$ is generated in degree one (in the total gradation of $k[\mathbf{x}, \mathbf{y}]$), so it is a standard k-algebra. Note that

$$k[x_1 - y_1, \ldots, x_n - y_n] \simeq k[X_1 - Y_1, \ldots, X_n - Y_n]/(I, I(\mathbf{Y})) \cap k[X_1 - Y_1, \ldots, X_n - Y_n].$$

Composing the inclusion $k[X_1 - Y_1, \ldots, X_n - Y_n] \subset k[\mathbf{X}, \mathbf{Y}]$ with the augmentation homomorphism $k[\mathbf{X}, \mathbf{Y}] \to k[\mathbf{X}]$ that maps $X_i \mapsto X_i$, $Y_i \mapsto 0$, gives an isomorphism $k[X_1 - Y_1, \ldots, X_n - Y_n] \simeq k[\mathbf{X}]$. Via this k-isomorphism, the above contracted ideal gets mapped to a homogeneous ideal $\mathrm{Sec}\, I \subset k[\mathbf{X}]$. Then the secant variety to V is $\mathrm{Proj}\,(k[\mathbf{X}]/\mathrm{Sec}\, I)$. Note that, by construction, $\mathrm{Sec}\, I \subset I$ as needed, as the secant to V should contain V as a closed subvariety.

Remark 3.7 Secant theory constitutes an example where the computation is not only effective but often obtainable, whereas the theoretical side of the subject carries difficult questions, at least from the geometric viewpoint. There are many open questions regarding secant varieties ever since Severi raised their importance for classification problems. Of course, the computation involves elimination (hence finding a Gröbner basis), but in this case one is eliminating not towards a subring generated by a general change of variables - which is a hard knuckle in the topic of projections - but a subring generated by differences of two variables. Again, experience seems to point in the direction of moderate complexity, however the question deserves a precise query.

We now go back to specifics, namely, to quadrics. In the preceding subsection, we developed the theory of exchangeable modules coming from quadrics of bidegree $(1, 1)$. Presently, we show that if one of the "side" modules, say, E is on itself the cokernel of the (transposed) jacobian matrix of a set of quadrics in the **X**-variables alone then a kind of self-duality takes place between E and its exchanged module F. We show, namely, that they are isomorphic after identification $k[\mathbf{X}] = k[\mathbf{T}]$. This is not surprising if one comes to think of it, since it is one among the multiple ways in which the correspondence between quadratic forms and symmetric matrices reveals itself.

What it turns to be slightly surprisinghas a grip is that this simple phenomenon triggers a substantial information on the ideal Sec I, where $I \subset k[\mathbf{X}]$ is generated by the given set of quadratic forms.

As a matter of notation, given a set of quadrics in $k[\mathbf{X}]$, the cokernel of the (transposed) jacobian matrix of these quadrics is a module which depends, up to isomorphisms, only on the ideal I generated by the quadrics. Accordingly, we denote this module by $\mathcal{J}(I)$ and call it the *jacobian module* of I. One has, namely:

Theorem 3.8 ([6]) *Let char $k \neq 2$ and let $I \subset k[\mathbf{X}]$ be an ideal generated by quadrics such that* depth $k[\mathbf{X}]/I > 0$. *Then* $(0) :_{k[\mathbf{X}]} \mathcal{J}(I) \subset \operatorname{Sec} I$.

The proof consists, very roughly, in showing that the polynomials belonging to the annihilator $(0) :_{k[\mathbf{X}]} \mathcal{J}(I)$ can be lifted back by polarization, which in turn can be translated in terms of the existence of a map between the symmetric algebra of the exchanged module F and a certain blowup algebra.

So, the following questions arise immediately:

(1) Can one effectively compute the annihilator?
(2) Is the annihilator (0) too often?
(3) How big a chunk of the secant ideal is the annihilator?

The first question is easily answered: "yes", since the module is given as a cokernel of a matrix - in fact there are standard routines to do it (cf. the scripts in *Macaulay* [1]). The second question is by far more interesting. It is equivalent to asking when is the 0th Fitting ideal of $\mathcal{J}(I)$ trivial (also see next subsection). The third question is the most difficult of the three since we lack any bare estimates as to what should be the codimension of the annihilator - although we do have from theory some "expected value" for the codimension of Sec I (namely, ht Sec I " $=$ "edim $k[\mathbf{X}] - 2 \dim k[\mathbf{X}]/I$).

We will survey a few structured classes of ideals I whose secant ideals Sec I can be computed and whose arithmetic properties can be guessed - among these, some classes of catalecticant (Haenkel) matrices. As a rule, the topic is very much open and has recently enticed many unexpected connections.

3.3   Jacobian modules of quadrics and polynomial relations

Looking at ideals generated by quadrics, one can basically distinguish three cases from the point of view of the theory of Gröbner bases. Precisely, let as before $I \subset k[\mathbf{X}]$ be an ideal generated by forms of degree two such that depth $k[\mathbf{X}]/I > 0$. Let $L(I) \subset k[\mathbf{X}]$ stand for the ideal generated by the leading terms of elements in I in the lexicographic order of $k[\mathbf{X}]$ (the so-called *initial ideal* of I). Here are the three cases:

1. $L(I)$ is generated by squarefree terms that correspond to the edges of a bipartite graph with vertices corresponding to $\mathbf{X} = X_1, \ldots, X_d$;
2. As above, but the corresponding graph is not bipartite;
3. $L(I)$ cannot be all generated by squarefree terms.

Case (3) is by far the most difficult and we lack as yet a good theory to deal with it. Case (2) is very close to the theory in the preceding section; thus, very roughly, the corresponding jacobian module tends to have nonzero annihilator and gives some boost to secant theory.

In this subsection we deal with case (1) - by far the one we have some reasonably complete results. The philosophy here is that the jacobian module in this case tends to have trivial annihilator, hence produces no impact in secant theory (in opposite to the previous subsection). In fact, quite frequently, in this case $\text{Sec}\, I = (0)$.

To remedy the situation one brings into the picture the polynomial relations of the quadrics $f_1, \ldots, f_n$ that generate I. Thus, let $J \subset k[\mathbf{T}] = k[T_1, \ldots, T_n]$ be the ideal of all these polynomials relations - which means that $k[\mathbf{T}]/J \simeq k[f_1, \ldots, f_n]$. Let $\mathcal{D}(I) := \ker\left(\sum_{i=1}^{d} k[\mathbf{X}]dX_i \to \mathcal{J}(I)\right)$ be the module generated by the differentials of the quadrics generating I. We call $\mathcal{D}(I)$ the *derived module* of I.

We consider the k-linear map $\lambda : \mathcal{S}(\sum_{j=1}^{n} k[\mathbf{X}]\, T_j) = k[\mathbf{T}] \to \sum_{j=1}^{n} k[\mathbf{X}]\, T_j$ given by $\lambda(F) := \sum_j \frac{\partial F}{\partial T_j}(f_1, \ldots, f_n)\, T_j \in \mathcal{Z}$, where $\mathcal{Z}$ denotes the module of first syzygies of the derived module $\mathcal{D}(I)$. By the usual rules of composite derivatives, if $F \in k[\mathbf{T}]$ then $\sum_{j=1}^{n} \frac{\partial F}{\partial T_j}(f_1, \ldots, f_n)\, df_j = 0$. This means that restriction induces a map $J \to \mathcal{Z}$. Any element in the image of J will be called a *polar syzygy*.

Furthermore, using the derivative rules, one can check that the latter induces a quotient map

$$\bar{\lambda} : J/(\mathbf{T})J \to \mathcal{Z}/(\mathbf{X})\mathcal{Z}. \tag{1}$$

The map (1) will be called the *polar map* (associated to I). One says that $\mathcal{J}(I)$ is *polarizable* if the polar map is an isomorphism. It is our belief that a characterization of sets of quadrics whose jacobian module is polarizable ought to be given in terms of the initial ideal of the ideal generated by these quadrics (in a suitable order), signaling to us that the question migth be of computational nature rather then expressable in terms of invariants of commutative algebra. In the talk we will discuss the underlying difficulties, some technical, some deeper.

As a guide through this seemingly disconnected question, we will give a class of ideals generated by quadrics whose jacobian module is polarizable.

Theorem 3.9 *Let there be given a bipartition of the set of variables* $\mathbf{X}$ *(so $k[\mathbf{X}]$ is naturally bigraded, bistandard). Let $I \subset k[\mathbf{X}]$ be an ideal generated by quadratic monomials of bidegree* $(1,1)$ [2]. *Then $\mathcal{J}(I)$ is polarizable.*

---

[2] Actually, one needs an additional *connectedness* condition to be explained in the talk

We will give hints on the proof, which is strongly algorithmic in the sense that we actually produce for any given syzygy of $\mathcal{D}(I)$ (whose coordinates are monomials) its explicit expression as module-combination of polar syzygies. The question of the uniqueness of these expressions is obviously translatable in terms of the existence of nonzero second syzigies. This leads us naturally to look at the whole free resolution of the jacobian module $\mathcal{J}(I)$. As it turns, one gets corollaries with a combinatorial feature.

To see this, first note that an ideal such as in the theorem defines a simple (connected) bipartite graph G. A cycle (necessarily with an even number of edges) of G is said to be *primitive* if no chord of the cycle is an edge of G. The number of primitive cycles of G will be denoted $\mathrm{prank}\,(G)$ to be distinguished from the usual $\mathrm{rank}\,(G)$. One has always $\mathrm{prank}\,(G) \geq \mathrm{rank}\,(G)$.

Corollary 3.10 *Let G be a connected bipartite graph and let $I \subset k[\mathbf{X}]$ be its ideal of edges as above. Let*

$$0 \to k[\mathbf{X}]^{b_r} \xrightarrow{\Theta_r} \cdots \to k[\mathbf{X}]^{b_2} \xrightarrow{\Theta_2} k[\mathbf{X}]^{b_1} \xrightarrow{\Theta_1} k[\mathbf{X}]^d \to \mathcal{J}(I) \to 0$$

be a minimal free resolution of its jacobian module over $k[\mathbf{X}]$, where Θ_1 is the (transposed) jacobian matrix of the generators of I corresponding to the edges of G. Then:

(i) $b_2 = f\mathrm{rank}\,(G)$.

(ii) *The following conditions are equivalent:*

(a) $\mathcal{J}(I)$ *has homological dimension at most* 2

(b) $\mathrm{prank}\,(G) = \mathrm{rank}\,(G)$

(c) $k[I_2]$ *is a complete intersection, where I_2 is the k-vector space of quadratic forms in I.*

4 Functions vanishing to higher order and jacobian matrices

The purpose of this section is to show an effective and reasonably fast method to compute the symbolic powers of an ideal $I \subset R = k[\mathbf{X}]$ in an iterative way. It is based on certain aspects of the so-called Zariski theorem on holomorphic functions, hence it will assume that k has characteristic zero and that I is radical.

Although one is mainly interested here in the computational aspects of the problem, there are some theoretical byproducts as well, such as a differential embedding of the symbolic associated graded ring of I ("infinitesimal normal cone") into a polynomial ring.

One defines two seemingly different notions, the first due to W. Krull.

Definition 4.1 Let $r \geq 0$.

(a) The *rth symbolic power* of I is the ideal

$$I^{(r)} := \{g \in R \mid \exists s \in R \setminus P \,(\forall P \in \mathrm{Ass}\,(R/I)), \text{ with } sf \in I^r\}.$$

(b) The *r th infinitesimal power* of I is the ideal

$$I^{<r>} := \{g \in R \mid \frac{\partial^\alpha f}{\partial \mathbf{X}^\alpha} \in I, \forall \alpha, |\alpha| \leq r - 1\}.$$

Taking a minute to look at these concepts, one is immediately striken by its different features even beyond the mere formal objects involved in their definitions. Thus, the first one is existencial, hence by and large non-effective; nevertheless, it is very much in the heart of commutative algebra and is an important theoretic tool thereon. The second concept is effective - inasmuch as the ideal membership problem is and we know the latter boils down to a Gröbner basis computation; however, its relation to typical reasoning in commutative algebra is far from obvious. So, wouldn't it be pleasant to switch between them at will?

This is indeed a basic result, due to Zariski and Nagata, stating that these concepts are the same provided char $k = 0$ and I is a radical ideal (for an indication of the proof, see [2, 3.9], where other references are quoted). The two propositions described below depend strongly on this result. Thus, in this subsection one assumes that char $k = 0$ and I is a radical ideal.

Given polynomials $\mathbf{f} := \{f_1, \ldots, f_m\} \subset R$, one denotes by $\Theta(\mathbf{f})$ the transposed jacobian matrix of $\mathbf{f}$. For an integer $r \geq 0$, $\Psi^{(r)}(\mathbf{f})$ will denote a lifting to R of the relation matrix of $\Theta(\mathbf{f})$ over the ring $R/I^{(r)}$ and $((\mathbf{f}) \cdot \Psi(\mathbf{f}))R$ will denote the ideal generated by the entries of the product matrix $(\mathbf{f}) \cdot \Psi^{(r)}(\mathbf{f})$. If r is fixed throughout the discussion, one shortens $\Psi(\mathbf{f}) := \Psi^{(r)}(\mathbf{f})$.

Proposition 4.2 ([5]) *Let $r \geq 0$. If $\mathbf{f} := \{f_1, \ldots, f_m\}$ is a set of generators of $I^{(r)}$ then*

$$I^{(r+1)} = ((\mathbf{f}) \cdot \Psi(\mathbf{f}))R + II^{(r)}.$$

The second result will only compute syzygies of matrices over the fixed ring R/I throughout for all powers $I^{(r)}$. The price to pay is that one has to use a kind of variable *jacobian* matrix, based on higher differential forms, which one now explains.

Recall that if F is a free R-module with basis $\{T_1, \ldots, T_n\}$, the rth symmetric power of F, denoted $\mathcal{S}_r(F)$, is a free R-module with basis the set of monomials of degree r on $\{T_1, \ldots, T_n\}$. Then the symmetric algebra of F is $\mathcal{S}(F) = \sum_{r \geq 0} \mathcal{S}_r(F) \simeq R[T_1, \ldots, T_n]$.

Thus, let $\mathbf{f} := \{f_1, \ldots, f_m\}$ be a set of generators of $I^{(r)}$. Define a map of free modules $R^m \to \mathcal{S}_r(\sum_{i=1}^n RdX_i)$

$$e_j \mapsto d^{(r)} := \sum_{u_1 + \ldots + u_n = r} \frac{1}{u_1! \ldots u_n!} \frac{\partial^r f_j}{\partial X_1^{u_1} \ldots \partial X_n^{u_n}} dX_1^{u_1} \ldots dX_n^{u_n}, \quad (2)$$

where e_j denotes the jth vector of the canonical basis of R^m. Let $\Theta^{(r)}(\mathbf{f})$ denote the matrix of this map relative to the canonical basis of R^m and the monomial basis of $\mathcal{S}_r(\sum_{i=1}^n RdX_i)$ coming from $\{dX_1, \ldots, dX_n\}$. Finally, let $\Psi^{(r)}(\mathbf{f})$ denote a lifting to R of the relation matrix of the matrix $\Theta^{(r)}(\mathbf{f})$ over the (fixed!) ring R/I and $((\mathbf{f}) \cdot \Psi^{(r)}(\mathbf{f}))R$ is the ideal generated by the entries of the product matrix $(\mathbf{f}) \cdot \Psi^{(r)}(\mathbf{f})$.

Proposition 4.3 ([5]) *Let $r \geq 0$. If $\mathbf{f} := \{f_1, \ldots, f_m\}$ is a set of generators of $I^{(r)}$ then*

$$I^{(r+1)} = (\, (\mathbf{f}) \cdot \Psi^{(r)}(\mathbf{f})\,)R + II^{(r)}.$$

Proposition 4.3 is useful for computing the defining equations of the so-called *symbolic associated graded ring*:

$$\mathrm{gr}_{(I)}(R) := \sum_{r \geq 0} I^{(r)}/I^{(r+1)}.$$

The result needed is yet another facet of the Theorem of Zariski-Nagata which we for convenience choose to state separately.

Lemma 4.4 (char $k = 0$) *Let $I \subset R$ be a radical ideal and let $r \geq 0$. The map (2) induces an injective R/I-module homomorphism*

$$I^{(r)}/I^{(r+1)} \rightarrow \mathcal{S}_r \left(\sum_{i=1}^{n} (R/I)\, dX_i \right). \tag{3}$$

Theorem 4.5 ([5]) (char $k = 0$) *Let $I \subset R$ be a radical ideal. The maps (3) induce an injection of R/I-algebras*

$$\mathrm{gr}_{(I)}(R) \rightarrow R/I[dX_1, \ldots, dX_n]. \tag{4}$$

Among all possible embeddings of $\mathrm{gr}_{(I)}(R)$ into polynomial rings over R/I, the preceding is a distinctive one. Suggestively, we call it the *differential embedding* of $\mathrm{gr}_{(I)}(R)$.

The differential embedding identifies $\mathrm{gr}_{(I)}(R)$ with the R/I-subalgebra of the polynomial ring $R/I[dX_1, \ldots, dX_n]$ generated by the set of polynomials $\{d^{(r)}f \mid f \in I^{(r)},\, r \geq 0\}$. Of course, in general this subalgebra is infinitely generated (Nagata–Rees). However, if it is finitely generated (typically, but not necessarily, if $I^{(r)} = I^r,\, \forall r \geq 0$) then we may obtain in an explicit way its *differential presentation ideal*, i.e., the kernel of the homomorphism of R/I-algebras $R/I[T_{r,j_r}] \rightarrow R/I[dX_1, \ldots, dX_n]$ obtained by assigning $T_{r,j_r} \mapsto d^{(r)}(f_{r,j_r})$, where $\{f_{r,j_r}\}$ generate $I^{(r)}$ and r is suitably bounded. Note that the computation, though involved, is effective provided one has a grip on a bound for r.

A point of relevance is to compare the advantages of the differential method to others. For instance, one could first compute the Rees algebra $\sum_{r \geq 0} I^{(r)}t^r \subset R[t]$. For each generator $f_{rj_r}t^r$ of $I^{(r)}t^r$ one picks a variable T_{rj_r} over R. Then, one computes the kernel of $T_{rj_r} \mapsto f_{rj_r}t^r$. The resulting ideal $J \subset R[\mathbf{T}]$ is the (differential) presentation ideal of the Rees algebra. To pass to the corresponding presentation ideal of $\mathrm{gr}_{(I)}(R)$ one should *not* as in the case of the normal cone just take the generators of J modulo I; rather, one has to compute for each generator $f_{rj_r} \in I^{(r)}$, its expression as an element of the preceding power $I^{(r-1)}$. Then one takes the ideal J modulo the ideal of R generated by the coefficients of all these expressions.

As one sees, this procedure is quite involved, if not costlier. However, even if one needs the above expressions for the latter method, the method of the differential embedding of $\mathrm{gr}_{(I)}(R)$ gives them as a byproduct since the coefficients of each such expression will be exactly the coordinates of some column vector of the lifted syzygy matrix $\Psi^{(r)}(\mathbf{f})$ in Proposition 4.3.

We will compare some of these in the talk, hopefully by showing some examples.

References

1. D. Bayer and M. Stillman, **Macaulay**, A computer algebra system for computing in algebraic geometry and commutative algebra, 1989.
2. D. Eisenbud, *Commutative Algebra with a view toward Algebraic Geometry*, Graduate Texts in Mathematics, vol. 150, Springer-Verlag, 1995.
3. J. Harris, *Algebraic Geometry: A First Course*, Graduate Texts in Mathematics **133**, Springer-Verlag, Berlin Heidelberg New York Tokyo, 1992.
4. H. Matsumura, *Commutative Ring Theory*, Cambridge University Press, 1986.
5. A. Simis, Effective computation of symbolic powers by jacobian matrices, Comm. in Algebra, to appear.
6. A, Simis and B. Ulrich, On the arithmetic of embedded joins and secant varieties, in preparation.
7. A. Simis, B. Ulrich and W. V. Vasconcelos, Jacobian dual fibrations, American J. Math. **115** (1993), 47–75.
8. A. Simis, B. Ulrich and W. V. Vasconcelos, Tangent star cones, J. reine angew. Math, to appear.
9. A. Simis and W. V. Vasconcelos, On the dimension and integrality of symmetric algebras, Math. Z. **177** (1981), 341–358.
10. A. Simis and W. V. Vasconcelos, Krull dimension and integrality of symmetric algebras, Manuscripta Math. **61** (1988), 63–78.
11. W. V. Vasconcelos, Symmetric algebras and factoriality, in *Commutative Algebra* (M. Hochster, C. Huneke and J. D. Sally, Eds.), MSRI Publications **15**, Springer–Verlag, New York, 1989, 467–496.
12. W. Vasconcelos, *Arithmetic of Blowup Algebras*, Lecture Note Series **195**, LMS, Cambridge University Press, 1994.
13. W. Vasconcelos, *Computational Methods in Commutative Algebra and Algebraic Geometry*, to appear.
14. U. Vetter, Zu einem Satz von G. Trautmann über den Rang gewisser kohärenter analytischer Moduln, Arch. Math. **24** (1973), 158–161.

Rigid body dynamics and measure differential inclusions

David E. Stewart

Abstract. Rigid body dynamics with unilateral constraints, collisions and Coulomb friction has been the subject of investigation and controversy for the past century, due to the failure to obtain solutions in certain situations. Recent extensions to the theory of ordinary differential equations, notably J.J Moreau's measure differential equations can be used to give a clearer idea of the issues in these problems. Novel numerical methods have had to be developed to solve these problems, and to prove the convergence of the numerical trajectories to solutions of the true rigid body problem.

1 Rigid body dynamics

Rigid body dynamics without inequality constraints is, in principle, easy to formulate. Given a set of generalized coordinates q, and the corresponding velocities $v = dq/dt$, one finds the expressions for the *kinetic energy* $T(q,v)$, which is homogeneous and quadratic in v, and the *potential energy* $V(q)$. Then one writes down the Lagrangian, $L(q,v) = T(q,v) - V(q)$, and solves Lagrange's differential equations:

$$\frac{d}{dt}\frac{\partial L}{\partial v} - \frac{\partial L}{\partial q} = 0. \tag{1}$$

If $T(q,v) = \frac{1}{2}v^T M(q)v$, then this can be re-written as

$$M(q)\frac{dv}{dt} - k(q,v) = 0 \tag{2}$$

where $k_i(q,v) = -\frac{1}{2}\sum_{r,s}\left[(\partial m_{ir}/\partial q_s) + (\partial m_{is}/\partial q_r) - (\partial m_{rs}/\partial q_i)\right]v_r v_s - (\partial V/\partial q_i)$. Provided $M(q)$ is a smooth function of q, and is always positive definite, this is a smooth ODE for dv/dt, and so local existence and convergence of numerical methods can be shown; in addition, since the total energy $E(q,v) = T(q,v) + V(q)$ is constant for exact solutions, existence and convergence are global.

Alternatively, a Hamiltonian formulation could be used in terms of generalized coordinates q and the corresponding momenta p. While the Hamiltonian formulation has a number of very pleasant properties, they are not to so useful in the presence of unilateral constraints or friction.

If there are some additional *bilateral* constraints $f(q) = 0$, then the Lagrangian formulation can be modified by means of Lagrange multipliers λ:

$$\frac{d}{dt}\frac{\partial L}{\partial v} - \frac{\partial L}{\partial q} - \lambda \nabla f(q) = 0 \tag{3}$$

where $\lambda = \lambda(t)$ is chosen to ensure that $\nabla f(q(t))\, q'(t) = \nabla f(q(t))\, v(t) = 0$. The value of λ represents the generalized forces necessary to maintain the constraint $f(q) = 0$. Provided f is smooth, and $\nabla f(q) \neq 0$ whenever $f(q) = 0$, λ can be solved for in (3), and again a smooth ODE can be found, and the theory goes through as expected.

However, unilateral constraints $f(q) \geq 0$ are different. At first it might seem (from an optimization point of view) that all that needs to be done is to use *Kuhn–Tucker conditions* [12, 11]:

$$\frac{d}{dt}\frac{\partial L}{\partial v} - \frac{\partial L}{\partial q} - \lambda \nabla f(q) = 0, \tag{4}$$

$$\lambda, f(q) \geq 0, \tag{5}$$

$$\lambda f(q) = 0. \tag{6}$$

(For those worried about such things, we can assume a constraint qualification, such as $\{\,\nabla f_i(q) \mid f_i(q) = 0\,\}$ is linearly independent for all q. More general constraint qualifications could be used, but they add little to the discussion.)

Note that there are an infinite number of constraints: one for each value of t. Further, the constraint is placed on q, for which (4) is a second order equation. If at some time t, $f(q(t)) = 0$, but $\nabla f(q(t)) \cdot v(t^-) < 0$, then there is no way of avoid penetration into the forbidden region if v is a continuous function: $f(q(t')) < 0$ for any $t' > t$ sufficiently close to t.

Therefore, the velocity function must contain jump discontinuities in general. The possible forces should then include Dirac-δ functions. From a functional-analytic point of view, λ needs to belong to a sufficiently large dual space. A simple and suitable choice is to let λ belong to the set of 0th order distributions, which are the Borel (vector) measures.

Frictionless problems can be treated in this framework, both analytically and computationally. Friction makes things more difficult, as the simple Lagrange multiplier formulation is no longer sufficient.

The type of impact is still undefined: impacts can be inelastic, partly elastic, or (perfectly) elastic. After inelastic impacts, the relative normal velocities of the two bodies at the contact point is zero. After perfectly elastic impacts, the relative normal velocities of the bodies at the contact point are the negatives of the the relative normal velocities before the impact. Partly elastic collisions result in velocities intermediate to these two possibilities. Care must be taken in interpreting these conditions, as they can give rise to physically impossible behavior with energy increasing in complex collisions. While there has been much discussion in the literature of the relative merits of Newton's and Poisson's impact laws [17, 25], or the new formulations of Stronge [23], the methods described here avoid much of this discussion, but still manages to obtain dissipation of energy for the solutions to rigid body dynamics problems.

1.1   Coulomb friction

Various approaches have been made to modeling Coulomb friction, from Coulomb's original work on the subject in 1781 [9, pp. 319–322], through the 19th century and early this century (see Painlevé [20], Delassus [8], and others including, e.g., F. Klein) to very recent approaches (Baraff [3, 2], Lötstedt [14, 15], J.J. Moreau [18], Trinkle, Pang, Sudarsky and Lo [24], and others).
      The basic principles of Coulomb are as follows:

1. The magnitude of the frictional forces between two bodies in contact is μ (the coefficient of friction) times the magnitude of the normal contact force provided there is relative motion between them at the point of contact, and
2. The direction of the friction forces maximizes the rate of loss of energy amongst all possible contact forces.

If the friction is isotropic, then item (2) implies that the friction force is directed opposite to the direction of relative motion, which is a more common expression of this principle. For anisotropic friction (as in, e.g., ice-skating), it is important to use the *maximum work* or *dissipation* principle in item (2), rather than the "opposite direction" formulation.
      While these principles seem simple enough, they give rise to discontinuous differential equations. As with discontinuous ordinary differential equations, it is better to work with a corresponding differential inclusion (see Filippov [10]). It is usual to construct a *friction cone FC(q)* which contains all possible contact forces (normal + friction) for a given configuration q. This is a closed, convex cone containing the negative of the normal cone: $\nabla f(q) \in FC(q)$.
      For a particle with isotropic friction, it is easy to describe the friction cone:

$$FC(q) = \{\, n(q)c_n + c_f \mid c_n \geq 0,\ c_f \perp n(q), \qquad \|c_f\| \leq \mu c_n \,\} \qquad (7)$$

where $n(q) = \nabla f(q)$. For more general systems, it becomes more difficult to describe since the forces in $\mathbb{R}^3$ must be transformed into generalized forces, which can include torques and other quantities.

1.2   The impossible problem

It was discovered last century by Painlevé [20] that Coulomb's friction law could lead to an apparent contradiction with rigid body dynamics. The example is a simple one, consisting of a rod on a table, as shown in Fig. 1. The generalized coordinates are the (x, y) coordinates of the center of mass, together with θ, the angle of the rod with respect to the horizontal.
      If the moment of inertia J is sufficiently small, and the coefficient of friction μ is sufficiently large, then existence can apparently fail as follows: since the rod tip is sliding left, the friction force F must have magnitude μN, directed to the right, where N is the normal contact force. If there is zero angular velocity, and no contact forces, the rod will penetrate the table, which is forbidden. On the other hand, the friction force induces a torque which (for small J, large μ, and

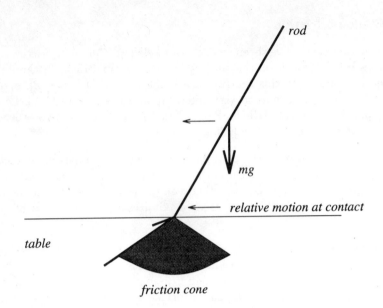

rod

mg

relative motion at contact

table

friction cone

Fig. 1. The impossible problem

suitable θ) drives the rod *into* the table. No matter how large the contact forces are, the tip of the rod is accelerated into the table.

The reason why the failure is only apparent, is that the tip is assumed to be sliding left. With impulses allowed, it is possible for the tip to be brought to rest instantaneously, and the friction force is no longer required to be μN in magnitude. This additional freedom enables the problem to be solved for the contact impulses and the velocities. This is an example of impulsive contact forces without a change in the contact state, which are *shocks*.

2   Measure differential inclusions – formulations

In order to use measure differential inclusions (MDI's), we need to understand what

$$\frac{dv}{dt} - k(q,v) \in K(q) \subset \mathbb{R}^n \tag{8}$$

means if v is a function which merely has bounded variation, and q is continuous. The definition should extend the definition of solution for ordinary differential inclusions (see, e.g., Aubin and Cellina [1], Clarke [4], Deimling [7], or Filippov [10]), which in turn extend Caratheodory's definition of solution to ordinary differential equations [5]. In keeping with the spirit of differential inclusions, it is assumed that $K(\cdot)$ is a set-valued function with convex values and a closed graph. (Nonconvex differential inclusions have been studied, but the properties of such differential inclusions are much more delicate.)

Equation (8) is understood in terms of Borel vector measures, and the differential measures of BV functions. Both of the formulations presented here are new formulations, although the strong formulation is closely related to the formulation used by J.J. Moreau [18, 19] and Monteiro-Marques [16]. The detailed proofs can be found in Stewart [21].

The space $BV[0,T]$ of functions of bounded variation on an interval $[0,T]$ can be identified with the dual of the space of continuous functions $C[0,T]^*$ via the duality pairing $\langle \mu, \phi \rangle = \int \phi(t)\,\mu(dt)$. Since v has bounded variation, dv is a Borel measure, and can be defined in terms of Riemann–Stieltjes integrals $\phi \mapsto \int_0^T \phi(t)\,dv(t)$.

If "t" is the independent variable (representing the identity function on $\mathbb{R}$), then "dt" represents the ordinary Lebesgue measure on $\mathbb{R}$. By abuse of notation, the MDI (8) can then be re-written in differential measure form

$$dv - k(q,v)\,dt \in K(q)\,dt. \tag{9}$$

At this point, there are two paths that can be taken; one is more direct, and leads to a *strong* formulation, and another which involves integration against sufficiently "nice" functions, which leads to a *weak* formulation. Under mild conditions the two are equivalent. The strong formulation is good for proving properties of the solutions, while the weak formulation is good for proving existence and convergence properties of solutions.

2.1 Strong formulation

The measure dv can be decomposed into an absolutely continuous part, and a purely singular part by the Lebesgue decomposition theorem:

$$dv = a(\cdot)\,dt + \nu_s \tag{10}$$

where the support of the Borel measure ν_s is a Lebesgue null set. Off the support of ν_s we should require that

$$a(t) - f(q(t), v(t)) \in K(q(t)) \tag{11}$$

for Lebesgue almost all t.

Since the support of $d\nu_s$ is a Lebesgue null set, bounded or L^1 functions multiplying the Lebesgue measure "dt" should be ignored in the support of ν_s. This removes $f(q(t), v(t))\,dt$, and the bounded part of $K(q(t))$.

If $K(q)$ is unbounded, then ν_s is not necessarily zero; in that case, we need ν_s to have values whose directions lie in $K(q)$ "at infinity". To extract specific values, we take Radon–Nikodym derivatives of ν_s with respect to its absolute value measure $|\nu_s|$.

Then the singular measure ν_s should satisfy

$$\frac{d\nu_s}{d|\nu_s|}(t) \in K(q(t))_\infty \tag{12}$$

where L_∞ is the *asymptotic cone* of the closed convex set L. This can be defined as in [13, §III.2.2]: $L_\infty = \{\lim_{i \to \infty} t_i z_i \mid t_i \downarrow 0,\ z_i \in L\} = \bigcap_{t>0} t(L - \bar{x})$ where $\bar{x}$ is any point in L.

If v satisfies (10), (11) and (12), then it solves the MDI (8) (or (9)) in the *strong* sense.

2.2   Weak form

An alternative approach is to consider (9), and to integrate against a non-negative continuous function ϕ with compact support:

$$\int \phi[dv - k(q,v)\,dt] \in \int \phi(t)K(q(t))\,dt \subset \overline{\mathrm{co}} \bigcup_{\tau:\phi(\tau)\neq 0} K(q(\tau)) \int \phi(t)\,dt. \quad (13)$$

Note that the corresponding discrete property

$$\sum_{i=1}^{n} \phi_i K_i \subset \left(\mathrm{co} \bigcup_{i=1}^{n} K_i\right) \sum_{i=1}^{n} \phi_i$$

holds for finite sequences of convex sets K_i nonnegative numbers ϕ_i. If the function ϕ is not identically zero, we can divide (13) by $\int \phi(t)\,dt$ to give

$$\frac{\int \phi[dv - k(q,v)\,dt]}{\int \phi(t)\,dt} \in \overline{\mathrm{co}} \bigcup_{\tau:\phi(\tau)\neq 0} K(q(\tau)). \quad (14)$$

If (14) holds for all continuous non-negative functions ϕ with compact support, not identically zero, then v is a solution in the *weak* sense.

2.3   Equivalence

The strong formulation is essentially that used by J.J. Moreau [19] and Monteiro-Marques [16]. The weak formulation is developed in Stewart [21].

   The weak formulation (10,11,12) and strong formulation (14) are equivalent under the following conditions:

1. $K(\cdot)$ has a closed graph, and has convex values.
2. $\min\{\,\|x\| \mid x \in K(q)\,\}$ is locally bounded as a function of q.
3. $K(q)_\infty$ is a *pointed* cone for all q. That is, $K(q) \cap -K(q) = \{0\}$ for all q.

   The need for $K(q)$ to be pointed is clear for the weak form of solution as can be made clear by considering a case where $K(q(t))$ is a line in the plane whose angle changes with t. Then $\bigcup_{\tau:\phi(\tau)\neq 0} K(q(\tau))$ will usually be the entire plane, for any ϕ not identically zero. However, the strong formulation still appears to be meaningful in this case. Then *any* measure $dv - k(q,v)\,dt$ satisfies the MDI in the weak sense. Whether the property that weak* limits of strong solutions are also strong solutions of the MDI for non-pointed $K(q)$ remains to be seen.

3   Numerical schemes

We consider here the one-contact case only. Multiple contacts are discussed in Stewart and Trinkle [22].

   Since $FC(q) = \{0\}$ if $f(q) > 0$, there needs to be a way to "switch on" and "switch off" contact forces. This is done here through complementarity conditions. Also, the friction cone $FC(q)$ for $f(q) = 0$ is approximated by a polyhedral cone

$$\widehat{FC}(q) = \operatorname{cone}\{\, n(q) + \mu d_i(q), \ i = 1, \ldots, m \,\},$$

where $n(q) = \nabla f(q)$, and let $D(q) = [d_1(q), \ldots, d_m(q)]$. All functions of q are assumed to be smooth, and that for each i there is a j where $d_j(q) = -d_i(q)$.

Let $t_l = l\,h$ be the time after l time-steps. Then the time stepping algorithm consists of finding $q^{l+1} \approx q(t_{l+1})$, $v^{l+1} \approx v(t_{l+1})$ and the auxiliary quantities β, c_n and λ, given $q^l \approx q(t_l)$, $v^l \approx v(t_l)$ satisfying the following equations and complementarity conditions:

$$
\begin{aligned}
M(q^{l+1})(v^{l+1} - v^l) &= n(q^{l+1})c_n + \\
& \quad D\,(q^{l+1})\beta + hk(q^l, v^l), \\
q^{l+1} - q^l &= hv^{l+1}, \\
\lambda e + D(q^{l+1})^T v^{l+1} \geq 0 \quad &\perp \quad \beta \geq 0, \\
n(q^{l+1})^T q^{l+1} - \alpha_0(q^{l+1}) \geq 0 \quad &\perp \quad c_n \geq 0, \\
\mu c_n - e^T \beta \geq 0 \quad &\perp \quad \lambda \geq 0.
\end{aligned}
\tag{15}
$$

Note that e is the vector of 1's of the appropriate size. Also, $\alpha_0(q) = n(q)^T q - f(q)$.

Note that c_n represents the normal contact force (or impulse) over the interval $(t_l, t_{l+1}]$, while $D(q)\beta$ represents the friction force (or impulse) over the same time interval. The inequality $n(q^{l+1})^T q^{l+1} - \alpha_0(q^{l+1}) \geq 0$ ensures that the discrete trajectory remains feasible at the discrete time values. The inequality $c_n \geq 0$ ensures that there is no adhesion. The complementarity condition between them ensures that there are contact forces only when there is contact.

The inequalities $\beta \geq 0$, and $\mu c_n - e^T \beta \geq 0$ ensure that the total contact force $n(q)c_n + D(q)\beta$ lies within the approximate friction cone $\widehat{FC}(q)$. The complementarity condition between $\lambda e + D(q)^T v^{l+1} \geq 0$ and $\beta \geq 0$ ensures that the friction vector is in the direction of maximum dissipation. And the final complementarity condition ensures that if there is relative motion at contact, then the contact force lies on the boundary of the approximate friction cone.

By writing q^{l+1} and v^{l+1} in terms of β, c_n and λ, and ignoring the dependence of n, α_0, M and D on q, (15) can be put in the form of a Linear Complementarity Problem (LCP) [6], for which solutions can be found by means of Lemke's algorithm [22].

The Nonlinear Complementarity Problem (NCP) (15) also has solutions under mild conditions, although it appears necessary to invoke Brouwer's fixed point theorem in some form to show this.

4   Convergence of numerical trajectories

The proof of the convergence of the numerical trajectories to a solution of the measure differential inclusion, and satisfying the additional conditions for inelastic impacts etc., is beyond the scope of this paper. Results on convergence will be published elsewhere. Several points about the proof should be made:

1. It is important in theory and practice that the friction cone $FC(q)$ be pointed in the space of generalized forces. If $FC(q) \cap -FC(q) \neq \{0\}$, then the object may be *wedged* and the problem is statically indeterminate. (That is, if all external forces are known, there are many possible admissible contact forces which keeps the body at rest. Physically, this corresponds to the common problem of a body being wedged between two other immovable objects.)
2. The numerical method automatically gives inelastic impacts. No known simple modification of the above formulation gives elastic or partially elastic collisions, which ensures that $f(q^l) \geq 0$ for all l.
3. The weak formulation of the MDI immediately shows that weak* limits of solutions $dv^h \rightharpoonup dv$ are also solutions — which also requires the pointedness of $FC(q)$.

Nevertheless, it appears that at long last, rigorous mathematical theory and corresponding computational techniques are available for general rigid body dynamics with Coulomb friction and unilateral constraints. What is surprising is the level of mathematical sophistication that has been needed to provide a complete resolution of this classical problem.

References

1. J.-P. Aubin and A. Cellina. *Differential Inclusions*. Springer-Verlag, Berlin, New York, 1984.
2. D. Baraff. Issues in computing contact forces for non-penetrating rigid bodies. *Algorithmica*, 10:292–352, 1993.
3. D. Baraff. Fast contact force computation for non-penetrating rigid bodies. In *Proceedings, SIGGRAPH '94*, pages 23–34. ACM, ACM, 1994.
4. F. H. Clarke. *Nonsmooth Analysis and Optimization*. SIAM Publ., Philadelphia, PA, 1990. Originally published by the Canadian Math. Soc., 1983.
5. E. A. Coddington and N. Levinson. *Theory of Ordinary Differential Equations*. Tata McGraw–Hill Publishing, New Delhi, 1972. Original edition McGraw–Hill, 1955.
6. R. W. Cottle, J.-S. Pang, and R. E. Stone. *The Linear Complementarity Problem*. Academic Press, Boston, San Diego, New York, 1992. Series on Computer Science and Scientific Computing.
7. K. Deimling. *Multivalued Differential Equations*. Walter de Gruyter, Berlin, New York, 1992. Series on Nonlinear Analysis and Applications #1.
8. E. Delassus. Considérations sur le frottement de glissement. *Nouv. Ann. de Math. (4ème série)*, 20:485–496, 1920.
9. R. Dugas. *A History of Mechanics*. Éditions du Griffon, Neuchatel, Switzerland, 1955. Original in French; translation to English by J. R. Maddox.
10. A. F. Filippov. Differential equations with discontinuous right-hand side. *Amer. Math. Soc. Tranlations*, 42:199–231, 1964. Original in Russian in *Math. Sbornik*, 5, pp. 99–127, (1960).
11. R. Fletcher. *Practical Mathods of Optimization*. J. Wiley and Sons, Chichester, New York, 2nd edition, 1987.
12. J. Franklin. *Methods of Mathematical Economics*. Springer-Verlag, New York, Heidelberg, Berlin, 1980.

13. J.-B. Hiriart-Urruty and C. Lemaréchal. *Convex Analysis and Minimization Algorithms I and II*. Number 305 in Grundlehren der mathematischen Wissenschaften. Springer–Verlag, Berlin, Heidelberg, New York, 1993.

14. P. Lötstedt. Coulomb friction in two-dimensional rigid-body systems. *Z. Angewandte Math. Mech.*, 61:605–615, 1981.

15. P. Lötstedt. Mechanical systems of rigid bodies subject to unilateral constraints. *SIAM J. Appl. Math.*, 42(2):281–296, 1982.

16. M. D. P. Monteiro Marques. *Differential Inclusions in Nonsmooth Mechanical Problems: Shocks and Dry Friction*, volume 9 of *Progress in Nonlinear Differential Equations and Their Applications*. Birkhäuser Verlag, Basel, Boston, Berlin, 1993.

17. M. T. Mason and Y. Wang. On the inconsistency of rigid-body frictional planar mechanics. In *Proceedings, IEEE Int. Conf. Robotics Automation*, pages 524–528, 1988.

18. J. J. Moreau. Frottement, adhésion, lubrification. *Comptes Rendus, Série II*, 302:799–801, 1986.

19. J. J. Moreau. Unilateral contact and dry friction in finite freedom dynamics. In J. J. Moreau and P. D. Panagiotopoulos, editors, *Nonsmooth Mechanics and Applications*, pages 1–82. Springer–Verlag, Vienna, New York, 1988. International Centre for Mechanical Sciences, Courses and Lectures #302.

20. P. Painlevé. Sur le lois du frottement de glissemment. *Comptes Rendus Acad. Sci. Paris*, 121:112–115, 1895. Following articles under the same title appeared in this journal, vol. 141, pp. 401–405 and 546–552 (1905).

21. D. E. Stewart. Asymptotic cones and measure differential inclusions. *J. Math. Anal. Appl.*, 1996. Submitted.

22. D. E. Stewart and J. C. Trinkle. An implicit time-stepping scheme for rigid body dynamics with inelastic collisions and Coulomb friction. *Internat. J. Numer. Methods Engineering*, 39(15):2673–2691, 1996.

23. W. J. Stronge. Rigid body collisions with friction. *Proc. Royal Society of London*, A431:168–181, 1990.

24. J. C. Trinkle, J.-S. Pang, S. Sudarsky, and G. Lo. On dynamic multi-rigid-body contact problems with Coulomb friction. submitted to *Zeitschrift fur Angewandte Math. Mech.*, 1995.

25. Y. Wang and M. T. Mason. Two-dimensional rigid-body collisions with friction. *ASME J. Appl. Mech.*, 59:635–642, 1992.

Linear decision lists and partitioning algorithms for the construction of neural networks

György Turán* and Farrokh Vatan*

Abstract. We consider the computational power of neural networks constructed by partitioning algorithms. These neural networks can also be viewed as decision lists with tests evaluating linear functions. An exponential lower bound is proved for the complexity of an explicit Boolean function in this model. The lower bound is extended to decision trees of bounded rank. We also discuss the relationship between these models and the hierarchy of threshold circuit complexity classes.

1 Introduction

Computational models with non-Boolean features can provide new methods for the computation of Boolean functions. For example, one can use general neural network construction techniques to build neural networks that compute or approximate a given Boolean function.

An interesting approach was proposed in the papers Ruján and Marchand [21], Marchand, Golea and Ruján [14], Marchand and Golea [15] (see also [10]). Their *partitioning algorithms* construct a neural network of linear threshold units with one hidden layer. A special case of this approach is described in Jeroslow [11].

Given a Boolean function f, the cube $[0, 1]^n$ is partitioned into regions by hyperplanes in such a way that f is constant in each region containing Boolean vectors. The halfspaces defined by these hyperplanes correspond to the threshold units in the hidden layer. The construction of the halfspaces implies that one can add an output unit in such a way that the resulting neural network indeed computes the function f.

The hyperplanes are determined sequentially. First, we find a hyperplane such that f is constant on one of its sides. The points in the corresponding halfspace are removed and we continue with the remaining points in a similar manner, until the set of remaining points becomes empty. Thus each halfspace is used to cut off a set of points with identical function values from the remaining set of points.

Let the halfspaces constructed by the algorithm be $H_1, \ldots, H_k$ and define u_i to be 1 (resp. -1) if f is 1 (resp. 0) on the region cut off by H_i for $i = 1, \ldots, k$.

* Partially supported by NSF grant CCR–9208170.

Then adding an output unit with threshold 0 and weight $u_i 2^{k-i}$ for the edge leaving the hidden unit corresponding to H_i, we get a neural network computing f. We assume that linear threshold units produce an output 0 or 1.

In a restricted version of the approach, called *regular partitioning* [21] it is also assumed that the hyperplanes do not intersect in $[0, 1]^n$.

This description only gives the general principle of partitioning algorithms. Particular implementations depend on the way the next hyperplane is selected. Experiments indicate that partitioning algorithms are successful in the sense that they construct small neural networks efficiently ([12], [14], [15], [21]).

Another way to view this process, without any reference to neural networks, is to consider a sequence of halfspaces $H_1, ..., H_k$. For each halfspace H_i we specify the value a_i of f for those points in H_i that are not contained in any of the previous halfspaces. Thus, for every Boolean vector $\mathbf{x}$ it holds that $f(\mathbf{x}) = a_i$, where i is the smallest index such that $\mathbf{x}$ is in H_i. It is assumed that H_k contains $\{0, 1\}^n$, thus i is always defined. This model is called a *neural decision list* [14] or *linear decision list*. In the original decision list model introduced by Rivest [19] the decisions are made by evaluating individual variables, instead of linear functions as above.

In what follows we also use the terminology of circuit complexity theory. Thus neural networks of linear threshold units with one hidden layer will also be referred to as depth 2 threshold circuits.

The previous discussion indicates that the following three computational models are equivalent (Marchand and Golea [14], also noted by Anthony [1]):

- neural networks constructed by partitioning
- depth 2 threshold circuits where the final gate has threshold 0 and its i'th incoming edge has weight $\pm 2^i$
- linear decision lists.

In this paper we consider the computational power of these models. In order to discuss their relationship to the hierarchy of threshold circuit complexity classes, let us give a brief overview of some lower bounds for threshold circuits.

There are exponential lower bounds known for depth 2 threshold circuits with integer weights bounded by some polynomial function of the number of input variables. These can be extended to depth 2 threshold circuits where *only the weights of the edges entering the final gate are assumed to be polynomial* and the other weights can be arbitrary (Hajnal, Maass, Pudlák, Szegedy and Turán [8], Krause and Waack [13], Nisan [17]). It is shown in Goldmann, Hastad and Razborov [5] that depth 2 threshold circuits with unrestricted weights can be exponentially more powerful than those with polynomial weights. No superpolynomial lower bounds are known for depth 2 threshold circuits with unrestricted weights. For depth 3 threshold circuits lower bounds are only known in restricted cases (Hastad and Goldmann [9]). Surveys of results on small depth threshold circuits are given by Razborov [18] and in Roychowdhury, Siu and Orlitsky [20].

Thus proving an exponential lower bound for the models considered in this paper is of interest as it would apply to a class of depth 2 threshold circuits with large, although quite restricted, weights allowed at the final gate.

In Section 2 we prove such a lower bound for the *mod 2 inner product of two Boolean vectors*. The bound is formulated in terms of linear decision lists. The argument is a variant of the lower bound proofs in Gröger and Turán [6],[7] for linear decision trees and threshold circuits.

The lower bound is extended to *linear decision trees of bounded rank* in Section 3. The rank of a binary tree T is the maximal depth of a complete binary tree embedded in T. This notion was introduced by Ehrenfeucht and Haussler [4] in computational learning theory. Decision lists have rank 1 and general decision trees can have rank n. Linear decision trees of bounded rank form an intermediate class between linear decision lists and linear decision trees. We obtain a superpolynomial lower bound for the complexity of the same Boolean function as long as the rank is $o(n/\log n)$.

From the point of view of threshold circuit complexity, the interesting property of neural networks produced by the partitioning algorithm is that they have large weight edges entering the output unit. Are these large weights necessary in general ? This problem is open. In Section 4 we review a result of Ruján and Marchand [21] showing that for nonredundant regular partitions the answer is *no*, and even ± 1 coefficients suffice. For completeness, we give a more detailed proof than the argument in [21]. In Section 5 we formulate some open problems.

2   Linear decision lists

A *linear test* L over the variables $z_1, \ldots, z_m$ is of the form $\sum_{i=1}^{m} \gamma_i z_i \geq t$, where $\gamma_1, \ldots, \gamma_m \in \mathbb{R}^m$ are the *weights* and $t \in \mathbb{R}$ is the *threshold*.

A *linear decision list* D over the variables $z_1, \ldots, z_m$ is a sequence

$$(L_1, a_1), \ldots, (L_k, a_k),$$

where L_i is a linear test and a_i is 0 or 1 for $i = 1, \ldots, k$. It is assumed that L_k, the last linear test is true for all Boolean vectors. The *length* of D is k. The Boolean function f_D computed by D assigns to every $\mathbf{z} \in \{0, 1\}^m$ the value a_i, where L_i is the *first* linear test in the list that is satisfied by $\mathbf{z}$. Every Boolean function can be computed by some linear decision list. For example, a disjunctive normal form with s terms can be represented by a linear decision list of length $s + 1$.

The *inner product mod 2* of two Boolean vectors $\mathbf{x} = (x_1, \ldots, x_n)$ and $\mathbf{y} = (y_1, \ldots, y_n)$ is defined by

$$IP2_n(\mathbf{x}, \mathbf{y}) = (x_1 \wedge y_1) \oplus \ldots \oplus (x_n \wedge y_n).$$

The following lower bound implies that the neural network constructed for $IP2_n$ by *any* partitioning algorithm must be of exponential size.

Theorem 1 *Every linear decision list computing* $IP2_n$ *has length greater than* $2^{n/2} - 1$.

Proof. First we formulate a general property of linear tests.

Lemma 2 *Let* $X, Y \subseteq \{0,1\}^n$, $|X| = |Y| = m$, *let* $v \in \mathbb{R}$ *and let* L *be a linear test over the variables* $x_1, \ldots, x_n, y_1, \ldots, y_n$. *Then either there are subsets* $X' \subseteq X$, $Y' \subseteq Y$ *such that* $|X'| = |Y'| \geq v$ *and* L *is true on* $X' \times Y'$, *or there are subsets* $X' \subseteq X$, $Y' \subseteq Y$ *such that* $|X'| = |Y'| \geq m - v$ *and* L *is false on* $X' \times Y'$.

Proof. Let the linear test L be of the form $\sum_{i=1}^{n} \alpha_i x_i + \sum_{i=1}^{n} \beta_i y_i \geq t$. Order the elements of X in decreasing order according to the value of $\sum_{i=1}^{n} \alpha_i x_i$, and order the elements of Y in decreasing order according to the value of $\sum_{i=1}^{n} \beta_i y_i$. Let $\mathbf{x}$ and $\mathbf{y}$ be the elements of rank $\lceil v \rceil$ in these orderings.

If L is true for $(\mathbf{x}, \mathbf{y})$ then it is also true for the whole set $X' \times Y'$, where X' and Y' are the elements of rank at most $\lceil v \rceil$ in X and Y. Similarly, if L is false for $(\mathbf{x}, \mathbf{y})$ then it is also false for the whole set $X' \times Y'$, where X' and Y' are the elements of rank at least $\lceil v \rceil$ in X and Y. $\square$

The following combinatorial lemma is the basis of several lower bounds for $\mathrm{IP2}_n$.

Lemma 3 (Lindsey, see [2]). *For every* $X, Y \subseteq \{0,1\}^n$

$$\left| \left| \{ (\mathbf{x}, \mathbf{y}) \in X \times Y : \mathrm{IP2}_n(\mathbf{x}, \mathbf{y}) = 1 \} \right| - \left| \{ (\mathbf{x}, \mathbf{y}) \in X \times Y : \mathrm{IP2}_n(\mathbf{x}, \mathbf{y}) = 0 \} \right| \right|$$

$$\leq \sqrt{|X||Y|2^n}.$$

As a direct corollary one obtains

Lemma 4 *For every* $X, Y \subseteq \{0,1\}^n$ *such that* $|X| = |Y| = s$ *and the function* $\mathrm{IP2}_n$ *is constant on* $X \times Y$ *it holds that* $s \leq 2^{n/2}$.

Now let $L = (L_1, \ldots, L_k)$ be a linear decision list computing $\mathrm{IP2}_n$. We show that for every $i < k$ there are sets $X_i, Y_i \subseteq \{0,1\}^n$ with $|X_i| = |Y_i| \geq 2^n - i(2^{n/2} + 1)$ such that $L_1, \ldots, L_i$ are all false on $X_i \times Y_i$.

This follows by induction on i. For $i = 0$, let $X_0 = Y_0 = \{0,1\}^n$. For the induction step, consider the linear test L_{i+1} and apply Lemma 2 to X_i, Y_i and $v = 2^{n/2} + 1$. If the first case holds in Lemma 2 then there are subsets $X' \subseteq X_i$, $Y' \subseteq Y_i$ such that $|X'| = |Y'| \geq v$ and L_{i+1} is true on $X' \times Y'$. But then, by the definition of linear decision lists, $\mathrm{IP2}_n$ must have value a_{i+1} on $X' \times Y'$. This cannot happen in view of Lemma 4. Hence the second case must hold in Lemma 2. Thus we obtain sets $X_{i+1} \subseteq X_i$ and $Y_{i+1} \subseteq Y_i$ such that $|X_{i+1}| = |Y_{i+1}| \geq |X_i| - (2^{n/2} + 1) \geq 2^n - (i+1)(2^{n/2} + 1)$ and L_{i+1} is false on $X_{i+1} \times Y_{i+1}$.

In order to complete the proof, note that $\mathrm{IP2}_n$ must be constant on $X_{k-1} \times Y_{k-1}$. Hence, again by Lemma 4 it follows that $2^n - (k-1)(2^{n/2} + 1) \leq 2^{n/2}$, implying the theorem. $\square$

3   Linear decision trees of bounded rank

A *linear decision tree* over the variables $z_1, \ldots, z_m$ is a binary tree T such that its inner nodes are labeled by linear tests, its leaves are labeled by 0 or 1 and its edges are also labeled by 0 or 1. The *depth* of T is the number of tests on a longest path leading from the root to a leaf. The Boolean function f_T computed by T assigns to every $\mathbf{z} \in \{0,1\}^m$ the label of the leaf arrived at, by evaluating the tests in the nodes for $\mathbf{z}$.

The *rank* $r(T)$ of a linear decision tree T is defined inductively. If T is a single leaf then $r(T) = 0$. Otherwise let T_l and T_r be the left and right subtrees of T. If $r(T_l) \neq r(T_r)$ then $r(T) = \max(r(T_l), r(T_r))$. Otherwise $r(T) = r(T_l) + 1$.

Linear decision lists are linear decision trees of rank 1 . A minor difference in the definitions is that in order to view a linear decision tree of rank 1 as a linear decision list, the rightmost leaf has to be replaced by an identically true test with the same label. Thus the complexity measures for lists and trees differ by 1 .

The following lower bound generalizes Theorem 1 from linear decision lists to linear decision trees of bounded rank.

Theorem 5 *Every linear decision tree of rank r computing* IP2$_n$ *has depth greater than*

$$(2^{n/2} - 1)^{1/r} - 1.$$

Proof. We prove a more general bound. A linear decision tree is said to compute IP2$_n$ on a *subset* of $\{0,1\}^n$ if it produces the correct result for inputs belonging to this subset.

Lemma 6 *For every $X, Y \subseteq \{0,1\}^n$ such that $|X| = |Y| = m$, every linear decision tree of rank r computing* IP2$_n$ *on $X \times Y$ has depth greater than*

$$\left(\frac{m}{2^{n/2} + 1} \right)^{1/r} - 1.$$

Proof. We prove the lemma by induction on r and m. The proof for the case $r = 1$ is identical to the proof of Theorem 1, with the minor difference remarked preceding the theorem. For every r, the statement is obvious if $(m/(2^{n/2} + 1))^{1/r} \leq 1$, i.e. if $m \leq 2^{n/2} + 1$. Thus we assume $r > 1$ and $m > 2^{n/2} + 1$.

Let T be a decision tree of rank r computing IP2$_n$ on $X \times Y$, and let the left and right subtrees of T be T_l and T_r. The definition of rank implies that at least one of the subtrees have rank at most $r - 1$. Let us assume *w.l.o.g.* that $r(T_l) \leq r - 1$. Let the linear test at the root of T be L.

Now let $v := m^{(r-1)/r}(2^{n/2} + 1)^{1/r}$. Apply Lemma 2 to X and Y, with v and L.

If in Lemma 2 the first case holds then it follows that T_l computes IP2$_n$ on a product of two sets of size $\lceil v \rceil$. Then $r(T_l) > 0$, as $v \geq 2^{n/2} + 1$. By induction on r, the depth of T is greater than

$$\left(\frac{v}{2^{n/2} + 1} \right)^{1/(r-1)} - 1 = \left(\frac{m}{2^{n/2} + 1} \right)^{1/r} - 1,$$

so the bound follows.

If in Lemma 2 the second case holds then it follows that T_r computes IP2$_n$ on a product of two sets of size $\lceil m - v \rceil$. Now we can apply induction on m ($v > 1$). Taking into account that L adds 1 to the depth, the depth of T is greater than

$$\left(\frac{m - v}{2^{n/2} + 1} \right)^{1/r}.$$

The concavity of $x^{1/r}$ implies that

$$m^{1/r} - (m - v)^{1/r} \leq \frac{1}{r} m^{-\frac{r-1}{r}} v < (2^{n/2} + 1)^{1/r},$$

hence

$$(m - v)^{1/r} \geq m^{1/r} - (2^{n/2} + 1)^{1/r}.$$

Thus

$$\left(\frac{m - v}{2^{n/2} + 1} \right)^{1/r} \geq \left(\frac{m}{2^{n/2} + 1} \right)^{1/r} - 1$$

and the bound again follows.   □

The theorem now follows by putting $m := 2^n$.   □

As direct corollaries we obtain a lower bound and a trade-off between rank and depth.

Corollary 7 *If $r = o(\frac{n}{\log n})$ then the depth of every rank r linear decision tree computing* IP2$_n$ *is superpolynomial.*

Corollary 8 *If T is a linear decision tree of rank r and depth d computing* IP2$_n$ *then it holds that $r \log d = \Omega(n)$.*

4   Restricted partitions

In neural networks constructed by the partitioning algorithm, edges entering the output unit have exponentially large weights. Therefore it is of interest to know that when additional restrictions are imposed on the partitioning process, the weights of the edges entering the output unit can be chosen to be *small* integers. In particular, it was shown by Ruján and Marchand [21] that for nonredundant *regular* partitions these weights can actually be chosen to be ±1. We reformulate their result in a form implicit in [21], that also strengthens a result of Nilsson ([16], Theorem 6.1). A related topic is the total unimodularity of matrices (see e.g. Schrijver [22]).

A *linear threshold unit* with input variables $z_1, \ldots, z_m$, weights $\gamma_1, \ldots, \gamma_m$ and threshold t outputs 1 if $\sum_{i=1}^{m} \gamma_i z_i \geq t$ and it outputs 0 otherwise. Neural networks of linear threshold units considered here have a hidden layer of such units and an output unit. Edges go from the inputs to the hidden layer and from the hidden layer to the output unit.

Let POS and NEG be finite subsets of $\mathbb{R}^n$. We would like to construct a neural network that accepts POS and does not accept NEG. In the framework used in the rest of the paper $POS = \{\mathbf{x} \in \{0,1\}^n : f(\mathbf{x}) = 1\}$, $NEG = \{\mathbf{x} \in \{0,1\}^n : f(\mathbf{x}) = 0\}$. A set of halfspaces $H_1, \ldots, H_k$ forms a *correct partition* of $\mathbb{R}^n$ if none of the regions formed by the hyperplanes determining these halfspaces contain points from both POS and NEG. We assume *w.l.o.g.* that none of these hyperplanes contains a point from POS or NEG. A region is nonempty if it contains some points from POS or NEG. A correct partition is *nonredundant* if deleting any of the hyperplanes, the new partition is not correct.

Theorem 9 (Ruján and Marchand [21]). *Let $H_1, \ldots, H_k$ be a nonredundant correct partition such that $k + 1$ regions are nonempty. Then there is a neural network separating POS and NEG with linear threshold units corresponding to $H_1, \ldots, H_k$ as its hidden layer and with weights ± 1 for the edges entering the output unit.*

Proof. Let the nonempty regions in the partition be $R_1, \ldots, R_{k+1}$. We refer to these regions as positive and negative, depending on whether they contain points from POS or NEG.

Let us form a graph G with edges labeled $\{1, \ldots, k\}$ having these regions as its vertices. There is an edge (R_i, R_j) labeled l in G if R_i is positive, R_j is negative, and after H_l is deleted, R_i and R_j form one region. As the partition is nonredundant, there is an edge labeled l for every $l = 1, \ldots, k$.

We claim that G is a spanning tree with exactly one edge for each label.

Let us select an edge e_l labeled l for every $l = 1, \ldots, k$. Assume that some of these edges form a cycle. Then along the edges of this cycle we can go from a region back to itself, crossing the hyperplanes corresponding to these edges exactly once, which is a contradiction. Thus the edges $e_1, \ldots, e_k$ form a spanning tree T of G. Now let us assume that G has another edge besides those in T. Adding this edge to T, we get a cycle. Along this cycle there is an edge label that occurs exactly once, and we obtain a contradiction as above. This completes the proof of the claim.

Let us consider for each region R_i the vector $\mathbf{r}_i \in \{0,1\}^k$ produced at the hidden layer. It has to be shown that the sets $P := \{\mathbf{r}_i : R_i \text{ is positive}\}$ and $N := \{\mathbf{r}_i : R_i \text{ is negative}\}$ can be separated by a hyperplane with coefficients ± 1.

We assume *w.l.o.g.* that $\mathbf{r}_1 = \mathbf{0}$ and that R_1 is a negative region. (In the description of the algorithm we assumed that the last hidden unit always outputs

1, thus $\mathbf{r}_1 \neq 0$. The proof for the general case is the same, but requires more notation.) Fix R_1 as the root of T. Along each path starting at the root, assign weight ± 1 to the edges of the tree alternatingly, starting with a $+1$. This defines the weights of the edges in the neural network leading into the output unit: the weight w_i of the edge leaving H_i is the weight assigned to e_i. Let $\mathbf{w} := (w_1, \ldots, w_k)$. The threshold of the output unit is set to $1/2$.

We assumed that $\mathbf{r}_1 = \mathbf{0}$, thus R_1 is not contained in any of the H_i's. Going from R_1 to R_i in T, we cross over to the other side of exactly those hyperplanes that occur as edge labels along the path. Hence for every region R_i, $\mathbf{r}_i$ is 1 exactly at components corresponding to the labels along the path in T leading from R_1 to R_i. We also assumed that R_1 is a negative region. Thus if R_i is a positive region then $\mathbf{w} \cdot \mathbf{r}_i = +1 - 1 + 1 - \ldots + 1 = 1$, and if R_i is a negative region then $\mathbf{w} \cdot \mathbf{r}_i = +1 - 1 + 1 - \ldots - 1 = 0$. Thus the hyperplane $\mathbf{w} \cdot \mathbf{y} = 1/2$ separates P and N. □

As nonredundant regular partitions have the properties required in Theorem 9, one obtains

Corollary 10 (Ruján and Marchand [21]). *Neural networks obtained by the regular partitioning algorithm [21] can be assumed to have weights ± 1 for the edges entering the output unit.*

A special case of regular partitioning covers the standard threshold circuits for symmetric Boolean functions, where all hyperplanes are parallel.

5   Open problems

Razborov [18] presents a hierarchy of complexity classes for depth 2 threshold circuits. Where do polynomial size linear decision lists fit in this picture? In particular, are they incomparable with polynomial size depth 2 threshold circuits with polynomial weights? Are they weaker than polynomial size depth 2 threshold circuits with unrestricted weights? With other words, is there a Boolean function that is computed by a small neural network with one hidden layer, but for which every neural network constructed by the partitioning algorithm is large?

The computational power of linear decision trees of bounded rank could also be studied in greater detail. Are the bounds of Section 3 tight? We note that linear decision trees in general can be simulated by threshold circuits of depth 3. The learnability of linear decision trees is studied, for example, by Brent [3].

As noted in the introduction, it would be interesting to know if exponentially large weights are necessary for edges entering the output unit in neural networks constructed by the partitioning algorithm.

Another open problem is to prove an exponential lower bound for algebraic decision lists, where the tests can be higher degree, e.g. quadratic, polynomials. Results about algebraic decision trees computing Boolean functions are given in Nisan [17] and Vatan [23].

Acknowledgement   We thank Vwani Roychowdhury for helpful comments about the paper and Matthias Krause for a useful discussion about this topic at the Neural Computing Workshop in Dagstuhl, in November 1994.

References

1. M.Anthony: Threshold functions, decision lists and the representation of Boolean functions, NeuroCOLT Tech. Rep. Ser. NC-TR-96-028, January 1996.
2. L.Babai, P.Frankl, J.Simon: Complexity classes in communication complexity, *Proc. 27th FOCS* (1986),337-347.
3. R.P.Brent: Fast training algorithms for multilayer neural nets, *IEEE Trans. on Neural Networks* **2** (1991),346-354.
4. A.Ehrenfeucht, D.Haussler: Learning decision trees from random examples, *Proc. COLT'88* (1988),182-194.
5. M.Goldmann, J.Hastad, A.A.Razborov: Majority gates vs. general weighted threshold gates, *Computational Complexity* **2** (1992),277-300.
6. H.D.Gröger, Gy.Turán: On linear decision trees computing Boolean functions, *Proc. 18th ICALP, Springer LNCS 510* (1991), 707-718.
7. H.D.Gröger, Gy.Turán: A linear lower bound for the size of threshold circuits, *Bulletin of the EATCS* **50** (1993),220-222.
8. A.Hajnal, W.Maass, P.Pudlák, M.Szegedy, Gy.Turán: Threshold circuits of bounded depth, *J.Comp.Syst. Sci.* **46** (1993),129-154.
9. J.Hastad, M.Goldmann: On the power of small-depth threshold circuits, *Computational Complexity* **1** (1991), 113-129.
10. J.Hertz, A.Krogh, R.G.Palmer: *Introduction to the Theory of Neural Computing.* Addison-Wesley, 1991.
11. R.G.Jeroslow: On defining sets of vertices of the hypercube by linear inequalities, *Discrete Math.* **11** (1975),119-124.
12. S.A.J.Keibek, G.T.Barkema, H.M.A.Andree, M.H.F.Savenije, A.Taal: A fast partitioning algorithm and a comparison of feedforward neural networks, *Europhysics Letters* **18** (1992),555-559.
13. M.Krause, S.Waack: Variation ranks of communication matrices and lower bounds for depth two circuits having symmetric gates with unbounded fan-in, *Proc. 32nd FOCS* (1991),777-782.
14. M.Marchand, M.Golea: On learning simple neural concepts: from halfspace intersections to neural decision lists, *Network: Computation in Neural Systems* **4** (1993),67-85.
15. M.Marchand, M.Golea, P.Ruján: A convergence theorem for sequential learning in two layer perceptrons, *Europhysics Letters* **11** (1990),487-492.
16. N.J.Nilsson: *The Mathematical Foundations of Learning Machines.* Introduction by T.J.Sejnowski and H.White. Morgan Kaufmann, 1990.
17. N.Nisan: The communication complexity of threshold gates, *Combinatorics, Paul Erdős is Eighty, Volume I*, D.Miklós, V.T.Sós, T.Szőnyi, Editors, Bolyai Math. Soc. (1993),301-315.
18. A.A.Razborov: On small depth threshold circuits, *Proc. 3rd Scandinavian Workshop on Algorithm Theory, Springer LNCS 621* (1992),42-52.
19. R.L.Rivest: Learning decision lists, *Machine Learning* **2** (1987),229-246.
20. V.Roychowdhury, K.-Y.Siu, A.Orlitsky, Editors: *Advances in Neural Computation and Learning.* Kluwer, 1994.

21. P.Ruján, M.Marchand: A geometric approach to learning in neural networks, *Proc. IJCNN 1989*, Vol. II., 105-110. Also: *Complex Systems* **3** (1989),229-242.
22. A.Schrijver: *Theory of Linear and Integer Programming.* Wiley, 1986.
23. F.Vatan: Some lower and upper bounds for algebraic decision trees and the separation problem, *Proc. 7th Structure in Complexity Theory* (1992),295-304.

Ill-Posedness and Finite Precision Arithmetic: A Complexity Analysis for Interior Point Methods

Jorge R. Vera

Abstract. In this paper we analyze the effect of ill-posedness measures on the computational complexity of an interior point algorithm for solving a linear or a quadratic programming problem when finite precision arithmetic is used. The complexity is analyzed from the point of view of the number of iterations required to achieve an approximately optimal solution, as well as from the point of view of the numerical precision required in the computations. This work gives a view of computational complexity based on more "natural" properties of the problem instance.

1 Introduction

In this paper we consider some computational complexity questions related to optimization problems of the form

$$\hat{z} = \min\{f_0(x) \ : \ Ax \le b\} \ , \tag{1}$$

where the objective is taken to be linear of the form $f_0(x) = c^T x$ or quadratic of the form $\frac{1}{2}x^T Q x + c^T x$. Here $A \in \mathbf{R}^{m \times n}$, $b \in \mathbf{R}^m$, $c \in \mathbf{R}^n$, $Q \in \mathbf{R}^{n \times n}$, $m \ge n$. We are interested on anlyzing the computational complexity of solving this problem, but in the context of finite precision computations. Because of this, our approach differs from the traditional one in the classical bit complexity theory, where it is assumed that problem instances are given in rational numbers, and exact rational arithmetic is used in the computations. In our setup we allow for finite precision arithmetic in some computations and hence, the consideration of rounding errors is important. Not being able in this case to find the exact solution to the problem, we look for a solution whose objective value does not differ from the true optimum in more than a given acceptable tolerance. Our interpretation of complexity is dual: we want to estimate the number of arithmetic operations needed to achieve the goal (this is consistent with the traditional approach), as well as the numerical precision required in the computations to run an appropriate algorithm to completion. This kind of complexity analysis is closer to the one we would like to use when analyzing performance of numerical computations, and it is in the line of a suggestion by Smale[13].

Central to the analysis is how the complexity, in the sense defined above, is affected by "conditioning" or "ill-posedness" characteristics of the problem instance. Conditioning in optimization has been studied before using different approaches, but most of them correspond to "local" analysis to measure the sensitivity of the solution or objective value of the optimization problem with respect to changes in the data. A good example of this research is the work by Mangasarian[7]. In recent years a new approach to the analysis of ill-posedness in mathematical programming has been initiated by Renegar[10]. In his work, he shows that the distance from the problem instance to some set in the data space corresponding to the "ill-posed" instances can be interpreted as a condition measure. This condition measure plays a central role on explaining the complexity of certain algorithms for deciding existence of solutions to linear programs, as has been shown by Renegar and by Vera[14, 15]. It also gives a measure of sensitivity of solutions (as a "traditional" condition number), as shown in Renegar[11] and Vera[16].

We now briefly review this new notion of ill-posedness in optimization. The first observation is that the notion of conditioning is problem specific and depends on the setup adoted. For instance, for the problem of deciding whether a given system of linear inequalities of the form $Ax \leq b$ is consistent or not, we consider the set

$$\mathcal{F} = \{(A, b) \in \mathbf{R}^M / \exists x \in \mathbf{R}^n \ : \ Ax \leq b\} \ ,$$

where M is the appropriate dimension of the data space (in this case $M = m \times n + m$). This set corresponds to the collection of all consistent problem instances. It is only for instances in the interior of this set that the problem is well defined when no exact knowledge of the instance is available, or rounding errors may affect the computations: for an instance in the boundary of $\mathcal{F}$ exact data is needed to make a decision. This follows from the fact that in this case, any neighborhood of the instance will contain instances which are consistent and instances which are inconsistent.

In this paper we want to analyze the effect of this notion of ill-posedness on the complexity of an interior point method to find an approximate solution to the problem. Given this, the problem instance will be well-posed if it is feasible and can be solved by an interior point method. This requires inmediately that the feasible region be bounded, and for the quadratic problem, that the objective function be convex. This means that the sets of well posed instances can be defined as:

$$\mathcal{F}_L = \{(A, b, c) \in \mathbf{R}^M / Ax \leq b \text{ is consistent and bounded}\}$$

for the linear objective, and

$$\mathcal{F}_Q = \left\{ \begin{array}{c} (Q, A, b, c) \in \mathbf{R}^M / \ Ax \leq b \text{ is consistent and bounded,} \\ Q \text{ is positive definite} \end{array} \right\}$$

for the quadratic objective. We denote by $\mathcal{F}$ the sets $\mathcal{F}_L$ or $\mathcal{F}_Q$ indistinctively. Let as asume that α is an instance of the problem and $\alpha \in int\mathcal{F}$. We define $\rho = dist_\infty(\alpha, \mathcal{F}^C)$. This is the distance from the problem instance to the set of ill-posed instances. Renegar[11] has defined $C(\alpha) = \|\alpha\|_\infty / \rho$ as the condition measure of the instance α. Further investigations on the notion of ill-posedness are Filipowski[1, 2] where the effect of additional knowledge is considered, and Freund and Vera[3] who characterize extensively the condition measure in terms of more "computable" problems.

In this paper we will analyze the effect of this condition measure on the computation of an approximate solution to the optimization problem when finite precision arithmetic is used in the computations. The paper condenses and generalizes results previously developed in Vera[17, 18]. As we mentioned, we use a classical logarithmic barrier interior point method to approximate the solution. A central step in the convergence analysis of the procedure is to estimate how well Newton's method can be used to follow a certain central trajectory. This has been studied extensively in recent years, even for very general classes of barrier functions, as in the work of Nesterov and Nemirovskii[8]. We use in our analysis the estimates developed by Den Hertog[5, 6] which can be extended to handle the presence of rounding errors in the computations. Den Hertog analysis simplifies some of the treatment by Nesterov and Nemirovskii. The basis of the analysis is, however, the same: the central element is to study the behavior of Newton's method when used on these algorithms. In our setup we will assume that floating point arithmetic affects only some critical computations, namely the calculation of the Newton direction. A full complexity result implies the consideration of rounding errors on every single arithmetic operation of the algorithm, but the critical effect of rounding errors is in the solution of the linear system required by Newton's method at each iteration and that is why we have restricted the analysis this way.

To simplify the treatment we assume that the instance has been normalized so that $\|(A, b, c)\|_\infty = 1$ (and $\|(Q, A, b, c)\|_\infty = 1$ in the quadratic case). Under this assumption $\rho \le 1$ and the condition measure is simply $C(\alpha) = 1/\rho$. There is no lost of generality on assuming this.

In this paper we will summarize general results for problem (1), but whenever needed we give particular forms for the linear or the quadratic objective. We first establish the general approach and the elements to analyze convergence under finite precision arithmetic, which concentrates basically on the analysis of Newton method.

2   General Setup

We begin reviewing a logarithmic barrier interior point method, and define some of the notation we use. We assume that an interior point $x^0 \in P(A, b)$ is available. The algorithm proceeds through the sequential minimization of a barrier function of the form

$$\Phi(x, \mu) = \mu f_0(x) + \sum_{i=1}^{n} \log(b_i - \alpha_i^T x) \ ,$$

where α_i is the i-th row of A and $\mu > 0$ is a penalty parameter. For a given μ we solve the penalized problem

$$
P_\mu) \qquad
\begin{array}{l}
\min \Phi(x,\mu) \\
\text{s.t.} \\
x \in intP(A,b) \; .
\end{array}
$$

As the objective function of this problem is strictly convex, P_μ has a unique solution. Let $x(\mu)$ denote the optimal solution to this problem. The set

$$
\mathcal{C} = \{x \in \mathbf{R}^n \; : \; x = x(\mu), \mu \in [0, +\infty)\}
$$

is known as the central trajectory. In particular, $x(0)$ is the analytic center of the polyhedra. Most interior point algorithms developed up to now attempt to follow this central trajectory. Following closely the notation in Den Hertog[6] we define:

$$
\begin{aligned}
g(x,\mu) &= \nabla\Phi(x,\mu) = \mu\nabla f_0(x) + A^T D(x)^{-1} e_m, \\
H(x,\mu) &= \nabla^2\Phi(x,\mu) = \nabla^2 f_0(x) + A^T D(x)^{-2} A \; ,
\end{aligned}
$$

where $D(x)$ is the diagonal matrix whose diagonal elements are $b_i - \alpha_i^T x, i = 1, \ldots, m$ and $e_m = (1, \ldots, 1) \in \mathbf{R}^m$. For each P_μ we use Newton's method to obtain an approximate solution to the problem. For a given (x,μ) the Newton direction is given by

$$
p(x,\mu) = -H(x,\mu)^{-1} g(x,\mu) \; . \tag{2}
$$

One of the main concerns in the analysis is to have good estimates for the proximity of the current iterate to the central trajectory. In this paper, as in others, we measure this proximity using a particular norm based on the Hessian at the current iterate. For a given z define

$$
\|z\|_{H(x,\mu)} = |z^T H(x,\mu) z|^{1/2} \; .
$$

The usual measure of proximity is actually the norm of the Newton direction, $\|p(x,\mu)\|_{H(x,\mu)}$.

The typical algorithm is as follows (here θ and τ are operational parameters):

INTERIOR POINT ALGORITHM
$x^0 \in intP(A,b)$, $\mu^0 \in \mathbf{R}_+$, s.t. $\|p(x^0,\mu^0)\|_{H(x^0,\mu^0)} \leq \tau$,
$k = 0, 0 < \theta < 1$.
Do while $(k < \bar{k})$
$\qquad y = x^k;$
$\qquad$ **Do while** $(\|p(y,\mu^k)\|_{H(y,\mu^k)} > \tau)$
$\qquad\qquad$ solve approximately:
$\qquad\qquad \min\{\Phi(y + \lambda p(y,\mu^k)), \lambda \geq 0\};$
$\qquad\qquad$ let $\bar{\lambda}$ be the approximate solution;
$\qquad\qquad y = y + \bar{\lambda} p(y,\mu^k);$
$\qquad$ **End while;**

$$x^{k+1} = y;$$
$$\mu^{k+1} = \mu^k(1 + \theta);$$
$$k = k + 1;$$

End while.

$\bar{k}$ is chosen to guarantee an ϵ-optimal solution, that is $x^{\bar{k}}$ satisfies $|f_0(x^{\bar{k}}) - \hat{z}| \leq \epsilon$, where $\hat{z}$ is the optimal value of (1). It is well known that, under appropriate selection of the parameters, this algorithm converges to the optimal solution of the problem. The convergence result is enough to transform the algorithm into a polynomial time procedure for linear and for quadratic programming, in the sense of the classical bit complexity theory. In this paper we analyze a "short step" version of the algorithm, where τ and θ are chosen in such a way that only one Newton step is needed per each major iteration. This simplifies the analysis a little. However a "long step" algorithm can be analyzed as well, as it was done in Vera[17] for linear programming. The main idea behind the analysis is that all estimates developed to prove convergence of interior point algorithms attempt to get bounds on the proximity to the central trajectory. Now, if floating point arithmetic is used and rounding errors are introduced, those estimates are valid as long as enough working precision is used. In this way convergence will still follow.

3   Estimates for the Approximate Direction

One of the central steps in an interior point algorithm is the solution of a system of linear equations for computing the Newton direction. We analyze the error in the solution of this system when finite precision arithmetic is allowed. The analysis relies on the results for the Cholesky factorization obtained by Wilkinson[19]. It can be proved (see, for example, Golub and Van Loan[4]), that when the process is applied to a system of the form $Bx = d$, with $B \in \mathbf{R}^{l \times l}$ positive definite, then the computed solution $\tilde{x}$ solves exactly the perturbed system $(B + E)\tilde{x} = d$, with $\|E\|_2 \leq c(l)u\|B\|_2$, where $c(l)$ is a polynomial in l, and u is the machine precision, in this case equals to $2^{-\bar{t}}$, where $\bar{t}$ is the number of binary digits used. Moreover, if u satisfies $q(l)u\kappa(B) \leq 1$, where $q(l)$ is another polynomial and $\kappa(B)$ is the condition number of B (in the linear algebra sense), then the process runs to completion with no catastrophic numerical failure (in this case, square root of a negative number). From this it follows that u has to satisfy

$$u \leq \frac{1}{\kappa(B)q(l)} .$$

We can show that u can be further reduced in order to obtain an small relative error in the computed solution. The main results are the following:

Proposition 1 *Consider the application of Cholesky factorization to the solution of the linear system*

$$[H(x,\mu)]p = -g(x,\mu) = -\nabla\Phi(x,\mu)$$

using floating point arithmetic. Suppose we use $\bar{t}$ binary digits of precision, where

$$\bar{t} = O\left(\log f(n,m) + \log \kappa(H(x,\mu))\right)$$

and $f(n,m)$ is a polynomial. Then, the algorithm runs to completion and if $\tilde{p}$ is the computed solution, we have:

$$\frac{\|p - \tilde{p}\|_2}{\|\tilde{p}\|_2} \leq r(n) \ ,$$

where $r(n)$ is a rational function of the form $c(n)/q(n)$, where c and q are polynomials.

The next corollary says that in fact, we can obtain as good a direction as we want by "simply" increasing the working precision.

Corollary 2 *Under the same assumptions of the previous proposition, if we use $\bar{t} + \log(1/\beta)$ digits, where $0 < \beta < 1$, then*

$$\frac{\|p - \tilde{p}\|_2}{\|\tilde{p}\|_2} \leq \beta f(n) \ ,$$

where f is a polynomial.

The precision requirement depends on the conditioning of the Hessian at the current iterate, so an important part of the analysis is to estimate a bound on that conditioning. Given the structure of both the linear and quadratic problem, it is easy to see that this conditioning will depend on the regularity of the matrix $A^T A$, as well as on the proximity of the current iterate to the boundary of the polyhedron. In the quadratic case, a further effect is introduced by the positive definiteness of Q. Let us define

$$\nu(x) = \min_i\{b_i - \alpha_i^T x\} \ .$$

This number measures the separation from x to the boundary of the polyhedron.
    The bound on the conditioning of the hessian is given by the following result.

Proposition 3

$$\kappa(H(x,\mu)) \leq \frac{O(n^2 m)\beta^s}{\rho^r \nu(x)^2} \ ,$$

where s and r are integers and β is a real number . If f_0 is linear then, $s = 2$, $r = 6$ and $\beta = 1$. If f_0 is quadratic, then $s = 1$, $r = 2$ and $\beta = \frac{1}{\mu}$.

The result condenses bounds obtained separately for the linear and quadratic case in Vera[17, 18]. As can be seen, the bound depends on the distance to ill-posedness and on the separation of the current iterate from the boundary of the polyhedron. A general estimate for the minimal precision is then:

$$\bar{t} = O\left(f(m,n) + \log\left(\frac{1}{\rho}\right) + \log\left(\frac{1}{\nu(x)}\right) + \log\beta\right). \tag{3}$$

With the bound on the error in the Newton direction given by Proposition 1, it is possible to analyze the efect of the finite precision computation on the estimates for the convergence. In Vera[17, 18] a detailed analysis is performed for both the linear and quadratic case, but also a general analysis can be done, and that will lead to the following result regarding the centrality of the next iterate of the algorithm. In the proposition, λ_{min} and λ_{max} denote respectively the smallest and largest eigenvalue of a matrix.

Proposition 4 *Let* (x,μ) *be the current iterate. Lest* $\bar{p}(x,\mu)$ *be the computed direction and suppose that* $\bar{p}(x,\mu) = p(x,\mu) + h(x,\mu)$. *Let* $\mu^+ = (1+\theta)\mu$ *and* $\tilde{x} = x + \bar{p}(x,\mu)$. *Suppose* $\|\bar{p}(x,\mu)\|_{H(x,\mu)} \leq \frac{1}{8}$. *Then, if* $\theta = \frac{1}{30\sqrt{n}}$,

$$\|\bar{p}(\tilde{x},\mu^+)\|_{H(\tilde{x},\mu^+)} < \omega + R$$

where ω *is a constant value such that* $\omega < \frac{1}{8}$, *and*

$$R = O(\|h(x,\mu)\|_2^2 \lambda_{max}(H(x,\mu)) + \|h(x,\mu)\|_2 \lambda_{min}(H(x,\mu))) \ .$$

This result says that the proximity criteria ($\|\bar{p}(x,\mu)\|_{H(x,\mu)} \leq \frac{1}{8}$) will be satisfied for the next iterate as long as $\|h(x,\mu)\|_2$ is small enough, and that can be achieved by using enough precision in the computation of the Newton direction. Specific bounds on the eigenvalues of the Hessian can be estimated and this bounds depend, as expected, on the distance to ill-posedness. The following result summarizes the usual "proximity" bounds for the interior point iteration:

Proposition 5 *Suppose* $\|\bar{p}(x,\mu)\|_{H(x,\mu)} \leq \frac{1}{8}$ *and* $\|p(x,\mu)\|_{H(x,\mu)} \leq \frac{1}{8}$, $\theta = \frac{1}{30\sqrt{n}}$, $\tilde{x} = x + \bar{p}(x,\mu)$, $\mu^+ = (1+\theta)\mu$, *and* β *is chosen so that*

$$\beta = \Omega\left(\frac{\rho^r \beta^s}{f(m,n)}\right) \ ,$$

where r *and* s *are integers, and* β *is real. If* f_0 *is quadratic, then* $r = 12$, $t = 2$ *and* $\beta = \frac{1}{\mu}$. *If* f_0 *is linear,* $r = 6$ *and* $\beta = 1$. *Under this setup, the following bounds hold:*

$$\|p(\tilde{x},\mu^+)\|_{H(\tilde{x},\mu^+)} \leq \frac{1}{8} \ , \tag{4}$$

$$\|\bar{p}(\tilde{x},\mu^+)\|_{H(\tilde{x},\mu^+)} \leq \frac{1}{8} \ , \tag{5}$$

$$|f_0(\tilde{x}) - f_0(x(\mu^+))| \leq C\sqrt{n}\left(\frac{1}{\mu^+}\right) \ , \tag{6}$$

where $C = O(1)$.

$$\|\tilde{x} - x(\mu^+)\|_{H(\tilde{x},\mu^+)} \leq \frac{1}{3} \ . \tag{7}$$

The result says that appropriate proximity to the central trajectory (in terms of norm of the Newton direction, geometric distance and objective value), will be preserved from one iteration to the next, as long as sufficient working precision is used.

4   Convergence of the Algorithm

Convergence of the algorithm to an ϵ-optimal solution follows now from the previous results, under sufficient working precision and an appropriate stopping iteration $\bar{k}$.

Theorem 6 *Let $x^0 \in P(A, b)$ and $\mu_0 > 0$ be such that*

$$\|p(x^0, \mu_0)\|_{H(x^0, \mu_0)} \leq \frac{1}{8} \ , \ \|\bar{p}(x^0, \mu_0)\|_{H(x^0, \mu_0)} \leq \frac{1}{8} \ .$$

Suppose we use $\hat{t}$ digits of precision, where

$$\hat{t} = \Omega\left(f(m, n) + \log\left(\frac{1}{\rho}\right) + \log\left(\frac{1}{\epsilon}\right) \right)$$

with $f(m, n)$ a polynomial function. Then, the algorithm achieves an ϵ-optimal solution after at most $\bar{k}$ main iterations, where

$$\bar{k} = O\left(\frac{1}{\theta}\left\{ \log n + \log\left(\frac{1}{\mu_0}\right) + \log\left(\frac{1}{\epsilon}\right) \right\} \right) \ .$$

5   Initialization of the Algorithm

In our results we have assumed that a suitable starting point, centered enough is available. This simplifies the analysis a little, but it is not a restriction. It is well known that the same interior point algorithm can be used to start from any point $\tilde{x}$ in the polyhedron and iterate in a finite number of steps until obtaining a point satisfying the centrality condition. In Vera[17] we have done that analysis for the linear programming case, obtaining a complexity bound which now also depends on the starting point: it is proportional to $\log(1/\nu(\tilde{x}))$. This quantity can be interpreted as a measure of centrality of the starting point and it represent our ability to use the knowledge of the problem to obtain a good guess. The analysis can be generalized to the setup presented here and will lead to a complexity result of the form

$$k = O\left(m\left\{ f(m, n) + \log\left(\frac{1}{\rho}\right) + \log\left(\frac{1}{\nu(\tilde{x})}\right) \right\} \right)$$

for the number of iterations, and

$$t = O\left(f(m, n) + \log\left(\frac{1}{\rho}\right) + m\log\left(\frac{1}{\nu(\tilde{x})}\right) \right)$$

for the precision requirement. The iteration complexity is consistent with previous results by Nesterov and Nemirovskii and by Renegar[12] in a more general setup of linear operators in functional spaces.

6   Conclusions

We see from the result that the "precision complexity" appears dependent on the logarithm of the condition measure as well as on the precision goal for the approximate solution. Moreover, this dependency is linear. The proportionality constants (which most surely can be improved) have been hidden by the "Ω" notation, but what is really important is the proportionality: if the problem gets "twice as much" ill-posed, "no more" than double precision is needed to cope with that. This results tells us, in a sense, that interior point algorithms are "well-behaved" from a numerical point of view. An specific estimate of precision requirements has been presented which depends on a measure of "conditioning" for the instance. It is easy to realize that the analysis we have performed is general for any convex objective function, except for the fact that the bound on the condition number is obtained based on the specific form of the problem. Anyway, what this means is that the analysis can be extended to include finite precision arithmetic complexity for general convex problems. The connection with a measure of conditioning for a general convex problem, however, can only be achieved after full understanding of the concept of ill-posedness in this case. Right now the analytical techniques used depends heavily on the linear algebra structure of the problem, and that is why the analysis has been restricted to linear and quadratic programming. A more general analysis of conditioning has been performed by Renegar[12] in a functional space context where linear cones can be used to represent more general constraints. Also Peña and Renegar [9] have analyzed sensitivity of the central trajectory in this context. This can be an starting point for a further analysis of conditioning in general convex optimization.

7   Acknowledgements

This research has been supported in different stages of its develeopment by funding from FONDECYT, the chilean fundation for science and technology, (contract 1930948), Fundación Andes, and by a research fellowship from CORE, the Center for Operations Research and Econometrics of the Catholic University of Louvain, Belgium.

References

1. S. Filipowski, *On the Complexity of Solving Systems of Linear Inequalities Specified with Approximate Data and known to be Feasible.* Preprint, Dept. of Industrial and Manufacturing Systems Engineering, Iowa State University, 1994.
2. S. Filipowski, *On the Complexity of Solving Linear Programs Specified with Approximate Data and known to be Feasible.* Preprint, Dept. of Industrial and Manufacturing Systems Engineering, Iowa State University, 1994.
3. R. Freund and J. Vera, *Some Characterization and Properties of the "Distance to Ill-Posedness" and the Condition Measure of a Conic Linear System.* Working paper, November 1995.

4. G. H. Golub, C. Van Loan, Matrix Computations, second edition, Johns Hopkins University Press, 1989.
5. D. Den Hertog, C. Roos, T. Terlaky, *On the Classical Logarithmic Barrier Function Method for a Class of Smooth Convex Programming Problems.* JOTA, vol 73, No. 1, pp. 1-25.
6. D. Den Hertog, *Interior Point Approach to Linear, Quadratic and Convex Programming*, Kluwer Academic Publishers, 1994.
7. O. L. Mangasarian, *Lipschitz continuity of solutions of linear inequalities, programs and complementarity problems*, SIAM J. Control and Optimization, Vol. 25, No. 3, May 1987.
8. Nesterov, Y., A. Nemirovskii, Interior-Point Polynomial Algorithms in Convex Programming. SIAM Studies in Applied Mathematics No. 13, SIAM, Philadelphia, 1994.
9. Peña, J., J. Renegar, *On the Condition Numbers of the Linear Equations Arising in Interior Point Methods*, Fifth SIAM Conference on Optimization, Victoria, Canada, May 1996.
10. J. Renegar, *Incorporating Condition Measures into the Complexity Theory of Linear Programming*, SIAM Journal on Optimization, Vol. 5, No. 3, 1995.
11. J. Renegar, *Some Perturbation Theory for Linear Programming*, Math. Programming 65 (1994), pp 73-91.
12. J. Renegar, *Linear Programming, Complexity Theory and Elementary Functional Analysis*. To appear in Mathematical Programming.
13. S. Smale, *Some Remarks on the Foundations of Numerical Analysis*, SIAM Review, 32 (1990) pp. 211-220
14. J. Vera, Ill-Posedness in Mathematical Programming and Problem Solving with Approximate Data. Ph.D. Dissertation, Cornell University, Ithaca, N.Y. 1992.
15. J. Vera, *Ill-Posedness and the Complexity of Deciding Existence of Solutions to Linear Programs*, SIAM Journal on Optimization, vol. 6, No. 3, 1996.
16. J. Vera, *Ill-Posedness and the Computation of Solutions to Linear Programs with Approximate Data*, working paper.
17. J. Vera, *On the Complexity of Linear Programming under Finite Precision Arithmetic*, Working paper, Dept. of Industrial Engineering, University of Chile, 1994.
18. J. Vera, *Ill-Posedness in Quadratic Programming: Stability and Complexity Issues*, Working paper, Dept. of Industrial Engineering, University of Chile, 1995.
19. J. H. Wilkinson, *The Algebraic Eigenvalue Problem*, Oxford Univ. Press, 1965.

Iterated Commutators, Lie's Reduction Method and Ordinary Differential Equations on Matrix Lie Groups

Antonella Zanna and Hans Munthe-Kaas

Abstract. In the context of devising geometrical integrators that retain qualitative features of the underlying solution, we present a family of numerical methods (*the method of iterated commutators*, [5, 13]) to integrate ordinary differential equations that evolve on matrix Lie groups. The schemes apply to the problem of finding a numerical approximation to the solution of

$$Y' = A(t, Y)Y, \qquad Y(0) = Y_0,$$

whereby the exact solution Y evolves in a matrix Lie group G and A is a matrix function on the associated Lie algebra $\mathfrak{g}$. We show that the method of iterated commutators, in a linear setting, is intrinsically related to Lie's reduction method for finding the fundamental solution of the Lie-group equation $Y' = A(t)Y$.

1 Introduction

In many applications of practical (and theoretical) interest, a given problem is often reduced to the solution of a differential equation or a family of differential equations evolving on some prescribed manifold. In the framework of devising structural discretization methods for Ordinary Differential Equations (ODEs) on manifolds (cf. [6, 7] for a discussion on such matters), we describe *the method of iterated commutators* for those manifolds that posses an underlying Lie-group structure. Such manifolds are interesting *per sé*, as for instance the *orthogonal group*, the *unitary group* etc., but also because, once we have Lie-group invariant discretization methods, we can solve ODEs on *homogeneous spaces* whilst retaining the mathematical structure of the underlying solution [11]. For the sake of completeness, we mention that other Lie-group invariant schemes are also available. To our knowledge, the first such methods, essentially of a Runge–Kutta-type, were introduced by Crouch and Grossman [4] and a similar approach is currently followed also by Owren and Marthinsen [8], who have presented systematic analysis of order conditions. A slightly different approach is being currently pursued by Munthe-Kaas: his schemes are also of a Runge–Kutta-type, but the order analysis in made in the algebra instead of the group. Thus, his methods can be obtained as modification of classical schemes with the introduction of correction functions on the Lie algebra $\mathfrak{g}$ [9, 10].

2   The Method of Iterated Commutators

The method of iterated commutators was introduced by Iserles in [5] to solve linear systems with variable coefficients,

$$Y' = A(t)Y, \qquad Y(0) = Y_0, \tag{1}$$

where $Y \in \mathbb{R}^d$ and $A(t) \in \mathbb{R}^{d \times d}$ is an analytic function. Iserles proved that the solution $Y(t)$ of (1) can be obtained as the infinite product

$$Y(t) = \lim_{m \to \infty} e^{B_0(t)} \cdots e^{B_m(t)} Y_0, \tag{2}$$

where the matrix functions $B_i(t), i \in \mathbb{Z}^+$, are evaluated in the following iterative manner. First of all, we let $A_0(t) = A(t)$ and for $i = 0, 1, \ldots$, we evaluate the integrals

$$B_i(t) = \int_0^t A_i(\tau)\, d\tau. \tag{3}$$

Next we construct the sequence of commutators

$$C_{i,0}(t) = A_i(t), \qquad C_{i,k+1}(t) = [C_{i,k}(t), B_i(t)], \quad k \geq 0, \tag{4}$$

and, letting

$$A_{i+1}(t) = \sum_{k=1}^{\infty} \frac{k}{(k+1)!} C_{i,k}(t),$$

we set

$$B_{i+1}(t) = \int_0^t A_{i+1}(\tau)\, d\tau.$$

The importance of (2) comes to light when we truncate the infinite product to a finite number $M \in \mathbb{Z}^+$ of terms,

$$Y(t) \approx e^{B_0(t)} \cdots e^{B_M(t)} Y_0. \tag{5}$$

If we denote by $\| \cdot \|$ the Frobenius norm on matrices, then the order of approximation of the scheme (5) is ensured by following result.

Theorem 1 (Theorem 3, [5]). *Let* $Y^{[M]}(t) = e^{B_0(t)} \cdots e^{B_M(t)} Y_0$, *where the matrices* $B_i(t), 0 \leq i \leq M$, *are evaluated as explained above. Then, if* $A(t)$ *is sufficiently smooth function of* t, *we have*

$$\| Y^{[M]}(t) - Y(t) \| = \mathcal{O}\left(t^{2^{M+2}-1}\right). \tag{6}$$

Therefore, for every integer $M \geq 0$ *the formula (5) is an approximation of order* $p = 2^{M+2} - 2$.   □

The procedure that we have presented requires the construction of fairly complicated functions. In order to implement the schemes numerically, we replace the integrals $B_i(t)$ by quadrature formulas and truncate the infinite series $A_{i+1}(t)$ to a finite number of terms. As long as these quadrature formulas are of sufficiently high order and we have a sufficient number of terms in the series, the order analysis in Theorem 1 remains valid.

We will assume that the matrix exponentials are evaluated exactly. Thus, with a single exponential we can approximate up to order two, with two up to order six, etc. The order increases exponentially with the number of exponentials we use.

When A commutes with its derivative, we have

$$B_1(t) = \cdots = B_M(t) = O$$

and the order of approximation depends only on the quadrature formula that we employ in approximating the integral $B_0(t)$.

The numerical implementation of the method of iterated commutators is described in [5, 13]. In [13], we prove that the method of iterated commutators for Lie-group equations (8) produces numerical approximations in the group G and that the algorithm can be implemented in a nonlinear setting.

3   Group Actions on Manifolds and Lie's Reduction Method

In this section we intend to show that Iserles's method of iterated commutators is in fact related to the well known method of reduction for *equations of Lie type* due to Sophus Lie. We refer the reader to [1] for theory and notation in the current section.

Let G be a Lie group and $\mathcal{M}$ a manifold. A *left action* of G on $\mathcal{M}$ is an operation $\lambda : G \times \mathcal{M} \to \mathcal{M}$ such that

$$\lambda(e, m) = m,$$
$$\lambda(g_1 g_2, m) = \lambda(g_1, \lambda(g_2, m)), \qquad g_1, g_2 \in G, \quad m \in \mathcal{M},$$

where e denotes the identity element in G. When there is no danger of confusion, we will denote the left action by means of a dot, that is

$$\lambda(g, m) = g \cdot m, \qquad g \in G, \quad m \in \mathcal{M}.$$

Definition 2 *Let $\mathcal{M}$ be a manifold endowed with a left action λ, $\mathfrak{X}(\mathcal{M})$ the Lie algebra of vector fields on $\mathcal{M}$, $\mathfrak{g}$ the Lie algebra of the group G. If $\lambda_* : \mathfrak{g} \to \mathfrak{X}(\mathcal{M})$ is the Lie algebra homomorphism induced by the left action, then a curve $\gamma : t \in \mathbb{R} \to \mathcal{M}$ on the manifold $\mathcal{M}$ obeys the differential equation*

$$\gamma'(t) = \lambda_*(A(t))(\gamma(t)) \tag{7}$$

where $A : \mathbb{R} \to \mathfrak{g}$ is a curve on the Lie algebra. The equation (7) is called an equation of Lie type.

All the solutions of (7) can be characterized in terms of the following theorem, due to Sophus Lie.

Theorem 3 (Prop. 3, [1]). *On any manifold $\mathcal{M}$ endowed with a left G-action, the solution of the equation of Lie type (7) with initial condition $\gamma(0) = m \in \mathcal{M}$, is given by*

$$\gamma(t) = S(t) \cdot m,$$

where $S : \mathbb{R} \to G$ is the fundamental solution of the Lie-group equation

$$S'(t) = Y_{A(t)}(S(t)), \quad S(0) = e, \tag{8}$$

and $Y_{A(t)}$ is the right-invariant vector field.

In the matrix case, $Y_{A(t)}(S(t)) = A(t)S(t)$, and (8) reduces merely to the Lie-group equation

$$S'(t) = A(t)S(t), \quad S(0) = I, \tag{9}$$

where I denotes the identity matrix.

Therefore, having integrators that solve differential equations on Lie groups means, at the same time, that we are able to construct $\mathcal{M}$-invariant integrators for all those manifolds which are endowed with a left G-action.

As an example, we consider the following group action of the special orthogonal group $SO(d)$ on the manifold $\mathcal{M}_{[d]}$ of $d \times d$ real matrices. The action is given by

$$U \cdot L = ULU^T \quad U \in SO(d), \quad L \in \mathcal{M}_{[d]}.$$

The corresponding equation of Lie type is

$$L' = [B(L), L], \quad L(0) = L_0, \tag{10}$$

where $B(L)$ maps L into a skew-symmetric matrix. Recall that the Lie algebra of $SO(d)$ is denoted by $\mathfrak{so}(d)$ and it consists of all $d \times d$ skew-symmetric matrices. The equation (10) is well known, since it characterizes *isospectral flows*. The solution $L(t)$ of (10) is similar to the initial condition L_0, as it can be evinced by the group action, hence both $L(t)$ and L_0 have the same spectrum. The numerical solution of these flows has been discussed with greater generality in [2], to which we refer the reader for further references.

The knowledge of the fundamental solution $U(t)$ that solves the Lie-group equation (8),

$$U' = B(L)U, \quad U(0) = I, \tag{11}$$

allows us to obtain the solution to (10). We have

$$L(t) = U(t) \cdot L_0 = U(t)L_0U(t)^T.$$

In this light, the approach proposed in [2] to solve numerically (10) whilst retaining isospectrality, can be depicted as finding the fundamental solution to (11) by means of an $SO(d)$-invariant numerical method. This allowed the authors to introduce *isospectral* numerical methods, whereas classical methods for ODEs fail to retain the isospectrality of (10).

3.1   Lie's Reduction Method

The main problem reduces to find the *fundamental solution* of (9) where S evolves in a Lie group G. Assume that S can be written in the form

$$S(t) = T_1(t)S_1(t), \tag{12}$$

whereby $T_1(t)$ is an arbitrary curve in G. Differentiation and substitution in (8) yields the following differential equation for S_1:

$$S_1' = \left(T_1^{-1}(AT_1 - T_1')\right)S_1, \tag{13}$$

together with the initial condition $S_1(0) = I$. The equation (13) is still in the Lie group G, but in particular S_1 evolves in a proper subgroup of G, see [1] for details.

The choice of Iserles for the arbitrary curve $T_1 \in$ G corresponds to

$$T_1(t) = \exp\left(\int_0^t A(\tau)\,d\tau\right) = \exp(B_0(t)), \tag{14}$$

accordingly to the notation introduced in Sect. 2. A long calculation[1] leads to the following formula for the derivative of T_1,

$$\frac{d}{dt}T_1(t) = \exp(B_0(t))\sum_{k=0}^{\infty}\frac{1}{(k+1)!}C_{0,k}(t),$$

and therefore, substitution in (13) yields

$$S_1' = T_1^{-1}\left(AT_1 - T_1'\right)S_1 \tag{15}$$

$$= \exp(-B_0(t))\left[A\exp(B_0(t)) - \exp(B_0(t))\sum_{k=0}^{\infty}\frac{1}{(k+1)!}C_{0,k}(t)\right]S_1$$

$$= \left[\exp(-B_0(t))A\exp(B_0(t)) - \sum_{k=0}^{\infty}\frac{1}{(k+1)!}C_{0,k}(t)\right]S_1$$

$$= \left[\sum_{k=0}^{\infty}\frac{1}{k!}C_{0,k}(t) - \sum_{k=0}^{\infty}\frac{1}{(k+1)!}C_{0,k}(t)\right]S_1$$

$$= \left[\sum_{k=0}^{\infty}\frac{k}{(k+1)!}C_{0,k}(t)\right]S_1.$$

Note that $\sum_{k=0}^{\infty}\frac{k}{(k+1)!}C_{0,k}(t)$ is just the function $A_1(t)$ introduced in Sect. 2. Next step consists in applying the same splitting to $S_1 = T_2 S_2$, where $T_2 = \exp(B_1(t))$ and finding the differential equation for S_2, $S_2' = A_2(t)S_2$ and so on. The iterated procedure yields exactly the method of iterated commutators described in Sect. 2.

---

[1] A similar expression appears implicitly in [5]. The general formula for the derivation of the exponential can be found in [12] and in [10]

3.2 Another Interpretation of Lie's Reduction Method

We have seen that the main task for solving equations on Lie groups and equations on manifolds on which there is a group action is to find the fundamental solution of the system (8). For this construction we require that G is connected and simply connected.

Assume that we can find a homomorphism $\phi : G \to H$ to some (possibly simpler) group H with Lie algebra $\mathfrak{h}$ on which it is possible to solve the corresponding fundamental equation more easily. Then our fundamental solution can be written in the form

$$S = T_1 S_1, \tag{16}$$

where $T_1 \in \phi^{-1}(\phi(S))$ and $S_1 \in \ker \phi$. An optimal situation is for instance when H is Abelian, therefore all ordinary differential equations of H can be explicitly solved by quadrature.

In passing we mention that a similar approach was proposed in [3] whereby the authors proposed the *method of diffeomorphisms*. Such method consists of reducing, whenever possible, the integration on some difficult manifold $\mathcal{M}$ to the integration on simpler manifold $\mathcal{N}$ numerically tractable with some classical scheme for ODEs. However, in this Lie-group setting we do not need the map ϕ to be one-to-one, since we are able to incorporate the contribution of its kernel, $\ker \phi$.

How to find the homomorphism ϕ and a simpler group H? Given the algebra $\mathfrak{g}$, we construct the linear space $[\mathfrak{g}, \mathfrak{g}]$ of elements that are commutators of some elements of $\mathfrak{g}$. This linear space is an *ideal* of $\mathfrak{g}$, since $[[\mathfrak{g}, \mathfrak{g}], \mathfrak{g}] \subset [\mathfrak{g}, \mathfrak{g}]$, therefore it is possible to construct the quotient algebra $\mathfrak{t} = \mathfrak{g}/[\mathfrak{g}, \mathfrak{g}]$. The elements of $\mathfrak{t}$ are equivalence classes $\langle u \rangle = u + [\mathfrak{g}, \mathfrak{g}]$, and each coset consists of those elements of $\mathfrak{g}$ of the form $u + [v, w]$ for some $v, w \in \mathfrak{g}$. If $\varphi_0 : \mathfrak{g} \to \mathfrak{t}$ denotes the canonical quotient mapping,

$$\varphi_0(v) = \langle v \rangle = v + [\mathfrak{g}, \mathfrak{g}],$$

then there exists a unique Lie-group homomorphism (cf. [1]), $\phi_0 : G \to \mathfrak{t}$, such that the induced Lie algebra homomorphism coincides with φ_0. The relation between ϕ_0 and φ_0 is that $\varphi_0 = \phi_0'(I)$. Thus, H = $\mathfrak{t}$. In fact, $\mathfrak{t}$ might be regarded as a group, and, being a vector space, its Lie algebra can be identified with $\mathfrak{t}$ itself. The advantage of this intricate construction is that $\mathfrak{t}$ is Abelian! For simplicity, we denote $\bar{\mathfrak{g}} = [\mathfrak{g}, \mathfrak{g}]$ and let $\langle u \rangle, \langle v \rangle \in \mathfrak{t}$. Then $\langle u \rangle = u + \bar{\mathfrak{g}}$ and $\langle v \rangle = v + \bar{\mathfrak{g}}$. Thus,

$$[\langle u \rangle, \langle v \rangle] = [u + \bar{\mathfrak{g}}, v + \bar{\mathfrak{g}}] = [u, v] + [\bar{\mathfrak{g}}, v] + [u, \bar{\mathfrak{g}}] + [\bar{\mathfrak{g}}, \bar{\mathfrak{g}}] = \bar{\mathfrak{g}}$$
$$= 0 + \bar{\mathfrak{g}} = \langle 0 \rangle,$$

since $\bar{\mathfrak{g}}$ is an ideal in $\mathfrak{g}$. If we denote $\gamma = \phi_0(S) \in \mathfrak{t}$, then γ obeyes the ordinary differential equation

$$\gamma' = R_\gamma'(\varphi_0 \circ A),$$

whereby R_γ' denotes the right multiplication induced on the algebra $\mathfrak{t}$. Abusing notation, we rewrite the above equation as

$$\gamma' = \langle A(t) \rangle \gamma,$$

whose fundamental solution, because of commutativity, is given by

$$\gamma = \exp\left(\int_0^t \langle A(\tau)\rangle \, d\tau\right) = \left\langle \exp\left(\int_0^t A(\tau)\, d\tau\right)\right\rangle.$$

We let $T_1 \in \phi_0^{-1}(\gamma)$ and we are allowed to choose

$$T_1 = \exp\left(\int_0^t A(\tau)\, d\tau\right),$$

then we need to find the differential equation for S_1 in $\ker \phi_0$, which is given exactly by (16). Letting $G_1 = \ker \phi_0$, its Lie algebra being denoted by $\mathfrak{g}_1$, the same procedure is repeated in an iterative fashion, to generate a sequence $\mathfrak{g} \equiv \mathfrak{g}_0, \mathfrak{g}_1, \mathfrak{g}_2 \ldots$, obeying the condition $\ldots \subset \mathfrak{g}_2 \subset \mathfrak{g}_1 \subset \mathfrak{g}_0$. If after some iteration k we have $\mathfrak{g}_k = \{0\}$, then the procedure terminates, the solution of (8) can be found exactly in a finite number of steps and the group is said to be *solvable*. In such cases, the method of iterated commutators, if the numerical approximation is carried with sufficient accuracy, samples the exact solution and the matrices $B_k(t)$ from a certain index k onwards are zero matrices.

4   Conclusions

We have evidenced that the method of iterated commutators for matrix ODEs can be viewed as a numerical implementation of the Lie's reduction method, which is genererally used to solve some special ODEs by quadrature. The method of iterated commutators, however, comes together with an order analysis, therefore it is possible to decide the order of accuracy for each scheme. Some numerical methods from this family are presented in [5] and in [13], where we consider application to a nonlinear case and establish that the order analysis is still valid, assuming sufficient regularity of the underlying problem. It is remarkable that Lie's procedure turns out to be a viable way to solve numerically ordinary differential equations.

References

1. Bryant, R.L.: An introduction to Lie groups and symplectic geometry. In Geometry and Quantum Field Theory, (D.S. Freed and K.K. Uhlenbeck eds) AMS IAS/Park City Math. Series, vol. 1, 1995.
2. Calvo, M.P., Iserles, A., Zanna, A.: Numerical solution of isospectral flows, DAMTP Techn. Rep. NA03/95, University of Cambridge, 1995.
3. Calvo, M.P., Iserles, A., Zanna, A.: Runge–Kutta methods on manifolds. In A.R. Mitchell 75th Birthday Volume, (D.F. Griffiths & G.A. Watson eds), World Scientific, Singapore, 57–70, 1996.
4. Crouch, P.E., Grossman, R.: Numerical integration of ordinary differential equations on manifolds, J. Nonlinear Sci., **3** (1993), 1-33.
5. Iserles, A.: Solving ordinary differential equations by exponentials of iterated commutators, Numer. Math., **45** (1984), 183–199.

6. Iserles, A.: Numerical methods on (and off) manifolds. This volume.
7. Iserles, A., Zanna, A.: Qualitative numerical analysis of ordinary differential equations, DAMTP Tech. Rep. 1995/NA05, to appear in Lectures in Applied Mathematics (J. Renegar, M. Shub & S. Smale eds), AMS, Providence RI, 1995.
8. Marthinsen, A., Owren, B.: Order conditions for integration methods based on rigid frames, to appear.
9. Munthe-Kaas, H.: Lie–Butcher Theory for Runge–Kutta Methods, BIT, **35** (1995), 572–587.
10. Munthe-Kaas, H.: Runge–Kutta methods on Lie groups. Submitted to BIT.
11. Munthe-Kaas, H., Zanna, A.: Numerical integration of differential equations on homogeneus manifolds. This volume.
12. Varadarajan, V.S.: Lie Groups, Lie Algebras, and Their Representations, Springer-Verlag, New York, 1984.
13. Zanna, A.: The method of iterated commutators for ordinary differential equations on Lie groups, DAMTP Techn. Rep. NA12/96, University of Cambridge, 1996.

Springer
and the
environment

At Springer we firmly believe that an international science publisher has a special obligation to the environment, and our corporate policies consistently reflect this conviction.

We also expect our business partners – paper mills, printers, packaging manufacturers, etc. – to commit themselves to using materials and production processes that do not harm the environment. The paper in this book is made from low- or no-chlorine pulp and is acid free, in conformance with international standards for paper permanency.

Springer

Printing: COLOR-DRUCK DORFI GmbH, Berlin
Binding: Buchbinderei Lüderitz & Bauer, Berlin